CRYOGENICS AND REFRIGERATION

—Proceedings of International Conference

May 22-26, 1989
Zhejiang University
Hangzhou, China

Edited by
Chen Guobang
Thomas M. Flynn

INTERNATIONAL ACADEMIC PUBLISHERS
(*A Pergamon − CNPIEC Joint Venture*)

Cryogenics and Refrigeration—Proceedings of International Conference

The book has been photographically reproduced from the best available copy. The papers were not refereed but were reviewed for their technical contents. Editing was restricted to matters of format, general organization and retyping.

The editors assume no responsibility for the accuracy, completeness or usefulness of the information disclosed in this volume. Unauthorized use might infringe on privately owned patents of publication right. Please contact the individual authors for permission to reprint or otherwise use information from their papers.

Publishers:
International Academic Publishers
Xizhimenwai Dajie, Beijing Exhibition Centre,
Beijing 100044, China

Distributors:
Domestic—International Academic Publishers
Foreign, outside the People's Republic of China—Pergamon Press
Headington Hill Hall, Oxford OX3 0BW, U.K.
(New York Frankfurt São Paulo Sydney Tokyo Toronto)

ISBN 7-80003-053-9/TB.3 International Academic Publishers
ISBN 0-08-0375340 Pergamon Press

SPONSORS

Zhejiang University, Hangzhou, China
University of Colorado, Boulder, USA
Ball Aerospace, Boulder, USA
Hangzhou Air Separation Plant Research Institute, China
Chinese Institute of Refrigeration (CIR)
International Institute of Refrigeration (IIR)

SUPPORTERS

Chinese State Education Commission
National Science Foundation of China
Ball Aerospace, USA
China National Air Separation Plant Corporation
Hangzhou Air Separation Plant Manufactory
China Kejian Corporation Ltd.
Department of Thermoscience & Engineering, Zhejiang University

ICCR CHAIRMEN

K. D. Timmerhaus	University of Colorado
Wang Qidong	Zhejiang University

PROGRAM COMMITTEE

Bian Shaoxing	Xi'an Jiaotong University, China
Chen Guobang	Zhejiang University, China
Douglas Stuart	Simmonds Precision, USA
Guo Fangzhong	Huazhong University of Science and Technology, China
Hua Zechao	Shanghai Institute of Mechanical Engineering, China
Huang Zhixiu	Zhejiang University, China
James Pierce	CVI Inc., USA
Jeff Bergen	Lake Shore Cryotronics, USA
Jin Xin	Nanjing University, China
Junya Yamamoto	Osaka University, Japan
Li Shimo	Zhejiang University, China
Li Yongkang	Hangzhou Air Separation Plant Research Institute, China
Lin Lihe	Zhejiang University, China
Peter Kerney	CTI Cryogenics, USA
R. J. Soulen	U. S. Naval Research Laboratory, USA
Ralph Longsworth	APD Cryogenics, USA
Ray Radebaugh	National Bureau of Standards, USA
Richard Blaugher	Westinghouse, USA
Ruan Yaozhong	University of Science and Technology of China
Shao Jian	Zhejiang University, China
Takasu Hashimoto	Tokyo Institute of Technology, Japan
Thomas Flynn	Ball Aerospace, USA
Venkatarao Ganni	Koch Process Systems, USA
Walter Swift	Creare, Inc, USA
Wu Peiyi	Xi'an Jiaotong University, China
Zhang Liang	Cryogenics Laboratory, Academia Sinica

ADVISORY BOARD

Chen Yiqiao	Hangzhou Air Separation Plant Manufactory, China
Hong Chaosheng	Cryogenics Laboratory, Academia Sinica
Wu Tingju	China National Air Separation Plant Corporation
Yu Chibin	Shanghai Jiaotong University, China
Zhang Zhiyou	Xi'an Jiaotong University, China

CONTENTS

1. LIQUEFACTION AND SEPARATION OF GASES

AIR SEPARATION PLANT TECHNOLOGY IN CHINA

Li Yongkang
Hangzhou Air Separation Plant Research Institute
Hangzhou, China

ABSTRACT

This is a review of the development on air separation plant technology in China. It describes large and small air separation plants; PSA-Pressure Swing Adsorption oxygen plants; recovery rare gases from air separation; aluminum plate fin heat exchangers; turbine expanders and cryogenic liquid containers and tankers. Up to now, various air separation plants with specifications less than that of 30000 m3/h (O2) could be designed and manufactured. A plate fin heat exchanger has been manufactured with maximum dimensions of 1000x 916x5430 mm. High speed turbine expanders with an air bearing have been running smoothly in the small air separation plants.

INTRODUCTION

An air separation plant is used to obtain oxygen, nitrogen, argon and other rare gases from air. The main methods of air separation are: cryogenic rectification, pressure swing adsorption and membrane osmosis. The cryogenic rectification method has been used for more than 80 years. This method is the most economic in the industry when oxygen and nitrogen are produced in large scales, and occupies a dominant position among air separation methods. In the early 1970'S, the pressure swing adsorption method began to step onto the stage, and after further improvements and perfections, it may soon compete with the cryogenic rectification method in small air separation plants. The membrane osmotic method is now in the process of turning laboratory achievement into practical production. The air separation plants manufactured in China are almost all based on the cryogenic rectification method. In the early 1980's, the pressure swing adsorption method began to be developed in small-scale air separation plants.

Before 1951, it was impossible for air separation plants to be manufactured in China. Only 89 sets of small-scale imported plants were operating in some major cities along the coast. The oxygen capacity of each set is 10-200 m3/h (O2). The total oxygen production capacity was about 3500 m3/h (O2).

30 m3/h (O2) air separation plants were designed in 1951 and 2 sets were produced in 1953. With the development of industry, we were able to produce a series of air separation plants which had capacities of oxygen of no more than 300 m3/h (O2) at the end of the 1950's. The technological process of the air separation plant was based on high pressure throttle Linde cycle, meduim pressure Claude cycle and high/low-pressure Linde Frankel cycle. We also developed 3350 m3/h (O2) air separation plant based on Linde Frankel cycle and the plant was put into operation in the steel industry in 1961.

In the 1960's a large number of tests and studies were done on the stone regenerator, reaction low pressure turbine expander, alumimum plate fin heat exchanger, carbon dioxide adsorption technology and cryogenic insulation. Many design data on developing the low pressure air separation plants were obtained. From 1966 to 1969, we produced 3200 m3/h (O2), and 6000 m3/h (O2) low pressure air separation plants with stone regenerators. With the successful development of plate fin heat exchangers, and Al welding technology, we were able to use Al to replace the expensive copper in making vessels, columns and pipelines in a cold box. The heat exchangers in the cold box all made use of the plate-fin structure. In 1972, we manufactured several air separation plants based on the reversing plate fin heat exchanger freezing process. The oxygen production per hour was 1000, 1500, 3200, 6000, 10000 m3 respectively.

Since 1978, we have introduced the technology of 10000 m3/h (O2) and 28000 m3/h (O2) air separation plants from Linde AG. of Federal Republic of Germany and the technology of medium pressure oxygen turbo-compressor from HITACHI Co. of Japan. Joint design and co-operative production of these new plants improved our ability to desige and manufacture large air separation plants ourselves. We have since produced air separation plant with recovery equipment for delivery of all rare gases such as argon, neon, helium, krypton and xenon. A 6000 m3/h (O2) air separation plant with a process of boost-turbine expander and molecular sieve adsorption is going to be erected. We now can design whatever the user demands. Computers are used to choose the optimum process and calculate the rectification process. The digital concentrating and dispersing system is adopted in the measurement and control system. A new generation of small scale air separation plants with capacities of less than 300 m3/h (O2) is developed . So far, we are able to design and manufacture air separation plants of various specifications with capacities below 30000 m3/h (O2). The products in series are shown in table 1.

Table 1

Oxygen Output (m3/h)	Oxygen Purity (%)	Nitrogen Output (m3/h)	Nitrogen Purity (%)
8	>99.2		
20	>99.2	40-80	>99.5
50	>99.2	100-120	>99.5
150	>99.2	300	>99.5
300	>99.5	600	>99.99
600	>99.5	1200	>99.99
1000	>99.5	1100	>99.99
1500	>99.5	1500-3000	>99.99
3200	>99.5	3200-6000	>99.99
4500	>99.5	4500-9000	>99.99
6000	>99.5	6000-12000	>99.99
10000	>99.5	10000-20000	>99.99
20000	>99.5	20000-40000	>99.99
30000	>99.5	30000-60000	>99.99

Meanwhile, air separation plants producing pure nitrogen for the petro-chemical enterprises with capacities of 720 m3/h (N2), 1000 m3/h (N2), 1500 m3/h (N2) and 3000 m3/h (N2) are also developed. The purity of the nitrogen is higher than 99.999%.

LARGE AND MEDIUM SCALE AIR SEPARATION PLANTS

The large and middle scale model KDON air separation plants in China are based on the low pressure cycle with turbine expanders. The freezing process with reversing heat exchanger, the purification process with molecular sieve adsorber at ambient temperature and the purification process of molecular sieve adsorber with boost-turbine expanders are used. The stone regenerator has a big volume, a heavy weight and a large heat capacity. The temperature difference on the warm end of the regenerator is also large. This requires more expanded air for refrigeration. So it has a low oxygen extract rate, a long starting time and a large electric energy consumption. It is now replaced by the reversing plate fin heat exchanger, which has a high efficiency and a light weight. The second generation of KDON air separation plant adopts a freezing process with reversing plate fin heat exchangers and produces oxygen and nitrogen in a ratio of 1:1. The nitrogen output is limited by the freezing and self-cleaning properties of the reversing plate fin exchanger. In the early 1980'S, nitrogen requirement increased rapidly with the development of metallurgical industry, chemical fertilizer industry and petro-chemical industry. In order to obtain more nitrogen, the molecular sieve adsorption purification process was developed. The ratio of oxygen to nitrogen is typically 1:2, although some achieved 1:3.

Two sets of KDON 6000/13000 air separation plants using the molecular sieve purification process were manufactured in 1982-1983, and they are now running normally at Shanghai Petro-Chemical Co-operation. Fig.1 is the photo of the plants and Fig.2 shows the process diagram.

The main characteristics of the plant are:

1) a process of purifying air with molecular sieve at ambient temperature is adopted. The ratio of O2 to N2 is 1:2.17.
2) using 13X molecular sieve clean out H2O, CO2 and C2H2 in the air.
3) using a high efficiency turbine expander with adjustable nozzles.
4) the main heat exchanger, condense-evaporator, subcooler are all

plate-fins.

5) a counter-flow-perforated-plate-rectification column without center column is used.
6) a small amount of liquid oxygen and liquid nitrogen can be obtained.
7) argon production equipment can be installed to produce argon.

The other advantages of the air separation plants with ambient temperature molecular sieve purification process are: less apparatus inside the cold box, easily and safely operated. But the energy consumption of per unit O2, N2 product is slightly higher. In order to reduce the energy consumption, a boost-turbine expander has been developed. It makes use of the work produced in the expansion process to drive the boost rotator, then increase the air pressure before entering the expander. So the pressure ratio of the expander is increased and more refrigeration capacity is obtained and less expansive air is processed. The amount of air entering the upper column decreases correspondingly. So the oxygen extraction rate increases and the energy consumption per unit product will decrease. This boost turbine expander was developed and is now operating in the KDON 6000 m3/h (O2) air separation plant. The energy consumption per unit oxygen output may be decreased by about 8%.

Fig.3 shows the process diagram of a 1000 m3/h (N2) air separation plant, which produces 920 M3/h gaseous nitrogen and 100 l/h liquid nitrogen (convert to 80 M3/h GN2). The purity of N2 is less than 10 ppm O2, the quantity of air processed is 3600 M3/h, air pressure is 0.91 MPa. The pressure of nitrogen out of the cold box is 0.83MPa. The main characteristics of the plant are:

1) oilless lubricate air compressor.
2) air precooled by a Freon refrigeration system.
3) molecular sieve absorber is adopted to clean air.
4) counter flow turbine expander is used, plate fin heat exchanger is used in hte cold box, a liquid nitrogen storage vessel is used to supply pure gaseous nitrogen steadily, the cold box is placed outdoors.

SMALL SCALE AIR SEPARATION PLANTS

In China, the main technological processes of small-scale air separation plants are medium-Claude cycle process, and counter flow nitrogen expansion cycle.

The former has higher air pressure, usually from 2 to 2.5 MPa, less air is pressured while more cold is produced. A small amount of cryogenic fluid may be obtained. The latter has a lower air pressure, usually below 1 MPa, and more air is processed. The amount of cryogenic fluid obtained is related to the air pressure and the amount of air being processed. It may have high speed turbine expanders and compact, highly efficient plate fin heat exchangers. In the 1980's, the renovated small-scale air separation plants have been improved in the following aspects:

1) The air compressor is oil free lubricated.
2) A Freon precool system is adopted.
3) 13X molecular sieves or double-layer beds consist of 13X molecular sieve and active aluminium oxide. They are used to purify the air in the adsorber.
4) The column and vessels in the cold box have aluminum structures.
5) High speed gas bearing turbine expanders are adopted.
6) Liquid oxygen pumps or oil-free oxygen compressors are used for filling oxygen to bottles.

Fig.4 shows the technological process of the K2ON-50/120 air separation plant. Fig.5 shows the K1-15 mobile air separation plant.

Fig.6 shows the schematic diagram of the K1-15 plant. This plant is based on the reverse-Sterling cycle. It can produce at a rate of 15l/h LO2 or 12m3/h GN2 or 15l/h LN2 or 10m3/h GN2. The purity of oxygen and nitrogen is 99.2% and 99.5% respectively.

PRESSURE SWING ADSORPTION METHOD IN OXYGEN PRODUCING PLANT

We began to study the oxygen producing plant using 5A molecular sieve adsorption in 1966. The adsorptive process includes pressure adsorption and vacuum desorption and regeneration. An industrial experiment was carried out in a scale of 100 m3/h (O2), and a 0.24 m3/h (O2) small plant was developed. But it was not well developed because of its low separation efficiency. In the 1970's, we began to study the PSA method which had succeeded in recovering hydrogen, and used it to separate O2 and N2 from air. The carbon molecular sieve was developed and experimental study on the efficiency of O2, N2 separation using 4A, 5A molecular sieves and carbon molecular sieves was also carried out. After that, we developed a series of 2-50 m3/h (O2) air separation plants using PSA method. The separation of nitrogen from air using PSA method is now under development. Fig.7 is the process diagram of a 50 m3/h (O2) PSA oxygen production plant. Oxygen is produced at 52 M3/h with a purity of 90%. The air processed is 750 m3/h and the air pressure is 0.32 MPa. The energy consumption is 0.91 KWh/m3 for "pure" oxygen.

RARE GASES RECOVERY

Equipment to extract the rare gases argon, neon, helium, krypton and xenon from air was trial-produced in 1962 and attached to a 3350 m3/h (O2) air separation plant. For argon, the copper catalyst with H2 to remove oxygen and the N2-Ar rectification process was used to obtain pure argon. For Ne and He, a process of high pressure condensation to remove nitrogen and using the liquid hydrogen to condense neon and separate helium from Ne-He mixture was adopted. For Kr and Xe, a 0.1% concentration of krypton and xenon mixture was extracted from product oxygen, to clean CH4 in a silver-aluminium catalytic stove, to absorb carbon-dioxide in alkaline solution, krypton and xenon mixture to be concentrated and separated and purification process followed. Although the technology was rather backward, we began to recover the rare gases and their production facilities on our own.

1. Argon Recovery from Air Separation

In order to meet the requirment of argon for the machinery and illumination industry, the YFS-2.5 type argon recovery equipment was trial-produced to attach to a 150 m3/h (O2) air separation plant in the middle of 1960'S. This was a catalytic-combustion system with hydrogen and cryogenic rectification to produce an output of 2.5 m3/h (Ar) with a purity of 99.99% Ar. We also succeeded in a new technological process which used molecular sieve to adsorb oxygen and nitrogen in argon at low temperature. A new purification technology of using 5A molecular sieve at 90K to adsorb nitrogen and 4A to adsorb oxygen was developed in the middle 1970's. Now it is applied in 150 m3/h (O2) and 600 m3/h (O2) air separation plants to obtain argon. The argon recovery equipment attached to 6000 m3/h (O2) and 10000 m3/h (O2) air separation plants was also manufactured in 1970's. Its advanced technology included: A Pd or Pt catalytic combustion system adding hydrogen to remove oxygen, cryogenic rectification to remove nitrogen, liquid argon pump pressurization and pipeline transportation to cylinder after evaporating the liquid argon. We now have eight sizes which are: 10, 20, 35, 45, 70, 95, 130, 190 m3/h (Ar). The purity is greater than 99.99% (Ar). After introducing the technology from Linde, we have produced 190 m3/h (Ar) and 250 m3/h (Ar) argon recovery equipment attached to 10000 m3/h (O2) air separation plants in the early 1980's. The purity is 99.999% (Ar) and is now in operation.

2. Neon and Helium Recovery from Air Separation

The method of cryogenic isothermal temperature adsorption was used to separate Ne and He from air at the end of 1970, and was soon applied in 3350 m3/h (O2) and 6000 m3/h (O2) air separation plants for recovery of Ne and He. This new technological process replaced the old one which was backward and dangerous because it used liquid hydrogen.

3. Krypton and Xenon Recovery from Air Separation

In the middle of the 1970's, Kr-Xe recovery equipment attached to 6000 m3/h (O2) air separation plant was developed. It extracts crude krypton from gaseous oxygen product, by using silver aluminium catalyst to remove CH4, molecular sieve to adsorb CO2, and finally concentrating and separating Kr-Xe, and purifying the Kr and Xe product. A new technology introducing from Linde includes: extract 0.3%(Kr+Xe) crude Kr from liquid oxygen, use Palladium(Pd) catalyst to remove methane (CH4), use molecular sieve adsorption to remove CO2, concentrating and separating Kr, Xe mixture, and purifying, then obtain the pure Kr and Xe. At the same time, the study on adopting the large scale chromatographic method to separate Kr and Xe was carried out. Results yielded Kr and Xe with purity of 99.995%. The extraction rate of pure Kr and Xe with 70-90% were obtained.

The method of argon recovery from the tail gas of a synthetic ammonia plant was also studied, and it is now applied in industry.

ALUMINIUM PLATE FIN HEAT EXCHANGER

The plate fin heat exchanger has a high heat transfer efficiency. It is compact, light, and suitable for 10-12 mediums heat transfer simultaneously. The heat exchanger surface per unit volume is up to 1500 m2/m3. It is used in main heat exchangers, reversing heat exchanger, condenser evaporaters, liquefiers, subcoolers, etc.

The plate fin heat exchanger was began to be developed in our country in 1963. It is now widely applied in various air separation plants and other aspects such as C_2H_4 cold boxes, oil field gas separation plants, helium liquefiers, oil-coolers and heat dispensers of internal combustion engines.

In order to guarantee the quality of the plate fin heat exchanger, we cooperated with manufacturers, institutes, and universities to study and develop the original material quality, the manufacturing precision of components, assemblage techniques, model design, brazing technology and product inspection. A set of specification rules regarding components cutting, assemblage, brazing, cleaning, argon-arc welding, general assembly and product inspection is laid down. A fatigue test to the plate fin reversing heat exchanger has been made for more than 1 million times. The exchanger is still in good condition. The burst experiment applied to the sample shows that the burst pressure is as high as a factor of 16 to 30 above the working pressure.

Our technology of fin-forming was transfered to Linde AG. of FRG in 1980. We also supplied the special purpose fin-punch machine and its mould.

With the development of technology of plate fin heat exchanger, the available pressure of plate fin heat exchanger applied in operating plants is increased to a level of 4MPa. The maximum dimension of the exchanger is 1000x916x5430 mm. The cross-section of the plate fin heat exchanger of 1000x900x3000 mm and 1000x1200x3300 mm have been batch produced. Fig.8 shows 1000x1200x3300 mm plate fin heat exchanger.

TURBINE EXPANDER

The radial-flow inward impulse turbine expander was first developed in a 3350 m3/h (O2) air separation plants in 1958. It is not available for low-pressure air separation plant, because its efficiency is only as low as 65%. The radial-flow reaction turbine expander was designed and tested in the early 1960's. A close cycle test rig was set up to trial-operate the expander under warm and cold conditions. The experiments of gas flow rate adjustment such as throttle adjustment before inlet to the expander, adjustment by nozzle width, rotary nozzle adjustment, and part gas-intake adjustment, the property of blower fan and electric motor brakes and the possibility to adopt a non-bladed nozzle in a cryogenic turbine expander were tested. In the early 1970's, the development of a turbine expander using a high speed gas bearing was carried out. Meanwhile the low pressure turbine expander for 600-10000 m3/h (O2) air separation plants, the medium pressure expander using gas bearing for 50-300 m3/h (O2) air separation plants, and the medium pressure nitrogen turbine expander for nitrogen liquefaction plant were built. In recent years, the design of three dimensional flow blade was studied, and the boost turbine expander as well as the wet turbine expander with 10% liquid were developed.

Table 2 lists the main performances of some cryogenic air turbine expanders.

Table 2 Performances of some cryogenic air turbine expanders in China

	1	2	3	4	5
Qe	0.385	0.790	0.90	1.44	2.52
Kg/s		0.932	1.08	1.62	2.88
				1.80	3.24
Pi,Mpa	0.58	0.52	0.583	0.583	0.55
Po,Mpa	0.137	0.135	0.137	0.137	0.135
Ti,K	132	122	124.5	130	128
n,rpm	41108	35000	30500	23190	19000
W,KW	12.8	24.5	30.5	50.4	87.5
D,mm	90	100	120	160	190

1--PLK192.2x2/4.8-0.37; e--expand air
2--ZLP33.2x2/4.7-0.35; i--inlet
3--PLK2500x2/4.83-0.37; o--outlet
4--PLK4500x3/4.83-0.37; D--diameter
5--PLK8000x3/4.5-0.35

CRYOGENIC LIQUID STORAGE AND TRANSPORTATION FACILITIES

Studies on cryogenic insulation for cryogenic fluid storage and transportation were carried out at the end of the 1950's in China. The first produced was a 15 liter Dewar for liquid oxygen using a double-wall ball-shaped structure with high vacuum insulation. The earliest produced 5 m3 liquid storage tank has a horizontal cylinder-shape and double-wall structure. The space between the two walls is filled with powder. The evaporation rate of this kind of storage was very high. The evacuated powder insulation was developed in the early 1960's and was first applied to horizontal liquid oxygen storage tank. The evacuated silica gel powder insulation with pressure of 1.33 Pa was adopted. Since 1970's, we have developed 300 liter to 100 m3 evacuated perlite insulation storage tanks. These tanks were used to store liquid oxygen, liquid nitrogen and liquid argon. The 0.3-15 m3 mobile storage tanks are developed to transport cryogenic liquid. This type of storage tanks is equipped with a presser to raise the pressure to 0.196 MPa or a cryogenic pump to raise the pressure to 1.57 MPa. Under special conditions the pressure can be raised to 34.3 MPa. The air-heated evaporator and water-heated evaporator fit with the cryogenic liquid storage are also developed. The evaporation capacity is 20 m3/h to 11000 m3/h and the evaporation pressure may be as high as 34.3 MPa. High vacuum multilayer insulation cryogenic containers with capacities of 2- 500 liters have been developed and the multilayer insulation structure is also used in transport pipes to reduce heat inleak. The developing trend of the cryogenic liquid vessel for liquid oxygen and nitrogen is large scale, high pressure and transportion of high purity liquid with minimum loss.

Figure 1 KDON 6000/13000 air separation plant

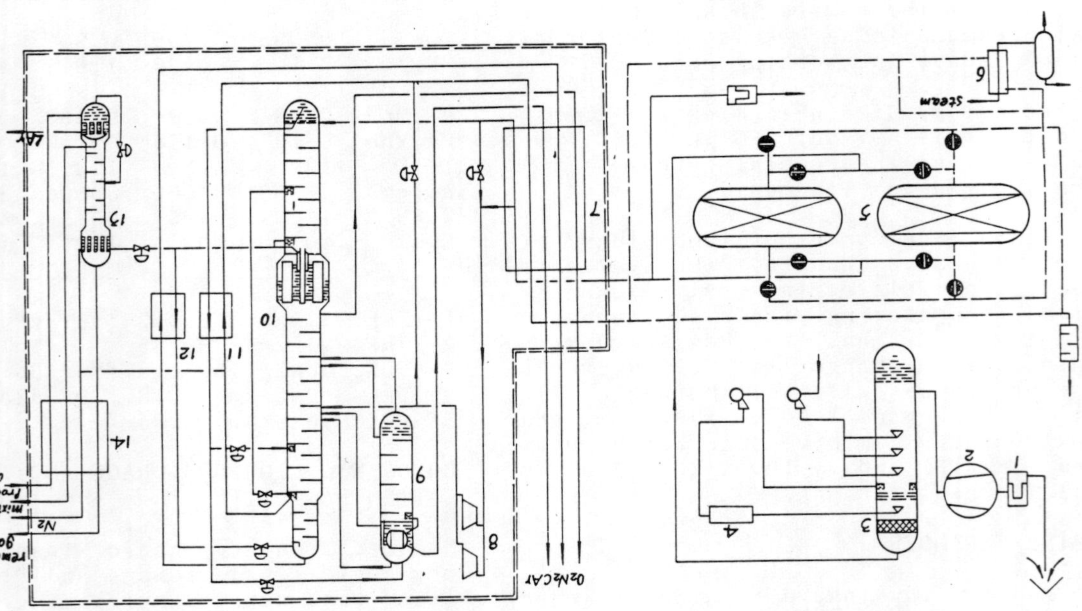

1 --air filter
2 --air turbo-compressor
3 --spray cooler
4 --Freon refrigerator
5 --molecular sieve adsorber
6 --steam heater
7 --main heat exchanger
8 --turbine expanders
9 --crude argon column
10--rectification column
11--oxygen enriched liquid air subcooler
12--liquid nitrogen subcooler
13--pure argon column
14--argon heat exchanger

Figure 2 Process diagram of model KDON 6000/13000 air separation plant with molecular sieve purification process

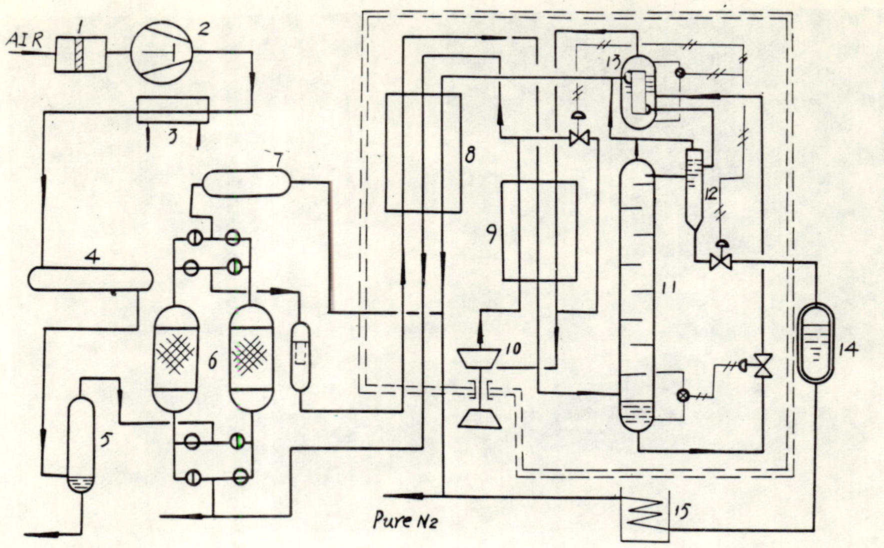

1 -- air filter
2 -- air compressor
3 -- air cooler
4 -- Freon refrigerator
5 -- water separator
6 -- molecular sieve adsorber
7 -- electroheater
8 -- main heat exchanger
9 -- liquefier
10-- turbine expander
11-- rectification column
12-- liquid separator
13-- condenser-evaporator
14-- liquid nitrogen tank
15-- vaporizer

Figure 3 Process diagram of KDN 1000 air separation plant

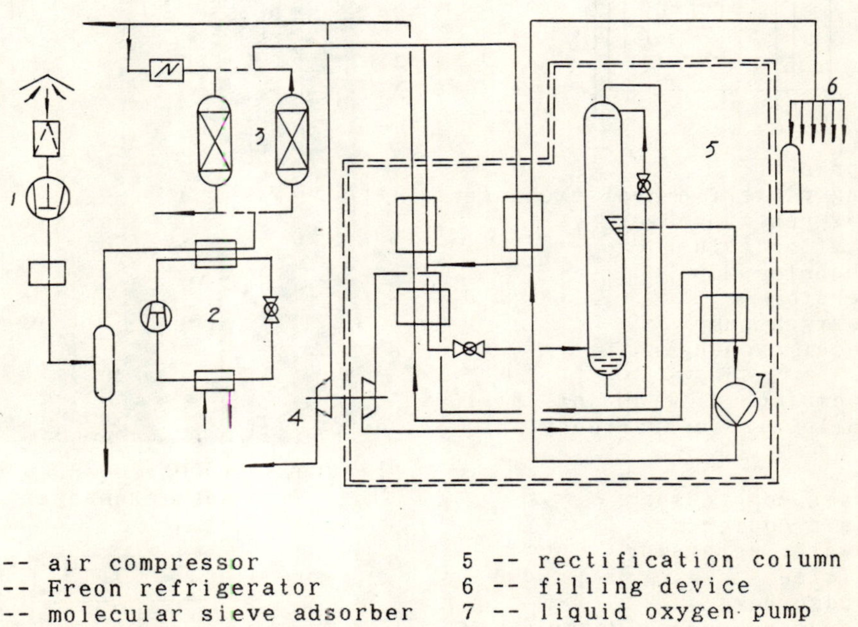

1 -- air compressor
2 -- Freon refrigerator
3 -- molecular sieve adsorber
4 -- turbine expander
5 -- rectification column
6 -- filling device
7 -- liquid oxygen pump

Figure 4 Process diagram of K2ON 50/120 air separation plant

Figure 5 K1-15 mobile air separation plant

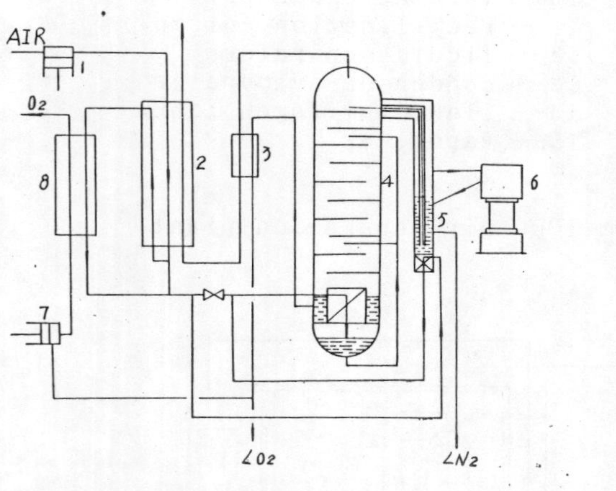

1 - air compressor
2 - reversing plate fin heat exchanger
3 - liquid oxygen subcooler
4 - rectification column
5 - vapour bubble pump
6 - cryogenerator
7 - liquid oxygen pump
8 - oxygen heat exchanger

Figure 6 Schematic diagram of K1-15 mobile air separation plant

Figure 8 1000X1200X3300 mm plate fin heat exchanger

1 -- air compressor
2 -- air cooler
3 -- water separator
4 -- filter
5 -- adsorbers
6 -- oxygen product vessel
7 -- silencer

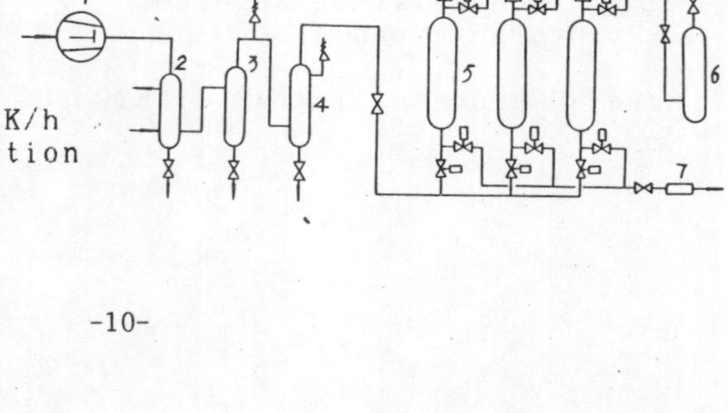

Figure 7 Process diagram of 50 m3K/h (O2) PSA oxygen production plant

SECOND GENERATION CRYOGENIC NITROGEN PLANTS

H. Cheung
Union Carbide Industrial Gases Inc., Linde Division
Tonawanda, New York 14151-0044

ABSTRACT

Three recently implemented second generation nitrogen production cycles designated as the Hybrid, MC/R Split Column, and MC/R Dual Column cycles are described and discussed with reference to past nitrogen production cycles. The new cycles represent patented processes which have been installed in the aluminum, chemicals, electronics, and petroleum industry. A versatile cryogenic nitrogen plant line is created utilizing these new energy efficient cycles for moderate to large capacities combined with the classical single column cycles for small capacities. This combination provides competitive operating cost and investment balance over plant capacities from 250 to over 200,000 NM^3/hr.

INTRODUCTION

On-site nitrogen production in the United States has more than doubled in the last decade to about 12.5 billion NM^3/yr. This has created the world's largest on-site nitrogen plant market for which UCIG has more plants than any other supplier. Major consumer of nitrogen capacity growth has been enhanced oil recovery (EOR) projects which commenced in 1979. However, the petroleum industry was not the only industry that increased nitrogen capacity. Other traditional industries such as the aluminum, chemicals, rubber, electronics, glass, textile, and steel industry also participated through expansions and new applications.

Nitrogen product requirements differ significantly from one industry to another. The electronics industry demands the highest product purity, often less than 1 ppm oxygen, and ultra product particulate cleanliness. Special particulate filters and purity maintenance precautions are common. Direct customer technical service and consultation concerning particulate technology (5) is often provided. Backup systems provide vaporized liquid nitrogen at the same ultra high purity to ensure product reliability and availability. Plant capacity is usually small to moderate. Product pressures are between 300 to 1000 KPa (43.5 to 145.0 psia), often without need for product compression. Conversely, enhanced oil recovery (EOR) for the petroleum industry represents largest plant capacity and highest product usage pressure in the order of 30,000 to 40,000 KPa (Approx. 5000 psia). Product purity requirement is a modest 10 ppm oxygen level to avoid corrosion. Short production interruptions are tolerable which allows minimum liquid nitrogen backup. Nitrogen product requirements for the other industries, aluminum, chemicals, rubber, glass, textile, and steel industry, are intermediate to the electronics and petroleum industry. Product purities are between 2 to 100 ppm oxygen; product pressures are between 135 to 3100 KPa (19.6 to 450 psia); and plant capacities vary from small to large. Some plants require liquid nitrogen backup to ensure uninterruptable supply. Computer control systems (6) are provided in some instances to match the particular industry customer demand and load patterns.

The multitude of nitrogen product requirements for the many industries with diverse applications result in the development of a cryogenic nitrogen plant line which possess adequate flexibility to fit as close as possible specific nitrogen customer usage conditions.

Foundation for the cryogenic nitrogen plant line comprises conventional small plants and novel second generation, energy efficient larger plants. This paper presents background information concerning past plant cycles and describes the three second generation plant cycles which have been implemented and are currently in service.

PAST PROCESS CYCLES

Past cryogenic nitrogen plants have been based upon three process cycles. The first and second cycle utilize a single air rectification column with waste expansion and air expansion turbine arrangements, (1,7) respectively. The third cycle is a double column oxygen process modified (8) to maximize nitrogen production and uses molecular sieve adsorption air prepurificaion.

(a) Waste Expansion Cycle

The single column, waste expansion cycle was developed about three decades ago by UCIG (7) and is the most widely used cycle for nitrogen plant capacities less than 6000 NM^3/hr. This cycle is illustrated in Figure 1.

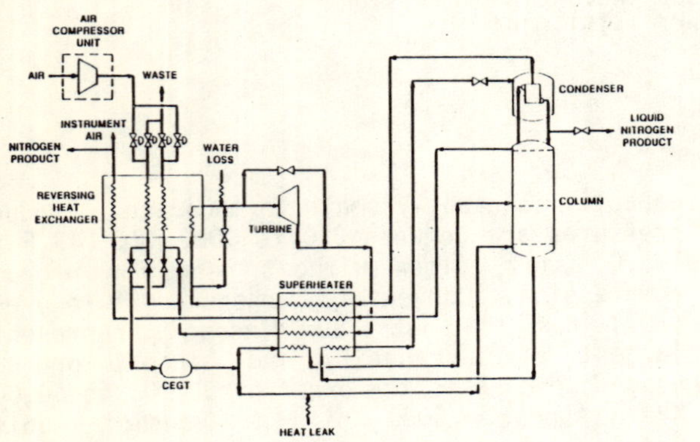

Fig. 1 Basic Single Column RHX Waste Expansion Cycle

Air is filtered, compressed (600 to 1100 KPa), cooled in an aftercooler, and the condensed moisture is removed in a condensate separator. Feed air then enters the reversing heat exchanger which cools the air feed to cryogenic temperatures and deposits on the heat exchange surface carbon dioxide and water vapor. By alternately switching or reversing the air feed and waste streams through the exchanger passages allows deposited contaminants to evaporate into the waste stream which exhausts to the atmosphere. This reversing also expels unevaporated liquid water which results in a water loss that constitutes additional refrigeration which must be generated by the turbine expander. The cooled air feed next enters a gel trap containing a filter to remove solid water ice and carbon dioxide carryover and silica gel to ensure primarily that acetylene and heavier contaminants do not enter the distillation equipment. After the gel trap, the air stream is fed to the distillation column where it is separated into an oxygen rich liquid at the bottom and pure nitrogen product gas at the top. The feed air pressure provides the driving force to accomplish the separation. The oxygen-rich liquid is subcooled and flashed into the reboiler/condenser where it is vaporized by condensing nitrogen at the top of the column to provide column reflux. This generated oxygen-rich vapor is the waste and is heated and then turbine expanded to generate plant refrigeration. Cold waste after expansion is re-warmed in the superheater and RHX and then vented to the atmosphere. Product nitrogen about 20 to 30 KPa less than the original air feed pressure is warmed also in the superheater and RHX. Product nitrogen often is delivered without the need for product compression.

The air prepurifier or molecular sieve adsorption precleanup version of the waste expansion cycle is shown in Figure 2. Here, a conventional primary heat exchanger (PHX) is applied; thus, designating the two versions as RHX and PHX versions. In the PHX version, there are no water losses and instrument air can be taken from the air feed prepurifier effluent while in the RHX version product nitrogen is used.

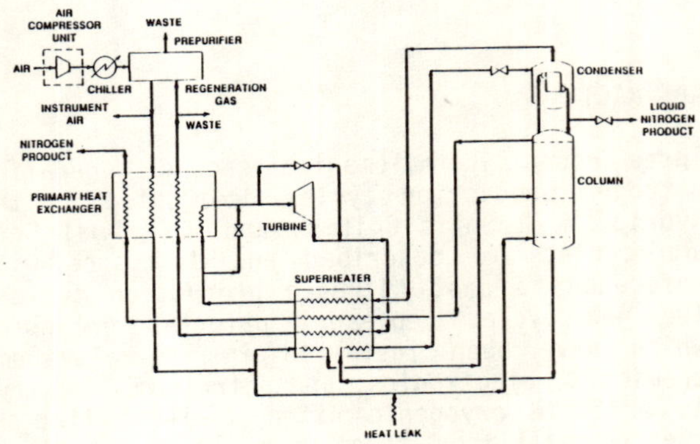

Fig. 2 Basic Single Column PHX Waste Expansion Cycle

The prepurifier removes water and carbon dioxide to replace the freeze-out benefit of the RHX and removes acetylene and heavier contaminants so that the gel trap can be eliminated. These differences simplify the cold box, but the prepurifier itself is a separate equipment system usually containing an air chiller, dual bed adsorption system, and a heating system for adsorbent regeneration. This results in the PHX version being slightly more complex, but both versions offer high reliability and availability.

The waste expansion cycle recovers 35 to 45 percent of the air feed as nitrogen. Delivery pressures range from 550 to 1000 KPa (80 to 145 psia) without product compression. The 550 KPa abs. product delivery pressure is indicative of operation at air supply pressure which provide no excess refrigeration to produce liquid nitrogen for plant backup in the event of plant shutdown. High product pressures result in air supply pressures sufficient to provide excess refrigeration for about 5 to 10 percent of the product as liquid nitrogen.

(b) Air Expansion Cycle

The single column air expansion cycle is similar to the waste expansion cycle except that cooled feed air is expanded to generate plant refrigeration, and the turbine discharge is then fed to the bottom of the separation column. Figure 3 depicts the RHX version and Figure 4 illustrates the PHX version. Equipment, plant availability and plant reliability is similar to the waste expansion cycle.

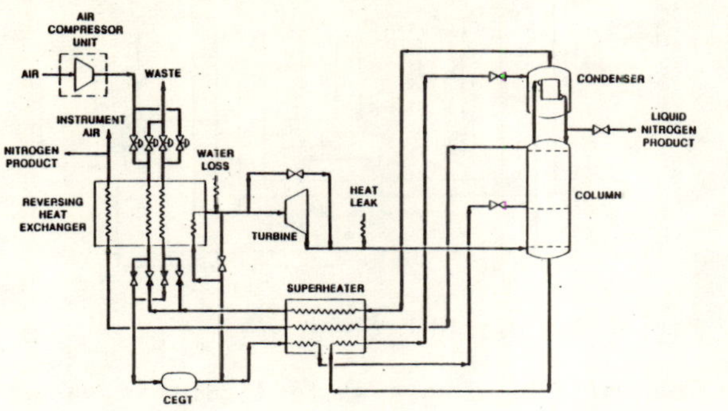

Fig. 3 Basic Single Column RHX Air Expansion Cycle

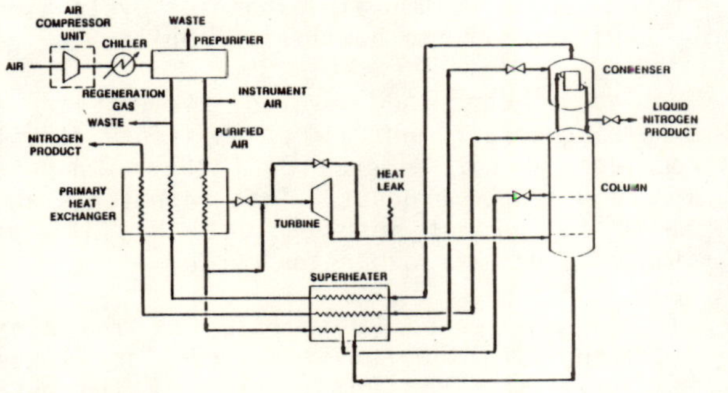

Fig. 4 Basic Single Column PHX Air Expansion Cycle

The air expansion cycle recovers about 45 to 50 percent of the air feed as nitrogen. The recovery is higher than the waste expansion cycle because the separation column pressure is significantly lower. This, in turn, results in waste liquid in equilibrium with the feed air to contain higher oxygen composition due to increased relative volatility. Nitrogen product pressures range from 275 to 450 KPa (40 to 65 psia). The higher product pressure or column pressure corresponds to the lower nitrogen recovery and the least pressure ratio across the turbine to provide refrigeration. Raising the air compressor discharge pressure increases the turbine pressure ratio and provides more refrigeration. Excess refrigeration can be generated to produce about 3 to 6 percent of the product as liquid nitrogen.

(c) Double Column Cycle

Double column cycles are classical in cryogenic air separation. Essentially, all oxygen plants utilize a double column process arrangement. Minor equipment and cycle modifications are necessary to maximize nitrogen production and discard oxygen which now is the exhausted waste. The double column nitrogen production cycle is shown in Figure 5.

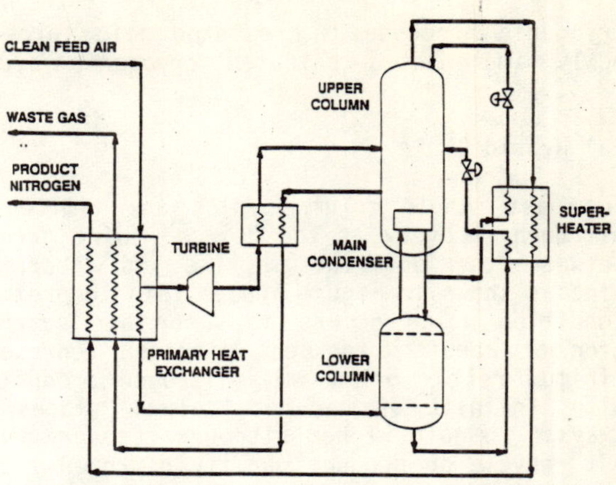

Fig. 5 Double Column Cycle

Air is filtered, compressed, chilled, and prepurified to provide clean feed air to the PHX at typically 278°K and 690 KPa (100 psia). At a PHX midpoint, about 8 to 10 percent of the feed air is removed and turbine expanded to provide plant refrigeration. The turbine exhaust air then is desuperheated against warming oxygen waste and is inserted into the low pressure upper column of the double column system. The main air feed is further desuperheated in the PHX and is fed to the bottom of the high pressure lower column where it is separated into a high purity nitrogen shelf liquid and an oxygen rich kettle liquid. The shelf liquid is subcooled against warming nitrogen product in the superheater, throttled, and inserted into the top of the upper column as reflux. Kettle liquid is subcooled also in the superheater, throttled, and fed into the upper column at a midpoint location. Upper column vapor flow is generated at the main condenser where upper column oxygen is boiled against condensing high purity, lower column nitrogen vapor. Total air separation occurs in the upper column in which oxygen rich vapor is taken from the bottom while high purity nitrogen vapor is provided at the top. Both waste and product are warmed in superheaters and the PHX. The waste is exhausted partly to the atmosphere and used also to regenerate the prepurifier adsorption beds. Nitrogen product is at about 140 KPa (20 psia).

The double column cycle recovers about 70 to 74 percent of the air feed as nitrogen. This high recovery eliminates the application of reversing heat exchangers which are limited to cycles which produce less than 60 percent of the air feed as clean products. The double column cycle favors large nitrogen plants (greater than 50,000 NM^3/hr capacity).

NEW PROCESS CYCLES

On-site nitrogen production growth has been both market and technology driven. This situation results in strong competition in the air separation industry to develop new cryogenic processes. Presented below are our newly implemented patented cryogenic cycles (2,4,9).

(a) Hybrid Cycle

The past single column cycles are limited in nitrogen recovery at the "pinch" point created between the equilibrium line and operating line as shown in Figure 6(a). This theoretical condition which occurs at vapor air composition of about 78 percent nitrogen represents minimum reflux or maximum nitrogen production with infinite number of column stages or trays. To gain higher nitrogen recovery without recycle or product quality nitrogen liquid addition, it is necessary to shift the equilibrium "pinch" point to higher nitrogen vapor compositions than in air. A liquid feed in addition to the existing vapor feed air to the column accomplishes this shift. The most convenient liquid for this purpose is liquid air provided from an auxiliary reboiler as depicted in Figure 6(b). Additional column liquid feed richer in nitrogen than air composition requires rectification with an auxiliary column (3) instead of just a reboiler. Theoretical maximum possible fractional nitrogen recovery from a unit of air fed into the column is one minus the slope of the equilibrium line at 100 percent nitrogen composition.

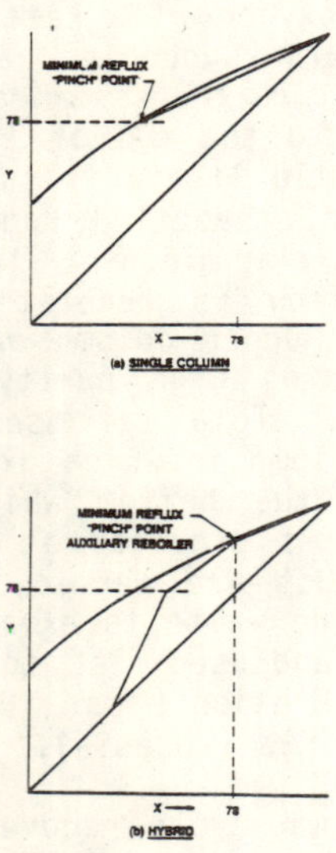

Fig. 6 X-Y Diagrams

The reversing heat exchanger version of the hybrid cycle applying the auxiliary reboiler is presented in Figure 7. If one compares Figures 7 with Figure 3 (the single column air expansion cycle), the process is similar except that a portion of the air feed is not turbine expanded and is condensed in the auxiliary reboiler to furnish column liquid air feed as well as to provide an approximate quantity of internal column stripping vapor to enrich the oxygen concentration of waste liquid.

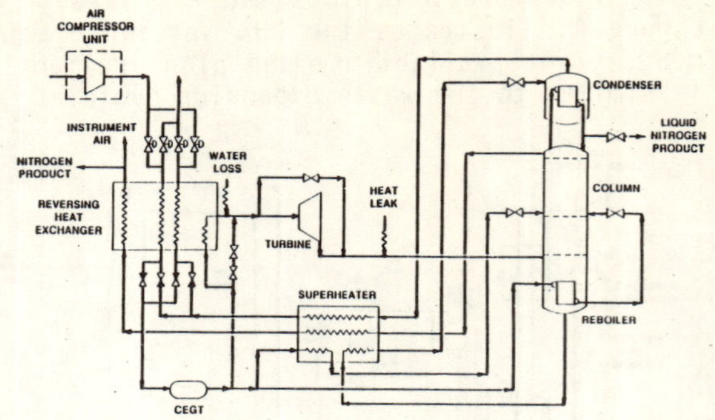

Fig. 7 RHX Hybrid Cycle

The enriched oxygen waste liquid is vaporized in the main condenser to provide column reflux and is warmed in the superheater and RHX and exhaust to the atmosphere. Liquid nitrogen production capability can be restored to that of the single column air expansion cycle simply by shutting off the auxiliary reboiler.

The above described hybrid cycle with air expansion arrangement and an auxiliary reboiler recovers 54 to 57 percent of the column feed air as nitrogen product. This represents about 10 to 15 percent more product than the past single column air expansion cycle.

The air precleaned PHX version of the hybrid cycle applying the auxiliary reboiler is shown in Figure 8. This PHX version of the hybrid cycle is comparable with Figure 4, the single column air expansion PHX version. Previous discussion of the RHX version comparison apply. A booster compressor is depicted in Figure 8 to provide the auxiliary reboiler air. The booster can be driven by the turbine in a coupled turbine/booster arrangement to provide a power savings.

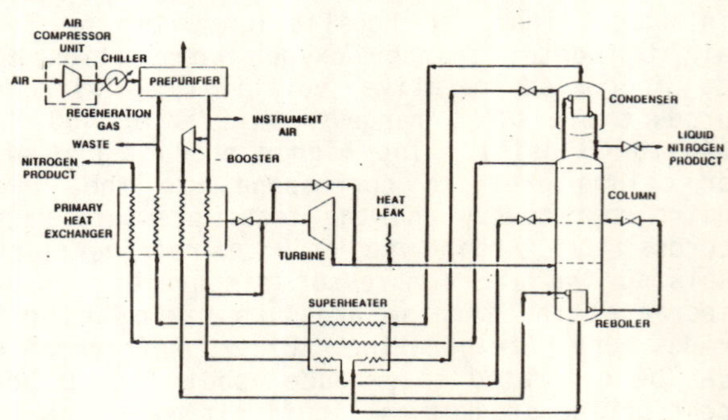

Fig. 8 PHX Hybrid Cycle

(b) MC/R Dual Column Cycle

The previously described double column cycle is very attractive from the nitrogen recovery point of view, but the nitrogen product is available at much lower pressure than the single column cycles. The multiple condenser/reboiler double column cycle designated as the MC/R Dual Column cycle is illustrated in Figure

9. This patented cycle achieves high nitrogen recovery at the expense of collecting product at two pressure levels in order to retain as much as possible the pressure energy of the compressed air feed. A second condenser/reboiler is added to provide upper column reflux rather than take liquid nitrogen directly from the lower column as done in Figure 5 of the double column cycle. Instead of transferring reflux liquid nitrogen, a high pressure nitrogen product is now taken.

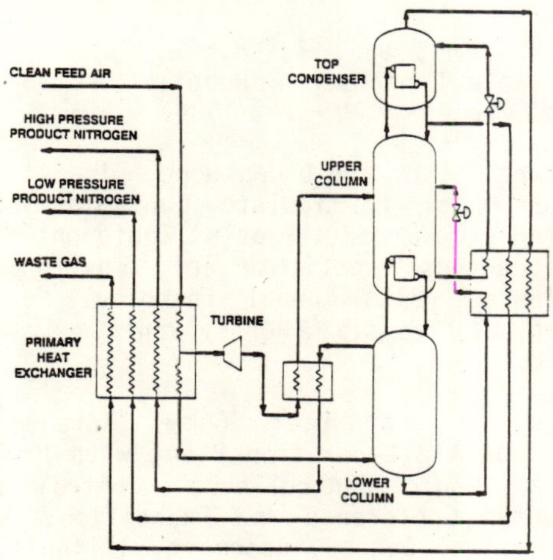

Fig. 9 MC/R Dual Column Cycle

In operation, clean feed air which is filtered, compressed to about 830 to 1030 KPa (120 to 150 psia), chilled and prepurified is inserted to the PHX at 278°K and is cooled by returning product streams and the oxygen rich waste stream. A portion of the feed air, about 8 to 10 percent, is taken at a midpoint of the PHX and turbine expanded to provide plant refrigeration. The air turbine exhaust is desuperheated against high pressure nitrogen product and is inserted into the low pressure upper column. The main air feed stream is cooled further in the PHX and is fed into the bottom of the lower column. Here, it is separated into the high pressure nitrogen product at the top of the lower column and a kettle liquid portion. The kettle liquid is subcooled, throttled, and fed into the upper column where together with turbine air feed is separated into the low pressure nitrogen product and an oxygen rich liquid waste which is subcooled and evaporated in the top condenser/reboiler to provide reflux for the upper column. The two warmed product streams are at about 760 to 830 KPa (110 to 120 psia) and 330 to 380 KPa (48 to 55 psia), respectively for the high pressure and low pressure products.

The MC/R Dual Column Cycle is capable of recovering 66 to 71 percent of the column system feed air as high plus low pressure products. This product recovery is slightly less than the 70 to 74 percent recovery achievable by the Double Column Cycle, but the preservation of pressure energy in the nitrogen products far outweigh the recovery loss.

(c) MC/R Split Column Cycle

The multiple condenser/reboiler Split Column Cycle is another patented dual nitrogen product pressure process developed simultaneously and with the same philosophy as the MC/R Dual Column Cycle. This column arrangement is comprised of two separate columns which both have its own main condenser or condenser/reboiler. In addition, the low pressure column is furnished with a reboiler. The process arrangement is depicted in Figure 10.

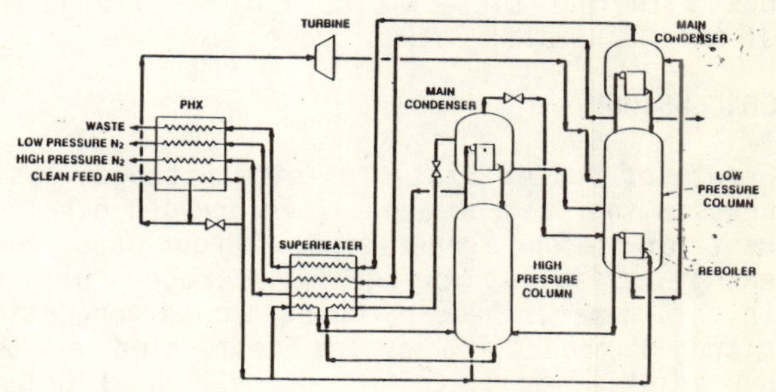

Fig. 10 MC/R Split Column Cycle

Filtered, compressed, chilled and prepurified clean air feed at 278°K between about 790 to 1000 KPa (115 to 145 psia) is fed to the primary heat exchanger, PHX, and is cooled by returning product streams and the oxygen rich waste stream. A 8 to 10 percent portion of the feed air is taken at a PHX midpoint and turbine expanded into the low pressure column to provide plant refrigeration. The main air feed stream is cooled further in the PHX and is split to provide a feed to the high pressure column and a drive for the low pressure column reboiler. High pressure nitrogen product is taken from the top of the high pressure column and is superheated and then warmed in the PHX. High pressure column kettle liquid comprised of column reflux and condensed air from the low pressure column reboiler along with a slight amount of condensed air from the superheater is subcooled, throttled and inserted into the top main condenser of the high pressure column. The evaporated high pressure column kettle liquid is fed into the low pressure column along with the unevaporated kettle liquid which is gravity drained into the low pressure column. The low pressure column feeds are separated into the low pressure nitrogen product and the oxygen enriched low pressure column, kettle liquid waste which is evaporated in the top main condenser of the low pressure column. The waste is superheated and then warmed in the PHX. The two warmed product streams are at about 730 to 800 KPa (106 to 116 psia) and 330 to 380 KPa (48 to 55 psia), respectively for the high pressure and low pressure products.

The MC/R Split Column Cycle is capable of recovering 65 to 70 percent of the column system feed air as high and low pressure products. This product recovery and high pressure product pressure are slightly less than that for the MC/R Dual Column Cycle, but the air compressor discharge pressure is slightly less. Consequently, the performance of both MC/R cycles tend to be similar and are very significantly better than the Double Column cycle.

The two multiple condenser/reboiler cycles complement one another. They offer some flexibility in choosing the optimum air compressor and construction choice of a tall cold box with the MC/R Dual Column cycle or a shorter cold box with the side-by-side column arrangement of the MC/R Split Column cycle.

CONCLUSION

Growth of United States on-site nitrogen business in the past decade has encouraged development of second generation, larger and more energy efficient cryogenic nitrogen plants. In this paper, three patented second generation nitrogen production cycles designated as the Hybrid, MC/R Split Columm and MC/R Dual Column cycles have been described and compared with past nitrogen production cycles. These cycles have been implemented at customer locations. Sufficient process advantages have been demonstrated to allow the Hybrid cycle to obsolete large capacity, past Single Column cycles and to permit the two MC/R cycles to replace the classical Double Column cycle for production of modest to high pressure nitrogen. On-site nitrogen plants based upon new second generation cycles have been installed in the aluminum, chemicals, electronics, and petroleum industry.

REFERENCES

1. Ayres, C. L., J. M. Alrudo and L. M. Himmelberger, "Developments in Nitrogen Generators," Cryogenic Processes and Equipment - 1984, The Fifth Intersociety Cryogenic Symposium, New Orleans, Louisiana, Dec. 9-14, 1984.

2. Cheung, H., U.S. 4,594,085, "Hybrid Nitrogen Generator with Auxiliary Reboiler Drive," June 10, 1986.

3. Cheung, H., U.S. 4,604,117, "Hybrid Nitrogen Generator with Auxiliary Column Drive," Aug. 5, 1986.

4. Cheung, H., U.S. 4,448,595, "Split Column Multiple Condenser - Reboiler Air Separation Process," May 15, 1984.

5. Cheung, S. D. and D. A. Hope, "In Situ Monitoring of Particulate Contamination in Integrated Circuit Process Equipment", 1988 Proceedings Institute of Environmental Sciences, 34th Annual Technical Meeting, King of Prussia, Pennsylvania, May 3-5, 1988.

6. Chow, K. Y. and C. M. Chow, "Advance Control of Air Separation Plant with Programmable Logic Controllers," Control Engineering Conference and Exposition, O'Hare Exposition Center, Rosemont, Illinois, May 19-21, 1987.

7. Dubs, M. A., D. Halberstam, M. S. Lipman and E. L. Malinowski, U.S. 3,210,947, "Process for Purifying Gaseous Streams by Rectification," April 3, 1961.

8. Pahade, R. F. and J. H. Ziemer, "Nitrogen Production for EOR," International Cryogenic Materials and Engineering Conference, St. Charles, Illinois, June 18, 1987.

9. Pahade, R. F., J. H. Ziemer and H. Cheung, U.S. 4,453,957, "Double Column Multiple Condenser - Reboiler High Pressure Nitrogen Process," June 12, 1984.

DEVELOPMENT OF NEW LIGHT-HYDROCARBON RECOVERY UNIT WITH PETROLEUM GAS LUBRICATED BEARING TURBOEXPANDER

Bu Cong
North China Petroleum Design Institute
Chen Chunzhen
Xian Jiaotong University
Zhu Renchu
Wu Xian Oxygen Machinery Plant

ABSTRACT

A small capacity light-hydrocarbon recovery modular unit collecting scattering associated gas source is introduced in this paper. A new type of externally-pressurized gas bearing with multiple-row tangential supply holes is used in the turboexpander set. The process determination, main equipment selection and development, unit and turboexpander operation test in site are also introduced. According to economy evaluation C_3° recovery of the unit is about 61%. The payout is one year and the profit is obvious.

I. INTRODUCTION

In a new-built single well or remote wells far from gathering and transportation system with high pressure and high gas-oil ratio, the associated gas always vents while only petroleum is produced. How to recover some hydrocarbons in the associated gas (including C_3 and heavier) is very important.

Therefore, a new packaged light-hydrocarbon recovery unit has been developed. The design principles are as follows.

(1) The $1.5 \times 10^4 Nm^3/d$ gas flow rate is chosen for one unit and $3 \times 10^4 Nm^3/d$ for two paralleled units. Because this unit is packaged, so it must be equipped with small size, light and handy expander, and a turboexpander (TE) is suitable. Two sets of TE-recompressor in parallel are used in this unit.

(2) There is no inlet gas compression system. The recovery of liquid hydrocarbon is only based on the refrigeration produced from gas expansion by means of the well itself pressure energy. If we want to recover the liquid hydrocarbon of low pressure associated gas, it is necessary to install a packaged natural gas compressor.

(3) Owing to the large fluetuations of working conditions at well head, it is suitable to use packaged unit. It is easy to transport and should be simple in technology with high recovery.

In the design course, we designed a case with ammonia refrigeration, however, it was abandoned because the recovery of C_3° is only about 20--30%, owing to the higher temperature level, and the equipments are huge and not easy to move. If the gas expansion alternate with ammonia refrigeration is used, the recovery of C_3° may be increased over 70%, but the compound flow sheet will be more enormous and not easily moveable. Based on the comparison results of some flow sheet alternates, the expansion refrigeration has been determined, since it is better than others.

(4) The unit products are light-hydrocarbon mixture C_3^+ and dry gas. The light-hydrocarbon mixture (LPG and NGLS) are sent to the stabilizer located in the central battery by tank, and are separated finally. If the final separation is realized in the package, the stabilizer must be installed and it is necessary to supply heat for reboiler at the bottom and cooling water for condenser at the top. In this case, the number of equipments in the system will be increased and the unit will become difficult to be moveable. This configuration is adopted.

The pressure of outlet dry gas is 0.4 Mpa. It may be used at worksite or for sales.

II. THE FLOW SHEET OF THE PACKAGED UNIT

The flow sheet of the packaged unit is shown in figure. The oil and gas from a well is 2.1 Mpa in pressure and 303K in temperature.

The two phase stream flow via the vertical and horizontal separator in series, the liquid is separated and sent to a measuring station. The gas passes through in turn to each of two purifiers filled with activated alumina. The purified gas whose dew point is about 213K, enters the main heat exchanger. The gas outlet temperature is 245.8K and the gas then goes to low temperature separator. The separated liquid phase is throttled to 1.04 Mpa, heated in the main heat exchanger and fed to the deethanizer. The separated gas phase at pressure 2.03 Mpa enters the turboexpander. The exit pressure of TE is 0.4 Mpa. The mixture phase at temperature 198.4K enters the second low temperature separator. The separated liquid is pumped to the top of deethanizer as a reflux. The gas enters the reflux condenser to cool the gas from the top of deethanizer. The cooled mixture of liquid and gas returns to the second low temperature separator and is separated again. The gas from the reflux condenser flows to the main heat exchanger, it is heated and then enters to the recompressor. The recompressor discharge gas, at 0.45 Mpa, is saled. The heat sources for switch regeneration in two cylinders purifier and for reboiler at bottom of deethanizer are supplied by gas cylinder heater.

The liquid products at the bottom of deethanizer are stored in tank and then sent to central battery for final separation.

III THE EQUIPMENT SELECTION

1. Medium Pressure Turboexpander-recompressor set

It is the key equipment of producing cold in the unit. The enthalpy drop of TE is based on the pressure energy of well head. The gas flow rate per day in the unit is 1.5×10^4 Nm^3. According to the analysis and comparison of flow sheet configurations, the gas pressure at the entrance and exit are 2.1 Mpa and 0.4 Mpa (or the lowest 0.2 Mpa) respectively. Because the expansion ratio of TE is larger (5--10) and the volumetric flow rate passed through the TE is smaller, the rotor speed must be higher (12×10^4 rpm) in order to obtain the higher efficiency, and it is necessary to adopt the gas bearing. It is well known that the rapid progress in gas bearing turboexpander of cryogenic engineering field in China has been achieved in recent years. It is in common use in small oxygen-producing plants and the rich experiences have been accumulated. Some of turboexpander have been operated for several ten thousand hours. The TE equipped with the gas bearings of multiple-rows of tangential feed holes and double-gas-film, which were developed by Xian Jiaotong University and Wu-Xian Oxygen Machinery Plant, has been widely used in China. These new types of gas bearing not only have higher stability, but also have not any limits for gas kinds and properties. Much design improvements and development based on the medium pressure TE in small oxygen-producing plant, have been anew made for this unit according to its technical requirements.The model of TE is PLYG-8.33/20-4. Its design parameters are as follows:

ET inlet pressure 2.1 Mpa, outlet pressure 0.4 Mpa, inlet temperature 245.8 K, flow rate 560 Nm^3/hr, adiabatic efficiency $\geqslant$ 65%; for recompressor inlet pressure 0.365 Mpa, inlet temperature 299K, outlet pressure 0.45 Mpa: rotation speed 12×10^4 rpm: bearing gas pressure 0.2--0.6 Mpa. A small part of plant petroleum gas is heated in the heat exchanger and supplied to the bearing near ambient temperature. The gas bearing type is multiple-rows of tangential feed holes and double-gasfilm. The bearing gas is exhausted to the blow-off pipe. The close circuit system is formed and there is not any external leakage. Many parts, such as the gas bearing and the case of TE, have been redesigned. The blower was adopted as a brake of TE in the initial test, in which the working fluid was air and exhausted outside after recompression. Even though it was simple in construction, the energy was not recovered for utilization and it was unsafe. Therefore, the blower was anew designed and changed to the recompressor in later stage of test. The working fluid is petroleum gas and the close circuit is formed as shown in the flow sheet. The energy is full utilized and it is better in safety of unit.

2. Plate-fin Heat Exchanger for Low Temperature Heat Exchanger and Condensation

The main heat exchanger located before the TE is a plate-fin type with 4 passes. The warm stream is the raw gas after purification, and the other three cold streams are (1) liquid separated before expansion; (2) gas after expansion and (3) bearing gas. The cold produced in the unit may be recovered sufficiently. The feed of deethanizer and bearing gas may be at ambient temperature. The loss of cold produced may be decreased.

The deethanizer overhead condenser is 2 passes plate-fin heat exchanger.

3. Purifier

The purification of water in raw gas is realized by absorption of active alumium oxide. As for the choice of absorbent, the performance comparison and on-the-spot operational check have shown that the performance of aluminium oxide is better. It has enough strength and is not easy to form powder. It may absorb a small amount of hydrocarbons.

It is strong and durable. The dew point of the gas having treated may be lower than 213 K. The requirements of the absorbent are well satisfied.

There are two cylinders for absorption. The heat is supplied by the heater with ϕ60 cylinder for switch regeneration heat sources.

(4) Deethanizer

In order to decrease the column height and increase the fractionating effect, the stainless sectionalized beds made of corrugated plates are used as column packing. The heat in the reboiler is supplied by gas heater. There is ambient temperature feed at the middle of column. The reflux on the top of column is supplied by the liquid separated in second low temperature separator after TE and by the liquid pumped with liquid-hydrocarbon plunger pump.

(5) The Liquid Hydrocarbon Plunger Pump Specifications

inlet pressure:	0.2--0.4 Mpa;
outlet pressure:	0.98--1.2 Mpa;
volume displacement:	50--126L/h;
rotation speed:	82--242 rpm;

The model of non-stage reducing gear: FPC11/1.97--3.1 x 1/10 R and motor explosion-proof class is KB.

IV. WORKSITE TEST OF TE

The gas bearing TE set installed in this unit has been jointly developed since 1983 by Xian Jiaotong University and Wu Xian Oxygen Machinery Plant. After completing the manufacture of the TE, experimental investigations had been repeatedly conducted in the test rig with air working fluid since the last half year of 1983, to see the operational stability. Then it was sent to the test worksite in North China Oil Field as a cold box block in 1984. In order to determine the actual performance of TE, the test had been made in the oil fields Quan No.36 and Zhe No.39 and we had got good results.

During the test in Quan No.36 well, the test for thermodynamic and mechanical performance had been made many times with petroleum gas under various working conditions, including normal design conditions,varying working conditions, and overspeed conditions. According to the actual measurement in worksite, the design requirements of the thermodynamic performance of the module TE had been met. The actual operational parameters at design condition are shown as follows.

TE inlet pressure	2 Mpa
TE outlet pressure	0.4 Mpa
nozzle outlet pressure	0.72 Mpa
sealing gas pressure	0.65 Mpa
bearing gas pressure	0.5 Mpa
TE rotation speed	12.4×10^4 rpm
TE inlet temperature	222K
TE outlet temperature	174K
TE flow rate	620 Nm^3/h

based on these actual data, the adiabatic efficiency calculated by computer is 0.68. It meets the design requirement ($\geq$ 0.65).

During the tests, we have measured the amplitude of rotor whirl under various operational conditions by means of measuring-vibration meter of electric eddy current type. The amplitude-frequency characteristic curves under various parameters have been obtained, and the orbits of shaft center under various working condition have been shown on the oscillograph. The results of actual measurements have shown that the operation of TE is very stable in all the course from starting to normal speed. The rotor amplitude near the bearing of TE side is 3.9μm, and it is 2.4μm for the charge side. In work site, the operation and starting are very handy. While the overspeed test is conducted, for example, during the increase of expansion ratio from 5--10, and the speed reaches 15×10^4 rpm, the end of the meter scale, the operation of TE is still stable. The test results have shown that the mechanical performance of this TE is very excellent and the normal operation of the unit can be guaranteed.

In order to further check the performance of model TE, the tests of small flow rate and moist gas have been especially carried out. When the inlet pressure of TE and the corresponding rotation speed are decreased, the operation is still stable. There are not any abnormal phenomena, such as charger pant, etc. When liquid drops are appeared at the flange of TE outlet during decreasing outlet temperature to less than -140°C in the worksite tests in Quan No. 36 well, the operation of TE is still stable too. According to the first analysis and calculation, the liquid amount in the gas more than 10% may be obtained.

Although all the test on the worksite have sufficiently shown that the performance indexes are very excellent and can meet the requirements of the unit, we have also found some problems to be improved in the first stage of test, and we have solved them in later stage of development. These problems are as follows.

(1) Because the TE always worked in moist condition in the test of Quan No. 36 well, and the turbine wheel made of aluminum alloy was not strong enough to bear the abrasion from liquid particles, the surface of turbine wheel was seriously worn and torn under the erosion of liquid particles. Therefore the outside surface of turbine wheel has been treated by hardening and wear-resisting in the last stage of development

(2) In the worksite test of Quan No.36 well, the blower of TE was still used as a brake. Not only the power produced by TE was not recovered and used, but also the noise was higher. In the last stage of development, the charger with the same shaft of TE has been used as braking device.

(3) If the exhaust pipe of TE drawn off at the upper part of cold box, the liquid might be accumulated at the TE outlet during the two phase formation in it. The pipe arrangement in the TE cold box has been improved. The exhaust pipe of TE draws off at the lower part of cold box.

(4) Line size bypass is added on inlet and outlet of the turboexpander. The explosion solenoid valve and rotating speed gauge interlock is set to control the turboexpander rotating speed. In the recompressor outlet a diaphragm operated valve is set to control flow rate by remote control or hand control

(5) The bearing gas outlet pressure is increased to guarantee the pressure differential between the inlet and outlet of static pressure gas bearing. The bearing gas which was vented originally is recovered and put into sale gas system.

(6) The recompress end shell has been changed into air-cooling from water cooling.

Above problems have been solved when the packaged unit was assembled.

V. ECONOMICS

The flow rate with two TE is 3 x 10^4 Nm^3/day. The gas composition was taken from the An No. 56 well, the C_3^o recovery of the unit is 0.55. The output value, capital cost and profit of the unit are shown in the following tables.

Table 1. Output Value

ITEM	RECOVERY	LPG	NGLS	C_3^o	value
UNIT	T/D	T/D	T/D	mol%	10^4yuan / year
value	6.166	4.332	1.834	61.0	86.6

Note: (1) Unit prices are calculated with 250 yuan/ton for liquified gas and 500 yuan/T for NGS.

(2) Working period every year --330 days.

Table 2. Capital Cost

item	main equip cost	material& instrument-ation cost	installation cost
unit	10^4 yuan	10^4 yuan	10^4 yuan
value	31.1	16	5

Table 2. (continuous)

item	profit	plant total funds	decrease gas expenses	operation cost
unit	10^4 yuan	10^4 yuan	10^4 yuan /year	10^4 yuan/year
value	10.4	62.5	3.3	3.98

Table 2. (continued)

item	wages	total production expenses
unit	10^4yuan/ year	10^4 yuan/year
value	1.92	9.2

Table 3. Profit

item	capital cost	operation cost	output value	payout
unit	10^4 yuan	10^4 yuan/ year	10^4 yuan/ year	month
value	62.5	9.2	86.6	10

VI. THE MODULAR PLANT

There are three modular block of the unit: (1) technology block; (2) storage tank block; (3) gas heater block.

Technology block is arranged as frame type and there are three storeys in it. The frame size of the bottom storey is 8 m (length) x 3.5 m (width) x 3.72 m (height). The equipments installed in it are as follows: the vertical and horizontal inlet separator of oil and gas, water cooler, purifier, low temperature separator, deethanizer and reboiler. The frame size of the second storey is 5 m (length) x 3.5 m (width) x 2.8 m (height). The equipments installed in it are second low temperature separator, main heat exchanger, powder filter, and cryogenic light hydrocarbon plunger pumps. The frame size of the third storey is 5m (length) x 3.5 m (width) x 2.8 m (height). The equipments installed in it are turboexpander--recompressor set reflux condenser, measuring and controlling room, and the room on duty. The deethanizer passes through the three storeys. The total height of the three storey frames is 9.32 m. Every frame is transitted to the worksite and then assembled together.

VII. CONCLUSIONS

A new light-hydrocarbon recovery unit with petroleum gas lubricated bearing turboexpander has been successfully developed in China. It has excellent performance and good profits. It can be widely used in oil fields. The idea of adopting natural gas bearing in turboexpander is successful. All the experiences in developing this unit may extend to other light-hydrocarbon recovery plant of large flow rate.

REFERENCES

1. Chen Chun-zhen "The Advance in Gas Bearing Application for the Expansion Turbines of Air Separation Plants in China" -- Proc. International Conf on Production and Purification of Coal Gas and Separation of Air. Oct, 1985.

2. Chen Chun-zhen et al "Stability Research of EP Gas Bearings for Cryogenic Expansion Turbines" Cryogenics, No.7, 1986.

3. Chen Chun-zhen et al "The Double-Gas-Film Bearing Equipped on the Cryogenic Expansion Turbine". --Proc. ICESR, 1986.

4. Chen Chun-zhen "The Gas Bearing Expansion Turbine of 150 Nm^3/hr Oxygen Plant". Cryogenic Technique 1984. No.6. (in Chinese).

PHOTO OF THE UNIT

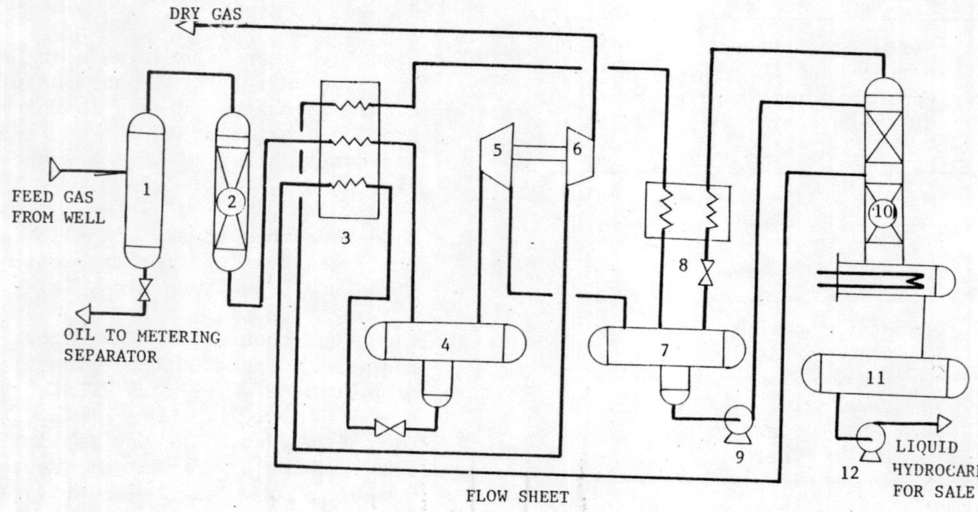

FLOW SHEET

+ 1. HIGH PRESSURE SEPARATOR
2. DEHYDRATOR
3. MAIN H EAT EXCHANGER
4. LOW TEMPERATURE SEPARATOR
5. EXPANDER
6. RECOMPRESSOR
7. REFLUX ACCUMULATOR
8. REFLUX CONDENSER
9. REFLUX PUMP
10. DEETHANIZER
11. LIQUID HYDROCARBON TANK
12. LIQUID HYDROCARBON PUMP

ETHYLENE RECOVERY FROM REFINERY FUEL GAS

Vijay A. Deshpande
Engineering Division, Texaco Inc.
Houston, Texas, U.S.A.

In the past three decades, ethylene has been established as the principal commodity petrochemical in the chemical industry. Based on annual production, ethylene is now ranked third, surpassed only by sulfuric acid and nitrogen (1). After some years of over-capacity and resultant lack of any new capacity coming on stream, the demand for ethylene has caught up with the supply. Plants are operating at capacity and prices have soared from 17-18 ¢/lb in the fall of 1987 to about 30 ¢/lb in the summer of 1988. While some grass-roots plants are being planned, producers are taking a very cautious approach to new mega-capacity because of past over-capacity problems and the relatively high investment required for a new cracker plant for ethylene. On the other hand, ethylene recovery from refinery fuel gas presents a highly promising approach.

Refinery Fuel Gas

Ethylene is a by-product of a Fluidized Catalytic Cracker (FCC) found in nearly all refineries. The ethylene, mixed with other light hydrocarbons from the FCC, is collected for use as fuel gas within the refinery. Typically, FCC off-gas contains 13 to 18 percent ethylene. Currently, ethylene priced as fuel gas is worth only about 5 ¢/lb, whereas high purity, polymer grade ethylene is worth about 30 ¢/lb. Even if the price was to drop and stabilize around 20-25 ¢/lb, there appears to be an economic incentive to separate and recover ethylene.

In recent years, many refiners have implemented energy conservation projects with an overall reduction of energy usage within a refinery. As a result, some refiners do not need all the fuel gas generated from various refining units and face an excess fuel gas situation. Excess fuel gas can be a serious problem because eliminating excess fuel gas through flaring negates energy conservation measures and is environmentally unacceptable. For these refiners, reduction of excess fuel gas through cryogenic recovery of light hydrocarbons and/or ethylene is a very attractive and profitable solution.

An ethylene recovery unit can serve a single refinery, or a number of adjacent refineries can combine fuel gas streams for a centrally-located recovery plant of modest size. In either case a recovery unit is likely to be relatively smaller compared to a world scale grass-roots ethylene plant. As a result, it can be brought on stream at a fraction of the cost and in a much shorter time. Additionally, a smaller recovery unit would reduce the risk of creating significant over-capacity.

Ethylene Product Specifications

Typical specifications for high purity ethylene are shown in Table-1. Note the stringent purity requirements.

TABLE-1

HIGH PURITY ETHYLENE
TYPICAL PRODUCT SPECIFICATIONS

Ethylene:	mol %	min	99.9
Impurities:			
Methane	mol %	max	0.1
Ethane	mol %	max	0.1
C_3+	wppm	max	10
Acetylene	wppm	max	5
Oxygen	wppm	max	5
Carbon monoxide	wppm	max	5
Carbon dioxide	wppm	max	15
Sulfur	wppm	max	2
Water	wppm	max	5
Hydrogen	vppm	max	5

FCC Off-Gas Composition

Table-2 shows a composition range for FCC off-gas. The off-gas is contaminated with many unwanted components present in ppm range. Some commonly found contaminants and their expected amounts are listed in Table-3.

TABLE-2
TYPICAL FCC OFF-GAS COMPOSITION

Component	mol %
Hydrogen	15-20
Nitrogen	5-10
Carbon Monoxide	0-2
Methane	30-40
Ethylene	13-18
Ethane	15-20
Propylene	2-7
Propane	1-5
C_4S	0-4
C_5S	0-3
C_6+	0-2

TABLE-3
POSSIBLE FCC OFF-GAS CONTAMINANTS

Contaminant	ppm
Acetylenes	10-300
Dienes	10-100
O_2	500-2000
CO_2	100-300
COS	5-20
SOx	5-20
H_2S	20-100
Mercaptans	5-20
HCN	5-20
NOx	1-5
Chlorides	1-5
Arsine	0.01-0.5
Ammonia	5-20
Benzene	100-500
Toluene	300-1500
Xylenes	200-800
Other Aromatics	500-1500

Removal of Impurities

The first step in ethylene recovery is to remove impurities which will affect final product specifications. Additionally, components of gas whose freezing point is above the lowest temperature attained in the process and are insoluble in the light hydrocarbon liquid mixture must be removed to prevent the eventual plugging of processing equipment with solids.

It is essential that FCC off-gas as feed to a recovery unit must be analyzed carefully. Special techniques for sampling, on-site testing, and laboratory analysis are necessary to detect trace amounts of impurities. Multiple sampling and testing are recommended to detect, identify, and quantify trace components.

It is, of course, necessary to essentially remove all the contaminants in order to produce a marketable product. Economical removal of impurities has been a formidable problem in the past and a major concern for refinery fuel gas processing. However, newly available technology, coupled with economic incentives, has made ethylene recovery from refinery fuel gas a reality. It is emphasized that removal of impurities is critically important for a successful ethylene recovery unit.

Acid gases such as H_2S and CO_2 are corrosive in the presence of moisture. These can be reduced to relatively low concentrations by well-established processes such as absorption by amine solutions. Amine treating can provide natural gas pipeline-quality feed gas. However, this still leaves enough residual acid gases to throw high-purity ethylene off specifications. Additionally, H_2S and CO_2 will freeze at the cryogenic processing temperature and must be removed. Essentially complete removal is obtained by multistage caustic washing. A caustic wash system will also remove mercaptans.

Any chlorides, cyanides, ammonia, or caustic are removed by careful water-washing. Removal of water is achieved by a well-known molecular sieve adsorption process.

Acetylene is also found in FCC off-gas. The temperature in the catalytic cracker is usually lower than in the steam cracker for ethylene, resulting in somewhat lower levels of acetylene. Nevertheless, acetylene must be removed to meet high-purity product specifications. Acetylene removal can be done by selective hydrogenation to ethylene, a well established process, commonly used in ethylene plants.

FCC off-gas contains trace amount of dienes. The dienes are separated with the C_3+ product and are therefore not a problem for the ethylene product. However, if high-purity propylene is to be produced in the recovery unit, dienes will have to be removed. The diene saturation can be accomplished with a selective hydrogenation process.

Arsine (AsH_3) is a troublesome contaminant. In the separation process for separating C_2S from higher hydrocarbons, arsine mixes with C_2S. Substituted arsines, such as the one formed by substituting a hydrogen with a methyl group, are also generated in the cat cracker and will probably be separated with C_3 and C_4 hydrocarbons. The arsines are potent and permanent poisons for catalysts used in the selective hydrogenation process for saturation of acetylenes and dienes. Therefore, arsine must be removed. A proven process is available for arsine removal.

Carbonyl sulfide (COS) is another problem contaminant. COS also affects catalyst used in ethylene processing. COS is hydrolyzed and removed by adsorption on activated alumina.

The heavier hydrocarbons, such as C_6+, may freeze out if not removed. Especially troublesome are aromatics, namely benzene, xylenes, and toluene. These can be easily separated, as long as their existence is recognized. Separation is achieved by providing several stages of vapor-liquid separators in the chilling train which provide convenient removal of heavier liquid hydrocarbons.

Ethylene Recovery

Ethylene is recovered by cryogenic separation and fractionation. Ethylene and heavier hydrocarbons are liquified and separated from methane and other "non-condensable" vapors. Ethylene is then recovered from the mixture of light hydrocarbons by low-temperature fractionation. The recovery process is similar to that used in natural gas processing plants designed for high level recovery of ethane and heavier hydrocarbons. However, in addition to conventional gas processing, some of the techniques practiced in liquefying natural gas are also used for an optimized ethylene recovery unit. As an additional benefit, cryogenic processing can be easily modified if hydrogen recovery is also desired.

Figure 1 shows a block flow diagram of an ethylene recovery unit.

FIGURE-1

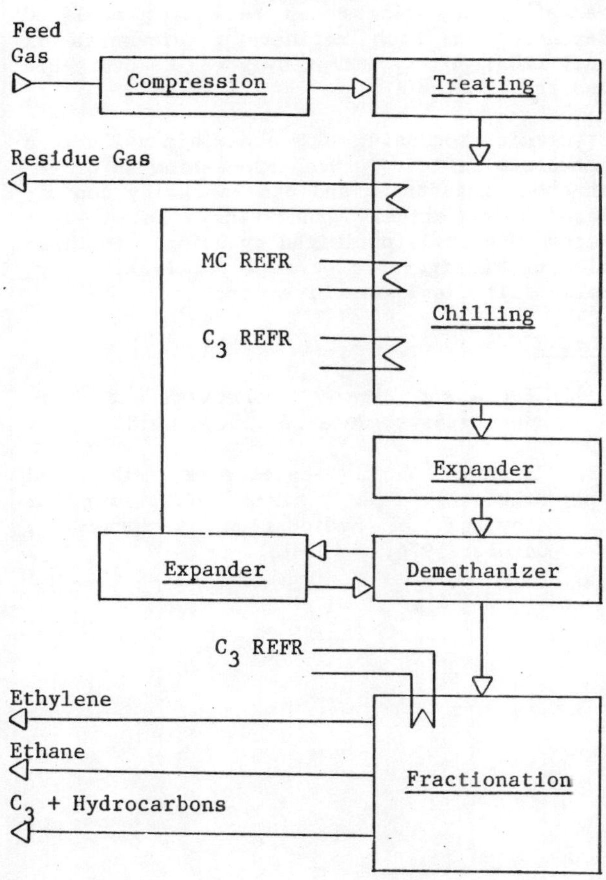

The feed gas is compressed in several stages. The heat of compression is removed at each stage where heavier components condense and are removed. The compression is followed by extensive treating steps to remove impurities as described above.

The purified feed gas is chilled in the chilling section. A combination of available refrigeration cycles is used to chill the gas to liquify ethylene and heavier components for separation from lighter components. An ethylene recovery unit uses closed loop single component, typically propane (or propylene) refrigeration cycle, to pre-cool the feed gas to -40 degrees C. The pre-cooling can be conveniently achieved in a conventional refrigeration unit, avoiding vacuum conditions at the compressor suction. A vapor-liquid separator separates heavier hydrocarbons liquified at these conditions.

A mixed component refrigerant (MCR) cycle is used to further cool the first cool separator vapor. A mixed component refrigerant (MCR) cycle has been analyzed (2). The MCR itself is condensed using propane or propylene refrigerant, thus limiting the range of MCR components. A second cold separator is used following MCR chilling.

The vapor from the second cold separator is further cooled by expansion through an expander. The separator products and expander outlet are fed to a Demethanizer column to fractionate methane and lighter components from ethylene and heavier components. The column is packed with several sections of packing to provide the required number stages for effective separation. To improve ethylene recovery, the tower overhead vapor is expanded further in a second expander to provide cold reflux to the column.

Thus, a combination of:

- Closed loop single component J-T cycle
- Closed loop mixed component J-T cycle
- Two refrigerant cascade cycle
- Expanders

provides an optimized unit capable of about 95 percent ethylene recovery. A great advantage of such a combination is seen in its reliability and flexibility.

A loss of the MCR compressor would still allow continued plant operation, simply reducing the ethylene recovery below the maximum design level. This is an advantage over a unit designed with single component refrigeration, such as ethylene refrigeration, found in many ethylene plants, where the loss of ethylene refrigeration compressor results in the loss of ethylene recovery. This advantage is in addition to the inherent advantages of the MCR system over ethylene refrigeration system such as carbon steel compressor instead of alloy compressor and a single plate and fin exchanger instead of multiple exchangers.

The expanders are provided with J-T valves in parallel. Therefore loss of an expander would still allow plant operation, affecting only the overall ethylene recovery. The plant reliability is further enhanced by multiple separators in the compression and chilling sections, successively condensing and removing liquids starting with the heaviest hydrocarbons. Of course, feed treating cannot be overemphasized, as even traces of impurities will result in mal-operation of the unit.

The demethanized liquid is routed to a low-temperature fractionation section where ethylene and ethane are separated from heavier components in a deethanizer. The ethylene-ethane mixture is separated in the ethylene fractionator where high-purity ethylene is drawn as a side-stream product. Both the fractionators are designed to operate with conventional propane or propylene refrigeration systems.

Cryogenics in Refinery Environment

A cryogenic unit for ethylene recovery from refinery fuel gas will be located inside or adjacent to a refinery, to avoid the large expenses associated with compression and shipping of fuel gas to outside processing. However, cryogenic processing with its associated specialized equipment will be a new concept for a petroleum refiner. Some of the "new" concepts to a typical refiner include:

Product - An ethylene recovery unit will be making a very high-purity (single component) product as compared to the typical, multicomponent blends produced in a refinery. The refiner needs to appreciate the design and operating requirements for such a unit.

Cryogenics - Simple freon or propane refrigeration systems producing moderately low temperatures are found in refinery applications such as gas recovery units, alkylation units, etc. However, applications involving cryogenics with attendant cascade cycles, mixed component refrigeration systems, and expanders are essentially non-existent.

Fractionation - Distillation is, of course, at the heart of petroleum refining. But, separation of close boiling components such as ethylene and ethane requiring super-fractionators will add a new dimension to a refinery's scope. These fractionators also present a unique opportunity for energy conservation through heat pump systems, generally not available for refinery fractionators.

Safety - Petroleum refiners are extremely safety-conscious and rightfully so. Cryogenic processing will add new safety situations and equipment such as the "cold" flare system for handling cryogenic reliefs.

Equipment - From a refiner's standpoint, high-speed expanders, high speed compressors, and plate-fin exchangers will be unconventional equipment. It must be recognized that the "non-refinery type" equipment has been designed, built, and commercially proven to be safe, operable, and reliable in cryogenic process industries such as natural gas processing, LNG plants, air separation, hydrogen recovery, and ethylene plants.

SUMMARY

A stand-alone cryogenic processing unit for ethylene recovery from FCC off-gas component of a refinery fuel gas system is a reality. Excess fuel gas situations at many refineries, coupled with tight supplies and increasing demand for ethylene, present a unique opportunity for refiners to enter the ethylene market. The process can be easily modified depending on each refiners requirements if only LPG recovery and/or hydrogen recovery are desired.

Cryogenic processing adds a new dimension to a petroleum refinery. Such processing is proven in other industries and can be easily assimilated into refinery processing. An ethylene recovery unit is optimized by using a combination of refrigeration cycles for flexibility, reliability, and ease of control.

REFERENCES

(1) "Facts and Figures", Chemical and Engineering News, June 20, 1988, p 38

(2) Kaiser, V., Becedelievre, C. and Gilbourne, D.M., "Mixed Refrigerant for Ethylene", Hydrocarbon Processing, October 1976, p 129-131.

A NEW METHOD FOR DEW AND BUBBLE POINT CALCULATIONS IN MULTICOMPONENT SYSTEMS

Feng Yanpu, Yu Jianping, Zhang Fengchu, Shao Jian
Refrigeration and Cryogenics Lab, Zhejiang University
Hangzhou, China

ABSTRACT

A modified approach for determining the dew or bubble point of multicomponent systems by using the GDSRK equation of state is presented here. In previous work, the convergence of the calculation depended on the initial estimates of the dew or bubble point. In this paper, we use pre-set initial value and new objective function to make the calculations converge successfully and quickly. The algorithm presented is applicable to the calculations using other equations of state.

Keywords: Bubble point; Computer program; Dew point; Equation of state; Hydrocarbon processing;

INTRODUCTION

Dew or bubble point calculations are basic problems involved in the stage-by-stage calculations of distillation processes, and they are often used in other areas. The algorithms for these problems are discussed in many treatises (1,2,3). The algorithms possess two main problems: the convergence of the calculation and the rate of convergence.

Dew and bubble point calculations fall into two types: one is to determine the dew or bubble temperature at a given pressure and composition of the mixture (the so-called "Dew (bubble) point--T problem"), and the other is to determine the dew or bubble pressure at a given temp. and composition (the so-called "Dew (bubble) point --P problem"). The first problem is discussed here because of its wide application. In these calculations, a crucial matter is to find out the compressibility factor Z in selected phase during iterative process. It has a bearing upon the initial value assumed at the beginning of the iteration. A poor initial assumption may lead to the failure of the iteration. Although self-adjusted method for initial value and temperature --disrupted method are raised in documents (1,2), they cannot ensure general convergence. Pre-set initial value is used in this work to remedy this. In addition, we need to calculate the dew or bubble point many times in some calculation processes, so the speed of convergence affects directly the entire calculation speed. We try to speed up the iterative process by using a preset initial value and by changing objective function.

FUNDAMENTAL CONSIDERATION

Many equations of state, such as Soave-Redlich-Kwong EOS(4), Peng-Robinson EOS(5), Chao-Seader EOS(6), and Matin-Hou EOS(7), are used to predict the vapor-liquid equilibria of multicomponent hydrocarbon mixtures. We chose the GDSRK EOS--- a modified version of the RK equation. According to Grabosri and Daubert(8), we have:

$$P = \frac{RT}{V-b_i} - \frac{\alpha_i a_i}{V(v+b_i)} \qquad (1)$$

where α_i, a_i, b_i are parameters defined as:

$$\alpha_i = [1+(0.048508+1.55171\, w_i - 0.15633\, w_i^2)(1-Tri)] \qquad (2)$$

$$a_i = 0.42747 R.Tci/Pci \qquad (3)$$

$$b_i = 0.8664 RTc_i/Pc_i \qquad (4)$$

In a multicomponent system, the mixing rules are as follows:

$$(\alpha a)_m = \sum_{i=1}^{c}\sum_{j=1}^{c} x_i\, x_j\, (\alpha_i \alpha_j a_i a_j)^{0.5} . (1.0-Kij) \qquad (5)$$

$$b_m = \sum_{i=1}^{c} Z_i\, b_i \qquad (6)$$

In the dew or bubble temperature calculations, the control equations (or convergent criteria) are fugacity equalities:

$$f_i^L = f_i^V \qquad i=1,2,\ldots c \qquad (7)$$

and the unity of the sum of compositions in liquid or vapor phase:

$$\sum_{i=1}^{c} y_i/K_i=1 \quad \text{for the dew temp. calculation}$$

$$\sum_{i=1}^{c} X_i \times K_i=1 \quad \text{for the bubble temp. calculation} \qquad (8)$$

In this paper we use secant method in an iterative process. Considering that the number of iterations can be reduced by choosing an objective function which is approximately linear. Here, we chose:

$$F(T)=\ln [\sum_{i=1}^{c} y_i /K_i] =0 \quad \text{for the dew temp.}$$

or

$$F(T)=\ln [\sum_{i=1}^{c} x_i \times K_i]=0 \quad \text{for the bubble temp.} \qquad (9)$$

as the objective function. It's more linear than that using formula (8) and we will see later its effect on the speeding up of the iteration.

ALGORITHM

In dew temperature calculations, the convergence of the iteration bears upon the initial value, especially for multicomponent systems with a narrow two-phase region. A great deviation of the initial value from the actual value may lead to the divergence of the problem. In previous method, an initial value is estimated before the iterative process begins. However, we can not foreknow the composition of the system in many calculations, giving difficulties in estimating initial value. The weighted average of the boiling point of each component is chosen as the initial value in our algorithm, or:

$$T_{set} = \sum_{i=1}^{c} y_i \, Tb_i \qquad (10)$$

where Tb_i is the boiling point at a given pressure and can be determined according to Antoine's equation:

$$\ln P_{vpi} = A_i - \frac{B_i}{Tb_i + C_i} \qquad (11)$$

Antoine's constants A_i, B_i, C_i for each component of hydrocarbon system are listed in Table-1. The calculation shows that Tset is close to the actual dew temperature.

Table. 1 Some Antoine's Constants

component	A_i	B_i	C_i
CO_2	22.5898	3103.39	- 0.16
N_2	14.9524	588.72	- 6.60
CH_4	15.2243	897.84	- 7.16
C_2H_6	15.6637	1511.42	-17.16
C_3H_8	15.7260	1872.46	-25.16
nC_4H_{10}	15.6782	2154.90	-34.42
nC_5H_{12}	15.8333	2477.07	-39.94
C_6H_{14}	15.8366	2697.55	-48.78
iC_5H_{12}	15.9454	2506.81	-31.81
iC_4H_{10}	15.5381	2032.73	-33.15
H_2	13.6333	164.90	3.19

An algorithm has been formulated that combines many of the features described above. Table-2 gives the steps involved in the algorithm for dew temperature calculation. The algorithm includes two loops: the fugacity equality is in inner loop (K-cycle) and formula (9) are checked outside (T-cycle). With the initial assumed temp., the fugacity of each component in the vapor phase (f), and the ideal K-values are determined, then the process enters the inner loop. Having estimated the composition of the mixture in the liquid phase based on the ideal K-values, we can determine the fugacity of each component in the liquid phase (f_i^v) at given T, P, x_i. Then the following conditions are checked:

$$|f_i^L / f_i^v - 1| \leq e_1 \qquad i=1,2, \ldots c \qquad (12)$$

If they are not satisfied, K-values are adjusted as:

$$K = \frac{f_i^L / x_i}{f_i^V / y_i} \qquad i=1,2,\ldots c \qquad (13)$$

where x_i is the present composition of the mixture in the liquid phase, y_i is the given component in the vapor phase. The calculation returns to step 4 with the adjusted K-values until all of the components satisfy the conditions (formulae 12). Here, the K-cycle is carried out at a given temperature. Whether the assumed temperature is correct or not is checked according to formula (9). If not, an initial value adjustment should be made. For comparatively complex GDSRK equations, it's better to use the secant method which needn't calculate derivative. The iterative formula are:

$$T_k = T_{k-1} + \delta T \qquad (14)$$

$$\delta T = - \frac{(T_{k-1} - T_{k-2}) \times F(T_{k-1})}{F(T_{k-1}) - F(T_{k-2})} \qquad (15)$$

where the subscript k is the iterative number in the cuter loop.

Table-2 Steps in the algorithm for the dew temp. calculation

Step No.	Process
1.	Assume an initial dew temp. T_{set}, then set k=1;
2.	Calculate f_i^V at given T,P, y_i;
3.	Calculate ideal K-values K_i^o;
4.	Calculate the composition, fugacity of each component in the liq. phase & new K-values
5.	Repeat step 4 until $\|(f_i^L/f_i^V)-1.0\| < e_1$;
6.	Calculate the value of $F(T_k)$;
7.	If $\|F(T_k)\|$ e_2 goto step 4 else k=k+1;
8.	If k > 2 goto step 9 else adjust T_k as $T_k = 0.999\ T_{k-1}$ goto step 2;
9.	Adjust T_k according to formula (14) (15);
10.	If $\delta T < e_3$ goto step 11 else goto step 2;
11.	convergence achieved: STOP.

As an example of how the algorithm is performed, consider the dew temperature calculation of a five-component mixture, the composition of which is listed in Table-3, at high pressure, 25.0 ata.

As an illustration, the results using different methods are summarised in Table 4. We can see from Table 4 that method C is the best for its smooth and quick convergence while the other methods may fail in iterative process with a bad initial estimate and their calculating speed are much lower than method C. The two-phase region is relatively narrow (about 10 K), so the algorithm is more applicable to the systems with a wide two-phase region.

Table 3 Composition of the Test Mixture

Component	Mol. Fraction %
Propane	65.00
n-Butane	28.00
i-Butane	5.70
i-Penlane	0.50
n-Hexane	0.80

Table 4 Comparision of Calculating Results

Method	initial Temp.	Iterative No.	Dew Temp.
Method A*	1. 65.00	11	95.195
	2. 92.00	7	95.195
	3. 130.00	disvergence	95.195
Method B**	T_{set}	6	95.195
Method C***	T_{set}	4	95.195

*: Method according to document (1);
**: Method with pre-set initial value, but the objective function not changed.
***: Method with pre-set initial value and new objective function.

CONCLUSIONS AND SIGNIFICANCE

In this work it is shown that the real factors affecting the convergence are: (1) the initial estimate for the dew or bubble temp.; (2) the choice of objective function. A new algorithm which uses a pre-set initial temperature and choses a more linear objective function has been presented. The calculating results of test example demonstrate that this simple-structure algorithm leads to smooth convergence and a quick calculation. The proposed algorithm can be used to perform dew and bubble point calculations for systems with a wide or narrow two-phase region. The GDSRK EOS is used for illustration and the algorithm is general and can be used with any other EOS including the BWR type of equation.

NOMENCLATURE

A_i, B_i, C_i--- Antoine's Constant;
a_i,b_i ------- Constant in GDSRK EOS;
c------------ Number of components;
e_1,e_2,e_3----- Controlling accuracy;
f_i^L, f_i^V ------ Fugacity in liquid or vapor phase;
F(T) -------- Objective function of Temp.;
K_i----------- Equili. Constant of component i;
K_i^o ---------- Ideal K-value of component i;
K_{ij} --------- Binary interaction coefficient;
P ----------- Pressure;
P_{ri}---------- Reduced pressure of component i;
P_{vpi} -------- Vapor pressure of component i;
R ----------- Gas constant;
T ----------- Temperature;
T_{bi} --------- Boiling point of component i;
T_k ---------- Temperature of No. k iteration;
T_{ri} --------- Reduced temp. of component i;
T_{set} -------- Assumed initial temperature;
v ----------- Special volume;
x_i ---------- Composition of component i in li-

quid phase;

y_i ----------- Composition of component i in vapor phase;

z_i ----------- Given composition of component i;

REFERENCES

1. Guo, T.M. 'Multicomponent Vapor-Liquid Equilibria & Distillation'. 1st ed.(1983), Chemical Engineering Press.

2. Khoury, F.M., Hydrocarbon Process, Vol.57, No.12, (1978) pp 155-159.

3. Shao, J. etc. 'Calculating Thermodynamic Properties of Petroleum Natural Gases by GDSRK Equation of State'. Acta Petrolei Sinica (Petroleum Precessing Section),

   Vol. 2, No.4 (1986) pp 67-77.

4. Soave, G., Chem. Eng. Sci., Vol.27 (1972) pp 1197-1203.

5. Peng, D.Y., Robinson, D.R. Ind. Eng. Chem. Fundam., Vol.15 (1976) p 59.

6. Chao, K.C., Seader, G.D., AIChE J., Vol. 7 (1961) p 598.

7. Martin, J.J., Hou, Y.C., AIChE, Vol. 5

   (1955) pp125-134.

8. Grabosri and Daubert, Ind. Eng. Chem. Proc. Des. Dev. vol.35 (1978) pp 444.

TECHNOLOGICAL ADVANCES FOR LARGE AIR SEPARATION UNITS

K. B. Wilson, J. M. Abrardo, and D. L. Bennett
©Air Products and Chemicals, Inc.
Allentown, Pennsylvania 18195, U.S.A.

ABSTRACT

The oil-price shocks of the 1970s provided a powerful incentive for the development of energy efficiency in many fields. Very substantial technological improvements in energy intensive processes, such as air separation, have resulted.

These oil-price increases also encouraged the development of major O_2 and N_2 consuming projects, such as oxygen for synthetic fuels and oxygen and nitrogen for the enhanced recovery of oil and natural gas. But the relative decline of oil prices during the 1980s has slowed the implementation of these projects.

Several recent major industrialization projects in east Asia have required large air separation facilities. Most of these have been for the steel industry - our company alone has provided almost 7000 T/D of oxygen capacity for these projects. Considerations such as low energy valuation, capital availability, and project schedule have generally precluded the use of our most efficient technology for these applications. But energy prices can be expected to rise substantially during the next two decades, justifying the use of high-efficiency processes and equipment.

ECONOMIC INFLUENCES

The oil-price shocks of the 1970s affected the world economy and the development of technologies in ways that are still being felt. This paper traces some of the effects on cryogenic technology for large air separation units during the past 15 years.

The sudden oil-price rise in 1973 caused immediate but modest changes in air separation plant design. Because most air separation plants use electric motors as prime movers, the effect of increased oil prices was not experienced immediately, but was soon reflected in a dramatic increase in projected future costs for electricity. Air separation plants designed in the mid-1970s had to use energy valuations which were two or three times higher than those for previous plants. (Incremental savings of energy which had been worth \$1000/kW, were suddenly valued at \$2000 to \$3000/kW.) As a result, all of the energy/capital optimizations undertaken during plant design were focused on decreasing energy consumption at the expense of additional capital investment. Abrardo et al. (2) describe in some detail - from process line sizing to variations in distillation tray count - the ways in which these optimizations are effected. Because energy cost can account for more than half of the value of products from an ASU, reoptimization of designs for high cost energy resulted in significantly more expensive plants.

For the United States, the increased cost of imported oil encouraged accelerated development of alternative energy sources. Three of these sources involve the use of large air separation units:

- Almost all of the processes for making synthetic liquid or gaseous hydrocarbons involve a gasification step which must be supplied with high pressure oxygen. Scharle and Wilson (2) reported estimates that up to nine million Nm^3/h of O_2 could be needed for a U.S. synfuels industry by the early 1990s.

- One of the promising techniques for recovering heavy oils involves in-situ combustion with oxygen. North America has vast quantities of heavy oil which may be amenable to this technique, and total oxygen requirements up to two million Nm^3/h were forecast.

- Other techniques for enhanced recovery of oil and natural gas have been developed which require large quantities of high pressure nitrogen; nitrogen generators with capacities up to 110,000 Nm^3/h were built for this market in the late 1970s and early 1980s (3).

The prospect of very large markets for oxygen and nitrogen, coupled with the need for more energy-efficient air separation technologies led to increased efforts in R&D by the major suppliers of air separation equipment. For example, Air Products' annual spending in this area increased by an order of magnitude in the 1970s and continues today at a high level.

TECHNOLOGICAL RESULTS

The drive toward increased energy efficiency for cryogenic air separation can be undertaken in two basic ways: by increasing the efficiency of individual pieces of equipment or by using more efficient process cycles. In fact, a third route is also available; some cycles can only be utilized if more efficient equipment is available. Examples of all three types are described below.

Wilson et al. (4) present the following breakdown of energy consumption for a typical low-pressure (LP) cycle oxygen plant, as depicted in Figure 1:

Work of Separation	26%
Compression Losses	23%
Prepurification Losses	6%
Heat Exchange Losses	22%
Distillation Losses	16%
Miscellaneous Losses	7%

Each of these energy losses represents an opportunity for efficiency improvement. Significant technological advances have been made in each of these categories during the past decade. The nature of these improvements and their applicability to new ASUs are discussed below.

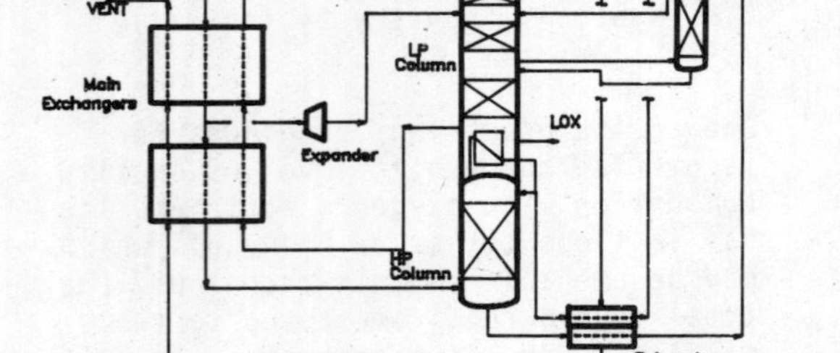

Fig. 1 Low pressure cycle oxygen plant

COMPRESSION

A substantial improvement in compression efficiency has taken place in the past 15 years, resulting from changes in aerodynamic design (such as improved impeller/diffuser designs or variable exit diffusers) and the optimization of intercooler pressure drops and temperature approaches. Opportunities for future improvements may lie in the use of turbo-compressor wheels with lower compression ratios, or from means to drive each wheel at close to its optimum speed. Most modern air compressors use three-dimensional, backward-leaning blade impellers; but the use of highly efficient axial compression stages is still limited to very large applications.

PREPURIFICATION

Most modern air separation units use molecular sieve adsorption for prepurification, although there are still some applications where reversing exchangers are preferred (5). Two principal opportunities for energy saving in prepurification lie in the means needed to precool the air and condense some of its moisture prior to adsorption, and the means used to preheat the adsorbent reactivation gas. Since many ASUs have substantial dry waste streams, it is frequently economic to recover high-level refrigeration with countercurrent humidifiers, thereby obtaining chilled water. This water can then be used to displace the mechanical refrigeration units previously needed for air precooling. In a somewhat analogous manner, the waste heat which is normally dissipated in the air compressor aftercooler can be used to displace electrical energy from the adsorbent reactivation gas heater.

The adsorption vessels used for large ASUs typically have thick walls with substantial heat capacities. Various proprietary designs have been developed to mitigate the waste of energy used in heating and cooling these walls during each reactivation cycle. Internal insulation, flow baffles and wall heaters have been demonstrated in various applications. Any design selected must be able to withstand cyclic variations in temperature and pressure, and must properly consider the detrimental effects of flow bypassing on adsorbent bed performance.

All of these techniques for saving prepurification energy will be most applicable when energy is most valuable; they may not be justifiable when low capital investment is given top priority.

HEAT EXCHANGE

Brazed aluminum core type heat exchangers are used in all modern large air separation units. They offer unparalleled efficiency of heat transfer between multiple gaseous streams. There have been only modest improvements to the performance of these exchangers during the past forty years, but the reduction in their cost per unit area has been substantial. As a consequence, the principal effect of increased energy cost has been to increase the optimum heat exchange area used in ASUs. Additional exchanger length reduces energy losses from temperature differences; additional exchanger cross section reduces pressure drop energy losses.

Special heat exchange concepts have been developed by Air Products to enhance boiling and condensing heat transfer (6). For energy values of about \$2500/kW, this type of exchanger can reduce the ΔT obtainable with conventional core type reboiler/condensers by 40%. But for energy values less than about \$1500/kW, use of these more expensive exchangers can seldom be justified.

DISTILLATION

The distillation columns are at the heart of the cryogenic process for air separation. Perhaps as a consequence, each of the major suppliers of air separation equipment has developed its own technology for this item. All suppliers probably used aluminum sieve trays with small tray spacing (100-250 mm) and with relatively small perforations (~1 mm). Bennett and Wilson (6) describe other aspects of tray design in some detail.

This type of design has achieved theoretical tray heights of 150-450 mm, theoretical tray pressure drops of 35-100 mm of liquid, and throughputs of 30,000 $Nm^3/h/m^2$. Recent developments in the use of ordered packing for distillation make it possible to achieve pressure drops which are an order of magnitude less, with only modest increases in column volume. The large number of theoretical trays used in a cryogenic ASU translates this reduction in pressure drop per theoretical tray into very substantial reductions in feed air pressure and air compressor energy requirements.

As with many other energy saving improvements, ordered packing is more expensive than the distillation trays that it replaces; it can only be justified when the benefits that it brings are adequately valued. In addition to energy efficiency, packing is superior to trays with regard to argon recovery and rangeability (turndown), but for applications where these factors are less important, efficient sieve trays are still appropriate.

The second major avenue for energy saving lies in improvements to process cycles. The widely-used LP cycle (Fig. 1) can produce large quantities of oxygen, nitrogen, and argon at low pressures. A nitrogen refrigeration cycle can be integrated with the LP cycle to produce up to 85% of the air feed as liquid product. The versatility of this cycle results from the presence of significant irreversibilities. If the cycle were modified to produce only the products needed for a particular application, some of these irreversibilities could be eliminated to yield a substantially more efficient cycle.

Most of the wasted energy in the LP cycle occurs because of excessive concentration driving forces in parts of the low pressure column. These driving forces can be reduced by providing for additional heat exchange with the distilling fluids. Wilson et al. (4) provide examples where this approach has led to energy reductions of up to 10%. Agrawal et al. (7) use the same approach to achieve substantial increases in product recoveries and pressures. (It should be noted that some of the processes cited in these papers are only practical with the low ΔPs and ΔTs obtainable with ordered packing and advanced reboiler concepts.)

Using nitrogen as the working fluid in the refrigeration cycle can also produce excessive driving forces in the distillation column. Unless much of the nitrogen is needed as liquid or as gas from the high pressure column, air may be a more efficient refrigerant. Roughly half of the advantage comes from reducing the heat transfer duty in the main heat exchangers, the balance from a more efficient distillation process.

Some of these cycle improvements are beneficial for any energy value; others, which obtain their efficiency improvements from more equipment or more stages of distillation, will need high value energy to justify their use.

OUTLOOK

Two additional consequences of the oil-price shocks were a worldwide economic slow down, a decline in oil prices, and a moderation in the rate of electricity price rise. As a result, the large, energy driven markets for oxygen and nitrogen have not yet materialized, and some of the energy efficient technologies developed for air separation are still too capital-intensive to be justified by today's energy cost.

Looking to the future, however, it seems clear that energy prices will continue to rise and efficient processes for air separation will ultimately gain wider use. Those investing in air separation plants during the next decade should be aware, therefore, that technology exists today which can insulate them from future increases in energy cost. By placing a realistic value on future energy costs, small incremental investments in better technology can be justified. These investments can result in substantial future benefits.

REFERENCES

1. Abrardo, J. M., A. R. Smith, and D. R. Vinson, "Optimization of Cryogenic Oxygen Plants," Proceedings, XVII International Congress of Refrigeration (1987), Vol. A, pp. 214-219.

2. Scharle, W. J., and K. B. Wilson, "Oxygen Facilities for Synthetic Fuel Projects," J. of Engineering for Industry (1981), Vol. 103, pp. 409-417.

3. Metzler, K. J., and L. J. Hvizdos, "Integrated Air Separation Facilities for Hydrocarbon and Sulfur Recovery," presented at the AIChE Summer National Meeting, Denver, CO, U.S.A. (1988).

4. Wilson, K. B., D. C. Erickson, and D. W. Woodward, "New, Low-Energy Processes for Cryogenic Air Separation," Proceedings, International Cryogenic Engineering Conference - 12 (1988), pp. 355-359.

5. Wilson, K. B., A. R. Smith, and A. Theobald, "Air Purification for Cryogenic Air Separation Units," Proceedings, XVI International Congress of Refrigeration (1983), Vol. A, pp. 33-39.

6. Bennett, D. L., and K. B. Wilson, "Optimization of Cryogenic Air Separation Column Designs," Proceedings, International Conference on Cryogenics and Refrigeration (1989), Hangzhou, PRC.

7. Agrawal, R., D. C. Erickson, and D. W. Woodward, "High Efficiency Processes for Cryogenic Air Separation," Proceedings, 7th Intersociety Cryogenic Symposium (1989), Houston, TX, U.S.A.

PROCESS ANALYSIS OF A KIND OF MOBILE AIR SEPARATION PLANT

Chen Lianshen and Lai Qiyao
Lin Yun Technical Research Center
Hangzhou, China

Cryogenerator was used as cold resource by the first kind of mobile air separation plant in China, which was provided with plate-fin heat exchangers, a molecular adsorber and a liquid nitrogen bubble pump. It has advantages of compact structure, flexible mobility, wide application, easy and safe operation and etc. Its flow process in shown in figure 1.

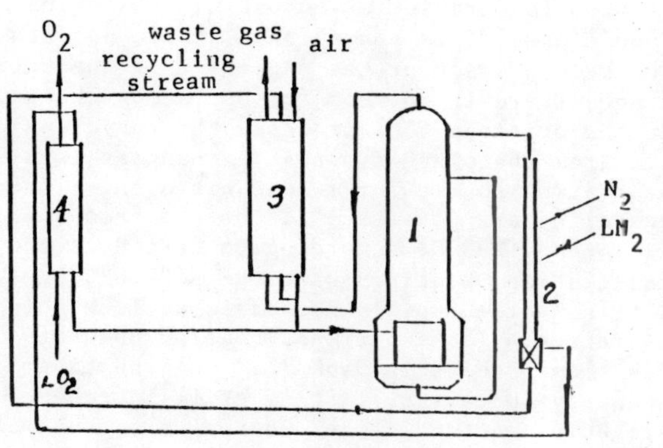

1 Column 2 Bubble pump
3 Plate-fin heat exchangers
4 Cold recoverage heat exchanger

Fig. 1 Flow process diagrammatic drawing

Through production and application, some imperfections have been discovered among this kind of plant which needs to be improved by technical renovation.

This paper will deal with problems in the process and some experiments improved.

I

What is the problems in this kind of air separation plant from view point of technological flow process?

A. Air into the column is on the low side

Discharge of air compressor provided is 180 m^3/h, reducing to 150 m^3/h after renovated. Design air capacity into the column is approx. 90 Nm^3/h. The ex-factory output of oxygen is 12 m^3/h. If delivery of air is increased, LN_2 level in LN_2 bubble pump will increase at once so as to damage the distillation condition. If the air into the column is not sufficient, the pressure of medium in cryogenerator in gas production condition must drop down, otherwise purity of oxygen reduces owing to excessive refrigeration. Since the working condition is limited by the air capacity into the column, the efficiency of air compressor and cryogenerator can not be given full play to and the output of the plant product remains on the low side.

B. Great cold loss because of lower oxygen temperature out of the column

After recoverage of cold by reflex in heat exchangers, oxygen temp. is 163°K which has a 140°K decrease in comparison with that of air into the column, i.e. it takes 6280-7540 KJ to produce 1 Kmol oxygen. So the lost is too great.

C. Unstable operation and difficult adjustment

Temperature in the middle of the platefin type regenerator is controlled by a recycling stream in the original flow process by which LN_2 in the bubble pump is heated to rise automaticly to the top of the column. Recycling flow is controlled by the opening of air adjusting valve of normal-direction flow so that approx. 173°K in the middle part of the regenerator can be ensured. This can be guaranteed easily. But the question occurs whether the bubble pump can work normally and whether LN_2 from the cryogenerator can be heated and whether all of LN_2 can rise to the top of the column by this recycling stream. In practical operation, the double task can not be fulfilled by one recycling stream. The temperature in the middle part of the regenerator is under control and the LN_2 level in the bubble pump is easily out of control and vice versa. It makes difficult to adjust operating conditions and keep stable operation.

D. High specific power consumption of the product

Specific power consumption of oxygen/Nm^3 is 6 KWH which is higher than that of ordinary small air separation plants, and is higher than that of Imitated-Soviet Type AK-12 Mobile Oxygen Plant.

II

Which is the causes of the existing problems?

A. Why does the LN_2 level immediately rise when a little more air is fed into the column?

There are two possibilities: One is excessive load so as to cause liquid flood in the column and down-flow liquid in rectification section is blocked so that LN_2 is detained at the top of the column; the other is that the poor function of the bubble pump so that the liquid from cryogenerator can not be lifted to the top of the column according to requirement.

Let us analyse the first possibility--liquid flood.

According to the hydraulic calculation under the original designed LO_2 condition (air into the column is approx. 90 Nm^3/h).

The min. perforation vapor speed without leakage of liquid:

ωo min =2.388m/s

Practical perforation vapor speed

ωo =1.172m/s

Tray vapor speed

ω =0.088m/s (all are data in rectification sector)

Since practical speed is low, tray distance is apparantly bigger than min. tray distance when there are no flood and mist of liquid mixed. Thus liquid flood will not happen without maloperation. In the experiment, we also make hydraulic calculation under gas production condition (air into the column approx. 120Nm^3/h),

ωo min =2.32 m/s
ωo =2.35 m/s (little over ωo min)
ω =0.125 m/s

While min. tray distance is 17.52mm, min.tray distance without mixed mist is 31.95mm, and actual tray distance is 50mm. Perforation vapor speed is on the low side, not on the high side according to air rate calculation of the experiment operation condition. In general, liquid flood is not the reason which causes LN_2 level to rise and influences the air into the column.

Analysis of the second possibility--problem of the bubble pump.

Liquid nitrogen from cryogenerator passes through the centre pipe to LN_2 tank. Liquid nitrogen is heated by air and part of it evaporates to create bubbles. LN_2 is lifted to the top of the column as reflux by the floating force of bubbles and the static pressure head of LN_2. The energy for lifting LN_2 comes from the heat released from air for heating. If we assume the increase of heat is divided into three stages, the lifting is increasing obviously while the heat is increasing in the first stage; the lifting nearly stops when the heat continues to increase to a certain value, i.e. the second stage. Because excessive bubbles enter the lifting pipe to decrease the liquid into the incoming pipe, the lifting remains basicly unchanging; After the second stage, if heat continues to increase, i.e. the third stage. This time heat is obviously excessive, the lifting rapidly decreases and the LN_2 level in the bubble pump increases speedily. So which stage our working point should be put on influences the performances strongly. If we choose the period at the end of the first stage or the beginning of the second stage, the lifting efficiency of the pump can be brought into full play and it is also favourable for adjusting of working conditions; If we choose that at the end of the second stage or the beginning of the third stage, there is no room for operation change. In the original flow process, the recycling air from the plate-fin heat exchangers is used as heating source for liquid nitrogen. The recycling air possesses 22% of the processed air, temp. 293°K; According to test data and calculation, heating quantity is 2788KJ/h, lifting value rapidly decreases while heating increases. The experiment heating quantity in the flow process is 916KJ/h. When heating quantity increases, lifting still increases slightly. According to our analysis, the working point at 2788KJ/h in the flow process may be situated at the beginning of the third stage; But that at 916KJ/h is at the end of the first stage.

From our view-point, the reason that excessive air can not be delivered is mainly the poor function of bubble pump.

B. Why are great losses caused from the low temp. of oxygen out of the column?

Cold of liquid oxygen is recovered by the recycling air after heat exchange with liquid nitrogen in the bubble pump in the original flow process. The temperature of recycling air is very high (293°K), but after heat-exchanged with liquid nitrogen, temp. lowers down to 193°K. The inlet temp. of cold recoverer is 108°K. The temp. differential of them is 85°K, i.e. the heat-exchange efficiency is poor. According to theoriginal calculation of the flow process, temp. differential ΔTh of warm end is 30°K, temp. differential ΔTc of cold end is 5°K, average log temp. differential

$$T_{ml} = \frac{30-5}{\ln\frac{30}{5}} = 13.95^{\circ}K$$

In experiment, the recycling stream ($293^{\circ}K$) was directly from the plate-fin regenerator to cold recoverer (not via bubble pump).

Based on measuring and calculation,

$T'h = 71^{\circ}K$ $T'c = 2.3^{\circ}K$

$$T'ml = \frac{71-2.3}{\ln\frac{71}{2.3}} = 20^{\circ}K$$

Heat-exchange efficiency in the experimental flow process was increased obviously and oxygen temp. out of the column rose to $222^{\circ}K$. The low temp. of heat stream (air) for recovering cold of liquid oxygen and the low efficiency of heat-exchange are one of the reasons for the low temp. of oxygen out of the column. Meanwhile this heat stream, i.e. the recycling stream from the plate-fin heat exchangers is used to treat air. 1 Nm^3 of air needs 0.22Nm^3 recycling stream for treatment. As for recovering of cold of 0.1684 Nm^3 of liquid oxygen, it is obviously not sufficient. This is another reason.

As for unstable operation, difficult adjustment and high specific power consumption of the mobile oxygen plant, the reasons were mentioned in the first part briefly. Here we are not going to describe further.

III

Some conclusions from the above analysis:

A. If you hope to increase oxygen production, air into the column must be increased; If you want to increase air into it, the question of the rising of LN_2 level in the bubble pump should be solved; If you wish to solve LN_2 level problem, heating quantity has to be reduced, the working point should be shifted to the end of the first stage, 2500-2930KJ/h should be reduced to 750-1000KJ/h.

B. If you want to reduce cold loss, temp.and flow rate of warm stream (air) in heat exchangers for cold recoverage should be raised. While cold loss reduces, reflux reduces, load of cryogenerator reduces and lifting capability of the bubble pump increases relatively. If the original reflux amount is maintained, air into the column can be increased.

C. If you want to improve operation reliability, air quantity for heating of liquid nitrogen in the bubble pump ought to be regulated separately, and will not be influenced by air quantity into the column.

All problems from the three sides concentrate on the line of the recycling channel outlet of plate-fin regenerator → the bubble pump heating body → cold recoverage heat exchangers. This provides a good condition to settle the problems and to improve the process at the least cost. flow process after revision is shown in figure 2.

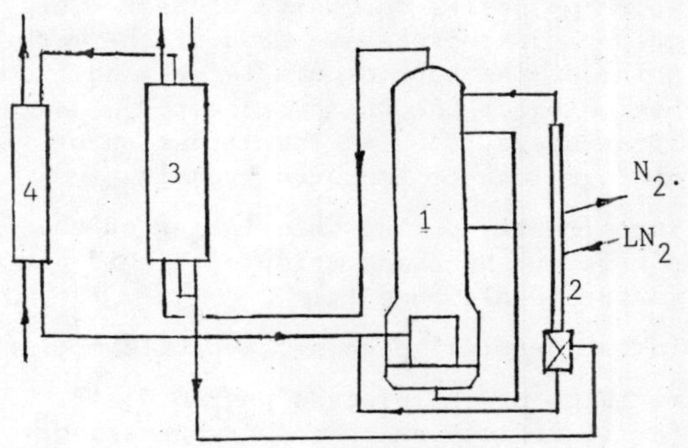

Fig. 2 Flow process diagrammatic drawing after revision

The main points for improvement:

A. The recycling stream ($293^{\circ}K$) from the plate-fin regenerator should be directly fed to the heat exchangers for cold recoverage of liquid oxygen. After reducing of temp., it is combined with direct flow air and enters condenser-evaporator.

According flow process calculation, enthalpy of oxygen out of the column in the original flow process:

$Io = 5036KJ/Kmol$

Enthalpy of oxygen out of the column after revision of the flow process:

$Ir = 9123KJ/Kmol$

Enthalpy difference

$I = 9123-5036 = 4087$ (KJ/Kmol)

This is the cold loss saved after revision of the flow process. H_2 working pressure of the cryogenerator is 2.5MPa, cold produced approx. 9604KJ/h before revision of flow process of the experimental model plant.

During measurement of revised plant, H2 presaure of the cryogenerator is 1.5MPa, cold produced approx. 6833KJ/h. The cold saved is 2771KJ/h. Production of oxygen is 15.76 Nm^3/h ($18.1m^3/h$) converting to 0.703 Kmol/h. Cold saved for production of 1 Kmol:

$$\frac{2771}{0.703} = 3942KJ/Kmol.$$

This figure conforms approximately to the enthalpy difference from the calculation according to the flow process.

B. Try to feed a part of lower temp. air to heat liquid nitrogen in the bubble pump.

As mentioned in the original flow process, air for heating liquid nitrogen is recycling stream, temp. $293^{\circ}K$, recycling flow 0.22 Nm^3

(unchangeable) is needed for treatment of 1 Nm^3 of air, releasing energy 2788KJ/h; After revision, heating air temp. is $113.5^{o}K$, flow 0.5 Nm^3 (adjustable, in general it is not needed to be adjusted), releasing heat 916KJ/h which possesses one third of heat energy in the original process. Thus if the working point of the bubble pump is shifted from the beginning of the third stage to the end of the first stage, the lifting situation of liquid nitrogen can be improved greatly.

Furthermore, during the lifting of the liquid nitrogen, the evaporation of liquid nitrogen in the bubble pump is:

In the original process, Avo=0.12Nm^3; After revision of the process, Avr=0.03Nm^3. Nitrogen for liquification in the cryogenerator:

In the original process:

Aco =0.436 Nm^3

After revision of the process:

Acr =0.53 Nm^3

Evaporation rate of liquid nitrogen:

$\frac{Avo}{Aco}$=0.277 $\qquad$ $\frac{Avr}{Acr}$=0.0566

From the equation, evaporation rate has been deoreased greatly.

While obviously reducing of evaperation of liquid nitrogen after revision, the rectification condition of the column is improved and the power consumption of cryogenerator reduced relatively.

C. The direction of air for heating of liquid nitrogen in the bubble pump heating body is now from the upper partto the lower part instead of from the lower part to the upper part. The purpose of doing so is to avoid air inlet from the centre pipe and so that the heating is even and slow to protect from air blocking.

D. Increase of process air. Air flow into the column after revision is 120Nm^3/h, before revision it is 80-90 Nm^3/h. After revision pressure in the column is raised from 0.04-0.045 MPa to 0.05-0.06 MPa.

IV

Summarising from the above, the crux of revising the flow process is to greatly increase air to the column under the same condition of the cold produced by the cryogenerator through reducing cold loss and improving working of the bubble pump so as to increase highly the oxygen output. Based on this basic summary, we revised the flow process for several scores of this kind of air separation plants,and have got good result from each of them.Oxygen output is raised from 10 Nm^3/h to 19Nm^3/h. H_2 pressure of cryogenerator is reduced from 2.5MPa. Power consumption is reduced from 6KWH/Nm^3 to 2.8KWH/Nm^3..

Merits after revision of the flow process:

A. Operation condition of rectification is more stable. Since operation performance of the bubble pump is improved, LN_2 level maintains stable no matter whether air into the column is increased or H_2 pressure in cryogenerator is raised (within a certain scope). When temp. in the middle deviats from the operation condition, it can be controlled within the regulated sphere by appropriate adjustment of the opening of direct-flow valve to keep the stability of temp. field because the recycling stream is not used to heat liquid nitrogen any more for specially adjusting of the temp. in plate-fin regenerator.

B. When adjustment is reduced, the operation is easier. After revision of the flow process, there is no need for adjustment of waste nitrogen discharge and liquid oxygen into the LO_2 pump. It is difficult to adjust LO_2 into the pump, if the operator is incompetent, the adjustment may influence the operation condition. After revision labor intensity is reduced and it is favourable to continuous stability of operation condition.

DEPHLEGMATOR PROCESSES FOR THE RECOVERY OF VALUABLE HYDROCARBONS

K. B. Wilson, D. P. Bernhard, and H. C. Rowles
©Air Products and Chemicals, Inc.
Allentown, Pennsylvania 18195, U.S.A.

ABSTRACT

High efficiency cryogenic processes using fractionating condensers, known as "dephlegmators," have been developed for the recovery of valuable hydrocarbons. These processes, associated with refinery and petrochemical operations, have been used to achieve high recovery of ethylene and heavier components from large flows of light gases. Such processes can be integrated with new or existing hydrocarbon recovery systems. Process flow diagrams for recovery of ethylene from cracked gas, light olefins from FCC unit off-gas, and MTBE precursors (butylenes) from dehydrogenation unit off-gas are presented.

The heart of these processes is the dephlegmator, which takes the place of the distillation column or phase separator in conventional processes. Liquid condensed from the feed gas rising in the dephlegmator flows back as reflux. Various refrigeration sources are employed for efficient condensation of the reflux. Brazed aluminum core type heat exchangers are used to provide efficient heat and mass transfer in compact cryogenic units.

INTRODUCTION

Dephlegmators have been proposed for many years to reduce energy consumption in gas separation processes. Some of the earliest cryogenic gas separation plants designed by Georges Claude used dephlegmators. The full potential of dephlegmation, however, has been hampered by the difficulty of predicting the performance of such equipment. Air Products has solved this problem and has been using dephlegmators commercially in a variety of cryogenic gas separation processes since 1970.

The dephlegmator, while maintaining simplicity of equipment and operation, provides greatly enhanced separation capability over conventional partial condensation processes, which rely on the equilibrium established between vapor and liquid after cooling and phase separation. A dephlegmator can also provide enhanced efficiency in many distillation processes, where it serves the dual function of rectifying column and partial condenser. This is particularly true where high recovery of heavy components is desired from gas mixtures having high concentrations of light components, and where the relative volatility between light and heavy key components is two or more. Integration of a dephlegmator process with conventional distillation, either for preseparation or postseparation, can increase the recovery of valuable hydrocarbon components.

DEPHLEGMATOR DESIGN AND OPERATION

Modern dephlegmators are brazed aluminum heat exchangers with special design features. These exchangers, which are commonly used in cryogenic service, consist of many layers or passages to transfer heat among several streams. Dephlegmators include enhancements that permit them also to behave as mass transfer devices. In particular, dephlegmator feed stream passages are designed to allow liquid flow downward, countercurrent to the flow of vapor.

Figure 1 illustrates how a dephlegmator functions. The feed vapor stream flows upward within some of the passages, and heat is removed by other process or refrigerant streams in adjacent passages. Some fraction of the feed condenses on the heat transfer surface, forming a liquid which flows downward, refluxing the rising vapor. The reflux becomes enriched with the less volatile components as the more volatile ones are revaporized.

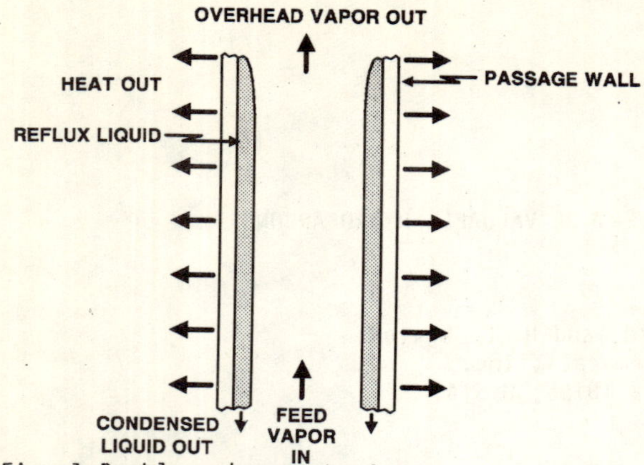

Fig. 1 Dephlegmator mechanisms

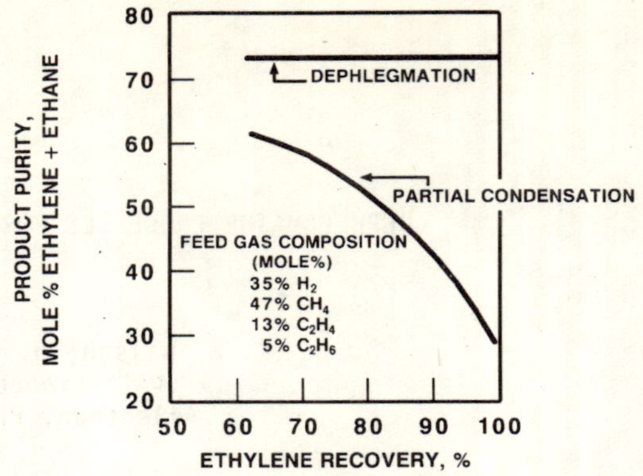

Fig. 2 Comparison of dephlegmator and partial condensation process performance

Unlike reflux liquid produced in a conventional distillation column condenser, dephlegmator reflux is generated throughout the mass transfer zone, over a wide temperature range. Conventional distillations have all the refrigeration supplied in the condenser, the coldest temperature point. Since much of the refrigeration can be supplied at warmer temperatures in a dephlegmator, energy savings are obtained.

In comparison to a partial condensation process, a dephlegmator process provides a different benefit. In a partial condensation process, a valuable hydrocarbon liquid is separated from the feed vapor at the coldest temperature. By comparison, a dephlegmator process recovers the liquid at a warmer temperature; this minimizes the quantity of light hydrocarbons and inerts dissolved in the liquid. Energy savings, or alternatively higher hydrocarbon recoveries, are reaped in downstream fractionation steps by reducing the concentration of lights.

Figure 2 compares the ideal performance of partial condensation to dephlegmation for separating ethylene from a feed gas laden with hydrogen and methane. The partial condensation process achieves ethylene recovery at the expense of purity. The dephlegmator process, on the other hand, achieves high ethylene recovery <u>and</u> purity. The flat dephlegmator curve is calculated for perfect equilibrium between the feed gas and the recovered ethylene-rich product. In practice, heat and mass transfer resistances limit the approach to equilibrium, but the deviation, which depends upon feed and refrigerant stream conditions as well as equipment configuration and size, is very slight.

Dephlegmator performance is determined by a latticework of interrelated transfer processes; accurate predictions of heat, mass, and momentum transfer resistances, as well as fluid thermophysical properties, are essential for equipment design. Air Products has conducted extensive bench-scale tests simulating dephlegmator operation, utilizing a variety of test systems, and covering a wide range of operating conditions. These tests have provided verification of: the thermodynamic data used for dephlegmator design, the heat transfer and mass transfer characteristics of dephlegmators, and correlations used to determine operating flow ranges and capacities. These tests, together with the results of successful commercial applications, provide highly reliable methods for predicting dephlegmator performance. Our thermodynamic data base, combined with rigorous cycle synthesis simulation techniques, has been shown to be extremely accurate in predicting the performance of dephlegmators.

APPLICATIONS

Dephlegmator processes are ideally suited for recovering valuable hydrocarbons such as ethylene from light gas mixtures. Three examples of hydrocarbon recovery processes are presented below to illustrate dephlegmator applications:

(1) ethylene recovery from cracked gas;

(2) olefins recovery from FCC off-gas; and

(3) C_4 hydrocarbons recovery from dehydrogenation unit off-gas.

These are only samples of the broad spectrum of gas separations to which dephlegmator technology can be applied. The same concept, for instance, can be used in processes to separate methane from H_2-N_2 and H_2-CO synthesis gases and C_{3}+ hydrocarbons from refinery and natural gases.

Example 1: Ethylene recovery from cracked gas

Figure 3 shows a block diagram of an ethylene plant. Following bulk ethylene recovery, lean cracked gas is conventionally processed to recover additional ethylene by partially condensing it at -130°C. A demethanizer, operated at 3.5 MPa, and subsequent fractionation equipment separate ethylene and other hydrocarbon products from the lean and concentrated C_{2}+ streams. Ethylene is lost to fuel in the uncondensed vapor leaving the preseparation step. More ethylene is lost with the substantial flow of methane that is rejected to fuel by the demethanizer.

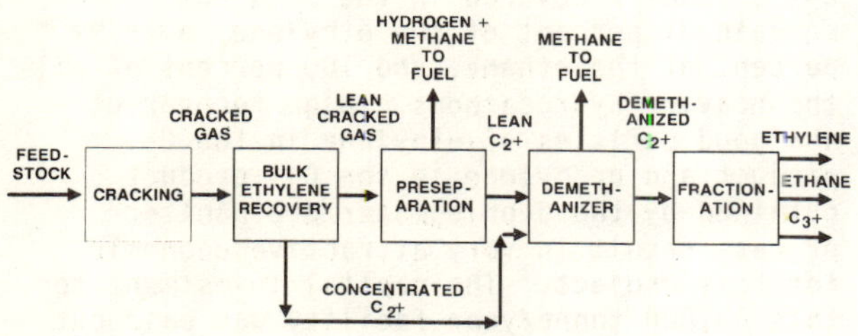

Fig. 3 Ethylene plant schematic

Dephlegmator processes have been developed to address both of these avenues for ethylene loss. These processes have been integrated with new and existing facilities to substantially increase ethylene production.

A dephlegmator preseparation process for ethylene recovery from lean cracked gas is illustrated in Figure 4 (2,3). When substituted for the partial condensation process, it enhances ethylene recovery, reduces demethanizer condenser refrigeration requirements, and/or increases the capacity of an existing plant. Because ethylene recovery takes place at a relatively warm temperature (-100°C or above), it contains little dissolved methane. This substantially reduces the loss of ethylene in the demethanizer overhead.

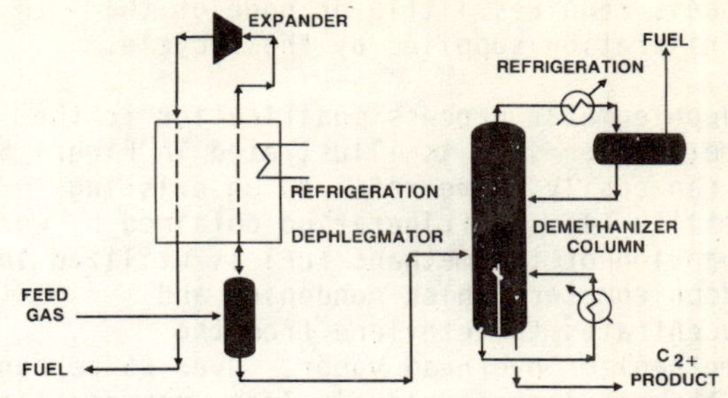

Fig. 4 Dephlegmator preseparation of ethylene from cracked gas

Table 1 compares dephlegmator preseparation to partial condensation preseparation for a new 450,000 tonne per year ethylene plant. A one percent improvement in recovery and power savings of 2000 kW, together worth over \$1.8 million per year, can be achieved with lower capital costs for demethanization and refrigerant compression, and a competitive capital cost for preseparation. A dephlegmator preseparation process, incorporated in an ethylene plant of this size, has continuously demonstrated these marked economic savings since it began operation in 1978.

Table 1. Comparison of Dephlegmation to Partial Condensation

Production Increase (1%) (a)	4,500 tonnes
Power Savings (10%)	2,000 kW
Net Benefit (b)	\$1,800,000

Notes:
(a) Basis: 450,000 tonnes per year production rate.

(b) Ethylene is valued at \$0.30/kg above the sum of its fuel value and the cost of production. Electricity is valued at \$8.30/GJ.

In an existing facility, a dephlegmator preseparation step can be used to debottleneck the plant. Typically, excess reflux, containing methane and a significant amount of ethylene, is condensed in the demethanizer condenser by evaporating ethylene refrigerant. The excess liquid is then evaporated to refrigerate the partial condensation preseparation process, and is subsequently recycled and compressed by the cracked gas compressor upstream of the bulk ethylene recovery section of the plant. Replacing the partial condensation preseparation step with a dephlegmator preseparation step can reduce or eliminate the recycle and allow increased throughput of cracked gas. To achieve the same fractional recovery of ethylene, the dephlegmator process requires little or none of the refrigeration supplied by the recycle.

A dephlegmator process modification to the demethanizer step is illustrated in Figure 5; it can easily be retrofit to an existing facility (4). Refrigeration obtained by work expansion of the methane fuel is utilized in a dephlegmator, which condenses and concentrates the ethylene from the demethanizer overhead vapor. Over 85 percent of the ethylene previously lost, representing almost one percent of plant production, can be recovered with no modification to the existing column or auxiliary equipment. Recovered ethylene is returned to the demethanizer. A recent study indicated that the investment in such a retrofit could be repaid in less than a year.

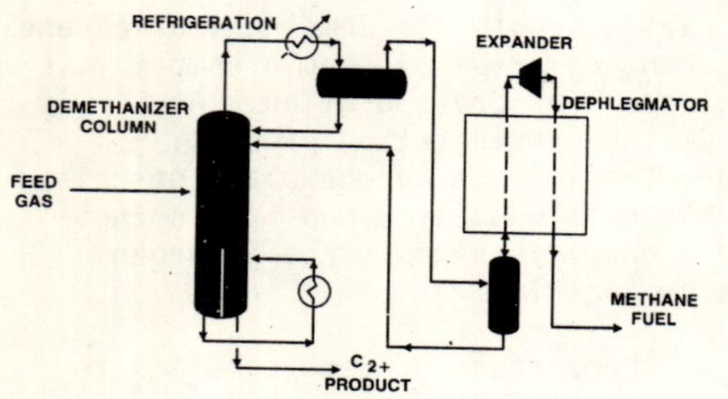

Fig. 5 Recovery of ethylene from demethanizer overhead gas

Example 2: Olefins recovery from FCC unit off-gas

The off-gas streams from fluid catalytic cracking (FCC) units are similar to the cracked gas streams produced in ethylene plants. Figure 6 is the flowsheet of a recently installed process in which C_2 and heavier hydrocarbons, previously burned as fuel, are recovered via preseparation and deethanization (5). The dephlegmator preseparation process recovers a concentrated C_2+ stream containing 12 percent methane and inerts, much less than the 36 percent lights content achievable by partial condensation. The partial condensation process would require additional capital expense for demethanization to meet product specifications. The deethanizer processes the concentrated C_2+ stream and recovers a C_2 fraction containing less than 2 percent propylene and propane and a liquid C_3+ fraction containing 400 ppm of ethane. The C_2 fraction is valuable in an ethylene plant as a supplementary source of ethylene. The C_3+ product is suitable for further fractionation to recover the individual hydrocarbon constituents, notably propylene.

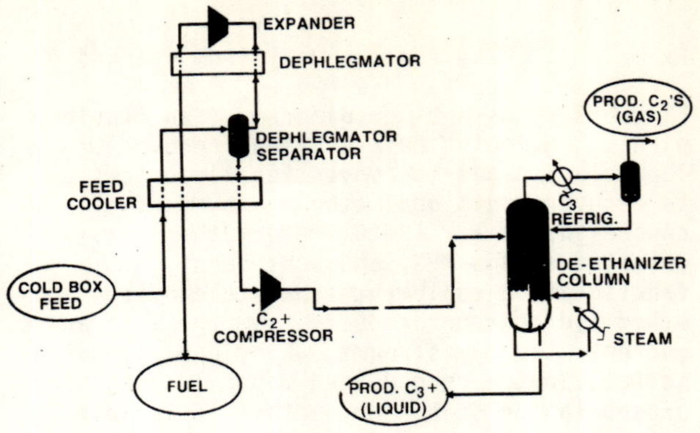

Fig. 6 Olefins recovery from FCC off-gas

Concomitant with the excellent purity of the concentrated C_2+ stream recovered by the dephlegmator process is high recovery. The C_2+ streams recovered in the cold box contain 91 percent of the ethylene, over 99 percent of the ethane, and 100 percent of all the heavier hydrocarbons. High recoveries and good purities of ethylene in the C_2 product and propylene in the C_3+ product obtained by the dephlegmator/deethanizer process result in very attractive economics for this project. The capital investment for this 58,000 tonne/year facility was paid out in 6 months.

Example 3: C_4 hydrocarbon recovery from dehydrogenation unit off-gas

One step in the production of the gasoline octane-enhancer MTBE involves recovery of C_4 olefins from dehydrogenation gas. Dephlegmator processes for recovery of C_4 hydrocarbons are similar to those for ethylene recovery. They provide very high levels of C_4 hydrocarbon recovery while minimizing the concentration of undesirable C_3 hydrocarbons and lighter components in the feed to the depropanizer downstream. The process flow diagram for a recently installed C_4 recovery unit is shown in Figure 7 (6). The unit recovers 25,000 tonnes/year of butylenes along with other C_4 hydrocarbons from 120,000 tonnes/year of feed gas containing 91 percent C_3 hydrocarbon and lighter impurities. All the refrigeration required for dephlegmator operation is provided by expanding the residual light gas from the top of the dephlegmator. Overall C_4 hydrocarbon recovery is 98.5 percent, and the liquid purity is 86 percent C_4 hydrocarbons.

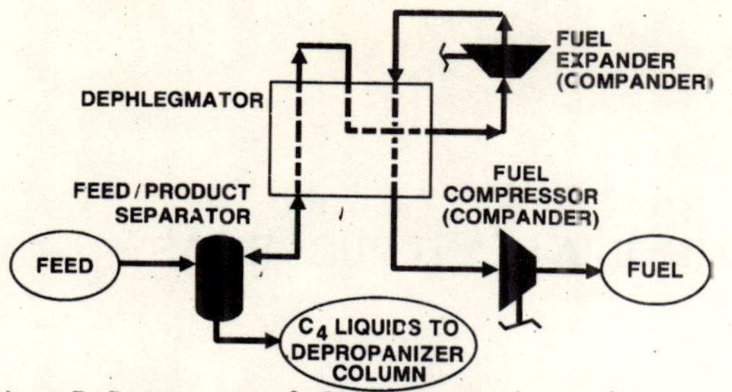

Fig. 7 Recovery of C_4 hydrocarbons from dehydrogenation off-gas

CONCLUSIONS

Dephlegmator processes have been used by Air Products in a wide variety of cryogenic gas separation applications. They are intrinsically simple and reliable, and yet efficient, both theoretically and practically. Their efficiency results from the effective use of available refrigeration to generate internal reflux.

Coupled dephlegmator and fractionation processes and equipment for hydrocarbon separations yield an excellent return on investment. Preseparation or postseparation steps can be added to existing facilities or integrated with new facilities.

Accurate thermodynamic data and valid correlations for heat/mass transfer and flow hydraulics, combined with rigorous process simulation techniques, provide the capability for design of highly efficient cryogenic gas separation systems.

REFERENCES

1 Lucadamo, G. A., D. P. Bernhard, and H. C. Rowles, "Improved Ethylene and LPG Recovery Through Dephlegmator Technology," Gas Separation and Purification 1 (1987), pp. 94-102.

2 Chiu, C. H., H. C. Rowles, D. Nahmias, and J. M. Geist, "Dephlegmator Cycle for Ethylene Plants Application," Proceedings of the 15th International Congress of Refrigeration (1979).

3 Pryor, J. A., and H. C. Rowles, "Recovery of C_2+ Hydrocarbons by Plural Stage Rectification and First Stage Dephlegmation," U.S. Patent 4,002,042 (1979).

4 Rowles, H. C., and T. R. Tsao, "Recovery of C_2 Hydrocarbons from Demethanizer Overhead," U.S. Patent 4,270,940 (1981).

5 Bernhard, D. P., J. A. Moss, J. L. Pickering, and H. C. Rowles, "Increased Olefins Production Via Recovery of Refinery Gas Hydrocarbons," presented at the NPRA Annual Meeting, San Antonio, U.S.A. (1988).

6 Bernhard, D. P., H. C. Rowles, and R. L. Reichman, "Process for Recovering C_4+ Hydrocarbons Using a Dephlegmator," U.S. Patent 4,519,825 (1985).

CRYOGENIC ARGON RECOVERY FROM AMMONIA PLANT PURGE GAS

W. S. Hwang and D. J. Kamrath
BOC Cryoplants, The BOC Group
Murray Hill, New Jersey 07974, U. S. A.

ABSTRACT

Ammonia plant purge gas is a viable source of argon. BOC Cryoplants has developed commercially proven processes to economically recover this argon from the purge gas itself or from the tail gas of existing cryogenic or membrane hydrogen recovery units.

INTRODUCTION

In many parts of the world, the demand for argon has exceeded its supply as a by-product of oxygen production. Ammonia plant purge gas containing relatively high concentrations of argon, (often in excess of 4%), is a viable argon source. BOC Cryoplants, the engineering and construction arm of The BOC Group, has developed processes to recover this argon economically for commercial sale (1, 2). Two commercial plants designed and constructed by BOC Cryoplants, both processing purge gas from two nominal 1,000 short tons per day (stpd) ammonia plants, are successfully operating in North America.

In addition to argon, the process can be adapted to recover hydrogen, nitrogen, and fuel gas as co-products for return to the ammonia plant to aid process economics. Krypton and xenon, which enter with the feed air to the secondary reformer of the ammonia plant and are concentrated with the argon, may also be recovered.

This paper will discuss the argon market, cycles for recovering argon from ammonia plant purge gas, and process economics.

THE NEED FOR ALTERNATE ARGON SOURCES

Since 1970, the demand for argon has been growing at an average rate of more than 10% per year worldwide. The United States alone now consumes about 1,600 stpd of argon. The gas has been steadily gaining importance in the production and fabrication of both ferrous and non-ferrous metals as well as the manufacture of electronic components. For instance, 90% of the stainless steel currently manufactured in the United States uses the argon oxygen decarburization (AOD) process.

Historically, the growth in argon demand in the United States has averaged about 8% per year while oxygen demand has grown at a much more modest rate of less than 2% per year (3). Because argon has traditionally been recovered as a by-product of oxygen production, the disparity in growth rates for the two gases has caused argon demand to exceed its supply. Ammonia plant purge gas, with its high inherent argon concentration, provides a cost effective alternate argon source.

HOW PURGE GAS IS PRODUCED

Figure 1 shows a typical ammonia process (4). High-pressure, catalytic reforming of natural gas is used to produce a raw synthesis gas consisting of hydrogen, carbon monoxide and carbon dioxide. Air is introduced to the secondary reformer to provide the stoichiometric quantity of nitrogen required for ammonia synthesis.

The reformer gas is shift converted, purified in subsequent steps, compressed to about 2,100 psia, and sent to the synthesis converter. Ammonia is recovered from the synthesis converter product stream by condensation. Unconverted synthesis gas is recycled back to the synthesis gas compressor for further processing. A purge gas stream is continually withdrawn to limit the concentration of inerts (primarily argon and methane) in the synthesis loop to about 16%.

The pressure and composition of the purge gas varies from plant to plant. Figure 2 shows the composition of a typical raw purge gas stream as it arrives at the feed gas dryers of the purge gas separation unit. This stream is of particular interest for argon recovery because the concentration of argon in the gas stream is several times that

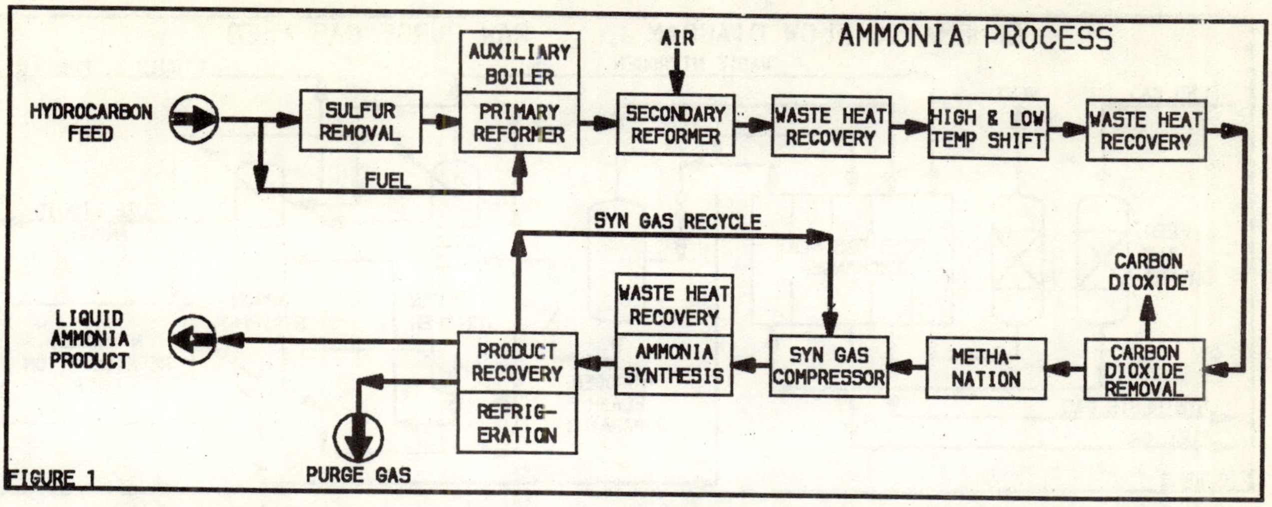

FIGURE 1

in air. The argon can be almost totally recovered as a useful product. The hydrogen and most of the nitrogen is recovered as a synthesis gas stream that is returned to the synthesis gas compressor. The methane is used as fuel gas for the primary reformer, and the argon is stored as a liquid for commercial sale.

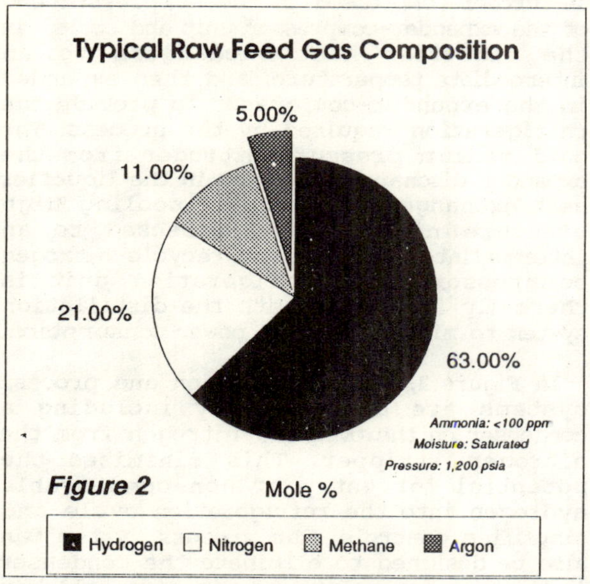

Figure 2

Complete recovery of the purge gas products from a 1,000 stpd ammonia plant will produce about 14 stpd of pure argon and increase ammonia production by about 60 stpd. It will also provide a methane-rich fuel gas stream containing approximately 30,000,000 Btu/hr of fuel energy. If this stream is utilized to generate saleable power using a combustion turbine, economics of the project will be greatly enhanced. It is estimated that 1,360 KW can be generated by using a simple power cycle or 1,900 KW using a combined cycle turbine-generator (5).

PURGE GAS PLANT CYCLES

The design of the argon recovery unit is tailored to the characteristics of its feed gas stream. This stream may be the raw purge gas from the ammonia synthesis loop or gas that has already been partially treated to recover hydrogen. The characteristics of three typical feed streams are shown in Table 1. BOC Cryoplants has developed cycles to process all three feeds. One BOC plant in North America processes raw purge gas feed while the second processes feed from a cryogenic hydrogen recovery unit.

TABLE 1
PURGE GAS FEED STREAM VARIATIONS

	Raw Purge Gas Feed	Cryogenic Hydrogen Recovery	Membrane Hydrogen Recovery
Feed Pressure, psia	1,500-2,000	45-90	1,500-2,000
Composition, %			
Hydrogen	60-63	10-15	9-30
Nitrogen	20-21	50-65	33-54
Methane	10-14	18-22	20-27
Argon	3- 6	4-12	7-12

CYCLE PROCESSING RAW PURGE GAS FEED

The BOC Cryoplants process separates purge gas from the ammonia plant into its constituent streams by distillation at cryogenic temperatures. A schematic of the process is shown in Figure 3 (1).

Ammonia in the purge gas feed is removed by absorption in water and returned to the ammonia plant. The saturated ammonia free purge gas stream is dried in a desiccant dryer, cooled and partially liquefied in the feed gas heat exchanger, and separated in a series of distillation columns.

The overhead product of the flash separator is the synthesis gas recycle stream. It contains 80-90% hydrogen and 10-20% nitrogen with a small quantity of argon. The synthesis gas stream is rewarmed in the feed gas heat exchanger to recover its refrigeration and returned to an intermediate stage of the

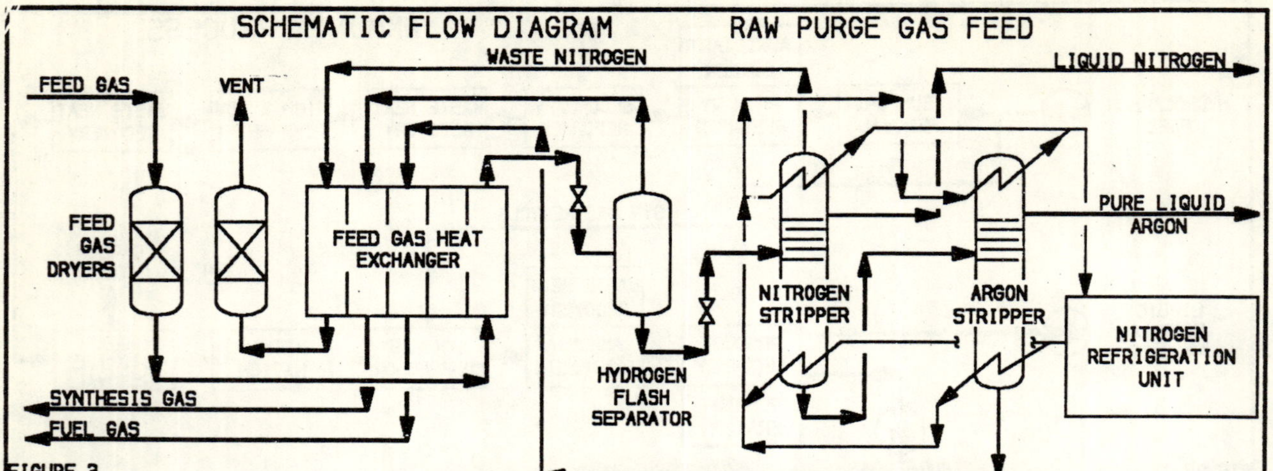

FIGURE 3

synthesis gas compressor. In this stream more than 90% of the hydrogen entering the plant is recovered.

All remaining hydrogen and nitrogen is removed from the process stream in the nitrogen stripper. The nitrogen content of the bottom stream leaving this column is maintained at less than 0.3 ppm to limit the nitrogen in the argon product to 1 ppm.

The nitrogen and hydrogen leave the column in an overhead waste nitrogen stream that is used to regenerate the feed gas dryers. A small pure liquid or gaseous nitrogen stream may also be withdrawn to provide make-up for the nitrogen refrigeration cycle machinery seal losses.

The final separation of argon and methane is performed in the argon stripper. Pure liquid argon containing less than 1 ppm nitrogen and less than 1 ppm methane is withdrawn as the overhead product; pure liquid methane is withdrawn as the bottoms product. The liquid methane is vaporized and rewarmed in the feed gas heat exchanger and sent to the ammonia plant fuel gas header. The process can recover up to 95% of the argon and essentially 100% of the methane entering in the feed to the dryers.

The nitrogen refrigeration cycle provides the cooling required to maintain the process at cryogenic temperatures and the energy needed to operate the distillation columns. High pressure nitrogen from the recycle nitrogen compressor is cooled to near saturation in the liquefier heat exchangers and condensed in the nitrogen and argon stripper reboilers. This condensation process provides the energy to drive the columns.

The liquid nitrogen thus produced is subcooled and sent to the column condensers. In the condensers, the liquid nitrogen is vaporized against condensing column vapors to provide column reflux. To close the cycle, the low pressure gaseous nitrogen leaving the condensers is warmed to ambient temperature in the subcooler and liquefier heat exchangers and recompressed to high pressure in the recycle nitrogen compressor.

Part of the high pressure nitrogen stream is further compressed in the compressor end of the expander-compressor unit and cooled in the liquefier heat exchangers to an intermediate temperature, and then expanded in the expander-compressor to provide the refrigeration required by the process. The cold medium pressure nitrogen from the expander discharge is warmed in the liquefier heat exchangers against the cooling high pressure nitrogen and directed to an intermediate stage in the recycle nitrogen compressor. The refrigeration unit is thermally integrated with the distillation system to minimize overall power consumption.

In Figure 3, the refrigeration and process systems are decoupled by including a condenser on the overhead nitrogen from the nitrogen stripper. This minimizes the potential for entry of non-condensable hydrogen into the refrigeration cycle and simplifies controls. The process system can also be designed to eliminate the condenser in the nitrogen column to minimize cold box height and capital investment. Low pressure nitrogen from the column is then fed directly to the refrigeration system and liquid nitrogen from that system is directly introduced to the column to provide reflux.

The process power consumption is basically that of the recycle nitrogen compressor. It requires about 460 KW to process a purge gas stream having the composition and pressure shown in Figure 2 from a 1,000 stpd ammonia plant and returning hydrogen to the ammonia plant at 400 psia. The plant can also be run to recover hydrogen only at a nominal power consumption.

BOC has built a plant of this basic design modified to achieve exceptionally high hydrogen recovery. Its performance is shown in Table 2. Generally the performance of the plant has exceeded expectations. Argon and hydrogen recoveries exceed design levels. Minimal operator attention is required even though feed flow and composition vary much more than originally anticipated.

TABLE 2
PERFORMANCE RESULTS
RAW PURGE GAS FEED

	Design	Actual
Argon Recovery, %	92.0	95.0
Hydrogen Recovery, %	98.5	99.0
Methane Recovery, %	100.0	100.0
Nitrogen in Argon, ppm	<1.0	0.3 avg
Methane in Argon, ppm	<1.0	Nil

CYCLE PROCESSING FEED FROM CRYOGENIC HYDROGEN RECOVERY UNITS

BOC Cryoplants has developed a process for recovering argon from the tail gas stream of a cryogenic hydrogen recovery unit (2). The cryogenic hydrogen recovery unit recovers hydrogen from the high pressure, dry, ammonia-free purge gas by taking advantage of available Joule-Thomson refrigeration from the feed. The recovered hydrogen-rich stream is returned to the ammonia synthesis loop at near feed gas pressure to enhance the ammonia production (6). The low pressure tail gas, containing almost all the argon and methane, the remaining nitrogen and residual quantities of hydrogen, becomes the feed to the argon recovery unit. Because the tail gas has been stripped of ammonia and dried upstream of the hydrogen recovery unit, no ammonia absorber and desiccant dryer is required for the argon recovery unit.

Figure 4 shows the schematic flow diagram of the argon recovery process. The tail gas is partially liquefied in the feed gas heat exchanger prior to introduction into two distillation columns. Operation of the nitrogen and argon strippers is similar to the cycle that processes raw purge gas.

The refrigeration required to maintain cryogenic temperatures and liquefy the argon product as well as the energy to drive the distillation columns is provided by a nitrogen refrigeration-vapor recompression unit consisting of a recycle nitrogen compressor, compressor-expander unit, and liquefier exchanger equipment. The refrigeration unit is optimized to provide the best fit for the process requirements.

The power consumption for the plant is essentially that of the recycle nitrogen compressor. Because the feed is low pressure tail gas from the cryogenic hydrogen recovery unit, little Joule-Thomson refrigeration is recoverable. Therefore, much more power is needed to process this tail gas than was required for the purge gas taken directly from the ammonia synthesis loop at high pressure. Approximately 880 KW is required to process a hydrogen depleted purge gas stream from a 1,000 stpd ammonia plant having the composition and pressure shown in Table 1.

BOC Cryoplants installed an argon recovery unit in Canada that processes the tail gas from cryogenic hydrogen recovery units on two nominal 1,000 stpd ammonia plants. The argon recovery unit had a trouble-free start-up and very successful performance run. The results of the performance test are tabulated in Table 3. The plant has demonstrated that it is able to operate stably under automatic control and meet product purity specifications and design argon recovery.

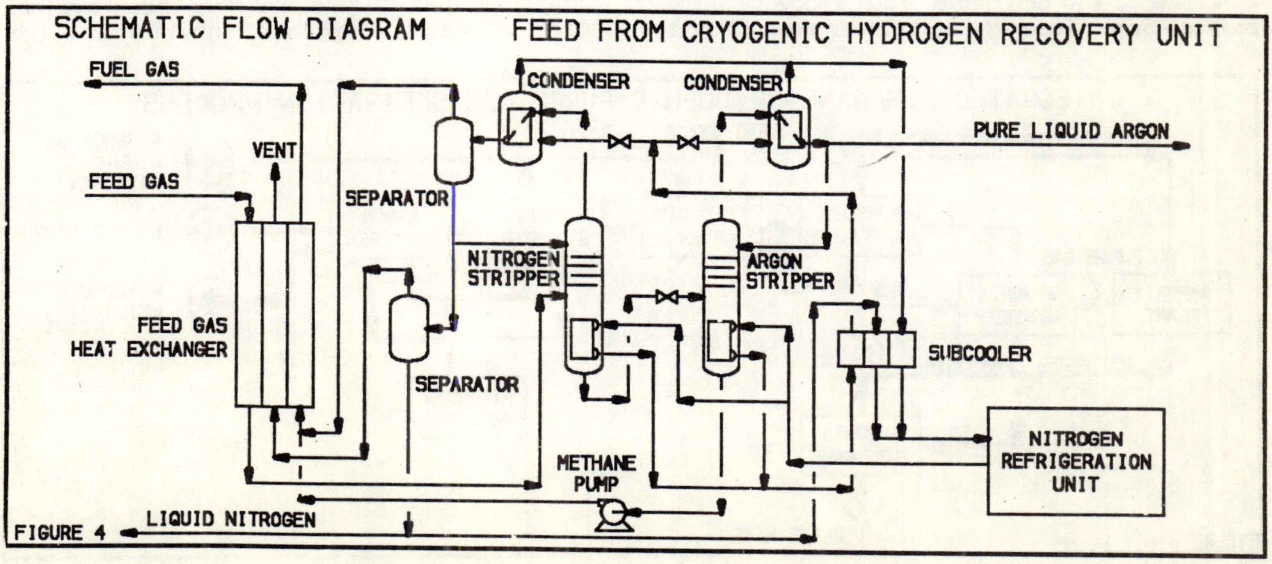

FIGURE 4

TABLE 3
PERFORMANCE RESULTS
CRYOGENIC HYDROGEN TAIL GAS FEED

	Design	Actual
Argon Recovery, %	95.0	95.2
Methane Recovery, %	100.0	100.0
Nitrogen in Argon, ppm	<1.0	0.37 avg
Methane in Argon, ppm	<1.0	0.27 avg

CYCLE PROCESSING FEED FROM MEMBRANE HYDROGEN RECOVERY UNITS

Typical tail gas composition and pressure ranges for a membrane hydrogen recovery unit are shown in Table 1. The membrane separation process utilizes the principle of selective permeation through a gas permeable membrane (7). The driving force for the separation comes from the partial pressure of the hydrogen in the purge gas stream. The process requires a high pressure feed, which is inherently available from an ammonia plant. Unlike a cryogenic hydrogen recovery unit, the hydrogen returns at a reduced pressure while little reduction in pressure is incurred in the tail gas stream.

Figure 5 is a block flow diagram of an integrated membrane-cryogenic purge gas separation process (1). Since hydrogen is already removed in the membrane separators, the purge gas unit does not require a hydrogen flash separator. The argon recovery unit process topology is similar to that of the cycle processing a tail gas from the cryogenic hydrogen recovery unit (2).

Since the membrane tail gas stream is at a pressure near that of the raw purge gas, significant Joule-Thomson refrigeration is available, resulting in low argon unit power consumption. Membrane removal of hydrogen upstream of the cryogenic argon recovery unit reduces argon recovery unit power consumption further. However, this effect is small because the high relative volatility between hydrogen and nitrogen allows cryogenic separation of these components with little energy input. Approximately 410 KW is required to process a membrane tail gas stream from a 1,000 stpd ammonia plant having the composition and pressure shown in Table 1.

TYPICAL PLANT ECONOMICS

The argon recovery unit is compact in design, requiring minimal site area. All cold process equipment is housed in a single shop fabricated cold box that only needs to be set and insulated on site. Other major equipment includes a recycle nitrogen compressor to provide energy for separation, an expansion turbine to provide refrigeration, a liquid storage tank and an enclosure to house analytical instruments. The argon recovery unit may be controlled remotely from the existing ammonia plant control room, if so desired, or from a local dedicated control room.

The capital investment for a plant processing raw purge gas installed in the United States and designed to recover recycle synthesis hydrogen, argon and fuel gas from a typical 1,000 stpd ammonia plant is approximately US $3,500,000. Beside normal maintenance costs for machinery and instrumentation, the major operating expense is the power cost for the recycle nitrogen compressor (about 460 KW to process the raw purge gas from a 1,000 stpd ammonia plant). The plant is fully automated, requiring minimal operator attention.

ADVANTAGE OF ALL CRYOGENIC ROUTE

Figure 6 shows the relative capital costs and power consumptions of argon recovery units designed to process the three typical feed streams. Capital requirements for

INTEGRATED MEMBRANE-CRYOGENIC PURGE GAS SEPARATION PROCESS

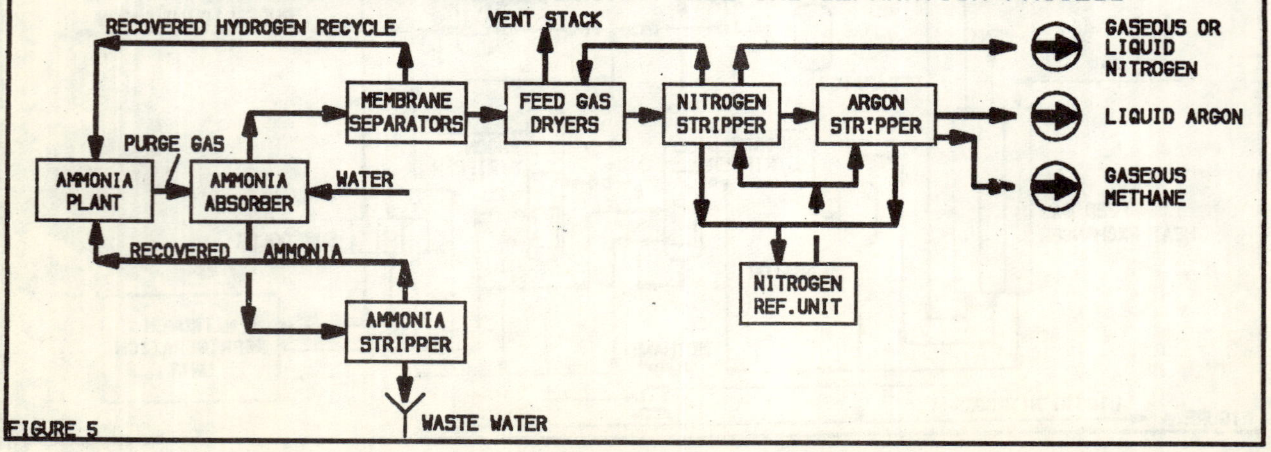

FIGURE 5

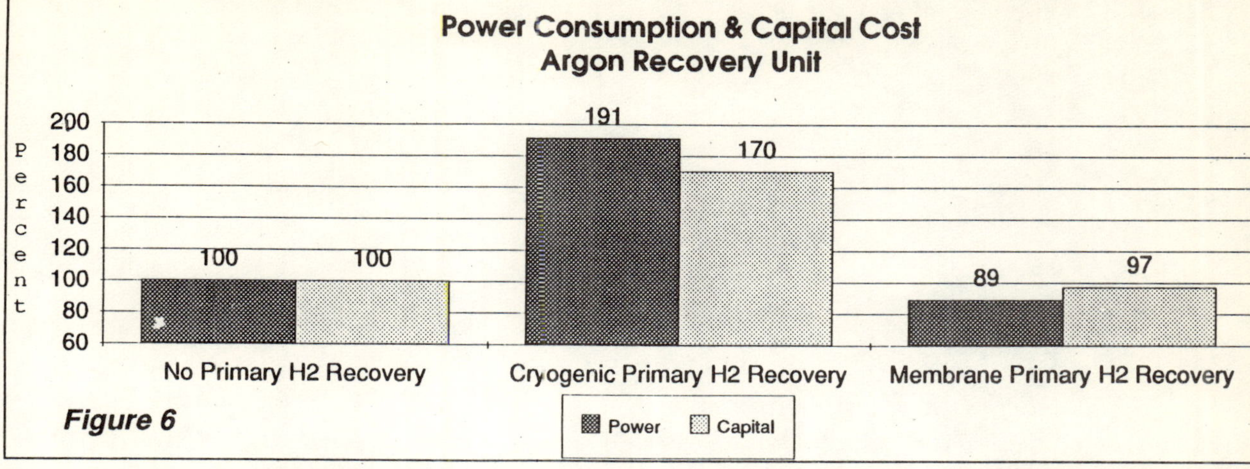

Figure 6

cryogenic or membrane primary hydrogen recovery equipment are excluded.

Figure 6 indicates clearly that there is both a capital cost and power disadvantage in decoupling cryogenic hydrogen and argon recovery. In this configuration, all available Joule-Thomson refrigeration is utilized by the cryogenic hydrogen recovery unit, resulting in a very low pressure tail gas stream as the feed to the argon recovery unit. All refrigeration for the argon recovery unit must then be provided mechanically, significantly increasing both capital cost and power consumption. Of course, the cost of the cryogenic hydrogen recovery unit must be added to the already high cost of the argon recovery unit, increasing the economic disadvantage of this approach even further.

The integration of the argon recovery unit with a membrane hydrogen recovery unit will reduce energy input up to 12%, but the additional cost of the membrane separators may offset the power savings.

CONCLUSION

Ammonia purge gas is an attractive source of argon. The argon concentration in the purge gas is relatively high and secondary products can be returned to the ammonia producer to defray part of the capital and operating costs of the purge gas separation unit.

With the successful commissioning of two commercial plants (35 stpd pure liquid argon capacity each) in North America, BOC Cryoplants has proven designs which will allow accurate estimation of costs and performance for future applications. The plants are easy to operate, require minimal operator attention and achieve high product purities and recoveries.

The lowest overall capital and operating costs for recovering both hydrogen and argon from the purge gas stream are realized by processing the stream in a single cryogenic recovery unit. For those ammonia plants that already have installed cryogenic or membrane primary hydrogen recovery units, special argon recovery unit designs have been developed that can still cost effectively recover the argon.

ACKNOWLEDGMENT

The permission to publish this paper by The BOC Group management is acknowledged.

REFERENCES

1 Gilmartin, B., Hwang, W., Kamrath, D. and Musicus, P., "Production of Argon From Ammonia Plants", The BOC Group Technology Magazine, No. 2, pp. 19-24, (1985).

2 Mishkovsky, V., Hwang, W. and Kamrath, D., "Process for the Recovery of Argon", U.S. Patent, 4,762,542, (1988).

3 Current Industrial Reports, Industrial Gases, M28C, U.S. Department of Commerce, Bureau of the Census.

4 LeBlanc, J.,"Make Ammonia with Less Energy", Hydrocarbon Processing, pp. 69-72, July 1984.

5 Kleinhaut, J., Private Communication, (1984).

6 Banks, R., "Hydrogen Recovery Unit Ups NH3-Plant Efficiency", Chemical Engineering, pp. 90-92, (Oct. 10, 1977).

7 Chae, Y., Legendre, G., and van Gelder, J., "Hydrogen Recovery from Ammonia Plant Purge Gas Using Hollow-Fiber Gas Separators", 4th Fert. Nitrogen Proc., Int. Conf. Fert. Technology., pp. 457-469 (1985).

2. CRYOGENIC MACHINERY AND EQUIPMENT

OPTIMIZING THE DISTILLATION COLUMN SYSTEM FOR AN AIR SEPARATION PLANT

D. L. Bennett and K. B. Wilson
Air Products and Chemicals, Inc.
Allentown, Pennsylvania, U.S.A.

ABSTRACT

The need for thermodynamic efficiency in air separation plants results in complex integrated distillation column systems. These systems can be optimized to meet the customer's needs, based on a clear understanding of product requirements and the relative values of products, capital and power.

Many factors need to be considered in this optimization, including the method of distillation, the type of packing or tray to be used, specific details to enhance distillation efficiency, maximum production requirements, turndown requirements, constraints on product purity and recovery at off-design rates, and the type and design details of the reboiler/condenser.

This paper discusses these considerations, provides some information about the various options and gives some guidelines for their application.

BACKGROUND

Distillation columns for cryogenic air separation are considerably more complex than columns usually encountered in the petro-chemical industry. Figure 1 illustrates this complexity for the low pressure cycle which produces nitrogen, oxygen, and argon products.

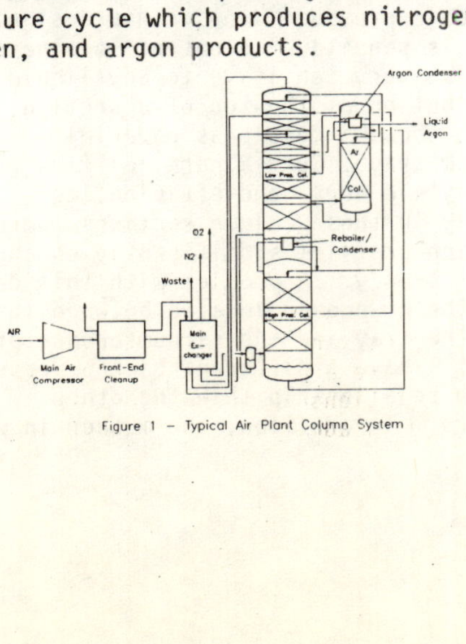

Figure 1 – Typical Air Plant Column System

The air compressor feeds air to the front-end cleanup portion of the plant where water vapor and carbon dioxide are removed. The purified air stream is then cooled to cryogenic temperatures and introduced into the high pressure column. A major function of this column is to generate the reflux streams for the low pressure column. The low pressure column illustrated in Figure 1 has six sections which allow a close matching of the operating line to the equilibrium curve. Reducing the composition driving forces in the column in this manner permits thermodynamically efficient separation. The crude argon column shown in Figure 1 separates argon from oxygen.

Two reboiler/condensers are illustrated in Figure 1. The crude argon condenser provides reflux for the argon/oxygen separation. The high and low pressure columns are thermally linked through the reboiler/condenser which provides reflux for the high pressure column, and boil-up for the low pressure column. The performance of this reboiler/condenser has a direct effect on the energy consumption of the air compressor.

Based on the relative values of capital, energy, and products, the process cycle and the design of the process equipment are selected to develop an optimized design for the customer's needs. The complexity of the design and the variety of available design options makes this optimization an important portion of the commercial package. This paper discusses the considerations required and the options available for the distillation sections and for the reboiler/condenser.

DISTILLATION

The need for thermodynamically efficient separation results in cycles and distillation column designs which use small driving forces for mass transfer. A large number of theoretical stages are needed to obtain near complete separation. The column illustrated in Figure 1 can have in excess of two hundred theoretical stages within the three-column distillation system. Cycles of higher efficiency may require even more stages. The pressure losses occurring in all these distillation stages have an important impact on plant energy requirements. Since the distillation equipment and the insulated cold box surrounding it make significant contributions to capital cost, economic distillation equipment will be compact and have low pressure drop. In addition, most plants must be designed to operate over a wide range of operating rates and with various product splits.

Method of distillation

Cryogenic distillation can be accomplished by using either trays or packing. Distillation trays generate interfacial area for mass transfer between the phases by bubbling streams of vapor through layers of fluid. A large amount of interfacial area can be generated by a properly designed tray. For mixtures with good transport properties, such as the air system, the gas-phase and liquid-phase mass transfer coefficients are high and excellent point efficiencies are obtained. For cryogenic distillation trays, the total pressure drop can range from 25 to 75 mm of liquid per tray.

Packing generates interfacial area by providing a structure which supports the liquid film. In general, the mass transfer area which is developed by packing is less per volume than that possible with trays. This results in somewhat greater heights for columns using packing than is possible with columns using distillation trays. However, since the packing structure supports most of the weight associated with the liquid, the pressure drop for packing can be as little as 10% of that achievable with tray designs.

Distillation columns which utilize packing will generally cost more and will consume less energy than equivalent tray columns. Consequently, they will be most appropriate for larger size plants, and when power and argon values are high.

Packing or tray type

Figure 2 illustrates examples of both random and structured or ordered packing. Various types of random packing have been available for years and have typically been used for relatively gross separations; those needing a small numbers of stages. Ordered packing is usually fabricated from corrugated sheets of metal. The packing matrix which results is between 2 and 7% metal by volume. The installation requires that neighboring layers be stacked at an angle to each other. This achieves considerable vapor and liquid mixing throughout the column section. The mixing benefits the overall mass transfer by substantially reducing the adverse impact of flow maldistribution. Random packings do not provide the same degree of mixing, and their cryogenic applications are limited.

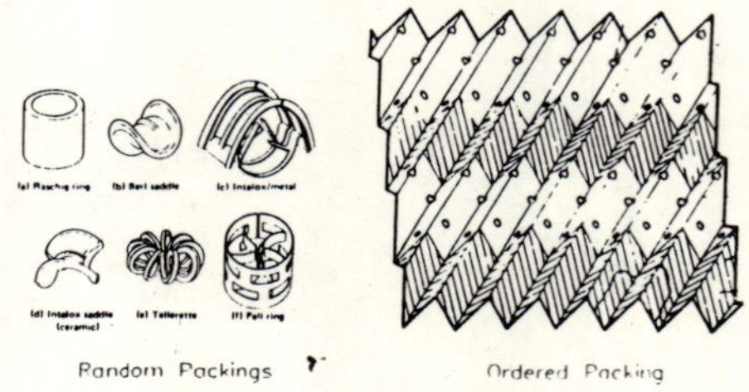

Figure 2 - Packing Types

Trays are used for most air distillation column designs. A number of types are possible and each tray type has somewhat different performance characteristics. Parallel flow trays have an infrequently recognized advantage for air separation distillation. The excellent transport properties of cryogens yield point efficiencies from 60 to 80%. Due to this high point efficiency, certain sections of the column system can have an overall section efficiencies which are substantially greater than the point efficiency. Under such conditions, tray designs which promote beneficial composition profiles in both the vapor and liquid are desirable. These trays are known as parallel flow trays. For such trays, liquid moves in a continuous direction as it flows from tray to tray.

Figure 3 illustrates several types that have been used in the cryogenic industry. For all these types, it is desirable to limit the liquid inventory and thereby reduce pressure drop. The outlet weir length needed to achieve low inventory requires the use of multiple passes for large diameter columns. Figure 3a, for example, shows a four-pass cross-flow tray. If properly designed, a relatively high efficiency can be obtained at a reasonable pressure drop. This design, however, is sensitive to out-of-levelness, and is also very sensitive to any liquid maldistribution at the top of a section. Figure 3b shows a multipass parallel cross-flow tray. Such a parallel flow type tray can yield very good efficiencies, especially in those column sections where composition profiles significantly enhance point efficiency. A problem with this design is that the downcomer located between the edge of the tray and the circumference of the column will have a different backup versus flow rate relationship than the other downcomers. In addition, the manner in which the liquid from this downcomer is introduced onto the subsequent tray can have a significant and adverse impact on the turndown characteristics of the tray. Figure 3c illustrates a racetrack design. An advantage of this design is that it has identical downcomers throughout. Its performance is also relatively insensitive to liquid maldistribution into the tray section. The parallel flow nature of the flow also gives it higher tray efficiencies than cross-flow trays. Its major problem is that about 10% of the total column area is unused. Figure 3d shows a radial design tray. This tray design has a tray efficiency comparable to that obtained with cross-flow trays. Like the cross-flow tray, its performance is sensitive to initial liquid maldistribution at the top of the section.

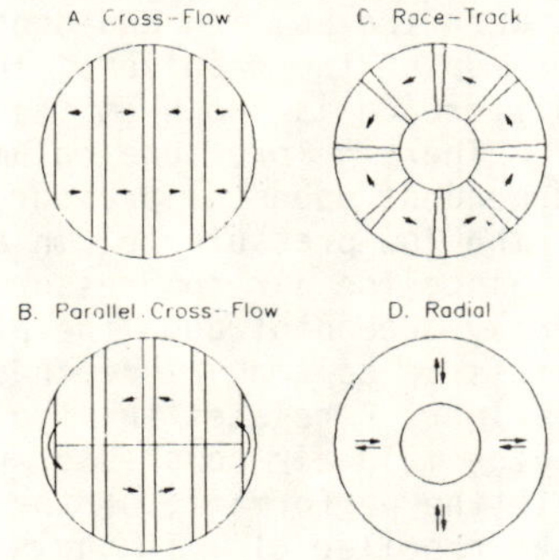

Figure 3 – Various Tray Types

Each of the above tray types can be used successfully if a thorough understanding of the fundamentals are incorporated in the design details. In general, each air plant supplier has optimized performance and costs for only one or two of these types.

Specific distillation details to enhance performance

Each design embodies a large number of design details. For packed columns, special attention and experience is necessary to develop a design which is compatible with the high purity oxygen environment within the distillation columns, and with the shipment of columns containing packing. Distributor designs are critical, requiring large-scale testing for verification of distributor concepts.

Tray designs require specification and optimization around the following parameters: specific vapor throughput, liquid inventory on the tray, inlet weir height, outlet weir height, number of downcomers, tray perforated area, perforation diameter, and tray spacing. Optimizations require accurate estimates of point efficiency, eddy diffusivity, tray efficiency, froth height, pressure drop, entrainment rate, weeping rate, and downcomer backup. Correlations for these parameters are usually based on comprehensive experimental cryogenic results and fundamental modeling. A number of references which address some of these areas are listed at the end of this paper.

Maximum design rate requirements

Typical designs will have a "design" case, a "maximum production" case, and a turndown requirement, or "minimum" case. Estimates of product rates and purities, together with energy consumption, may be required for each of these cases. The required operating range is an important aspect of overall column system design. In general, the wider the intended operating range, the more difficult the optimization. Research programs are designed to incorporate wide ranges of operating conditions around typical operating points. This information is needed to accurately estimate performance over all these operating conditions.

Flooding of the column may become a limiting factor for design at maximum rates. Figure 4 shows a distillation tray in elevation. The liquid enters the tray from the downcomer, is expanded into a froth by the vapor flowing through the perforations, and flows over the outlet weir into the downcomer before flowing onto the next tray. Flooding occurs when the liquid/froth within the downcomer starts to interfere with the froth flow over the outlet weir. This causes the liquid inventory on the tray to increase. This causes an increase in the tray pressure drop, and as tray pressure drop increases, backup increases, causing the tray to cascade into flooding.

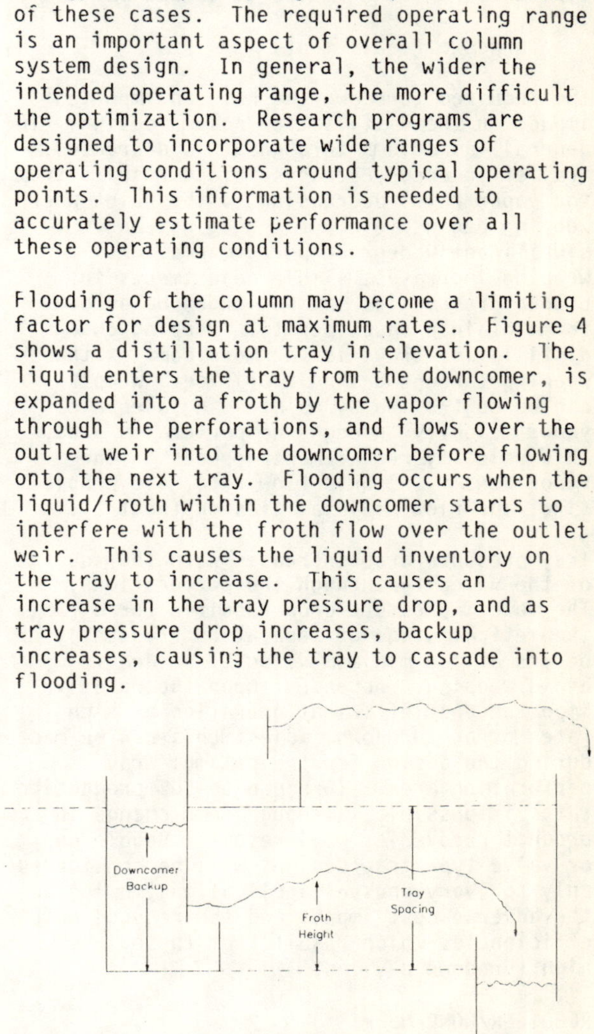

Figure 4 – Elevation of Distillation Tray

One of the most frequent causes of flooding of cryogenic trays is entrainment of liquid to the tray above. This increases the liquid traffic on the trays and in the downcomers, and it increases the tray pressure drop associated with the flow of the vapor and entrained liquid through the perforations. This increase in pressure drop increases backup, eventually causing flooding. This type of flooding can be controlled by choosing a tray spacing that permits maintenance of a vapor space above the froth. This concept is not always easy to apply, however, because the froth height can be ill defined. The froth height is dependent upon tray details and system pressure, composition and properties. Designing close to the expected flood point permits the utilization of the optimal number of trays for a given cold box height. Using design criteria based on comprehensive laboratory studies and full-scale plant performance permits guaranteed performance at throughputs within 5% of flood. Tray spacing typically ranges from 100 mm to 250 mm.

Turndown requirements

The required turndown can have an important impact on the distillation column design. In general, as column throughput is decreased, tray efficiency decreases. This results from the vapor rates decreasing to the point where weeping begins to occur. Weeping can significantly degrade performance. As weeping increases, simple adjustments in product recoveries may no longer be adequate to maintain product purity. Weeping on distillation trays is far more complex than weeping through a single orifice. Weeping can be controlled by maintaining adequate vapor velocity through the perforated area. This velocity requirement depends on the vapor density, liquid flow rate, sieve hole size, and flow nonidealities on the tray.

Tray pressure drop increases with the square of the velocity through the perforations. The lowest pressure drop required for a theoretical stage of separation frequently occurs at the minimum velocity needed to avoid weeping. Dependent upon the relative importance of efficient operation at high rates or at turndown, adjustments can be made during the design for the optimum tray perforation area. Turndown to 50% production rates is possible, although some change in product recoveries will result. Bubble cap or valve-type tray designs will be considered only for very unusual applications, owing to the adverse cost impact and the reduced point efficiencies which result from these high-turndown devices.

REBOILER/CONDENSER

The reboiler/condenser provides reflux for the high pressure column and boil-up for the low pressure column. Figure 5 illustrates its function. Most reboiler designs operate in a thermosyphon fashion. Typical designs incorporate brazed aluminum plate fin exchangers. Serrated or perforated fins are used in the high pressure nitrogen circuit. Perforated fins are typically used in the boiling passages. If proper design features are included, safe and reliable performance will result. Many hundreds of air separation plants have been built using these types of reboilers.

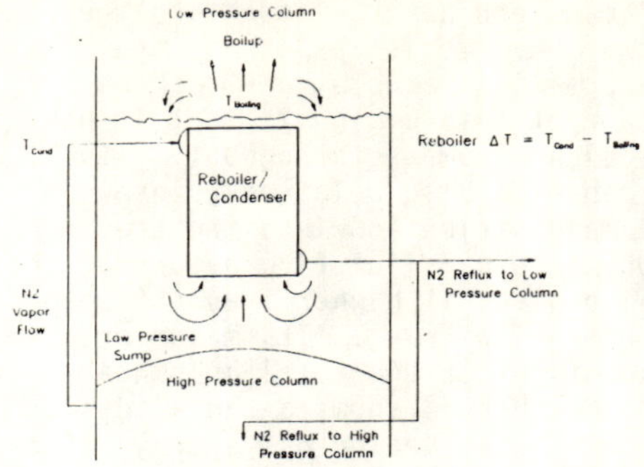

Figure 5 - Reboiler / Condenser

The performance of the reboiler can be most effectively characterized by the temperature difference between the boiling and condensing streams at the top of the reboiler. This temperature difference is known as the reboiler "ΔT". The high pressure column pressure is dependent upon the pressure at the bottom of the low pressure column and the reboiler ΔT. Since the air compressor power is directly dependent upon the pressure of the high pressure column, it depends on reboiler ΔT. Figure 6 relates the typical percentage of the main air compressor power associated with the performance of the reboiler to the reboiler ΔT. It shows that a reboiler design yielding a reboiler ΔT of 2°C will have a thermodynamic efficiency penalty corresponding to about 7% of the air compressor power. Clearly, the air plant efficiency can be substantially improved by using reboiler designs with small values of reboiler ΔT.

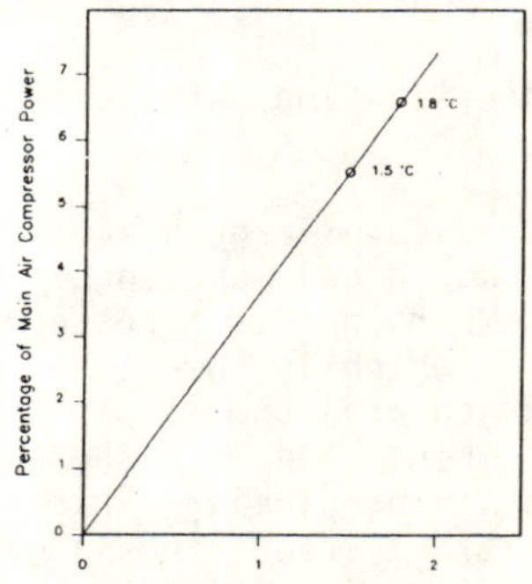

Figure 6 - Importance of Reboiler ΔT on Plant Efficiency

The reboiler ΔT is dependent upon the design of the reboiler and the vapor generation rate per cross-section or boil-up flux. For a given reboiler design, the reboiler ΔT can be decreased by decreasing the boil-up flux. The total column vapor generation rate is fixed by the capacity of the plant, so reboiler ΔT can be decreased by adding reboiler cross-sectional area. Generally, the most desirable location for the reboiler is in the sump of the low pressure column. Although dependent upon the specific layout and column diameter, a reboiler ΔT of about 1.8°C can be obtained with reboilers located within the column. Lower reboiler ΔT values will usually require the addition of more cross-sectional area than is available from within the column diameter. This can be done by installing an additional layer of reboilers (i.e., stacking), locally expanding the column diameter, or adding additional reboilers adjacent to the column. Even with these methods of adding additional reboiler cross-section, it is difficult to economically justify with conventional reboiler designs, reboiler ΔT values less than about 1.5°C. As seen in Figure 6, this reboiler ΔT still corresponds to 5% of the main air compressor power.

Novel and proprietary reboiler designs are required to further reduce the reboiler ΔT. Figure 7 shows a downflow reboiler in which the boiling and condensing streams flow in a downward direction. In the design which would result in the lowest reboiler ΔT, the generated oxygen vapor flux would be very low to reduce pressure gradients. This would require a high cross-sectional area and resultant high reboiler costs but could result in essentially constant temperatures in both the boiling and condensing circuits. Since thin film evaporation and condensation give high heat transfer coefficients, the net result would be a reboiler with a very low ΔT. Practical designs for downflow reboilers cause pressure gradients to develop in both circuits adversely effecting the reboiler ΔT.

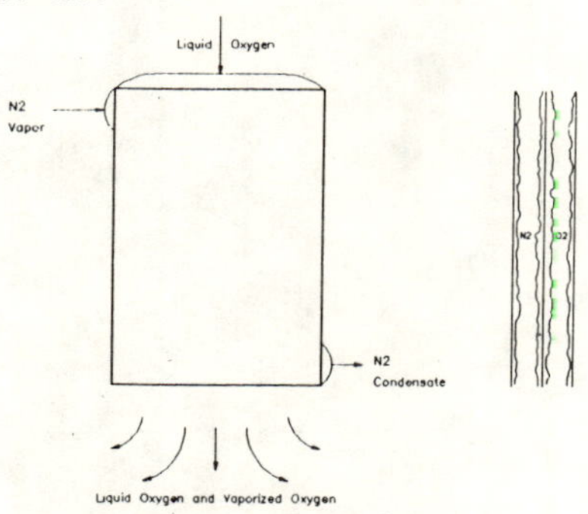

Figure 7 – Down-Flow Reboiler / Condenser Concept

Safety considerations require that all boiling surfaces must be wetted. Extremely careful attention to the design of downflow reboilers is required to yield adequate liquid distribution throughout all portions of the oxygen passages. The addition of a pump to provide adequate liquid oxygen flow to facilitate wetting introduces an additional refrigeration load on the plant. This will tend to dissipate the beneficial effects of the downflow reboiler on reboiler ΔT.

Figure 8 illustrates two related designs which can achieve very low values for reboiler ΔT. As in the case of conventional reboilers, the dual and multizone reboiler designs operate in a thermosyphon manner. This avoids the complex distribution design required for the downflow reboilers. In addition, the large liquid flow which results from the thermosyphon pumping action promotes safe operation. The key concept of these designs is the incorporation of a high pressure drop region in the bottom portion of the reboiler. This, in combination with the upper, low pressure drop region, provides a relatively constant temperature in the boiling zone. Since the condensing side is downflow, the temperature on the condensing side is also approximately constant. The resultant near-constant temperature difference throughout the boiling region makes this reboiler appropriate for air separation plants when reboiler ΔT values of less than about 1.5°C are economic.

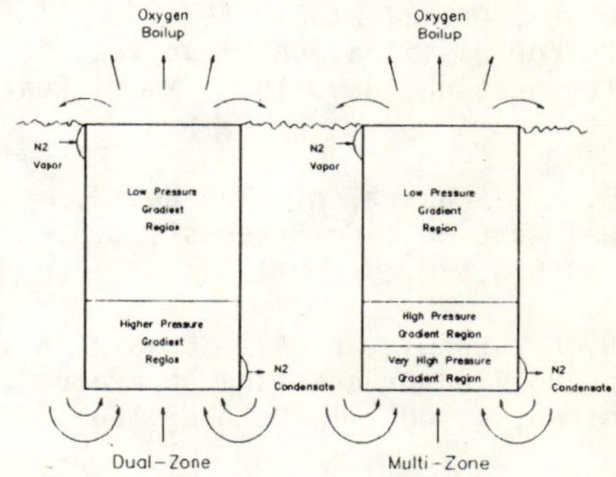

Figure 8 – Multi-Zone Reboiler / Condenser Concept

CONCLUSIONS

The distillation column system for a high efficiency cryogenic air separation plant is complex and integrated. This paper has presented some background information describing various features of these designs, the options which must be considered, and some guidelines for evaluation of these options.

The complexity of these designs and the number of options available make the optimization portion of the design an integral and important part of the technical package offered to the customer. The quality of the optimized design is dependent upon two items. Firstly, an open and complete understanding is required between the supplier and the customer about the relative values of energy, capital and products, and a thorough understanding of the range of desired production rates and product splits. Secondly, the quality of the optimized design depends on the understanding of the fundamentals of distillation and reboiler design by the supplier. This understanding should be based on comprehensive laboratory and full-scale plant performance tests which is translated into practice through thoroughly developed and fundamentally sound design methods.

The close cooperation between the customer and a knowledgeable and technically sophisticated supplier will result in an efficient, technically advanced air separation plant which will fully satisfy the customer's needs.

REFERENCES

Abrardo, J. M., Smith, A. R., Vinson, D. R., "Optimization of Cryogenic Oxygen Plants", Proceedings, XVII International Congress of Refrigeration (1987), Vol. A, pp. 214-219.

Bennett, D. L., Agrawal, R., Cook, P. J., "New Pressure Drop Correlation for Sieve Tray Distillation Columns", AIChE Journal (Vol. 29, No. 3) May 1983, pp. 434-442.

Bennett, D. L., Grimm, H. J., "Eddy Diffusivity for Distillation Sieve Trays", AIChE Winter Meeting, Nov. 1986, Miami Beach, U.S.A.

Bennett, D. L., Schwarz, A., Thorogood, R. M., "Dual-Zone Boiling Process", U.S. Patent No. 4,653,572, March 1987.

Bennett, D. L., Ludwig, K. A., Schwarz, A., "Multi-Zone Boiling Process and Apparatus", U.S. Patent No. 4,700,771, October 1987.

Garay, P. J., Bennett, D. L., "Single-Hole Bubble Dynamics During Weeping", AIChE Winter Meeting, No. 1985, Chicago, IL, U.S.A.

Thorogood, R. M., Sacks, R. E., "The Measurement of Fundamental Distillation Data with a New Pilot Scale Apparatus to Simulate a Commercial Cryogenic Tray", I. Chem. E. Symposium Series No. 104, Distillation and Absorption 1987, pg. B101.

Thorogood, R. M., Ying, D. H. S., "The Development of Modeling Techniques for the Determination of Tray and Column Performance from Point Efficiency", I. Chem. E. Symposium Series No. 104, Distillation and Absorption 1987, pg. A597.

Wilson, K. B., Abrardo, J. M., Bennett, D. L., "Technological Advances for Large Air Separation Units", Proceedings International Conference on Cryogenics and Refrigeration (1989), Hangzhou, PRC.

A STUDY OF GAS-PARTICLE FLOW PROPERTIES IN THE RADIAL CIRCULAR CASCADE

Gou Youyi, Ji Guanghua, Zhang Haoran, Bian Shaoxiong

Power Eng. Dept., Jiaotong University
Xi'an, China

ABSTRACT

A method is developed for calculating the flow properties of gas-particle in two phase flow flowing over blades in a radial inflow turbine cascade. Momentum and heat transfer between the gas and particles, as well as compressibility effects are considered. Experimental results of pressure distribution over a blade were obtained. Good agreement was acquired between the theoretical and experimental pressure distribution.

INTRODUCTION

A performance of turbomachinery operating under gas-particle in two phase flow conditions deterioated both structurally and aerodynamically. The discussion and analysis of the flow properties and pressure distribution of gas particle suspended over blade surfaces of a cascade nozzle were presented (1), (2). These studies were strongly limited to the axial flow turbomachinery. In this report the theoretical and experimental investigations concerning the effect of the properties over blades in radial inflow turbine are presented.

METHOD OF SOLUTION AND GAS-PARTICLE FLOW GOVERNING EQUATIONS

Firstly the pressure distribution over the blade without the particals is determined experimentally. Then considering the compressibility of the gas flow, the exit flow conditions are obtained. It is assumed that two stream tubes are to be existed in the flow field around the blade, one for the suction side.and the other for the pressure side. The gas flow without particles is used to get the ares of the stream tube as function of the given pressure distribution and inlet gas conditions of the particulate gas flow.

The following main assumption are to be made,

(1) The flow in the stream tube for a small small step is steady and one dimensional. And total mass flow rate of gas-particle mixture is W.

(2) The solid particles and gas properties are uniformly distributed over each cross-setion.

(3) The solid particles are in the shape of sphere with uniform diameter at constant temperatre with the same physical properties.

(4) The particles interact neither with one another nor with the boundaries, and occupy negligible volume.

(5) The gas is perfect and inviscid.

Besides of the drag force exerted by gas on the particles, the heat transfer by convection is only considered. The governing equations for particulate flow are formulated for compressible cases, and can be solved numerically to get the pressure distribution of particulate flow properties.

When the gas passes through a stream tube, the continuity equation of the momentum and energy as well as the equation of state can be manipulated to obtain the rate of the area change at the stream tubes.

$$\frac{dA}{dS} = \left(\frac{\rho U A}{p} - \frac{UA}{C_{pg}T} - \frac{A}{U} \right) \frac{dU}{dS} \tag{1}$$

The energy equation of particulate flow can be written as

$$\frac{dT_g}{dS} = \frac{T_g}{A}\frac{dA}{dS} - \frac{U_g}{R}\frac{\alpha}{1-\alpha}\frac{dU_p}{dS} + \left(\frac{T_g}{U_g} - \frac{U_g}{R}\right)\frac{dU_g}{dS} \tag{2}$$

The equation of momentum of gas-particle flow is

$$\frac{dp}{dS} = -\frac{W}{A}\left[(1-\alpha)\frac{dU_g}{dS} + \alpha\frac{dU_p}{dS}\right] \quad (3)$$

The continuity equations are as follows:

For gas phase, $\rho_g U_g A = (1-\alpha)W$ (4)

For particle phase, $\rho_p U_p A = \alpha W$ (5)

The state equation of gas is

$$p = \rho_g R_g T_g \quad (6)$$

The equation of state for the gas can be differentiated. And introducing Eqs. (2),(3) to the differentiated equation, the velocity rate of the flow is obtained,

$$\frac{dU_g}{dS} = \frac{\left\{-\frac{T_g}{A}\frac{dA}{dS} - \frac{\alpha}{1-\alpha}\left[\frac{C_{pp}}{C_{pg}}\frac{dT_p}{dS} + \left(\frac{dp}{C_{pg}} - \frac{U_g}{R}\right)\frac{dU_p}{dS}\right]\right\}}{\left[U_g\left(\frac{1}{C_{pg}} + \frac{T_g}{U_g^2} - \frac{1}{R}\right)\right]} \quad (7)$$

The equation for the heat transfer between solid sphere particle and the gas flow can be derived as follows:

$$U_p\frac{dT_p}{dS} = -\frac{6K_g}{\bar{\rho}_g C_{pp} d_p^2}\left[2 + 0.6Re^{0.5} p_r^{1/3}\right](T_p - T_g) \quad (8)$$

The equation of the drag exerted on the particles by gas flow is found to be

$$U_p\frac{dU_p}{dS} = \frac{18\mu_g}{\bar{\rho}_p d_p^2} \cdot \frac{28Re^{-0.65} + 0.48}{24Re^{-1.0}} \quad (9)$$

Eqs.(2),(3),(4),(5),(7),(8),(9) are the gas-particle suspension flow governing equations. These differential equations are solved numerically for obtaining the properties $\rho_g, \rho_p, U_g, U_p, T_g, T_p, p$ of a mixture flow.

TEST RESULTS

In order to examine that if the calculated results are correst, and to compare the properties of a gas flow with the properties of a gas-particle in two phase flow, a test is conducted. The test rig is showed in Fig.1. The tested cascade dimensions and pressure probe locations are showed in Fig.2 and table 1.

The area of the throat of the tested cascade is 19.5 8.1 mm. The rate of inlet gas is 37.5%. The arenaceous quartz with following specifications are used as the particles.

$$d_p = 180\mu m \quad C_{pp} = 585.2\ J/Kg.K,$$

$$\bar{\rho}_p = 2203\ Kg/cm^2,\ m_p = 25g/min,\ \alpha = 0.003$$

By regulating the valves 3,5,7,heater 4 and the compressor 9 of the secondare inlet flow, the inlet flow conditions are changed. The results of test are plotted in Fig.3 with Wg=0,134 kg/s, P =0,2112 Mpa, T=289 K, The maximum error is at about 5%.

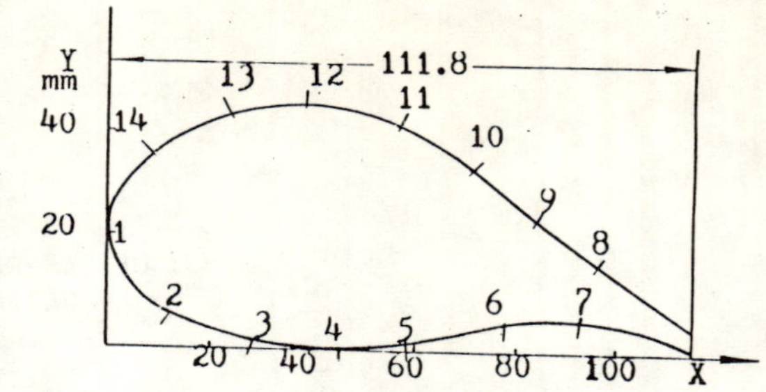

Fig.2 Cascade dimensions and locations of measurement on blade

Table 1 the locations of pressure measurement

	X mm	Y mm		X mm	Y mm
1	0.15	20.5	8	9.0	36.2
2	11.0	6.2	9	23.0	43.5
3	27.8	1.8	10	38.2	45.8
4	44.8	0.15	11	55.8	42.2
5	58.5	1.3	12	70.3	33.2
6	76.0	3.7	13	82.8	24.2
7	89.8	4.6	14	95.0	16.0

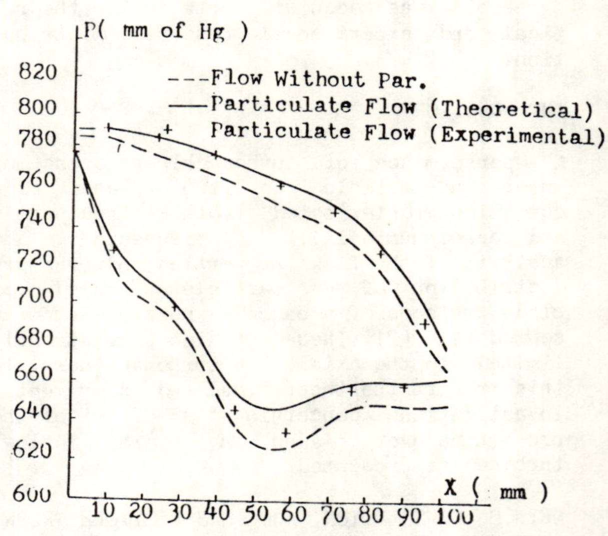

Fig.3 Pressure distribution

CONCLUSION

(1) The theoretical method developed here has good agreement with experiment at result in the compressible flows. Thus the method of estimating the forces on the blade in a radial cascade is provided for the designer.

(2) The results show that the entrance of the particles caused an increase in gas-particle flow pressure, a decrease in both gas and particle velocity, and an increase in gas and particle temperature. Therefore this

causes a decrease in the turbine adiabatic efficiency. The reduction of the efficiency comes with increase of the particle conentration and the particle mean diameter.

LIST OF MAIN SYMBOLS

A = cross-sectional area of stream tube, m.

= the ratio of the mass flow rate of particles to the total mass flow rate of the gas-particle mixture.

C_p= specific heat, J/kg K.

d_p= particle mean diameter, m.

K_g= coefficient of conductivity for the gas, J/s, m, k.

p = gas-particle suspension pressure.

ρ = density, kg/m^3.

R = gas constant.

T = absolute temperature K.

u = velocity m/s.

W = mass flow rate of gas particle mixture, kg/s.

Subscripts

g = gas.

p = particle.

REFERENCES

1 Tobakoff, W.Hussein, M.F."Effect of suspended solid particles on the properties in cascade flow'. AIAA Journal, Vol.8, No.8, August 1971.

2 Tobakoff, W.Balan, C.'Effects of solid particles suspended in fluid flow through a axial flow compressor stage'. Proceeding Airbreathing Engines, February 1981, Bangalore, India.

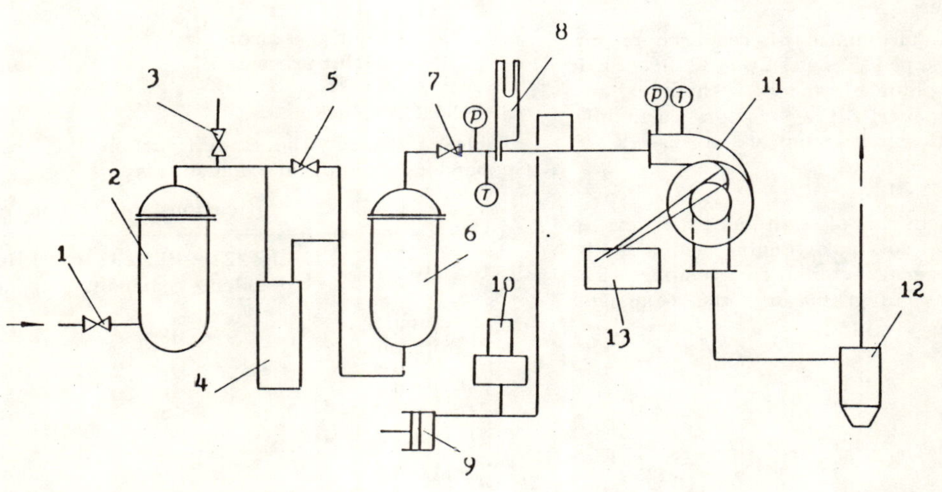

Fig.1 The test rig

1,3,5,7	Control valve	2,6	Stabilizer	4	Heater
8	Orifice meter	9	Compressor	10	Particle feeder
11	Tested nozzles	12	Particle recoverer	14	Pressure measuring unit.

THE DEVELOPMENT AND EXPERIMENTAL STUDY OF HELIUM EXPANDERS WITHOUT AN EXHAUST VALVE

Shi Jin

Hefei Institute of Cryogenics and Electronics
P. O. Box 19, Hefei--230061, China

ABSTRACT

The thermodynamic cycle, structure, performance and test results of 3 types of helium expanders without an exhaust valve are described. Five expanders have been mounted in an helium liquefiers. Reports from users affirm show. The expanders have the advantages of compact structure, reliability in working, steady performance and simple operation.

INTRODUCTION

Since the first liquefier using a helium expander was constructed by Kapitza at Cambridge, England, in 1934[1], there have been many improvements in helium expanders which are the most important part of helium liquefiers.

According to the requirements of researchers and users, we have developed Type PLHe 3240[2], PLHe 3846[3], PLHe 5250[4] helium expanders without exhaust valves. In this paper, the design principle, experiments, tests and improvement are discussed.

THERMODYNAMIC CYCLE

The thermodynamic cycle of helium expanders is similar to that of a two-stroke engine which is the Otto cycle of an internal combustion engine. The P(pressure)--V(volume) diagram of the expander without a valve is shown in Fig. 1.

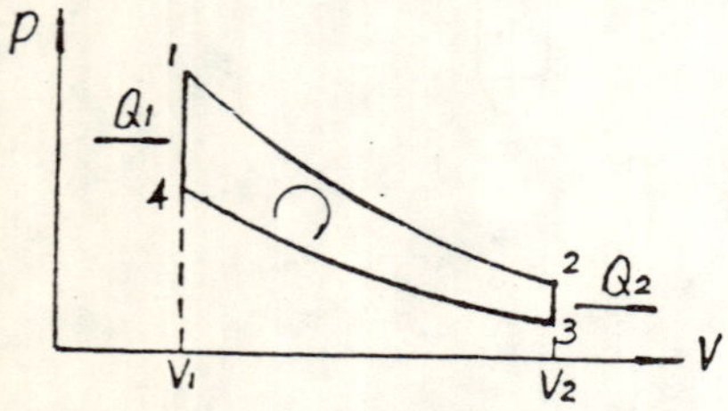

Fig. 1 P--V diagram of a helium expander without an exhaust valve

The theoretical cycle consists of four processes.
(1) Inlet process (from 4 to 1),
(2) Expansion process (from 1 to 2),
(3) Exhaust process (from 2 to 3),
(4) Compression process (from 3 to 4)

From Fig. 1, we can obtain[2]

$$\begin{cases} V_1=V_4 \\ V_2=V_3 \end{cases} \tag{1}$$

$$\varepsilon=V_2/V_1 \tag{2}$$

According to the thermodynamic calculations, the efficiency of a theoretical cycle can be expressed as[2]:

$$\eta=1-Q_2/Q_1=1-T_3/T_4=1-\varepsilon^{1-k} \tag{3}$$

where $k=C_p/C_v$ is the ratio of specific-heat. Because the cycle is irreversible, the isentropic efficiency is

$$\eta = \frac{A}{A_s} = \frac{1-\varepsilon^{1-k}}{1-(P_1/P_3)^{1-k/k}} \tag{4}$$

where

A is the work done by actual cycle;
As is the work done by a theoretical isentropic cycle;
P_1 is inlet pressure;
P_2 is outlet pressure.

SPECIFICATIONS

Table 1 shows the specifications of the helium expanders without an exhaust valve.

Table 1 The specifications of helium expanders

No	Items	PLHe3240 expander	PLHe3846 expander	PLHe5250 expander
1	cylinder dia. (mm)	32	38	52
2	piston stroke (mm)	40	46	50
3	piston length (mm)	250	250	250
4	revolutions (rpm)	1050	1050	750-820
5	inlet pressure (MPa)	2.8-3.3	2.8-3.1	2.8-3.1
6	outlet pressure (MPa)	0.02	0.02	0.02
7	efficiency	65%	60%	60%
8	LHe capacity (liter/hr)	5-10	15-20	20-30

STRUCTURE

The helium expanders consist of a crankcase, a crankshaft, a connecting rod, a crosshead, a connecting cylinder, a piston, a cylinder, an inlet valve and an inlet filter.

Each expansion of the helium expander is accompanied by two piston strokes. When the piston reaches the bottom of its stroke, the top of the piston meets the stainless steel ball of the inlet valve. The valve is then opened as it is fed with high-pressure helium gas at about 27 k inside the cylinder.

As the expansion stroke begins the entry of gas is cut off, the inlet valve is closed and the piston continues to rise to the top of its stroke, the 8-port which has a 4 or 5 mm diameter in the cylinder wall is opened, the expanded gas in the cylinder leaves at about 14 k. Then the piston goes down, the remaining gas in the cylinder is compressed, and the cycle is repeated.

1) Crankcase: The crankcase is sealed with a shaft seal which is made of nylon rings mounted in the flywheel side. Tests indicate that a good gas seal can be obtained both in pressure and vacuum.

2) Crankshaft: The crankshaft consists of a main shaft which is associated with subsidiary shaft carrying the crank, to which the crosshead is attached by the connecting rod, and the other side of the crosshead with the piston rod. The crankshaft is supported inside the crankcase by two ball bearings, the big end of the connecting rod is a sealed ball bearing, the other end which is in the crosshead is supported with a needle bearing.

3) Cylinder: The cylinder made of netarded stainless steel has a hardness about HRc56-62. It is joined to the crankcase by the connecting cylinder with two flanged unions.

4) Piston: The piston, made of stainless steel, has many special spiral grooves in the front of it, to improve the seal and steadiness in the cylinder. It is attached to the crosshead by the connecting rod by means of two ball-and-sockets flexible joints.

5) Inlet valve: The inlet valve contains a ball inlet valve and the inlet filter that is mounted on the cylinder head.

TEST RESULTS

The volume ratio is well known to be very important to the efficiency of the helium expander without an exhaust valve. The different ε values are obtained in the air tests at room temperature and the measured η-ε curve is shown in Fig. 2.

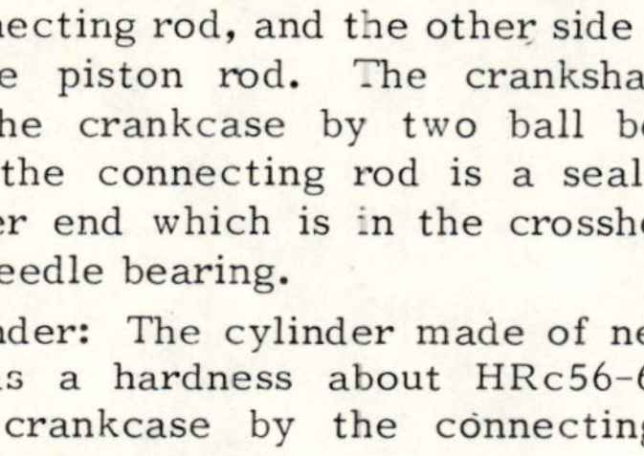

Fig. 2 η-ε curve

From Fig. 2 we can see that the optimal region of volume ratio is from 4.5 to 5.2. When the expander is mounted in a helium liquefier using a different δ (open height of inlet valve), L-δ curve is obtained.

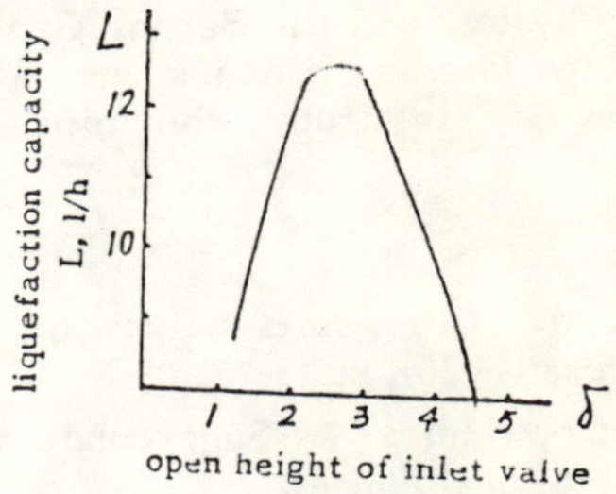

Fig. 3 L--δ curve

The phase diagram of the expander without an exhaust valve is as shown in Fig. 4.[8]

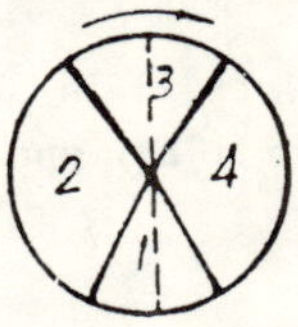

1. Inlet angle
2. Expansion angle
3. Exhaust angle
4. Compression angle

Fig. 4 The phase diagram of the expander

We can see that the phase diagram is symmetrical about the dotted line. The preinlet angle is equal to half of the inlet angle, and the expansion angle is equal to the compression angle. When the inlet angle is smaller or larger, the L will decrease. The open height of the inlet valve is 2.0-3.0 mm and the inlet angle is calculated in the region of $50^{o}\pm5^{o}$.

When p_1/p_3=25.4, p_1=3.0 MPa, the isentropical efficiency of the expander is 67.8% in the liquid Helium temperature region. The method of an optimal inlet angle is adopted in testing the helium expanders, so it is easy to determine the open height of the inlet valve and adjust the parameters of the expanders.

CONCLUSIONS

We have developed 3 types of helium expanders without an exhaust valve. Based on the tests, η-ε and L-δ curves are obtained. The optimal region of volume ratio is 4.5-5.2, the open height of the inlet valve is 2.0-3.0 mm, the phase diagram of the expander has been prepared and the optimal inlet angle is $50^{o}\pm5^{o}$. The isentropical efficiency is 67.8%. The method of the optimal inlet angle is adopted in testing the helium expanders without an exhaust valve. It is easy to determine the volume ratio and the open height of the inlet valve.

The valve can be integrated with the helium liquefier. The liquefication capacity is from 5 to 30 litre per hour.

Five helium expanders have been supplied to users where they have been mounted in the helium liquefiers. Reports from users show that our expanders without an exhaust valve have advantages of a compact structure, a steady performance, working reliability, simple operation and easy servicing. We are taking the further step of improving the performance to meet the requirements of research and users.

ACKNOWLEDGEMENTS

The author would like to thank the assistance and support of the Baoji Research Institute of Non-Ferrous Metals, the Lanzhou Institute of Physics, the Hefei Institute of Plasma Physics, the Shanghai Institute of Technical Physics, and the Beijing General Institute of Non-Ferrous Metals. Thanks are also given to the people in my Institute who took part in this work.

REFERENCE

1. Croft, A. J., Cryogenic Laboratory Equipment, Plenum Press (1970), pp27-73.
2. Shi Jin, Cryogenics & Superconductivity, No. 2 (1981), pp1-6 (in Chinese).
3. Shi Jin, Cryogenics & Superconductivity, No. 2 (1983), pp1-4 (in Chinese).
4. Shi Jin, Cryogenics & Superconductivity, No. 3 (1987), pp1-5 (in Chinese).
5. Shi Jin, Cryogenics, No. 4 (1982), pp1-5, in Chinese.
6. Shi Jin, INCONCRYO-88-II, India (1988), pp6.

HIGH-SPEED TURBOEXPANDERS FOR CRYOGENIC PLANTS

N.M.Grigorenko, A.B. Davydov,
A.Sh.Kobulashvili, V.V.Leonov
Scientific Industrial Association of
Helium Engineering Moscow, USSR

Our Association is traditionally oriented towards the use of turboexpanders in cryogenic plants; it began in the 1950s with the development of highly effective radial (centripetal) turboexpanders for various large air-separation plants operating at pressures of 0.5-0.6 MPa. A number of standard turboexpanders was produced that differed in capacity; the turbine stage was connected to the electric generator via a regulator. The expansion ratio in the turbine stage, $\delta = \frac{P_o}{P_2}$ (inlet pressure to outlet pressure ratio), was about 4, while the isentropic efficiency, η_s, was within 0.8-0.85 depending on the capacity.

The USSR also manufactured various and fairly small air-separation plants and nitrogen liquefiers operating at mean (4-7 MPa) and high (18-20 MPa) pressures. There were also cryogenic helium plants with operating temperatures of 4.5 and 20K (pressures up to 3.0 MPa). Those plants were equipped with reciprocal expanders which did not meet the growing requirements for such parameters as reliability, mass, size, simplicity of manufacture and maintenance. Turboexpanders successfully replaced reciprocal machines in many cryogenic plants under construction or development. Regulator-free, single-rotor turboexpanders with low flow rates and high speeds have been developed providing highly effective expansion of gas.

Note that the single-rotor design is a prospect for large-scale turboexpanders where the turbine stage serves as a drive for the centrifugal compressor final stage included in the cryogenic plant cycle. Such turboexpanders are designed for natural gas processing systems.

The diversity of cryogenic plants manufactured in the USSR requires the variety in the parameters of turbiexpanders:

- the expansion ratio in a single stage, δ, is 1.5-150;

- inlet volumetric flow rate (with inlet temperature and pressure), V_o, is 0.5-6000 m^3/h;

- the inlet temperature, T_o, is 300-5K.

These parameters are mainly found in turboexpanders with an impeller diameter of 10-800 mm and a rotor speed of $10\text{-}350.10^3$ rpm. Their power rating ranges from 2500 to 0.5kW.

Fig. 1 shows the applications of the turboexpanders developed in our Association.

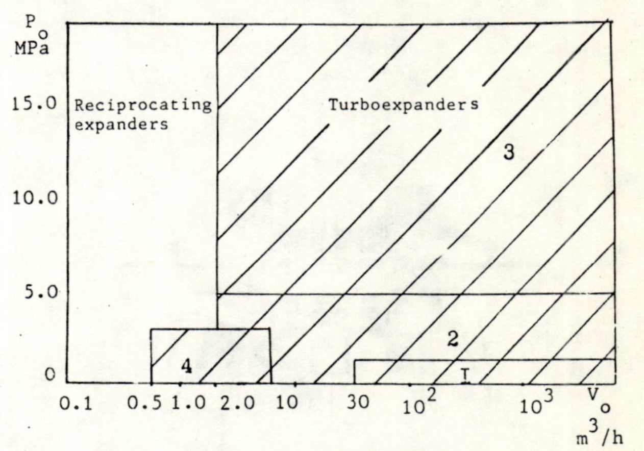

1 - low pressure air-separation plant; 2 - mean pressure air-separation plant; 3 - high pressure air-separation plant; 4 - cryogenic helium plant.

Fig. 1

Having in mind such diverse parameters we strictly differentiate our approach to the optimization of the turbine stages, minimization of heat inflow to the flow region, outer cold leaks through the shaft sealing and selection of adequate rotor suspension system.

We have got to use various designs for turboexpanders with different cooling capacity because known vibrationproof supports and brakes never possess universal properties that could equally satisfy a broad range of operating requirements.

All commercially manufactured turboexpanders are equipped with an effective centripetal reaction turbine with a radial-axial closed impeller for large machines and with a semi-open impeller for small machines. Large machines designed mainly for low pressure airseparation plants utilize nozzle blocks with variable-incidence blades. Small size machines use nozzle blocks with a constant flow section.

Consider most typical turboexpander designs for relatively small cryogenic plants of various applications.

HIGH PRESSURE TURBOEXPANDERS

Such turboexpanders are used in high pressure air-separation plants producing liquid products of air separation (from 200 to 2000 kg/h), as well as in nitrogen liquefiers and various cryogenic plants cooled by nitrogen. Depending on the type and purpose of a cryogenic plant, the expansion ratio, δ, is 33-133. The limit of the inlet temperature, T_o, is 300-233K. The flow rate of the existing machines, G, is in the range of 800-10000 kg/h with a capacity, N, of 22-350 kW.

Our centripetal turboexpanders provide for effective expansion of gas in a single stage at $\delta = 35$. So early turboexpanders operating at high pressures were single-rotor one-stage machines.

Today, improved single-rotor two-stage turboexpanders have been developed with turbine stages being positioned on the opposite cantilevers of the shaft (Fig. 2).

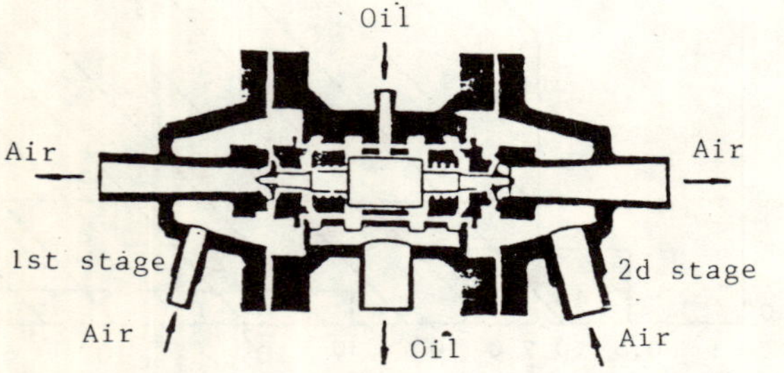

Fig. 2

This two-stage design makes full use of heat recovery and significantly shaves high axial forces off the rotor which increases the effectiveness and reliability of the turboexpander in comparison with a one-stage model. centripetal turbine stages, a loading hydrodynamic device positioned on the bulging portion of the rotor between the bearings, and the rotor suspension system are made as a single module which is connected via self-sealing bushes to the gas supply chambers and outgoing pipes. The turboexpander uses plain oil bearings. Labyrinth seals are placed between the bearings and turbine stages.

The comparative characteristics of high pressure two-stage and one-stage turboexpanders with a capacity of 800 and 5000 kg/h are shown in Figs 3 and 4. The effectiveness of two-stage turboexpanders exceeds that of two-stage turboexpanders although the metal content of the latter is 20 times higher.

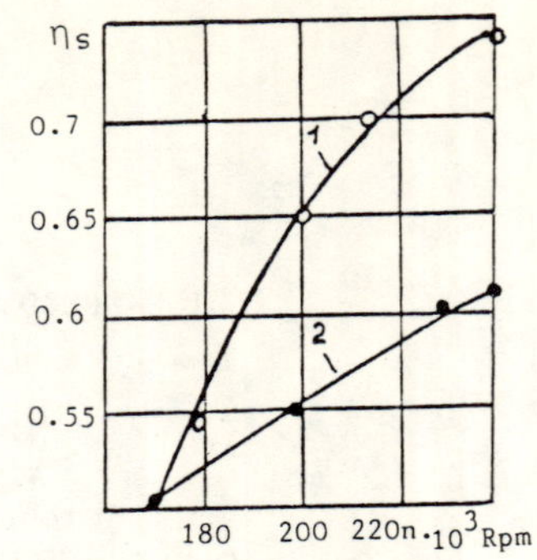

1 - two-stage;

2 - single-stage.

Fig. 3

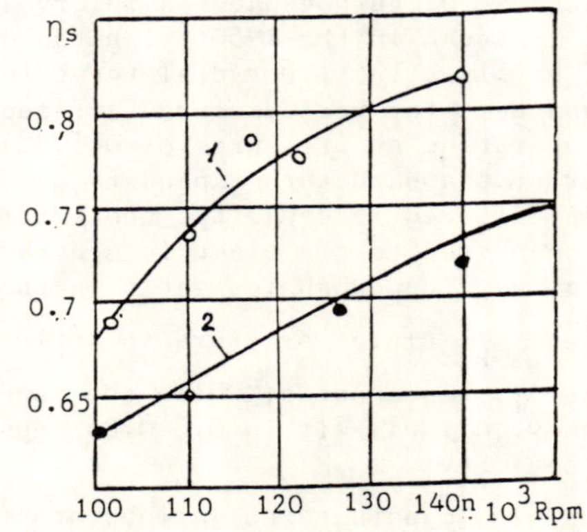

1 - two-stage;

2 - single-stage.

Fig. 4

Two- or three-stage turboexpanders are used in cryogenic plants that expand air or nitrogen from 20.0 to 0.15 MPa (expansion ratio $\delta = 133$). Due to a relatively low density of gas at the outlet, the dimensions of the final stage impeller become large enough, so the stage is made as a separate machine with an inlet pressure of 0.6-0.7 MPa. In this case, gas is expanded either in the two-stage single-rotor turboexpander shown above, or in a single-rotor one-stage two-flow turboexpander where impellers of the same size with the right-hand and left-hand profiles are placed on the opposite cantilevers of the rotor.

TURBOEXPANDERS FOR CRYOGENIC HELIUM PLANTS

Turboexpanders shown in Fig. 5 are designed for relatively large cryogenic helium plants operating at 3 MPa, producing 100-700 l/h liquid helium or refrigeration of 250-2000 W at 4.5K, as well as for helium refrigerators with a cooling capacity of 4-6 W at 20K.

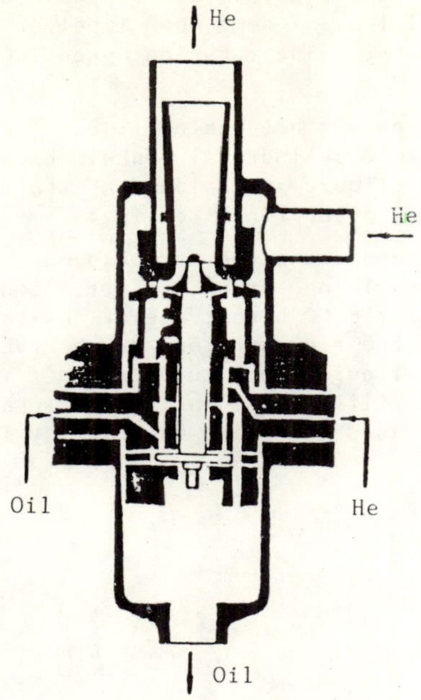

Fig. 5

Depending on the installation the turboexpander may have one, two or three self-sustained stages. The turboexpander uses a rotor suspension system with gas- and hydrostatic bearings and a shaft, the longitudinal net section of which is of a linear shape. The gas bearing placed directly at the turbine impeller partially provides for vibration resistance cf the rotor and its braking. Its main purpcse, however, is to pressurize the turbine stage from oil and oil vapor. Gas bearings are practically insensitive to temperature variations; in operation they release a very small amount of heat which makes a positive effect on the reduction of heat inflows to the cryogenic portion of the machine. The rotation is fully stabilized by an oil bearing that successfully withstands all radial and axial loads from the turbine side due to its good dynamic characteristics, Almost all the power of the turbine is consumed by the friction in the bearings and, consequently, by the heating of oil that circulates and cools in the pressurized oil supply system.

The cryogenic and warm portions of the turboexpander are separated along the shaft by a two-section labyrinth seal having an intermediate chamber for supplying pressurized warm helium in order to form an additional gas seal. The impeller and nozzle block, as well as the rotor suspension system and loading device are made as a single module. Thus, the manufacture of the turboexpander is greatly simplified; repairs consist in a ready change of the module without hreaking the leakproofness of the vacuum casing with thermal insulation.

The turboexpander bodv is installed vertically to ensure effective oil drainage.

Several standard sizes of such turboexpanders are available their capacities ranging from 1 to 10 kW and rotor speed from 200 to 350.10^3 rpm.

Within the whole speed range the rotation is stable and without precession. Turboexpanders are notable for their high reliability. In some cases, they operated failure-free for several years. Under design conditions the isentropic efficiency of turboexpanders is 0.7-0.8 depending on the size.

LIQUID-VAPOR TURBOEXPANDERS

Machines are called liquid-vapor turboexpanders if the line of the expansion process begins in the supercritical region or higher, then crosses the left-hand branch of the boundary curve and ends in the region of moist vapor. The process of pressure decrease (expansion) leads to the boiling of liquid in the flow portion. The level of humidity in a liquid-vapor turboexpander at the end of the expansion may vary from 60 to 100%.

A liquid-vapor turboexpander is essentially a cryogenic hydraulic turbine. The development of reliable and effective liquid-vapor turboexpanders opens new ways for novel and more economic cryogenic plants with reduced specific energy consumption. They also permit a relatively simple and inexpensive modernization of existing cryogenic plants, including air-separation devices, by replacing the throttle valve in the lower cooling stages with a proper turboexpander. It results in a substantial increase of cooling capacity.

Consider two examples.

For a sufficiently large cryogenic helium refrigerator-liquefier a liquidvapor turboexpander has been developed which is supplied with 400 kg/h of helium at 5.2-7K and 1.8-2.5 MPa. The turboexpander uses a centripetal turbine with a radial-axial impeller of 15 mm in diameter. The rotation apeed is 140-180.10^3 rpm. The turboexpander design is similar to the gas helium turboexpanders described above.

The liquid-vapor turboexpander replaced the throttling valve in the final cooling stage increasing the cooling capacity of the cryogenic plant by approximately 50%. Fig. 6 shows the experimental characteristics of the plant including refrigeration at 4.5K and the liquid helium temperature.

The turboexpander has been stably and reliably operating for more that two years now.

The other example refers to a high pressure air-separation plant producing about 1500 kg/h of liquid oxygen. The plant used a single-stage liquidvapor turboexpander which expanded about 4000 kg/h of air from 18.0 to 0.6 MPa at an inlet temperature of 110K. The turboexpander was placed instead of the throttle valve beyond the main heat exchanger.

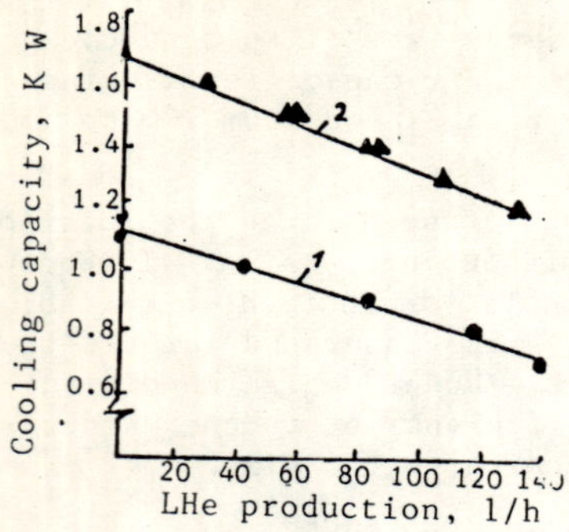

1 - throttling in final cooling stage;

2 - expansion in liquid-vapor expander.

Fig. 6

The turboexpander used a centripetal turbine with a radial-axial semi-open impeller of 25 mm in diameter. The rotation apeed 100.10^3 rpm.

The turboexpander had plain oil bearings and a hydrodynamic cylindrical brake. External cold leaks (liquid air) along the shaft was prevented by a labyrinth sealing.

Under all conditions the turboexpander demonstrated a stable operation. After a continuous operation for 8000 h no trace of cavitation effects on the flow portion was observed. The use of the liquid-vapor turboexpander instead of the throttling valve in the air-separation plant increased the plant capacity by 10%.

DEVELOPMENT OF A 250W HELIUM REFRIGERATOR

Wu Jianmei
Hangzhou Air Separation Plant Research
Institute China

ABSTRACT

This paper describes the application of a 14-15K helium refrigerator of 250W which provides refrigeration for the calibration and testing of an infrared remoted sensing device used in meterorological satellite. The Claude cycle with LN_2 precooling is adopted for the process, and an oil-free piston-type helium compressor and a gas-bearing turboexpander are used in the refrigerator so as to ensure its stable, reliable and continuous operation.

The main point of this article is to discusse and evaluate the selection of the process, design parameters and the types of heat exchanger, and to set forth the reasonable way to be taken for the necessary components and the key technology concerned. It has been proved that the system has stable parameters and good performance by trial runs. Experience has been gained for the long-term continuous operation of the helium refrigerator.

I. Intruduction

A Helium refrigerator of 250W refrigeraticn has been developed to provide refrigeraticn for the calibration and testing of an infrared remote sensing device used in meterorological satellite, which is needed for space simulation of ultrahigh-vacuum and very-low-temperature.

A continuous, reliable cold supply of 250W is available at 14-15K, one stream of which is set for radiation cold shroud so as to provide refrigeration for the infrared remote sensing device, and the other stream is used to maintain very-tow-temperature for helium plates of. the cold sink and cold shroud, so that high vacuum requirements for a calibration device can thus be met.

As described above, the selection of process should meet the requirements of temperature and refrigerating capacity, and at the same time the machines and components of the helium refrigerator should be for the purpose of long-term steady state continuous operation and easy handling.

II. Analysis of process

Two kinds of processes, which have already proved to be applicable for this refrigeration and temperature conditions, are as follows:

1. Single-stage expansion Claude cycle with LN_2 precooling (See Fig.1)

Helium is compressed in compressor C. A small fraction of helium leaving the compressor is cooled down by the returned nitrogen in heat exchanger I' while its major part is cooled down by returned helium in heat exchanger I". The two streams of preliminary cooled helium leaving exchangers are collected at point (2) and enters in to nitrogen bath II, where they will be further cooled down to state (3). Then it passes through heat exchanger III, expander E, which will result in a temperature drop at state (5). The cold content will be carried to a space simuration chamber by a cold-transfer line and flows to state (6), then it is reheated in heat exchangers III and I'. The incoming helium is cooled by returned helium at the same time.

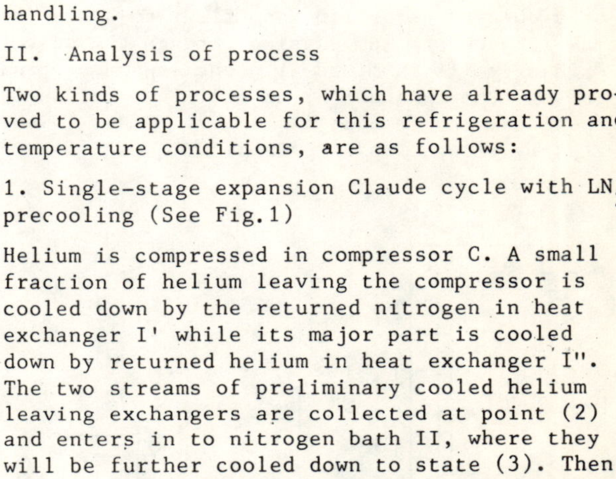

Fig.1 Single-slage expansion Claude cycle with LN_2 precooling

The advantage of singl-stage expansion helium refrigeration cycle with LN_2 precooling is simple and less moving parts. It is now widely used in small and medium sized helium refrigerators, but it consumes a quantity of liquid nitrogen.

2. Two-stage expansion Claude cycle (see Fig.2)

The helium is compressed by compressor C, and enters heat exchanger I, where it is cooled down by the returned helium. One part of the helium enters the first expander E_1, and expands to provide refrigeration and then mixes the returned helium at (6), while the other part of the incoming helium passes further through heat exchangers II and III to the second expander E_2 to provide refrigeration,with temperature drops to state (4). The cold contant is carried to a space simulation chamber by a cold-transfer line and flows to state (5). The returned helium passing through the heat exchanger III is mixed with the expanded stream coming from expander E_1 at (6). The mixture is then reheated in heat exchangers II and I and finally returns to compressor C.

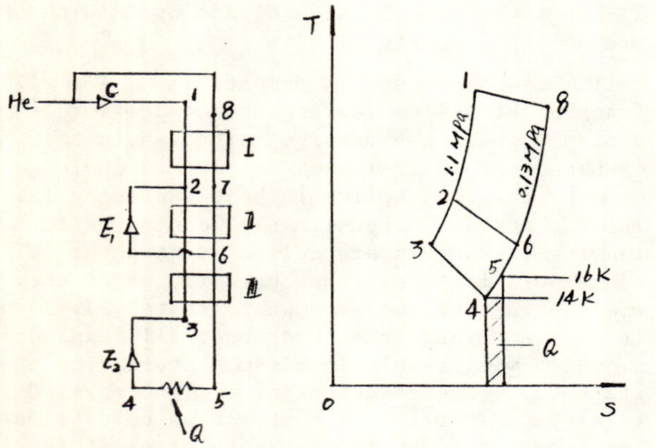

Fig.2 Two-stage expansion Claude cycle

Although no LN_2 precooling in this process confers same benefits, the process is more complicated and makes the manufacturing and maintenance more difficult. In actual practice, the complete experiment for calibration of an infrared detector needs a large quantity of liquid nitrogen. Therefore the single-stage expansion Claude cycle with LN_2 precooling has been introduced with the aim of simplifying process, easy handling and ensuring continuous steady operation as well. By doing so, the occurrence of unmatched coexistence of the two expanders can be avoided. In addition, it will also decrease production and maintenance costs.

III. Selection of design parameters

The process having been carefully chosen. the selection of design parameters will be taken into account, which are mainly related to the decision of pressure of the working medium as well as its amount of circulation. This will not only affect the efficiency and service life of the system as a whole, but also exert a direct influence upon the selection of the type and structural parameters of such major equipment as circulating compressors and refrigerating expanders as well as heat exchangers.

Derived from the formula of a compressor, the value of its power consumption (N) can be expressed as follows:

$$N \alpha G., \; L_n \; P_1/P_2$$

which means that power consumption is linear proportion to the amount of gas ciculation (G) and logarithmic proportion to pressure ratio (P). Assuming that the compressor has a definite efficiency and the temperature of the gas flowing into it is constant, it would be possible to calculate the amount of gas circulation by the equation:

$$G=\frac{Q}{i_6-i_5}$$

in which
Q-Refrigerating Capacity
i-Enthalpy

which indicates that the amount of working-medium circulation will be reduced as the pressure increases. If considered only from the aspect of reducing power consumption, it will become more advantageous as the pressure rises and the amount of gas circulation lessens [1]. Nevertheless, with regard to the structure of a compressor, the excessiveness of compression ratio will bring about many problems such as an increased number of stages for the compressure, shortening of service life of the oil-free sealing parts and an increase of leakage due to the fact that the exhaust temperature is too high, which will result in unfavourable factors to the operational reliability of the system.

In view of the turboexpander, after the value of U_1/C_0 (U_1-Peripheral velocity of impeller intet; C_0-Theoretical spouting velocity) having been reasonably chosen, the small diameter of the turbine wheel and the higher rotational speed are neceassary as a result of the high compression ratio and remarkable enthalpy drop, so the stability of bearings will make manufacture and operation more difficult.

As regard to the high-efficient and compact plate-fin heat exchanger, it is necessary to have reasonable arrangement of the flow passages for both of the incoming and returned fluids being heat-exchanged, to pass through. It will be more difficult to make such arrangements of passages if the compression ratio becomes too high, or else a special type of fins will have to be adopted.

To sum up, the refrigerator has been equipped with a three-stage, oil-free, piston-type compressor, whose moving mechanism is comparatively good, an excellent, reliable turboexpander of 4000 r.p.s. and compact heat exchanger. The gas circulation of $600m^3/h$, and the pressure of 1.1 Mpa. have been chosen. The power consumption based on calculation is accaptable.

IV. Selection of heat exchanger types

The heat exchanger of a helium refrigerator may be characterized as small quantily of working-medium flow and a remarkable drop in temperature so a compact heat exchanger with less resistance is required for a given duty.

Among many heat exchangers of this type, only the aluminium plate-fin heat exchanger or the copper fin-coiled tube heat exchanger can most likely meet the above requirements, the former of which is characterized as wide surface area for heat transfer per volumn, light weight and less resistance. However, great longitudinal thermal conductivity, which results in an increase of irreversible losses must be considered [2].

On the contrary the fin-coiled tube heat exchanger, compared with the aluminium plate fin heat exchanger, has less surface area for heat transfer per volumn and a heavier weight. In fact, its loss in longitudinal thermal concuctivity is relatively small, but it is still apt to bring about a phenomenon of short circuit of gas, which will affect the effectiveness of heat transfer on the fin-side, because the low-pressure channels for return stream demand high manufacturing precision that makes the fabrication exceedingly difficult [3].

The data, collected from calculation for the aluminium plat-fin type and the copper fin-coiled type heat exchangers, used as the first and the second stage heat exchangers in the process are compared in the following list:

		Plate-fin type	Fin-tube type
heat transfer area on the low-pressure side	M^2	67	22
Weight	Kg	202	660
Volum	M^3	0.184X0.18X3.6	φ0.324 L=8.2
Pressure drop	Pa	9870	6420

It shows that the aluminium plat-fin heat exchangers are more suitable for the requirements of the process.

V. Requirements for necessary components

1. Compresser

Two compressors in parallel are used with one in operation and the other ready for use by randon design. To prevent helium from being contaminated during operation, the compressors are required to meet the following requirements.

1) A semi-closed structure for the crankcase with its crankshaft sealed on the end-faces. Its total helium leak rate be less than $1X10^{-4}$ that of the amount of gas circulation;

2) The cylinder be oil-free lubricated;

3) The leaking helium from the piston-rod packing be inducted into the inlet of the first stage cylinder to prevent lubricating oil from contamicating helium;

4) No leakage from safety valves. The blowout gases be returned into the suction side and all the valve rods sealed by metal bellows;

5) All the pipes and vessels be made of stainless steel and fixed joints employed wherever possible. Operation experience has proved that the expected sealing requirements of the compressor are satisfied. The actual amount of leakage is considerably less than the design data. The compressor has been running steady with less wearing of its sealing parts.

2. Expander

Both the expander's turbine wheel and blower's brake wheel are of double-cantilever structure and the shaft is supported by static-pressure gas bearings with the thrustend set on the side of the brake wheel.

Two expanders are set in parallel with one in operation and the other ready for use, thus ensuring continuous and reliable operation.

After several periods of continuous running, it has been proved that the performance of the expander used is satisfactory and the refrigeration capacity required by the whole system can be satisfied.

3. Purification system and cold box insulation

The pressurized helium is purified by activated charcoal at liquid nitrogen temperature. The purifier will be cut off when the refrigeration system is in normal state. The refrigeration system will then be put into operation under the circumstances of no purification so as to reduce the losses of energy and gases.

All the components in the cold box are suspended form the top-cover which is supported by the outer casing. The cold box is vacuumized and insulated with multilayers.

VI. Experimental results

The performance test has been done through a 240 hour continuous operation. Later another four trial runs have been made with the result that all the expected design data are satisfied.

		Design data	Test data	Remarks
Refrigeration temperature	K	14-15	14.36	
Refrigerating capacity	W	250	251.6	Electrical measuring method is used
Start-up time	hr	3-6	1.5	
Leak rate of the working medium	$1X10^{-4}$	3	0.3	

References

1. С.С. Будневич РАСЧЕТ КРИОГЕННЫХ УСТАНОВОК 1979.

2. W.M.Kays and A.L.London, Compact Heat Exchanger 1964.

3. 渡辺正己. 化学装置 Vol6. No4. 1964.

THE CHARACTERISTICS OF ROLLING PISTON TYPE REFRIGERATION COMPRESSOR AND ITS APPLICATION IN CRYOGENICS

Zhou Zicheng
Shanghai College of Architectural And Municipal Engineering
71 Chi-Feng Road, Shanghai, China

ABSTRACT

A mathematical model of the rolling piston type refrigerating compressors has been developed, and the charecteristics of a 890W rolling piston type compressor are obtained either by computating or by testing. The refrigerant capacity, the indicated power and motor input power, the mass flow rate, the volumetic efficiency and the energy efficiency ratio of this compressor were obitained by varying the condensing and evaporating temperatures. The results show that this kind compressor is suitable not only for refrigeration use but also for cryogenics use.

1. INTRODUCTION

Rolling piston type compressors have many advantages such as lower components, small size, light weight, high efficiency and suited for mass production. These compressors have been used widely for room air-conditioners, unitary air-conditioners, domestic refrigerators and freezers.

Although the rolling piston type compressor and the reciprocating type compressor are all positive displacement compressors, but the constructions of them are quite different, and the working processes of them are also different. These lead to the differential charecteristics of each type compressor.

A computer simulation model has been developed to demostrate the charecteristics of a rolling piston type refrigerating compressor. The experimental measuring has also been completed. Comparing those charecteristics with reciprocating compressor, it is obvious that the application of the rolling piston type compressors in cryogenics will be widely.

2. THE COMPUTER SIMULATION MODEL

A computer simulation model for a rolling piston type refrigerating compressor has been made to demonstrate the charectristics of this type compressors. In the simulation computation, the cylinder working volume is divided into three control volumes, that is, the suction volume S, the compression volume C and the clearance volume V.

In order to analyze the various states of the gaseous refrigerant in each control volume, the working process of each rotation is divided into seven steps. Fig. 1 shows the geometrical shape of the cylinder working volumes and the seven positions on the cylinder wall O', $\theta_{v2}, \theta_{s1}, \theta_{s2}, \theta_{d1}, \theta_{d2}$ and θ_{v1}. The seven steps which refer to the rolling piston rotational angle θ are as follows:

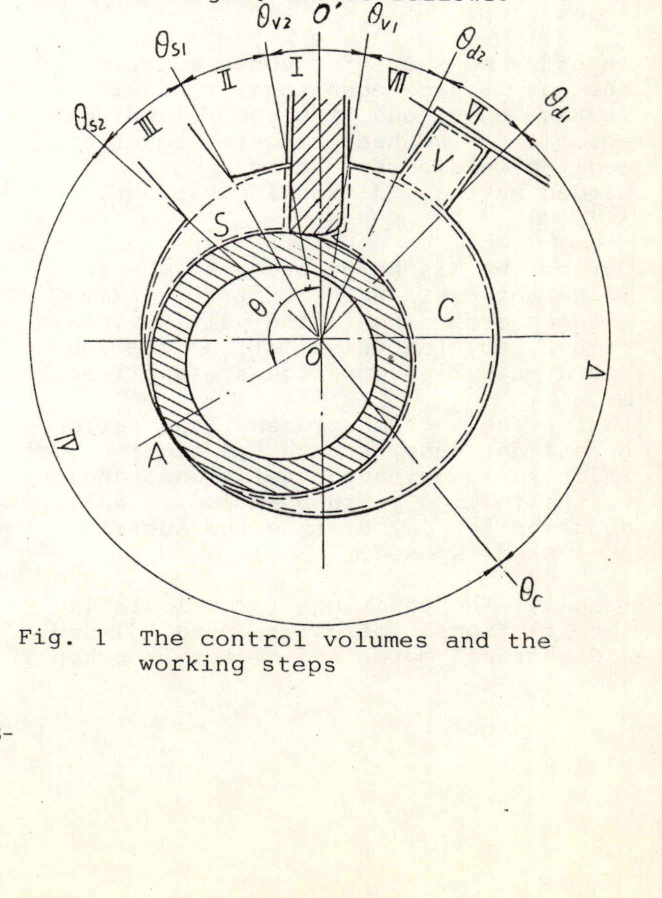

Fig. 1 The control volumes and the working steps

Step I : $-(2\pi-\theta_{V1})<\theta<\theta_{V2}$; The volume S is zero. There are only two volumes, C and V, and they are connected with each other.

Step II : $\theta_{V2}<\theta\leq\theta_{S1}$; The suction process begins in volume S.

Step III : $\theta_{S1}<\theta\leq\theta_{S2}$; The gaseous refrigerant flows from the suction pipe to the suction volume. A small part of the gaseous refrigerant flows back to the suction pipe.

Step IV : $\theta_{S2}<\theta<\theta_{c}$; The gaseous refrigerant in the volume C is being compressed and the suction volume is connected completely with the suction port. θ_c is the positional angle when the pressure in volume C reaches to the discharge pressure.

Step V : $\theta_c<\theta\leq\theta_{d1}$; The gaseous refrigerant in the volume C is in the discharge process, volume S is a continuation of the suction process.

Step VI : $\theta_{d1}<\theta\leq\theta_{d2}$; Volume V, volume C and volume S are connected, and a part of the gaseous refrigerant flows back from volume V to volume S.

Step VII: $\theta_{d2}<\theta\leq\theta_{V1}$; Volume C is so small that it can be neglected, and only volume S existes which is connected with the volume V.

The mathematical model is based on the energy balance and the mass balance. There are several equations for each of the three control volumes during the seven steps.

In order to solve the above equations, the gas leakage equations, the oil leakage equations, the control volume equations, the heat transfer equations and the refrigerant state equation are needed as the additional equations. (1) (2)

During the discharge process, the refrigerant mass in the control volume C changes according to the valve displacement, so, the valve motion equation and a mass flow equation are applied.

During the suction process, the refrigerant gas pass through the suction port which varies with the rotational angle of the rolling piston. an additional equation for calculating the suction port area is used.

Finally, the equations for calculating the frictional losses are used (3) and a electrical motor efficiency is taken into account.

3. THE COMPUTATION AND THE EXPERIMENTAL RESULTS

A rolling piston type refrigeration compressor of 890W motor power is used for computating and testing. The compressor capacity testing equipment is a calorimeter. The testing method is according to the chinese state standard GB 5773-86 and is equivalent to the international standard ISO 917-1974. The refrigerant is R502.

A comparison between the computation results and the experimental results is shown by figures 2 to 7 for condensing temperature from 35°c to 55°c and evaporating temperature from -30°c to 10°c. It can be seen that the computation results fit with the experimantal results over their entire scope.

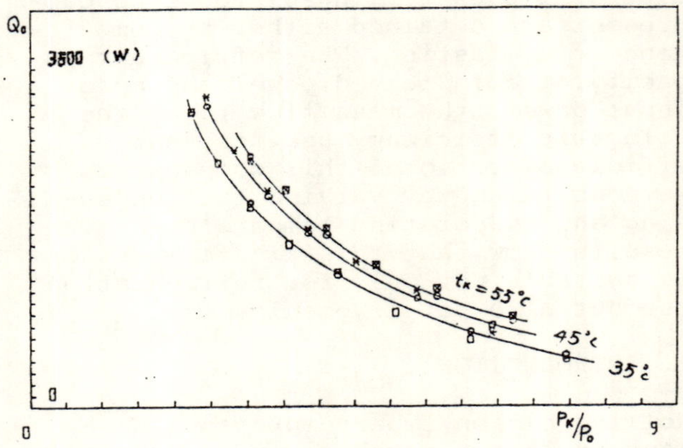

Fig 2. The refrigeration capacity versus the pressure ratio

Fig. 2 shows the refrigeration capacity versus the pressure ratio when the condensing temperature changes from 35°c to 55°c. It can be seen that the refrigeration capacity decreases slowly at the greater pressure ratio. It is the available charecteristic of the rolling piston type compressors for low temperature use.

Fig. 3 shows the curves of the motor input power versus the pressure ratio. There is a maximum power for each condensing temperature when the pressure radio changs from 3 to 8. so, for low temperature use, the el^ctrical motor should be desinged at the maximum power condition, not at the working condition.

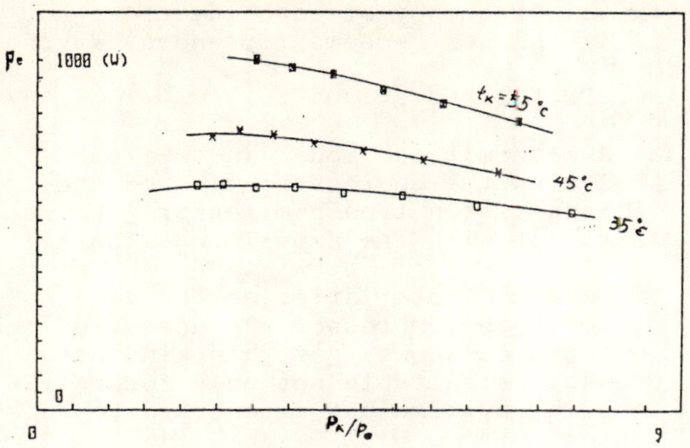

Fig. 3 the motor imput power versus the pressure ratio

Fig. 4 shows the E.E.R. of the rolling piston type refrigerant compressor versus the pressure during the condesing temperature changes from 30°c to 55°c. It can be seen that all points on the E.E.R-pressure ratio diagram will be so close that they appear as one curve.

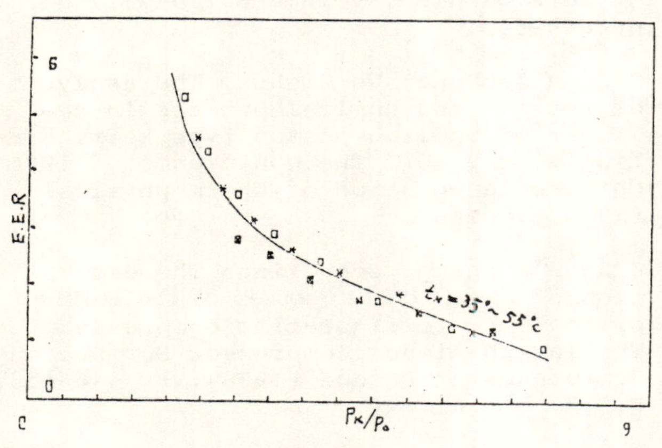

Fig. 4 The E.E.R. of compressor versus the pressure ratio

Fig. 5 Shows the indicated power of the rolling piston type compressor versus the pressure ratio. It is similar to Fig 3 and the data of this diagram are only calculated.

Fig. 6 Shows the mass flow rate of the rolling piston type compressor versus the pressure ratio. The mass flow rate changes slowly when the pressure ratio is become greater.

Fig. 7 Shows the volumetic efficiency of the rolling piston type compressor versus the pressure ratio during the condensing temperature at the range of 30°c to 55°c. In this diagram we can also see that all points are so close that they appear as one curve. The value of the volumetic efficiency at the greater pressure ratio is higher then the value of the reciprocating compressor.

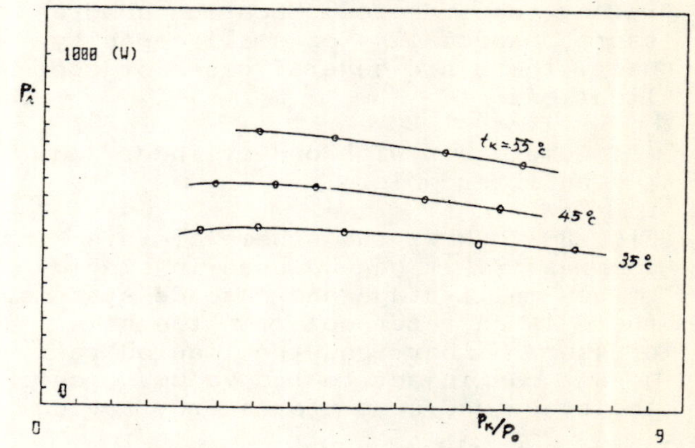

Fig. 5 The indicated power versus the pressure ratio

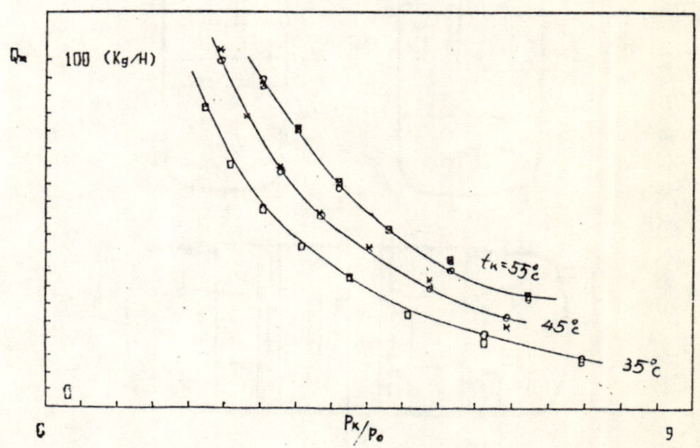

Fig. 6 The mass flow rate versus the pressure ratio

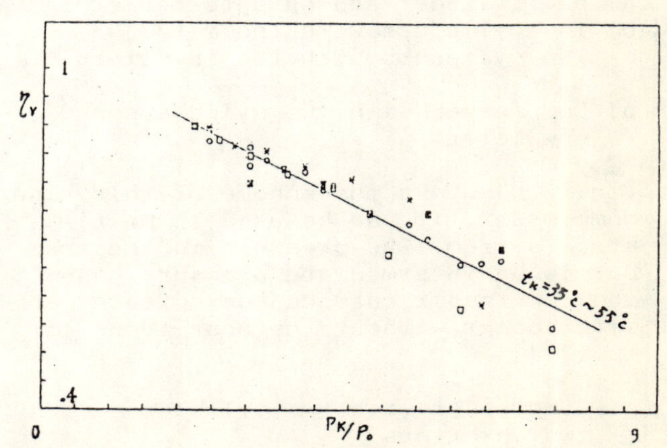

Fig. 7 The Volumetic efficiency versus the pressure ratio

4. THE APPLICATIONS OF ROLLING PISTON TYPE COMPRESSORS IN CRYOGENICS

From above analysis gives a clear idea that rolling piston type compressors are better for using in the cryogenics rather then the reciprocating compressors, espacially for small capacity unit. there are several cases of applications:

(a) Compressor with one cylinder and an intercooler

Fig. 8 a) Shows the scheme of this kind compressor. It can be used in single stage, multi-stage and cascade systems. there is an intercooler in the compressor chamber, and the liquiod refrigerant can inject to the volume C during the step Ⅳ and step Ⅴ for energy saving.

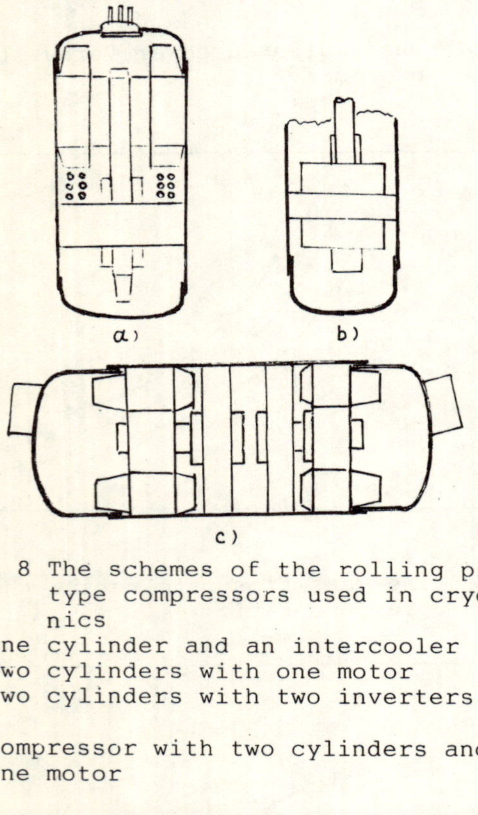

Fig. 8 The schemes of the rolling piston type compressors used in cryogenics

a) one cylinder and an intercooler
b) Two cylinders with one motor
c) Two cylinders with two inverters

(b) Compressor with two cylinders and one motor

Fig. 8 b) Shows the scheme of this kind compressor. It can be used in multistage system. The pressure in the chamber is an intermediate pressure, and the compressor case can be acted as an intercooler without the heat transfer tube.

(c) Compressor with two cylinders and two inverters

Fig. 8 c) Shows the scheme of this kind compressor. It can be used in multistage system. The displacement of each cylinder varies by changing the electrical frequency of each inverter. so, it is the best scheme for energy saving.

5. CONCLUSIONS

a) A mathematical model is developed for analyzing the charecteristics of the rolling piston type compressor. The results fit with the experimental data.

b) Some charecteristics of the rolling piston type compressor are analyzed, the performance shows that this kind compressor is suitable not only for refrigeration systems but also for cryogenics systems.

c) Some application cases in cryogenics are suggested and will save the energy comsumption.

6. REFERENCES

1 Zhou Zicheng, Wu Zichun, 'Computer simulation of the working processes for rolling piston type refrigerating compressors.' XVll international congress of refrigeraion. volume B (1987). pp 648-653.

2 Zhou Zicheng, Wu Zichun, 'The analysis of the thermodynamical process in cylinder of rolling poston type refrigerating compressor.' The conference of 1986 Chinese Association of Thermophysics. (1986), China.

3 Zhou Zicheng, Gong Ying, 'The estimation of the frictional losses of the rolling piston type refrigerating compressors.' The International Compressor Engineering Conference at Purdue (1988). PP 446-453. U.S.A.

A SIMULATION MODEL FOR A ROLLING PISTON TYPE ROTARY COMPRESSOR

Ma Guoyuan
Zhengzhou Institute of Light Industry

Yu Yongzhang
Xi'an Jiaotong University, Xi'an

ABSTRACT

Rolling piston refrigerating compressor is a new type of rotary compressor developed recently. Its construction and operation are different from conventional reciprocating compressor. Taking the working chambers as control volumes, this paper has developed a mathematical simulation model for rolling piston type rotary compressor, with the basis on the First Law of Thermodynamics and the Law of Mass Conservation. This model describes energy and mass transfer, leakage of working fluid and lubricating oil, suction gas heating loss, displacement of discharge valve and reexpansion of residual gas. The comporison of the calculated with the measured compressor performance has been made and has verified the validity and availability of the simulation model. By use of this model, the performance prediction has been made for the test compressor under the condition of variable rotation speed. The results demonstrate the possibility that this type compressor can be used in automobile air-conditioning system. Finally through the comparison of several design prejects have been obtained the optimized design data for rolling piston type rotary compressors, which can replace the reciprocating compressors in China-made refrigerators with the capacity of 90 and 160 watts respectively.

INTRODUCTION

Rolling piston type rotary compressors are widely used in air conditioners and domestic refrigerators. They have many dominant characteristics such as high efficiency and reliability, small size, light weight, small number of parts and large productivity, in comparison with conventional reciprocating compressors. These advantages prompt to switch from reciprocating compressors to rotary compressors.

A number of investigations to simulate rolling piston type rotary compressors have been carried out(1-4). Based on the previous investigations, a new model for a rolling piston type compressor was developed in this paper, with consideration of all factors influencing the performance of the compressor. To prove the suitability of this model, the calculated performance curves by the model for YZ-30R rolling piston compressor and the measured performance curves of the same compressor obtained in calorimeter tests were compared for various operating conditions. As the model proved to give good results, it can be used as a tool to predict the performance of the compressor.

SIMULATION MODEL

The Description of a Rolling Piston Type Compressor

Rolling piston type rotary compressors are similar to rotary-vane compressors in many respects. Fig.1 shows a schematic view of a rolling piston type rotary compressor. The roller is mounted on an eccentric of a shaft and divides a cylinder cavity into two chambers, a suction chamber and a compression chamber, separated by a vane. As the shaft rotates, the roller "rolls" over the inside surface of the cylinder, which causes suction in the suction chamber and compression and discharge in the compression chamber. If the tolerances were perfect, the roller would have a perfect rolling motion. However, in practice this does not happen and the roller has a complicated motion consists of rolling and the slipping motion relative to the cylinder.

Fundamental Law of Thermodynamics

The suction and compression chambers are considered as control volumes (shown in Fig. 2). The first Law of Thermodynamics and the Law of Mass Conservation are used for these control volumes. Neglecting the kinetic and potential energy differences of working fluid

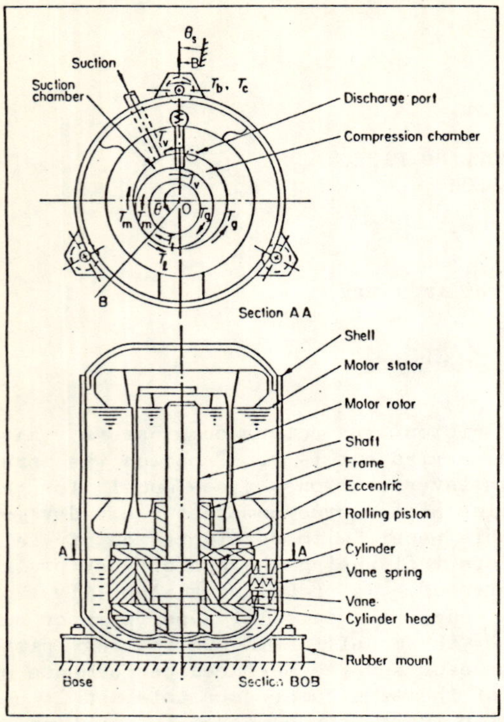

Fig. 1 Schematic view of rolling piston type rotary compressor

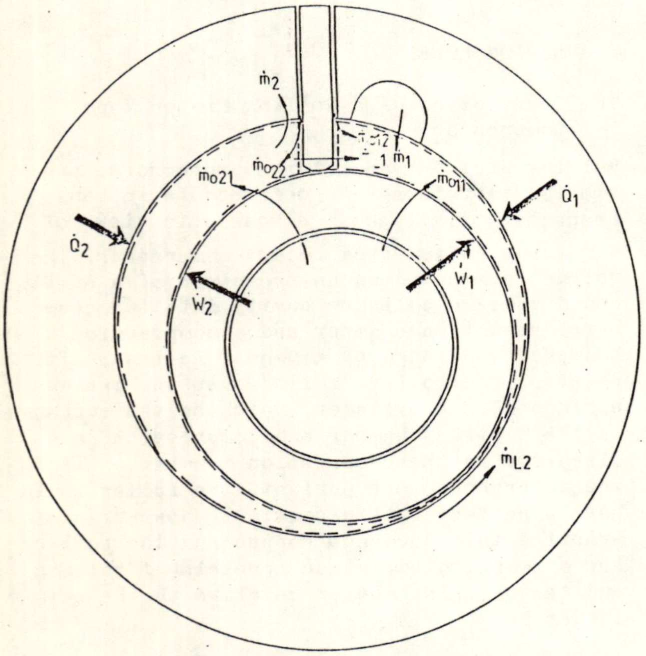

Fig. 2 Control volumes

between inlet and exit of the cylinder, the First Law of Thermodynamics can be expressed follows;

$$\frac{dU}{dt} = \dot{Q} + \dot{W} + \Sigma \dot{m}_{in} h_{in} - \Sigma \dot{m}_{out} h_{out} \quad (1)$$

For the suction chamber

$$\frac{dU_1}{dt} = \dot{Q}_1 + \dot{W}_1 + \dot{m}_1 h_1 + (\dot{m}_{L1} + \dot{m}_{L2}) h_2 - \dot{m}_b h_1 \quad (2)$$

Correspondingly, for the compression chamber

$$\frac{dU_2}{dt} = \dot{Q}_2 - \dot{W}_2 + (\dot{m}_2 + \dot{m}_{L1} + \dot{m}_{L2}) h_2 \quad (3)$$

The Law of Mass Conservation

For the suction chamber

$$\frac{dM_1}{dt} = \dot{m}_1 + \dot{m}_{L1} + \dot{m}_{L2} + a \cdot (\dot{m}_{011} + \dot{m}_{012}) - \dot{m}_b \quad (4)$$

And, for the compression chamber

$$\frac{dM_2}{dt} = -\dot{m}_2 - \dot{m}_{L1} - \dot{m}_{L2} + a \cdot (\dot{m}_{021} + \dot{m}_{022}) \quad (5)$$

State equation of working fluid

Thermodynamic parameters of halocarbon refrigerants can be described by Matin-Hou equation.

$$P = \frac{RT}{v-b} + \sum_{i=2}^{5} \frac{A(i) + B(i)T + C(i) \cdot \exp(-kT/T_c)}{(v-b)^i} + \frac{A(6) + B(6)T + C(6) \cdot \exp(-kT/T_c)}{\exp(av)[1 + c' \cdot \exp(av)]} \quad (6)$$

Where A, B, C are constant arrays and b, μ, v, a and c' are constants.

For applications in refrigeration and air condition, it is not necessary to use the full equation. The 6th term of the equation can be omitted only with less sacrifice in accuracy.

Other equations used to describe the halocarbon refrigerants are given in (5). From these equations specific enthalpy h can be derivated as a function of gas pressure p, temperature T and specific volume v.

$$h = h(p, T, v) \quad (7)$$

The definition of enthalpy is given by

$$h = u + pv \quad (8)$$

If u and v are known, the equation system (6)-(8) with three unknowns, p, T, and h, which can be calculated by iteration.

Volume-Angle Relationship

The volumes of suction and compression chambers are functions of time (or rotation angle). Referring to Fig.3, the following geometrical relationships can be easily established.

$$V1 = H(A_\theta - A_\gamma - A_T - A_v) \quad (9)$$

$$V2 = V_h(1+\varepsilon) - V_1 - 2A_v H \quad (10)$$

Where, $A_\theta = \frac{1}{2}\theta R_c^2$, $A_v = \frac{1}{2} H_e B_v - A_R$

The vane extension into the cylinder is

$$He = R_c - L \quad (11)$$

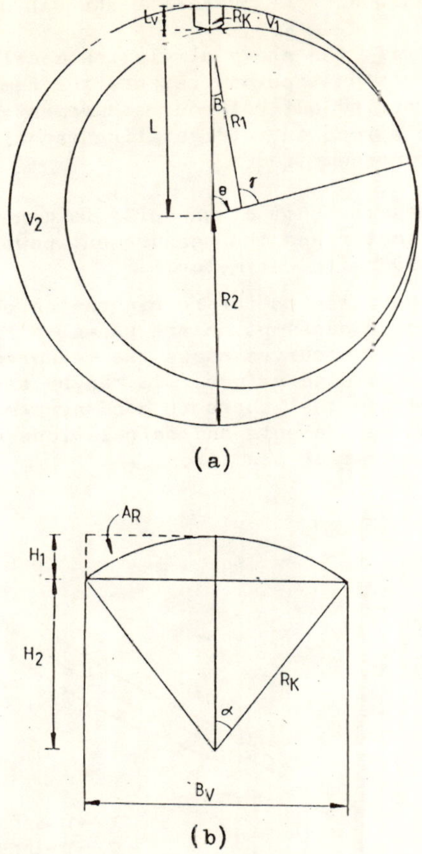

Fig. 3 Geometric values

Where, $L = \sqrt{R_r^2 + E + 2R_rE\cdot\cos\gamma}$

When rotation angle is θ, the heat exchange area inside the cylinder are

$$Ao1 = H(B_{\theta_1} - \tfrac{1}{2}B_v + B_{\gamma_1} + H_e) + 2\cdot\frac{V_1}{H} \quad (12)$$

$$Ao2 = H(B_{\theta_2} - \tfrac{1}{2}B_v + B_{\gamma_2} + H_e) + 2\cdot\frac{V_2}{H} \quad (13)$$

The shaft is driven by a motor. The angle of rotation is given by

$$\theta = \omega\cdot t \quad (14)$$

Compression Power

Assuming that the control volumes for the compressor chambers are all simple compressible thermodynamic systems, then the momentary power for these chambers are as follow.

$$\dot{W}_1 = -p_1\frac{dV_1}{dt} \quad (15)$$

For the compression volume

$$\dot{W}_2 = -p_2\frac{dV_2}{dt} \quad (16)$$

Leakage Loss

Refrigerant and lubricating oil leak through leakage paths from high pressure side to low side. The leakage is liquid vapor mixture of oil and refrigerant. To simplify the calculations, the leakage can be divided into two types: one is liquid leakage, another is gas leakage.

(A) Liquid Leakage

Leakage Past the Clearances of Rolling Piston Faces

Fig.4 shows an analytical model of leakage flow occurring radially outwards through the clearances on rolling piston faces. The balancing equation of the forces acting on the infinitely small element (shown in Fig. 4) expressed as follows

$$\tau r d\theta dr - (\tau + \frac{d\tau}{dy}) r d\theta dr = p r d\theta dy - (p + \frac{dp}{dr}dr)(r+dr)d\theta dy + 2p dr dy \frac{d\theta}{2} \quad (17)$$

$$\tau = \mu(\partial V/\partial y) \quad (18)$$

Resolving the above equation, the leakage mass flow rate can be calculated.

Leakage Through the Clearances Between Vane and slot

Liquid leakage occurs the clearances between vane and slot. The clearances have a very small height in relation to breadth and length, therefore, the leakage can be analysed as a lamina flow between two infinitely developed parallel plates.

(B) Gas Leakage

Leakage Past Pistion Radial Clearance

At piston radial clearance, the compressed gas leaks into suction chamber. As the leakage path is narrow and long compared with its height, it is necessary to analyse the leakage taking viscous effects into account.

The leakage path developed along the cylinder inner surface is modelled by the flow channel which consists of the compression chamber, a convergent nozzle, a straight channel including the viscous drag and the suction chamber.

Mass flow rate in the clearance is calculated as the adiabatic flow with fluid friction.

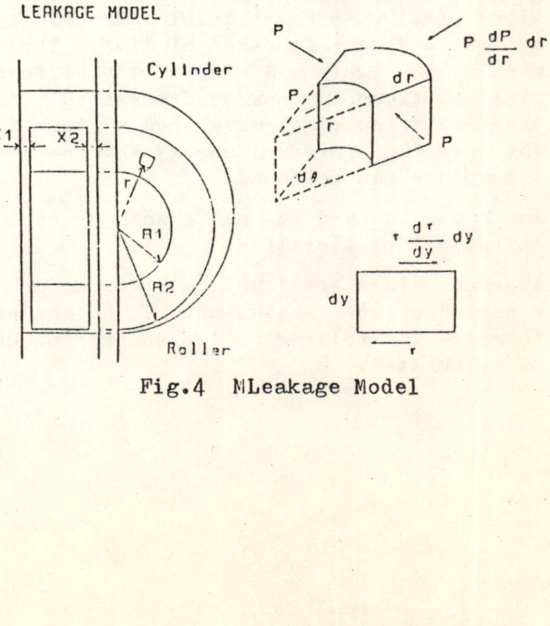

Fig.4 MLeakage Model

Leakage Past Vane Edges

The leakage past vane edge can be analysed as a flow through a convergent nozzle, under the assumption of adiabatic friction less, the flow rate through the nozzle per unit of area can be calculated.

Discharge mass flow rate

The discharge mass flow rate is dependent on the discharge valve lifting height, the pressure difference and discharge port area. Discharge valve for rolling piston compressor is reed valve. A reed valve model has been developed (6). Then, the discharge mass flow rate can be calculated by using this reed valve model.

Heat Transfer

Heat transfer in rolling piston compressor is very complicated. Up to date, there are little investigation in this field.

The heat transfer coefficient between working fluid and cylider walls can be calculated by using the McAdams correlation.

Suction Mass Flow

The suction mass flow is calculated from the pressure difference between suction pressure and the pressure in the suction chamber.

Re-expansion of Residual Gas

It is assumed that at the moment when the roller is at top dead point momentary re-expansion occurs. The residual gas is re-expansion in the whole suction volume. The re-expansion loss can be calculated by internal energy balance.

Return Flow

After re-expansion of the residual gas, compression chamber pressure will be larger than suction pressure. Therefore, there will be a return flow through the suction pipe during the first degrees θ_0 of rotation. The mass flow per unit of area is calculated as nozzle flow.

Final Calculations

Solution for the simulating differential equations (1-2)-(1-5) give total internal energy and mass as a function of time. The variables on right-hand side of the differential equations can now be determined from the simulation model developed above. Then, the specific internal energy and the specific volume can be found.

For known u and v, p, T and h can be calculated by iteration.

Average values are found by integration over a period of one revolution. Then, the performance for rolling piston compressor can be calculated.

COMPARISON OF MEASUREMENTS AND CALCULATION

To verify the above simulation model of rolling piston type compressor, the comparison was carried out between measurements and calculations on refrigerating capacity and power consumption.

The measurements of capacity is done with calorimeter and the measurement power consumption with wattmeter.

The characterics for refrigerating capacity and power consumption are shown as Fig. 5 and 6. The curves shows the measured and the calculated values. It can be seen from the curves that there is good agreement between measurements and calculations over a wide operating condition.

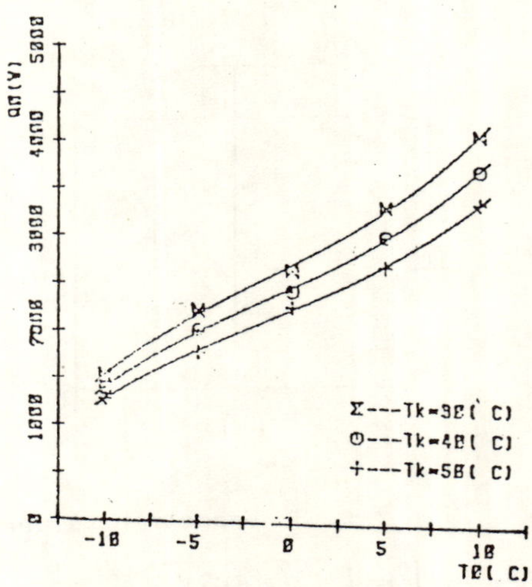

Fig.5 Refrigerating capacity comparison of measurements and calculations

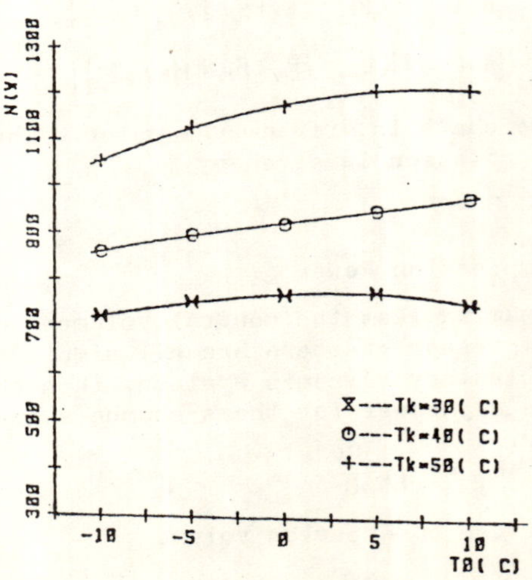

Fig.6 Power comparison of measurements and calculations

APPLICATION OF THE SIMULATION MODEL

As described above, the validity and availability of the simulation model have been verified. Therefore, this model can be used to prodict the performance of new machine.

(1) Performance prediction for a rolling piston type compressor in an automotive air conditioner.

We attempt to replace conventional swash plate compressor by the test rolling piston compressor for an automobile air conditioner, the data of the test compressor is shown in Table 1. Using the model developed in this paper, we had predicted the performance for the compressor under the condition of variable speed (shown in Fig.7). In Fig. 7,

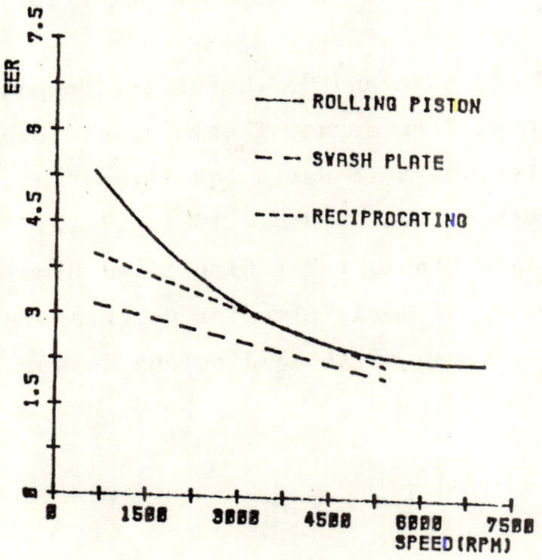

Fig. 7 EER of test compressor in comparison with two conventional compressors : reciprocating and swash plate compressor

the EER of the compressor is compared with that for two conventional compressors used in automobile air conditioner. When speed is below 2000 rpm or above 5500 rpm, the EER for the test compressor is significantly higher than the other two compressors. When speed is from 2200 to 5000 rpm, the EER of the test compressor is higher than that of reciprocating compressor, but approaches to that of swash plate compressor, this demostrates that the test compressor can be used in automotive air conditioner instead of these two conventional compressors.

(2) Performance prediction for the rolling piston type compressors in China made refrigerators

There are two commonly used refrigerators in China, which have capacity 90 and 160 watts respectively. EER of the reciprocating compressors in these refrigerators are about 1.

The performance prediction of the new rolling piston compressors is shown in Fig.8,

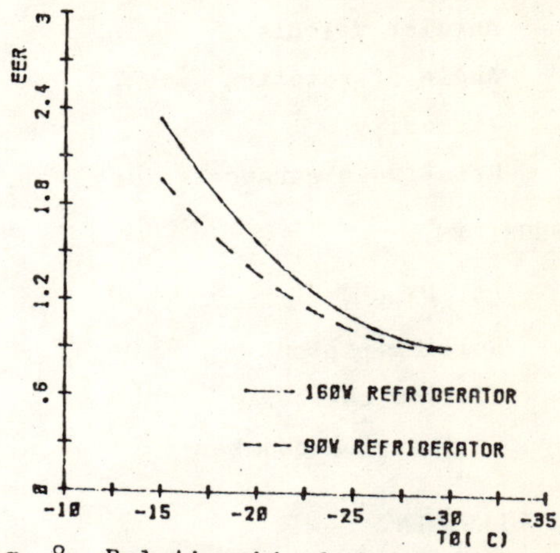

Fig. 8 Relationship between EER and evaporating temperature

EER of these two prejects (in Table 1) are about 1.2 and 1.3 respectively. This demonstrates that these two prejects have better preformance than the reciprocating compressors.

The performance calculated by the simulation model developed in this paper is in good correlation with the experimental one. This demonstrates that the model is valid. This model can be used to analyse working process, to predict the performance, and as a tool for optimization design. The three design prejects obtained by the aid of this model could be used in actual production.

LIST OR SYMBOLS

A	Area
a	Weight proportion of gas in the oil
H	Cylinder height
h	Specific enthalpy
M	Total mass
$\dot{m}$	Mass flow rate
p	Pressure
Q	Heat flow rate
R	Gas constant
	Radius
r	Radius
T	Temperature
t	time
U	Total internal energy
u	Specific internal energy

V	Volume
	Leakage velocity
v	Specific volume
W	Compression power
ω	Angular velocity
θ	Angle of rotation
μ	Viscosity
ε	Relative clearance volume

SUBSCRIPTS

c	Cylinder
r	Rolling piston
1	Suction chamber
2	Compression chamber

ACKNOWLEGEMENT

We are grateful to Mr. Liy Yong and Miss Li Li for giving us unselfish help in this work.

REFERENCES

1. P. Pandeya and W.Soedel: " Rolling Piston Type Rotary Compressors with Special Attention to Friction and Leakage" Proc. 1980 Purdue Compr.Tech.Conf.,p.209-216.
2. T.Yanagisawa and T. Shimizu: "Leakage Losses with a Rolling Piston Type Rotary Compressor. I.Radial Clearance on the Rolling Piston" Int.J Refrig.,Vol.8,No.2, 1985,p.75-84.
3. T.Yanagisawa and T.Shimizu: "Leakage Losses with a Rolling Piston Type Rotary Compressor.II.Leakage Losses Through Clearances on Rolling Piston Faces" Int. J Refrig., Vol.8,No.3, 1985,p.152-158.
4. F.Gyberg and H.Stentoft Nissen: "A Simulation Model For Fixed Vance Rotary Compressor Using Real Gas Properties" Proc. of ICECP, 1984, p.33-39.
5. C. Y. Chan and G.G.Haselden."Computer - Based Refrigerant Thermodynamic Properties, Part 1: Basic Equations "Int. J Refrig., Vol.4,No.1, 1981,p.7-12.
6. Huang Dianxum: " A Simplified Simulation Model of Reed Valve for compressors "M. D. Theme of Xi'an Jiaotong University, 1986.

Table 1 The design parameters of rolling piston compressors

Prejects	for automobile air conditioner	for China-made refrigerator with capacity 90 watts	for China-made refrigerator with capacity 160 watts
cylinder inner radius	27 mm	19 mm	27 mm
cylinder height	25 mm	10.6 mm	12 mm
rolling piston radius	21.5 mm	16.4 mm	23.9 mm
vane thickness	4.5 mm	3.6 mm	4.2 mm
eccentric radius	5.5 mm	11.5 mm	9.9 mm
discharge port diameter	7 mm	7 mm	9 mm
suction port diameter	11 mm	5 mm	6 mm
discharge port angle	15 degr.	10 degr.	13 degr.
suction port angle	350 degr.	352 degr.	350 degr.
relative clearance volume	1.0%	1.2%	1.0%
piston radical clearance	18 μm	10 μm	15 μm

3. CRYOCOOLERS

MECHANISM AND NUMERICAL ANALYSIS OF ORIFICE PULSE TUBE REFRIGERATOR WITH A VALVELESS COMPRESSOR

Wu Peiyi Zhu Shaowei
Power Machinery Engineering Department
Xi'an Jiaotong University, Xi'an, China

ABSTRACT

This paper presents a thermodynamic explanation of the pumping heat effect in pulse tube from the view point of thermodynamics of variable mass systems. A typical calculation by using numerical analysis for an orifice pulse tube refrigerator with a valveless compressor has been made. The results shown in the paper will reveal the processes occurring in the pulse tube in detail. The effects of some parameters are discussed as well.

INTRODUCTION

The pulse tube refrigeration discovered by Gifford in 1963 has some advantages over other cycle cryocoolers[1], such as only one moving part at room temperature, mechanical simplicity and inherent reliability. However, the basic pulse tube refrigerator developed by Gifford and his coworkers only achieved a temperature of 124 K with one stage, this temperature is usually too high for the most practical applications. In the recent few years Mikulin suggested a new type of the pulse tube called cryogenic pulse tube in which two innovative elements (a gas distributor and a reservoir) were added to the closed end of the tube. This kind of pulse tube refrigerator was able to achieve a low temperature of 105 K by using air in 1984[2]. Then Radebaugh at NBS made a modified version in which a valveless compressor and a needle valve placed between the heat exchanger and the reservoir were used (see Fig.1). This version called the orifice pulse tube achieved a low temperature of 60 K by using helium gas in 1985[3]. The achievement on the new type of the pulse tube is attracting considerable interest, many theoretical and experimential studies are being carried out[4].

This paper presents a discussion on mechanism from the view point of thermodynamics of variable mass systems[5] and a numerical analysis for an orifice pulse tube refrigerator, the processes occurring in the tube are revealed in detail and the effects of some parameters are discussed as well.

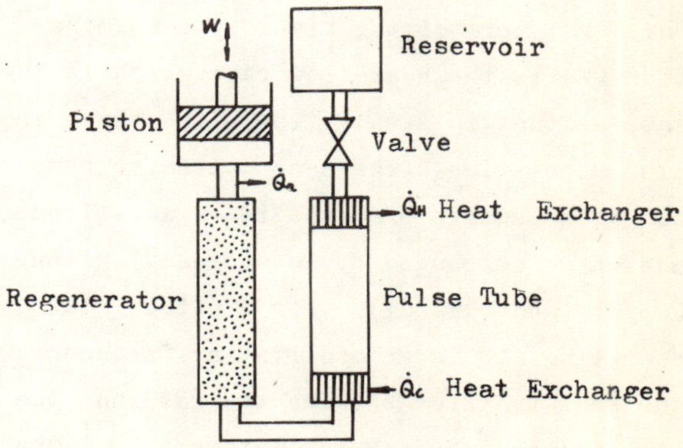

Fig.1 The orifice pulse tube refrigerator

THERMODYNAMIC FOUNDATION

Let us consider a thin and long tube shown

in Fig.2, the gas at lower temperature T_1 enters the tube from the section 1 and goes out of the section 2 at high temperature T_2 ($T_2 > T_1$). Fig.2 shows a positive mass flow. We take

the gas within the tube as a control volume, it is a variable mass system when the flow is unsteady. Assuming the working gas as ideal gas and uniform pressure in the tube, the conservations of energy and mass for the control volume are as follows:

$$\frac{dQ}{dt} + \dot{m}_1 h_1 = \frac{dU}{dt} + \dot{m}_2 h_2 \qquad (1)$$

$$\dot{m}_1 - \dot{m}_2 = m\left[\frac{1}{p}\cdot\frac{dp}{dt} - \frac{1}{T_m}\cdot\frac{dT_m}{dt}\right]$$

where T_m is the average temperature of the control volume. After rearrangement, we get

$$\frac{1}{\dot{m}_1}\frac{dQ}{dt} = (h_2 - h_1) + \frac{V}{k-1}\left[\left(1 - \frac{kT_2}{T_m}\right)\frac{1}{\dot{m}_1}\cdot\frac{dp}{dt} + \frac{KPT_2}{T_m^2}\cdot\frac{1}{\dot{m}_1}\cdot\frac{dT_m}{dt}\right] \qquad (2)$$

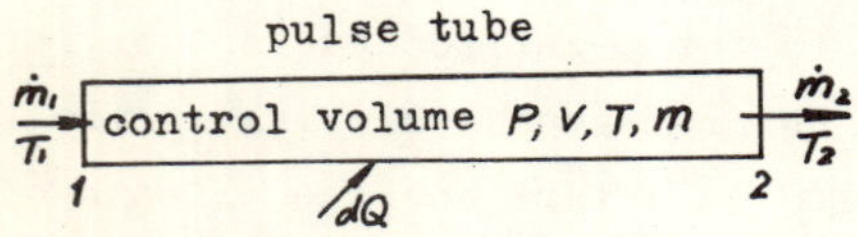

Fig.2 Unsteady flow in a pulse tube

The first term of the right side in equation (2) is the heat flow rate which is the same as that in steady flow, its values are always positive. The second term represented by the square brackets is an additional one which is caused by an unsteady flow (dp/dt and dT_m/dt). For a pulse tube operating at higher oscillating frequency the heat transfer between the gas and the wall of the tube may be quite small [3], the left term in equation (2) is approximately equal to zero, so the values of the square brackets in equation (2) should be negative. For a basic pulse tube, dp/p is usually much larger than dT_m/T_m, the values of $\frac{1}{\dot{m}_1}\cdot\frac{dp}{dt}$ should be positive as $kT_2 > T_m$. that is to say, when the pressure increases the gas should flow into the pulse tube from the cold end of the tube and vice verse. For an orifice pulse tube, as will be mentioned below, the average temperature of the gas in the tube will change with time greatly. However, a numerical calculation shows that the $\frac{1}{\dot{m}_1}\cdot\frac{dp}{dt}$ is also positive in the most time of a period. If the above conditions are satisfied, it is possible to pump heat in a tube by using an unsteady gas flow.

As for the hot end of the pulse tube, Radebaugh has mentioned by using an enthalpy flow analysis that the gross pumping heat rate is

$$\dot{Q}_H = \frac{c_p}{\tau}\oint \dot{m}_2 T_2\, dt = \frac{Ac_p}{R\tau}\oint p u_2\, dt \qquad (3)$$

Thus, $\dot{Q}_H$ is a maximum when p and u_2 (or approximately $\dot{m}_2$) is in phase and is zero when the phase difference between them is $\pi/2$. The phase shift by adding an orifice and a reservoir may be explained clearly by using Fig. 3. Because the reservoir volume is large, the pressure in it is almost no change with time and is nearly equal to the average pressure of the system, the direction of gas flow at the hot end of the tube depends on the pressure difference between

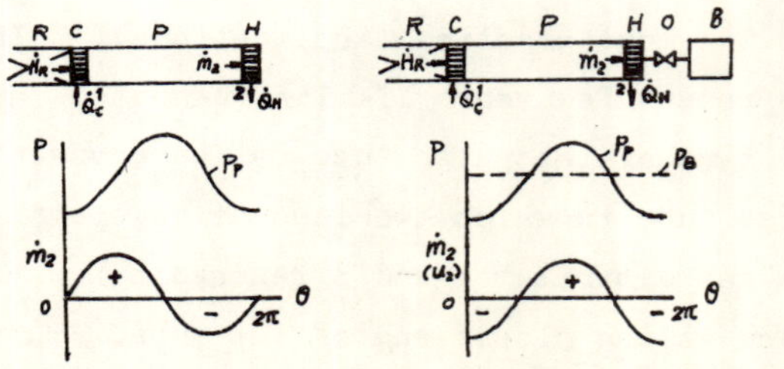

basic pulse tube   orifice pulse tube

Fig.3 Phase difference between p and $\dot{m}_2$

the tube and the reservoir for an orifice pulse tube instead of only on the pressure of the tube for a basic pulse tube.

NUMERICAL ANALYSIS AND RESULTS

In order to understand the processes occurring in the orifice pulse tube refrigerator in detail, a numerical analysis has been made. A mathematical model used here is to divide the whole refrigerator into many subsystems, each subsystem is considered as a uniform one which can exchange the work, heat and mass with the outside through its boundaries, and whose parameters including mass are changeable with time. For simplicity, the processes in the tube and the reservoir are considered as adiabatic and the pressure drop in the system is neglected except in the orifice. The working gas (He) is idea gas.

The basic equations are as follows:

For i-subsystem

$$\frac{dU_i}{dt} + (\dot{m}h)_{i+1} - (\dot{m}h)_i - \alpha_i A_i (T_{wi} - T_i) + p \cdot \frac{dV_i}{dt} = 0$$

$$\frac{dm_i}{dt} = \dot{m}_i - \dot{m}_{i+1} \qquad (4)$$

Note that the mass flow rate is defined as positive if the gas flows from the compressor to the reservoir and as negative during the opposite flow.

For the reservoir [5],

$$\frac{dm_B}{dt} = \frac{V_B}{KRT_H} \cdot \frac{dP_B}{dt} \qquad \text{when} \quad \frac{dm_B}{dt} > 0$$

$$\frac{dT_B}{dt} = \left(1 - \frac{T_B}{kT_H}\right) \cdot \frac{T_B}{P_B} \cdot \frac{dp_B}{dt} \qquad (5)$$

$$\frac{dm_B}{dt} = \frac{V_B}{KRT_B} \cdot \frac{dp_B}{dt} \qquad \text{when} \quad \frac{dm_B}{dt} < 0$$

$$\frac{dT_B}{dt} = \left(1 - \frac{1}{k}\right) \cdot \frac{T_B}{P_B} \cdot \frac{dp_B}{dt} \qquad (6)$$

For the matrix in the regenerator,

$$(mc)_{si} \cdot \frac{dT_{si}}{dt} + \alpha_i A_i (T_{si} - T_i) = 0 \qquad (7)$$

For the orifice, the mass flow rate through it is determined by using a formula for nozzle with a collection factor.

The above differential equations are rewritten to become a set of difference equations and then solved by an interative method [6]. A typical calculation has been made for an orifice pulse tube refrigerator whose structure is mainly taken from the reference [3], the tube is 12.7 mm O.D. by 237 mm long, the regenerator is 19 mm O.D. by 127 mm long, the reservoir volume is 1L. The average pressure is taken to be 0.97 MPa.

Fig.4 shows a set of changes in pressure, mass flow rate, temperature and velocity at two ends of the tube for three different values of d_o/D_p. It can be found that the $\frac{1}{\dot{m}_1} \cdot \frac{dp}{dt}$ is positive in the most time and p and $\dot{m}_2$ is almost in phase. In addition, when the d_o/D_p increases, the mass flow rate $\dot{m}_2$ (or velocity u_2) increases too, but the pressure fluctuation decreases, so there exists an optimum value of d_o/D_p which is shown in Fig.5. Now we have found that the effect of adding an orifice and a reservoir is not only possessed of phase shift but also of amplitude shift.

Fig.6 shows the results for three different values of frequency. The increase in frequency causes a larger fluctuation of p and $\dot{m}$, so the gross pumping heat rate $\dot{Q}_H$ will increase. However a larger mass flow rate will increase the heat load in the regenerator, as a result, the efficiency of regenerator will decrease and the enthalpy flow $\dot{H}_R$ will increase. Thus there also exists an optimum frequency for getting a maximum net refrigerating power $\dot{Q}_c$, which is shown in Fig.7.

Some internal processes occurring in the pulse tube for $d_o/D_p = 0.0643$ and f=9 are shown in Fig.8-11. Fig.8 shows an axial temperature distribution along the tube at different time. x = o and x = L represent the cold and hot end of the tube respectively. The temperatures of gas fluctuate with time greatly, the axial temperature distribution of the gas in an orifice pulse tube is not linear. Fig.9 shows how the mass flow rate and gas temperature change along the tube. Fig.10 and 11 are two trace maps for three gas particles, they show that there is a portion of gas in the tube which never moves out of the tube and oscillates within the tube with pressure oscillation. All the results given above may enable us to gain an important understanding of the working processes occurring in an orifice pulse tube.

CONCLUSION

1. The effect of adding a reservoir with an orifice to a basic pulse tube is not only phase shift but also amplitude shift.
2. There exist an optimum d_o/D_p and an optimum frequency for getting a maximum net refrigeration power or a maximum efficiency.
3. The local temperatures of the gas in the orifice pulse tube fluctuate with time greatly, the average temperature of the gas along the tube is not linear.
4. A portion of gas in the tube never moves out of the tube and oscillates in the tube with pressure oscillation.

LIST OF SYMBOLS

A	heat transfer area (m^2)
c_p	specific heat (J/Kg.k)
d_o	diameter of orifice (m)
D_p	diameter of pulse tube (m)
f	pulse frequency (Hz)
h	enthalpy (J/Kg)

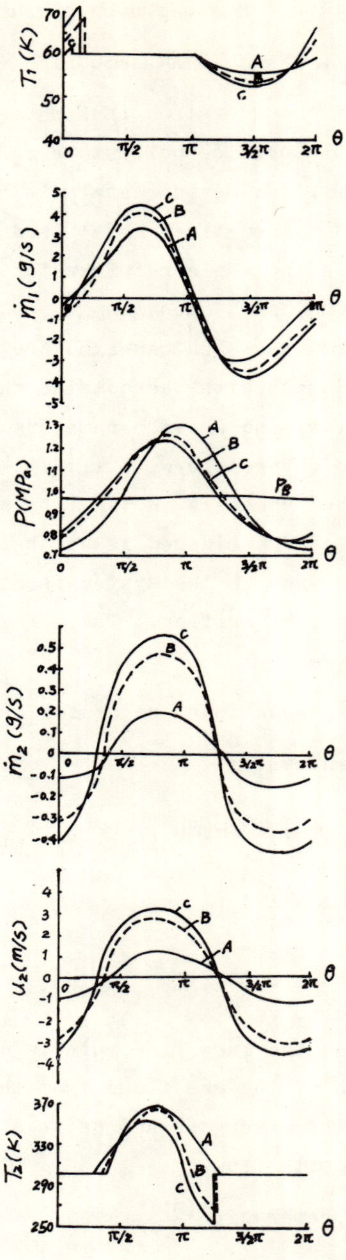

Fig. 4 Changes in P, ṁ, T and u_2

	A	B	C
d_o/D_p	0.0371	0.0643	0.0742

f = 9 Hz

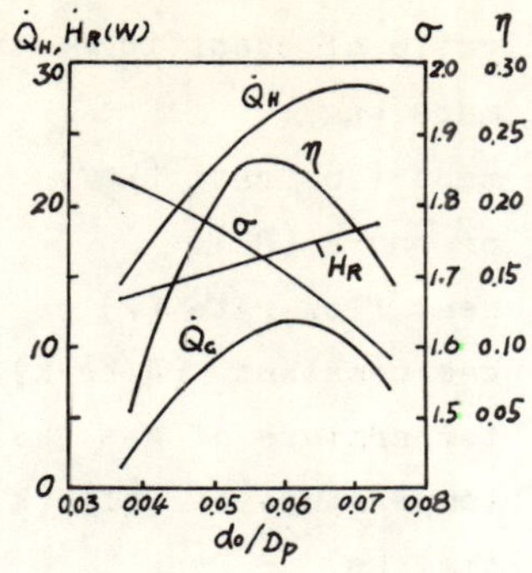

Fig. 5 Effects of d_o/D_p

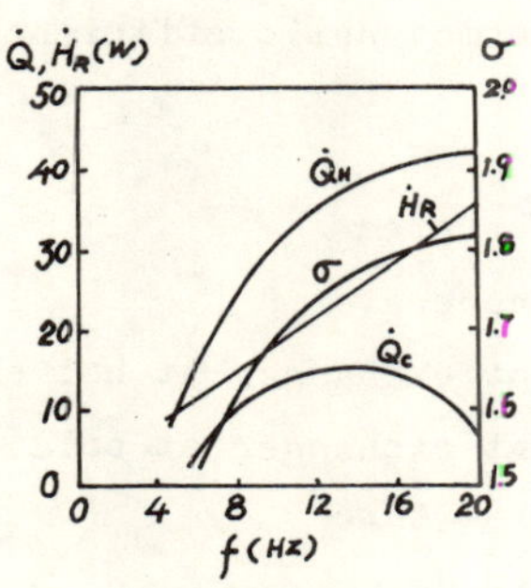

Fig. 7 Effects of frequency

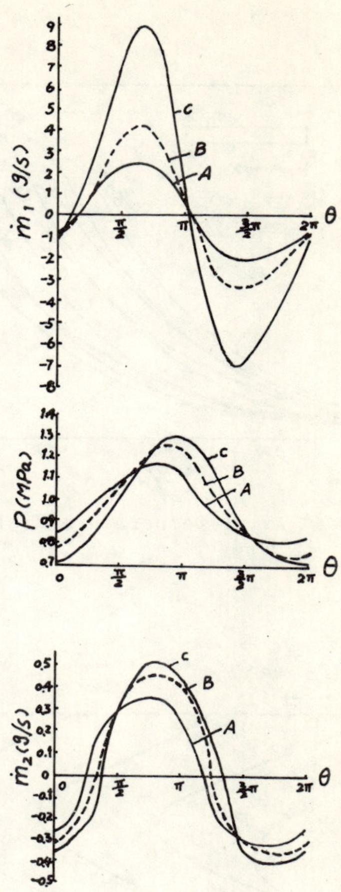

Fig. 6 Changes in P and $\dot{m}$

	A	B	C
f(Hz)	4.5	9	20

d_o/D_p = 0.0643

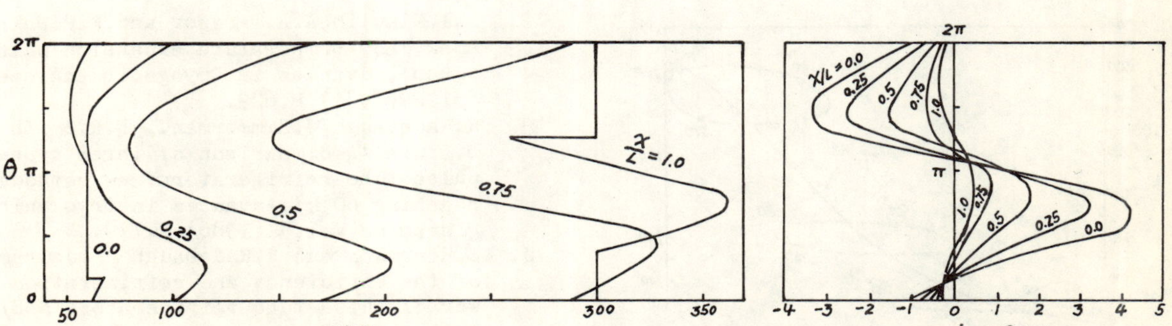

Fig. 9 Changes in T and $\dot{m}$ along the tube

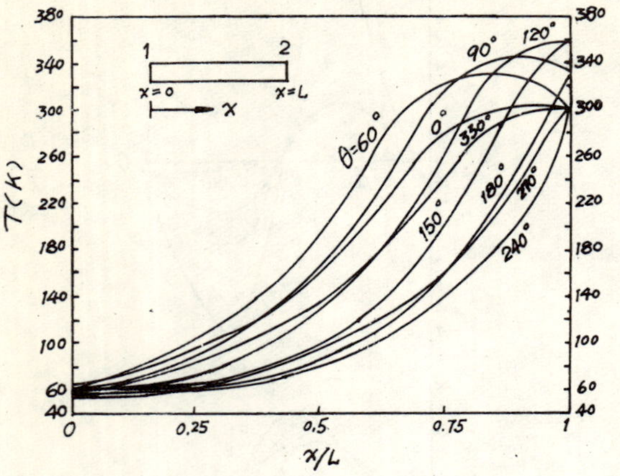

Fig. 8 Axial temperature distribution

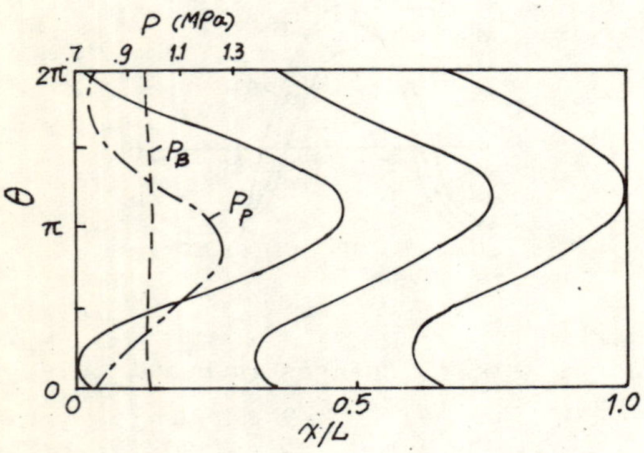

Fig. 10 Trace map (x-θ) of three gas particles

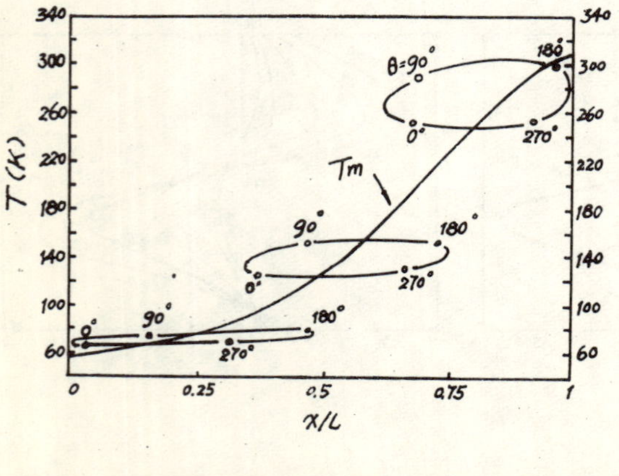

Fig.11 Trace map (x ~ T) of three gas particles

H	enthalpy flow rate (w)
k	ratio of specific heats
m	mass (Kg)
$\dot{m}$	mass flow rate (Kg/s)
P	pressure (Pa)
$\dot{Q}$	heat flow rate (w)
R	gas constant (J/Kg·k)
T	temperature of gas (k)
T_w	temperature of wall (k)
t	time (s)
U	internal energy (J)
V	volume (m^3)
α	heat transfer coefficient (w/m^2 k)
τ	period (s)
σ	pressure ratio (P_{max}/P_{min})
θ	crank angle (radian or degree)
η	thermodynamic efficiency

SUBSCRIPTS

i	subsystem
R	regenerator
H	heat exchanger at hot end
C	Heat exchanger at cold end
P	pulse tube
O	orifice
B	reservoir
S	solid
1	cold end of pulse tube
2	hot end of pulse tube

REFERENCES

1. W.E.Gifford and R.C. Longsworth,'Pulse tube refrigeration progress'. Advances in cryogenic Engineering,Vol.10(1965), P.69.
2. E.I.Mikulin,A.A.Tarasov and M.P.Shkrebyonock,'Low-temperature expansion pulse tubes'.Advances in Cryogenic Engineering, Vol.29(1984),P.629.
3. R.Radebaugh,J.Zimmerman., D.R.Smith and B.Louie 'A comparison of three types of pulse tube refrigerators:new methods for reaching 60 k'.Advances in Cryogenic Engineering,Vol.31(1986),P.779.
4. S.Herrmann and R.Radebaugh 'Measurements of the efficiency and refrigeration power of pulse-tube refrigerators'NBS/TN-1301, (1986).
5. Wu Peiyi and Ma Yuan 'Thermodynamics of variable mass systems and its applications'(in Chinese),the High Education Press,(1983).
6. S.V. Pantanker,'Numerical Heat Transfer and Fluid Flow', McGraw-Hill Press,(1983)

DEVELOPMENT OF A 5K GM REFRIGERATOR USING RARE EARTH COMPOUNDS AS A REGENERATOR MATRIX

T. Kuriyama, R. Hakamada, H. Nakagome, Y. Tokai, M. Sahashi, and O. Horigami
TOSHIBA Research and Development Center
Ukishima, Kawasaki, Japan

R. Li, O. Yoshida, K. Matsumoto, and T. Hashimoto
Dept. of Applied Physics, Tokyo Institute of Technology
Meguro, Tokyo, Japan

ABSTRACT

Nowadays, the application of two-stage GM refrigerators to various cryogenic systems is widely spreading. Below 15 K, regenerator efficiency drops rapidly, because of the low heat capacity of lead which is preferred as a 2nd regenerator matrix. That is why the lowest temperature of the GM refrigerator is limited to about 8 K and their refrigeration capacity below 10 K is very small. However, some rare earth compounds have large heat capacity, compared with lead in this temperature region. The present paper describes the experimental results obtained from the two-stage GM refrigerator, in which Er_3Ni is used as a regenerator matrix. The lowest temperature of 5.15 K was achieved and refrigeration capacity, such as 1 W at 6.35 K or 2 W at 7.20 K, was obtained. The average input power to the compressor unit for operating the GM refrigerator was about 2.8 kW.

INTRODUCTION

Small Stirling cycle and Gifford-McMahon(GM) cycle refrigerators are commonly spread for many cryogenic systems, because they have advantages of high reliability and simple mechanism. Especially, two-stage GM refrigerators are widely used in shield cooling for Magnetic Resonance Imaging (MRI) magnets and in cryopumps.

Pb (lead) is usually preferred as a 2nd regenerator matrix, because of its high specific heat for the temperature region from 77 K to 15 K. However, the specific heat for Pb decreases with decreasing temperature and, at lower temperature, the heat capacity of a regenerator matrix is no longer much greater than that of gaseous helium. Then, regenerator efficiency drops rapidly with decreasing temperature. That is why the lowest temperature achieved by commercial regenerative refrigerators, is limited to about 8 K.

To obtain lower no-load temperature, utilization of the magnetic specific heat of rare earth or of rare earth compound as a regenerator matrix has been proposed. Daniel et al.(1) achieved 7.8 K with EuS (Europium Sulfide) as a 3rd regenerator matrix in the triple-expansion Stirling-cycle refrigerator. Nakashima et al.(2) used GdRh (Gadolinium Rhodium) and GdErRh (Gadolinium Erbium Rhodium) as a regenerator matrix in the main Stirling-cycle refrigerator, which operated at from 30 K to 4-5 K with 0.14 MPa mean pressure. The maximum refrigeration capacity of 2.3 W was obtained at a temperature less than 5 K. Matsubara et al.(3) used GdRh powder with epoxy resin as a regenerative material for a final stage in the composite Vuilleumier cycle refrigerator. They achieved 5.4 K no-load temperature and 10 mW refrigeration power at 7 K was obtained. Nagao et al.(4) achieved 3.6 K no-load temperature by three stage GM refrigerator. GdRh and GdErRh were used as a third regenerator matrix. They also confirmed 5 cc/hr of helium liquefaction from room temperature. Stuart et al.(5) achieved the lowest temperature, 6.5 K, by three-stage GM refrigerator. They used Pb as both 2nd and 3rd regenerator matrices.

The authors focused to improve refrigeration performance below 10 K for conventional two-stage GM refrigerator. Because two-stage GM refrigerators are commonly spread for cooling superconducting devices or cryogenic systems and utilization of the improved two-stage GM refrigerator is very wide, such as MRI magnet or cryopumps. The two-stage GM refrigerator with rare earth compounds as a regenerator matrix was developed, which achieved the lowest temperature of 5 K level. Not only was the no-load temperature lowered, but both the refrigeration power and the refrigeration efficiency were also improved. This paper describes the experimental results obtained from the two-stage GM refrigerator.

REGENERATOR WITH RARE EARTH COMPOUNDS

(a) Rare earth compounds

To improve the regenerator efficiency in the low temperature region, several rare earth compounds have been developed(6)(7). The authors selected Er_3Ni (Erbium 3 Nickel) as a regenerator matrix. Figure 1 shows the volumetric specific heat for Er_3Ni and Pb. As a regenerator matrix in the low temperature region, Er_3Ni has several advantages. Below about 15 K, the volumetric specific heat for Er_3Ni is much larger than that for Pb. At higher temperature, volumetric specific heat for Er_3Ni is nearly the same as that for Pb. This characteristic is very convenient to enable using Er_3Ni as a 2nd regenerator matrix. The reason is that one may easily accept a hybrid structure with Pb and Er_3Ni in a 2nd regenerator or it is possible to use Er_3Ni only as a 2nd regenerator matrix. A sharp peak, determined by magnetic specific heat, is common in rare earth compounds, such as EuS, GdRh, or GdErRh, causing difficulty in initial cooling from room temperature to operating condition(8). Er_3Ni doesn't have such an obstacle and initial cooling should be done smoothly. The Er_3Ni cost is much less than that for GdRh or GdErRh.

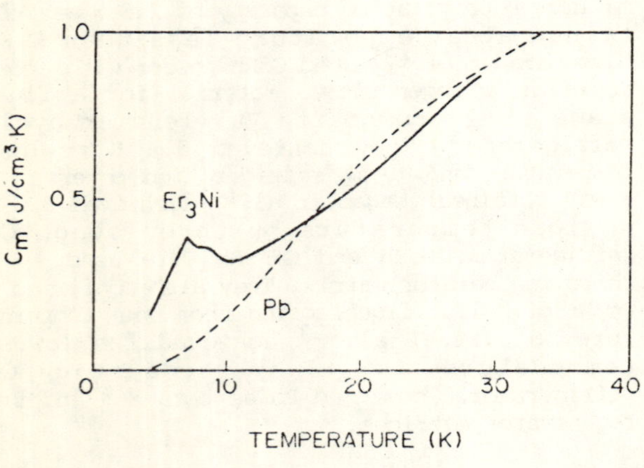

Fig. 1. Specific heat data for Er_3Ni and Pb

(b) Regenerator analysis

A hybrid structure with Er_3Ni and Pb is adopted for a 2nd regenerator. Then, the volume ratio for Er_3Ni and Pb, which assures the maximum regenerator efficiency, must be determined. For this purpose, thermal performance for a hybrid regenerator was calculated. There are many parameters for regenerator operation. Many thermal analyses have been reported in pertinent literature. Rios et al.(9) investigated the effect of the specific heat variation on the thermal performance of regenerators. Daney et al.(10) showed that the void volume fluid heat capacity could strongly influence the regenerator performance. Sahoo et al.(11) operated the governing differential equations with temperature dependent specific heat of a fluid. The effect of axial conduction in the fluid was calculated by Sarangi et al.(12). In this paper, the authors mostly follow Rios's method. Constant mass flow rate and constant gaseous helium properties are assumed for a simple one-dimensional model. Axial conduction of gaseous helium and matrix are neglected and transverse conduction is assumed to be infinite. In Rios's paper, the Debye function was used to change the specific heat of materials. In this paper, however, specific heat data for Pb and Er_3Ni were tabulated and regenerator efficiencies were calculated from this table. The governing equations are written as

$$m C_g \frac{\partial T_g}{\partial x} = \pm \frac{hA}{L} (T_g - T_m) \quad (1)$$

$$M \frac{\partial (C_m T_m)}{\partial t} = \pm hA (T_g - T_m) . \quad (2)$$

After solving Eq. (1) and (2) in the t and x directions, regenerator efficiency, η , is defined as

$$\eta = 1 - \frac{\int_0^\tau (T_{out} - T_c) dt}{\tau (T_h - T_c)} \quad (3)$$

Regenerator parameters, such as m, A, and L are fixed. Boundary conditions, Th and Tc, are also fixed. When a two-stage GM refrigerator with Pb regenerator is operated, no-load temperature is about 30 K for the 1st stage and 8 K for the second stage. Then, inlet helium flow temperature of 30 K at the warm end (T_h) and 8 K at the cold end (T_c) were determined. The only changing parameter is the volumetric ratio of Er_3Ni to a regenerator volume in the hybrid regenerator. The calculation model for the regenerator is shown in Fig. 2. Pb is filled in the high temperature part of the regenerator and Er_3Ni is filled in the low temperature part. The volumetric ratio for Er_3Ni, ε ,is defined as shown in Fig. 2. Figure 3 shows calculation results. Regenerator efficiency, η , increases with increasing Er_3Ni ratio, ε . The effect of Er_3Ni high heat capacity below 15 K is shown clearly. However, it is almost constant under the condition that ε is over 50 %. This is because the specific heat for Er_3Ni is almost same as that for Pb above 15 K. Then, the volumetric ratio of Er_3Ni to 50 % was determined.

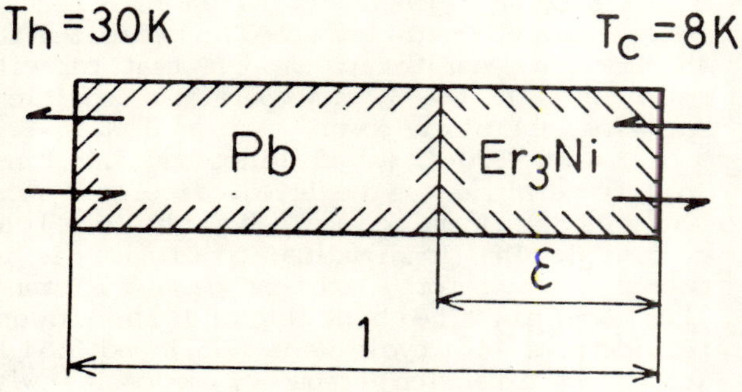

Fig. 2. Caluculation model for hybrid regenerator with Er_3Ni and Pb

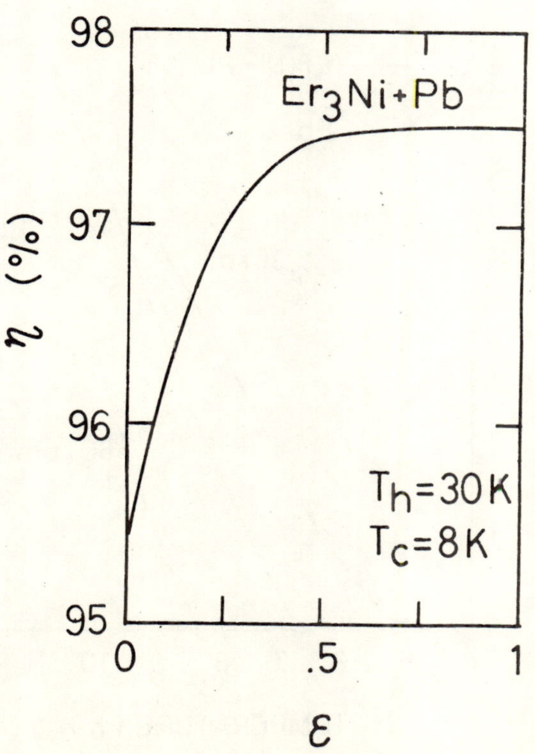

Fig. 3. Caluculated regenerator efficiency

(c) Regenerator construction

A hybrid regenerator with Er_3Ni and Pb was constructed. The half volume of the regenerator high temperature part is occupied by Pb balls (0.2-0.3 mm in diameter) and the latter half is occupied by Er_3Ni shots (0.2-0.4 mm in length). Figure 4 is a photograph of Er_3Ni shots used in the regenerator. These shots are selected by mesh screen after smashing Er_3Ni bulk into pieces.

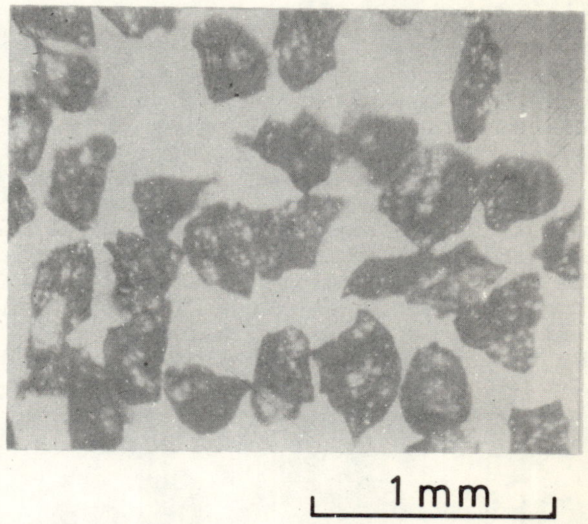

Fig. 4. Er_3Ni shots

EXPERIMENTAL APPARATUS AND PROCEDURE

The experimental apparatus consists of a two-stage GM refrigerator, compressor unit, and instrumentation. Figure 5 is a photograph of the refrigerator. The first stage expansion volume is 97.4 cm^3 and the 2nd stage is 25.4 cm^3. Phosphor bronze wire screen mesh is stacked in the 1st regenerator. Two kinds of 2nd regenerator were made. One is a hybrid Er_3Ni and Pb structure which is described in the above session. The other is made of Pb only as a comparison. Dimensions for these two regenerators are the same, except for the matrix. The displacer, which includes 1st and 2nd regenerators, is driven by AC motor. Since the AC motor is connected to the electrical frequency invertor, rotating speed can be easily changed and the displacer can reciprocate from 15 rpm to 80 rpm. Typical compressor unit operating condition is shown in Table. 1. The electrical input power to the compressor unit was measured by a watt meter. Carbon glass resistance thermometers were placed in contact with the cylinder at 1st and 2nd stages. Electrical heaters were also mounted on both 1st and 2nd stages to measure refrigeration capacity. Inlet and outlet helium gas pressure and mass flow ratio were also measured.

Table 1. Typical compressor unit operating condition

Supply pressure	1.8 MPa
Suction pressure	0.6 MPa
Mass flow ratio	2.0 g/s
Input power	2.8 kW

RESULTS AND DISCUSSION

Figure 6 shows the 2nd stage heat capacities obtained by the regenerator with Er_3Ni (60 and 36 rpm operation) and the regenerator with Pb (60 rpm operation). No heat load is applied to the 1st stage. The temperature of the 1st stage is in the 32 - 34 K range. Solid lines indicate results for the regenerator with Er_3Ni. Dotted line indicates results for the regenerator with Pb only. In the 60 rpm condition, no-load temperature for Er_3Ni regenerator (6.74 K) is much lower than that for Pb regenerator (8.8 K). In this paper, refrigeration temperature is defined as an average temperature of the highest and the lowest temperatures in a cycle. For example, 2nd stage temperature changes between 6.42 K and 7.07 K for Er_3Ni regenerator. Then, 6.74 K refrigeration temperature is calculated by determining the average for 6.42 K and 7.07 K. Heat capacity below 10 K was also improved by the Er_3Ni regenerator. At 10 K, 5 W heat capacity was obtained by Er_3Ni regenerator, but 3 W heat capacity by Pb regenerator. The Er_3Ni effect is shown clearly. Operating speed is optimized for the Er_3Ni regenerator to improve heat capacity below 10 K. In the 36 rpm operating condition, large refrigeration power, such as 1 W at 6.35 K or 2 W at 7.20 K, was obtained and 5.15 K no-load temperature was achieved. Figure 7 shows the temperature change in the no-load condition. The longitudinal-coordinate is an output voltage for a carbon glass resistance thermometer. The highest and the lowest temperatures in a cycle were 4.75 K and 5.54 K. Though an expansion frequency decreases with decreasing operation speed, mass flow ratio for helium gas flow decreases and the difference between suction and supply pressure for the compressor unit increases. Then, the pressure-volume diagram at expansion volume expands and refrigeration power for a cycle also increases

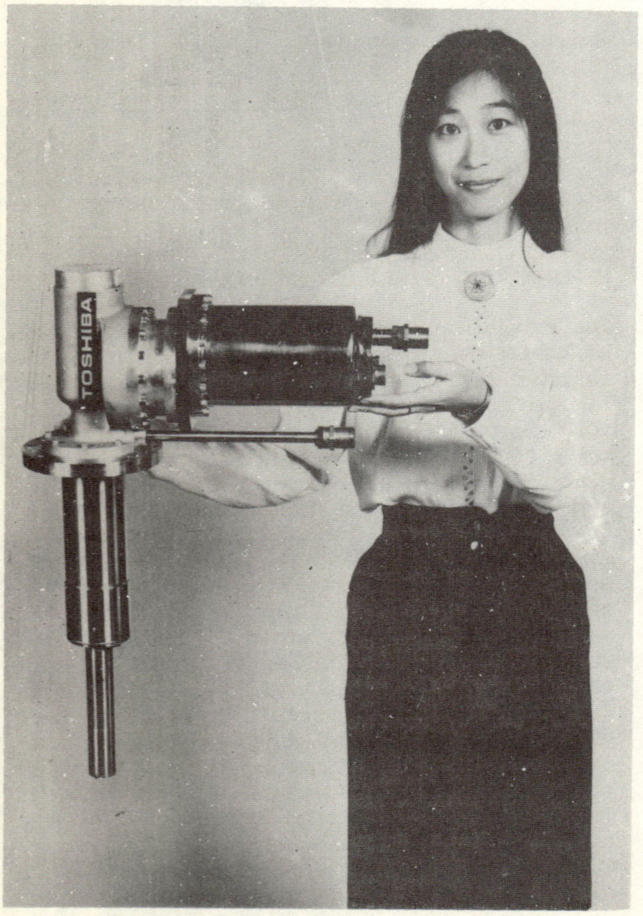

Fig. 5. Two-stage GM refrigerator

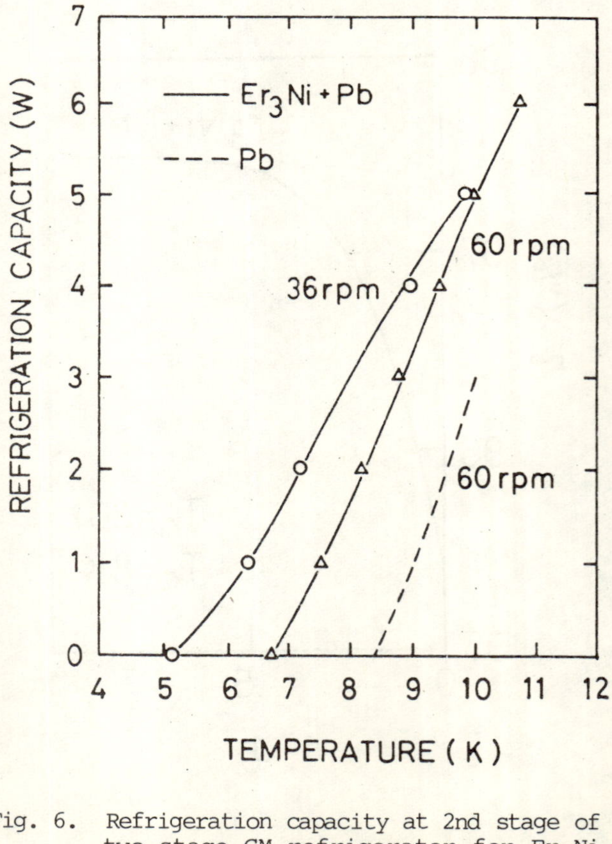

Fig. 6. Refrigeration capacity at 2nd stage of two-stage GM refrigerator for Er_3Ni regenerator and Pb regenerator

to some extent. That is why slow operation speed improves the refrigeration power and lowers the lowest temperature. These results were obtained with the same compressor unit. Though low pressure operation is valid for lowering no-load temperature, higher pressures are required to operate with the upper temperature at room temperature. The operating pressure is from 0.60 MPa to 1.80 MPa. This

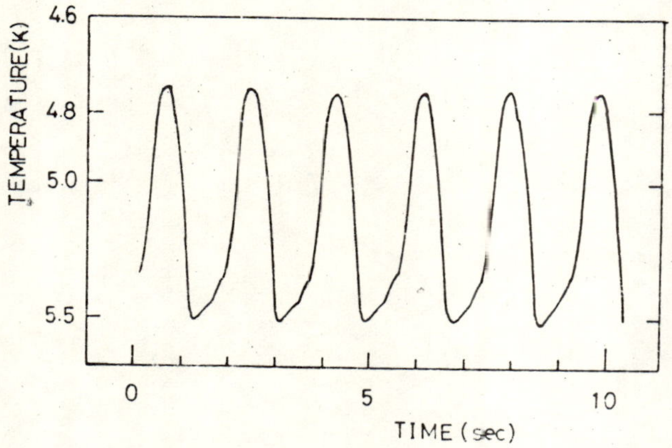

Fig. 7. Temperature change at 2nd stage for two-stage GM refrigerator with Er_3Ni regenerator

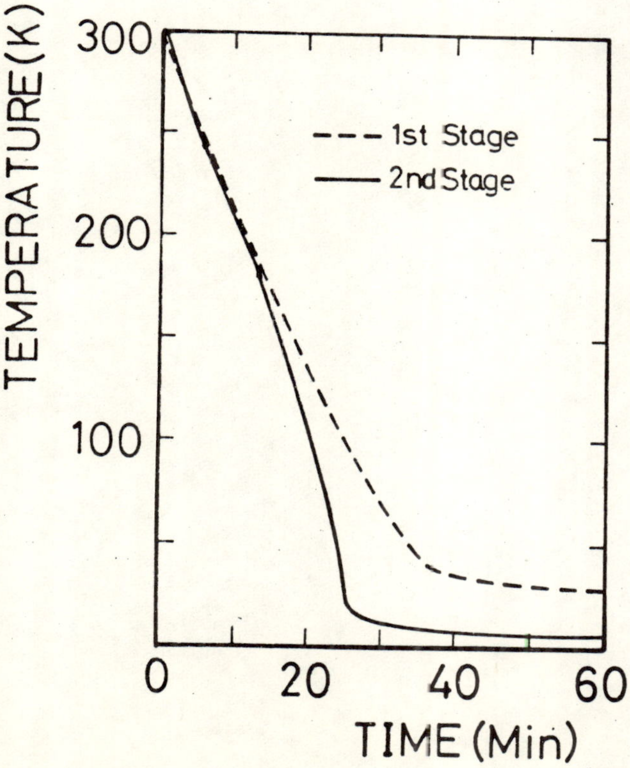

Fig. 8. Cooldown characteristics for two-stage GM refrigerator with Er_3Ni regenerator

pressure level is almost the same as that used in commercial GM refrigerators. Cooldown characteristics for the refrigerator with Er_3Ni regenerator are shown in Fig. 8. Solid line indicates for the 2nd stage and dotted line indicates for the 1st stage. The operation speed is 60 rpm. It takes about 60 minutes to achieve the lowest temperature of 6.74 K. This cooldown time is not as long as that of commercial two-stage GM refrigerators. An electric input power to the compressor unit is about 2.8 kW. The refrigerator efficiency, %carnot, is calculated from the input power, refrigeration capacity and refrigeration temperature. For the optimized operating condition of 36 rpm, 5.2 %carnot at 10 K is obtained. At 10 K, 3.0 %carnot was obtained for the Pb regenerator. The same efficiency is obtained at about 7.2 K for the Er_3Ni regenerator.

CONCLUSIONS

Experimental results, obtained for the 5 K level two-stage GM refrigerator in which rare earth compound is used as a regenerator matrix, are presented. A hybrid structure with Er_3Ni and Pb is adopted to the 2nd regenerator. Specific heat for Er_3Ni is greater than that for Pb below 15 K and is almost the same from 15 K to room temperature. A regenerator with Er_3Ni improves the refrigeration capacity below 10 K, compared with Pb regenerator. In the 36 rpm operation, 5.15 K no-load temperature was achieved and high refrigeration capacity, such as 1 W at 6.35 K or 2 W at 7.20 K, was obtained. The average input power to compressor unit was about 2.8 kW and refrigerator efficiency is extremely improved below 10 K. High refrigeration efficiency, 3.0 %carnot at 7.2 K or 5.2 %carnot at 10 K, was obtained. The refrigeration efficiency of 10 K level for Pb regenerator was obtained for Er_3Ni regenerator at 7 K level.

LIST OF SYMBOLS

A = heat transfer area of regenerator
L = regenerator length
T = temperature
C = specific heat
M = mass of regenerator matrix
h = convective heat transfer coefficient
m = mass flow ratio of helium gas flow
t = time
x = distance from inlet point of regenerator
τ = time between reversals
η = regenerator efficiency
ε = volumetric ratio of Er_3Ni in regenerator

subscripts
g = gaseous helium
m = matrix
h = warm inlet temperature
c = cold inlet temperature
out = outlet temperature at cold end

REFERENCES

1. Daniels,A. and du Pre, F.K. 'Triple-expansion stirling-cycle refrigerator.' Adv. Cryog. Eng. Vol. 16 (1971), pp.178-184.

2. Nakashima,H., Ishibashi,K., and Ishizaki,Y. 'Development of 4-5 K cooling Stirling cycle refrigerator.' Proc. 4th Int. Cryocoolers Cof. Sept. (1986),pp.263-271.

3. Matsubara,Y. and Kaneko,M. 'Vuilleumier cycle cryocooler operating below 8 K.' Proc. 3rd Int. Cryocoolers Cof. Sept. (1984),pp.234-239.

4. Nagao,M., 'Development of GM refrigerator using rare earth compounds.' Proc. 1988 Fall Meeting of the Cryogenic Society of Japan (1988),p.58 (in Japanese).

5. Sturat,R.W., Cohen,B.M., and Hartwig,W.H. 'Operation and application of a three-stage closed-cycle regenerative refrigerator in the 6.5K region.' Adv. Cryog. Eng. Vol. 15 (1970), pp.428-435.

6. Buschow,K.H.J., Olijhoek,J.F., and Miedema,A.R. 'Extremely large heat capacities between 4 and 10 K.' Cryogenics Vol.15 May (1975), pp.261-264.

7. Sahashi,M., Tokai,Y., Li,R., Ogawa,M., and Hashimoto,T. 'New regenerative R_3T system with extremely large heat capacity between 4 K to room temperature.' Proc. ICCR (1989), (to be published).

8. Walker,G. Cryocoolers Part 2, Plenum Press, New York and London.

9. Rios,P.A., and Smith,Jr.J.L. 'The Effect of variable specific heat of the matrix on the performance of thermal regenerators.' Adv. Cryog. Eng. Vol. 13 (1968), pp.566-573.

10. Daney,D.E. and Radebaugh,R. 'Non-ideal regenerator performance - the effect of void volume fluid heat capacity.' Cryogenics Vol.24 Sept. (1984), pp.499-501.

11. Sahoo,R.K. and Sarangi,S. 'Effect of temperature-dependent specific heat of the working fluid on the performance of cryogenic regenerators.' Cryogenics Vol.25 Oct. (1985), pp.583-590.

12. Sarangi,S. and Baral,H.S. 'Effects of axial conduction in the fluid on cryogenic regenerator performance.' Cryogenics Vol.27 Sept. (1987), pp.505-509.

AN APPROACH TO THE ESTIMATION OF THE SHUTTLE HEAT TRANSFER OF A PLASTIC DISPLACER

Sun Guohua, Chen Guobang
Cryogenics Laboratory, Zhejiang University
Hangzhou, China

ABSTRACT

In this paper, the factors influencing shuttle heat transfer of a plastic cryocooler are analyzed by introducing a semi-infinite sclid hypothesis based on the equation of heat conduction. The influence of temperature lag and convective thermal impedance on shuttle heat transfer is also considered. The total thermal impedance of the displacer with a sinusoidal movement is regarded as a vectorial summarization of five thermal impedances. Finally, a new equation of shuttle heat transfer has been developed, and the calculation results are given and compared with the experimental data (1). The calculation accuracy is significantly improved.

INTRODUCTION

The main heat losses of a plastic cryocooler with a gap regenerator include: radiated heat, axial thermal conduction through the plastic displacer and sleeve, imperfect regenerator loss, enthalpy deficit and shuttle heat transfer. When the refrigerating temperatures of a cryocooler are above 20K, the enthalpy deficit can be neglected and the imperfect regenerator loss is relatively small (2). By use of vacuum multilayer insulation, the radiated heat turns out to be a minor loss, the thermal conduction and shuttle heat transfer thus play a leading role in the total loss of a plastic cryocooler, in which the shuttle loss becomes the most important one. For example, a design analysis of a 77K modified Solvay-cycle expander indicates that the shuttle heat transfer is approximately 25% of the ideal refrigeration produced by the unit (3).

It is important to pay attention to the shuttle heat transfer in the optimum design and thermodynamic analysis of a cryocooler. Even though a few references have been reported on making an approach to shuttle losses for non-metal cryocoolers (1,4), the calculation values deviate from the experimental data with an error of 15% (1), due to many assumptions before derivation. It is not satisfactory for obtaining the optimum design of the cryocooler. In this paper, a new calculation formula of shuttle heat transfer is suggested by considering the effect of convective thermal impedances of the working gas. The calculation accuracy has been improved significantly.

MECHANISM OF SHUTTLE HEAT TRANSFER AND THE SEMI-INFINITE SOLID HYPOTHESIS

A. Mechanism of Shuttle Heat Transfer

Shuttle heat transfer is a form of convective heat transfer which occurs when a displacer with an axial temperature gradient reciprocates inside a cylinder with a similar axial temperature gradient. In a displacer-type cryocooler expander, this temperature gradient is typically established because of a temperature difference between both ends of the displacer and the cylinder. When the displacer is at the midstroke position, the temperature gradients in the cylinder wall and displacer are nearly matched, so the temperature at any point along the displacer is the same as that of the cylinder wall adjacent it. These temperature profiles are shown in Fig. 1(b). Figure 1(a) shows the temperature profiles with the displacer in top dead center. The entire length of the displacer and the cylinder wall is warmer than the displacer. Heat thus flows from the cylinder to the displacer,such as from point B to A in Figure 1(a).

Figure 1(b) shows the displacer back in the midpoint position with the temperatures again nearly matched and thus no heat transfer occuring. At the bottom dead center, Fig 1(c) shows that the temperature in the displacer throughout its entire length is greater than that of the adjacent cylinder wall. Heat thus flows from the displacer to the cylinder, such as from point A to C in Figure 1(b).

Not only is the heat flowing radially in and out of the displacer, but it is also being transported axially by the reciprocating motion of the displacer. Heat entering the displacer from point B on the wall is returned from the displacer to the wall at point C. Heat which originally enters the displacer at the warm end is eventually transported by this mechanism to the cold end (4,5). Obviously, heat is transported from the cylinder to the displacer through the gas inside the radial gap between the displacer and the cylinder, so the thermal impedance of the gas has a close relationship with shuttle heat transfer.

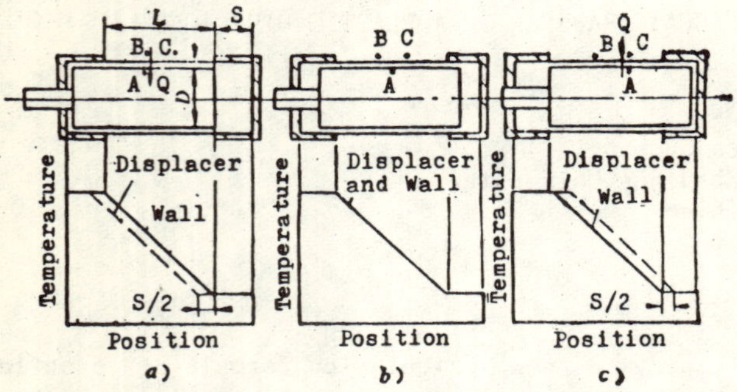

Figure 1. Mechanism of shuttle heat transfer

B. Semi-infinite Solid Hypothesis

There exist two kinds of unsteady-state conduction: periodic and aperiodic. One of the characteristics of periodic unsteady state conduction is that the temperature at any point of the body varies periodically with time. The conduction inside the displacer and cylinder of a plastic cryocooler with a sinusoidal movement can be regarded as periodic unsteady-state conduction. Due to the periodical variation of temperature, the heat flow inside the body changes periodically, and they are referred to as the temperature wave and the heat flux wave respectively. In the same way as with any other waves, the temperature wave will be attenuated to be zero at a certain depth while it is travelling in a body. The semi-infinite solid is so defined that it is wide enough to cause the attenuated temperature wave to be zero, and then, other surface conditions will have no effects on heat-transfer. The displacer and the cylinder thus are regarded as semi-infinite solids with two assumptions as follows:

(a) The surfaces of the displacer and the cylinder are separate planes.

(b) The thickness of the displacer and cylinder is sufficient to meet the semi-infinite assumption.

The gap between the displacer and the cylinder is much smaller than the diameter of the displacer and the cylinder, so assumption (a) is reasonable. Assumption (b) is acceptable only when the wall thickness of the displacer and the cylinder, δD is greater than Λ_T (where thermal penetration depth $\Lambda_T = \sqrt{K/Cv\dot{n}\pi}$) (6), and then, the other surface conditions are negligible.

ESTABLISHMENT OF FORMULAE

In order to derive the equations conveniently, we have introduced two assumptions relating to a semi-infinite solids to describe the conduction performances of the displacer and cylinder previously described. Some additional assumptions are now made here:

(1) A periodic variation of temperature exists on the surface of the displacer and the cylinder.

(2) There exists a temperature gradient running parallel to the surfaces of the displacer and the cylinder.

(3) The temperatures on the surfaces have been varying for a long time and have reached the limiting temperatures.

(4) The interaction of axial-conduction effects in the displacer and the cylinder are neglected.

(5) The effect of gas-pressure cycling is also neglected.

(6) The displacer runs sinusoidally.

We put emphasis on discussing the shuttle heat transfer of a plastic cryocooler which has reached the limiting temperatures, so assumption (3) can be neglected.

Within a solid body, the two dimensional unstable state differential equation which governs the heat flow is

$$\frac{\partial T}{\partial \tau} = a\left(\frac{\partial T^2}{\partial x^2} + \frac{\partial T^2}{\partial y^2}\right) \qquad (1)$$

If the surface temperature $Ts(\tau)$ is a harmonic function of time, we have

$$\delta T_s(\tau) = T_s(\tau) - T_\infty = \theta_0 e^{i2\pi\dot{n}\tau} \qquad (2)$$

where $\dot{n}$ is the frequency of the displacer

The temperature distribution in a semi-infinite solid is easily worked out by Carslaw and Jeager (7) to be

$$\delta T(x,\tau) = \theta_0 e^{-x/\Lambda_T + i(2\pi\dot{n}\tau - x/\Lambda_T)} \qquad (3)$$

where the thermal penetration depth is

$$\Lambda_T = \left(\frac{K}{\dot{n}\pi C_v}\right)^{\frac{1}{2}} \qquad (4)$$

The heat flow entering per unit area of the surface is

$$q = -K\frac{\partial T}{\partial x}\Big|_{x=0} = (1+i)\frac{K}{\Lambda_T}\delta T_s(\tau)$$
$$= e^{i\pi/4}(2\pi\dot{n}KC_v)^{\frac{1}{2}}\delta T_s(\tau) \qquad (5)$$

In terms of equations (2) and (5), we deduce that the surface heat flow is a periodic function of time, and the temperature phase lags $\pi/4$ behind the phase of heat flow. Due to a sinusoidal movement of the displacer, the total temperature difference in heat transfer is given by

$$\Delta T(\tau) = \frac{1}{2} S \frac{dT}{d\tau} \sin(2\pi \dot{n} \tau) \quad (6)$$

For a nylon displacer, the heat flow entering per unit area of the displacer surface is obtained from equation (5)

$$q = -K_N \frac{\partial T}{\partial x}\Big|_{x=0}$$

$$= e^{i\pi/4}(2\pi \dot{n} K_N C_{VN})^{\frac{1}{2}} \delta T_N(\tau) \quad (7a)$$

similarly, the heat flow entering per unit area of the G-10 cylinder surface is

$$q = -K_G \frac{\partial T}{\partial x}\Big|_{x=0}$$

$$= e^{i\pi/4}(2\pi \dot{n} K_G C_{VG})^{\frac{1}{2}} \delta T_G(\tau) \quad (7b)$$

So, we can see that the modulus of thermal impedances of the displacer and the G-10 cylinder which are caused by the the effects of thermal capacitances are as follows

$$R_N = (2\pi \dot{n} K_N C_{VN})^{-\frac{1}{2}} \quad (8a)$$

$$R_G = (2\pi \dot{n} K_G C_{VG})^{-\frac{1}{2}} \quad (8b)$$

The displacer moves inside the cylinder with a low frequency, the gas flowing inside the space between the nylon displacer and the G-10 cylinder thus remains laminar($Re < 500$), and the Nusselt number is assumed constant.

The thermal impedance of the gas is simply

$$R_g = \delta / K_g \quad (9)$$

and the convective thermal impedances of gas-nylon and gas-G-10 are

$$R_{Ng} = 2\delta / K_g Nu \quad (10a)$$

$$R_{Gg} = 2\delta / K_g Nu \quad (10b)$$

Thanks to the phase shift between thermal impedance and the temperature variation, the total thermal impedance of the shuttle heat transfer is a vectorial sum of the five thermal impedances above. Introducing the auxiliary variables

$$R_{gT} = R_g + R_{Ng} + R_{Gg} = \frac{\delta}{K_g}\left(1 + \frac{4}{Nu}\right) \quad (11)$$

$$R_{NS} = \sqrt{2}/2\,(R_N + R_G) \quad (12)$$

$$R_{com} = R_{NS} \cdot R_{gT} / (R_{NS} + R_{gT}) \quad (13)$$

and then

$$\vec{R_T} = \vec{R_{gT}} + \vec{R_G} + \vec{R_N} \quad (14)$$

So that the modulus of total thermal impedance is

$$|R_T| = \sqrt{(R_{NS} + R_{gT})^2 + R_{NS}^2} \quad (15)$$

and the vectorial angle is

$$\cos\beta = (R_{NS} + R_{gT}) / |R_T|$$

Finally, the formula for the shuttle loss is given by

$$Q_{shu} = \frac{\pi \dot{n}}{8} D S^2 \frac{dT}{d\tau} (2R_{NS} + R_{gT} - R_{com})^{-1} \quad (16)$$

On the basis of equation(3), Fig. 2 and Fig. 3 show the penetration of a temperature wave inside G-10 and nylon respectively. It can be seen that there are no temperature oscillations inside nylon and G-10 materials with the wall thickness exceeding four-fold thermal penetration depth. When the wall thicknesses of the G-10 cylinder and the nylon displacer are greater than 1.5 mm and 1.0 mm separately, the semi-infinite solid hypothesis is reasonable.

Fig. 4 indicates the relations of the thermal penetrations of nylon and G-10 with temperatures. Obviously, the lower the temperature, the greater the thermal penetration depth.

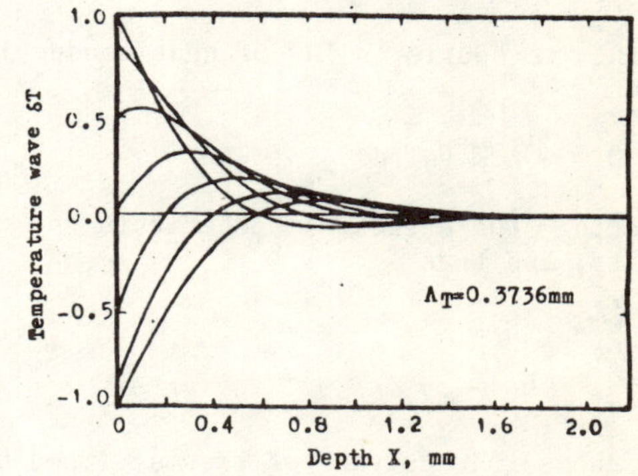

Figure 2. Penetration of temperature wave in infinite G-10 material with temperature 300K

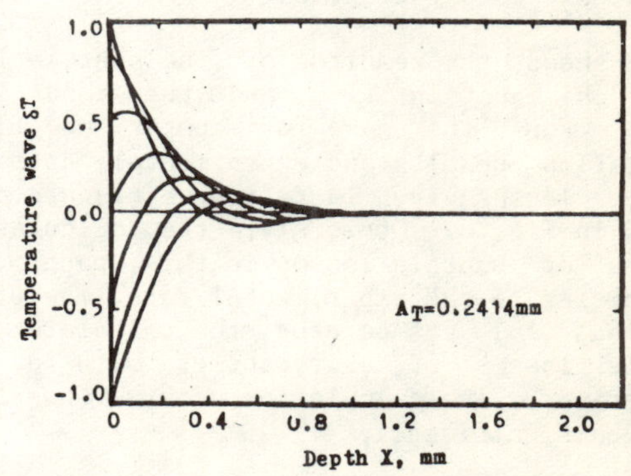

Figure 3. Penetration of temperature wave in infinite nylon material with temperature 300K

CALCULATIONS AND RESULTS

For the sake of comparison, the calculations are given with the sizes supplied by reference (1), that is, the value of the diameter and the length of the displacer is 0.95 cm and 30.3 cm

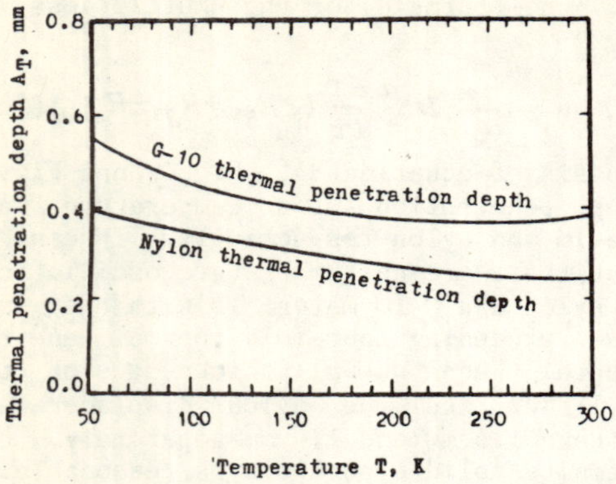

Figure 4. Penetration depthes of temperature wave as a function of temperature

respectively, the radial clearance between the displacer and the cylinder is about 0.1 mm at 145K and 0.06 mm at ambient temperature. The working fluid is helium.

According to Fourier's law of heat conduction, we have

$$Q = - A\ K\ (dT/dx) \tag{17}$$

Considering the effect of temperature on conductivity, and then

$$\frac{dT}{dx} = - B/K \tag{18}$$

where $B = \int_{T_c}^{T_h} K\, dT/L$

Fig. 5 shows four curves of thermal impedances as a function of temperatures. In Fig. 5, all kinds of thermal impedances increase with the decrease of temperature, and the thermal impedance of the gas varies sharply with temperature. The convective thermal impedance nearly approaches to that of nylon.

Fig. 6 shows the results of the shuttle loss based on equation (16) and the conditions above. For the sake of comparison, the calculation results and experimental data of shuttle losses given in reference (1) are also shown in Fig. 6. Obviously, the calculation results of shuttle losses in this paper are much closer to the experimental data. Therefore equation (16) can be used to calculate the shuttle losses of a cryocooler with a gap regenerator. The calculation accuracy has been improved significantly.

CONCLUSION

In a plastic cryocooler with a gap regenerator, the semi-infinite solid hypothesis has been successfully used to develop equation (16) for the shuttle heat transfer when the wall thicknesses of the nylon displacer and the G-10 cylinder are larger than 1 mm and 1.5 mm respectively.

Calculation results show that equation (16) is much more practical than that in reference (1). By considering the effects of convective thermal impedances and the lag of temperature in Fig. 6, the estimated maximum error of with the experimental data is about 10%. Therefore, equation (16) can be adopted in optimum designs and thermodynamic calculations for a plastic cryocooler.

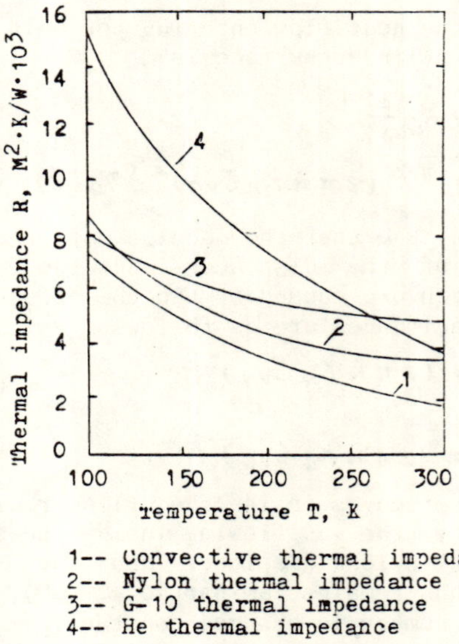

Figure 5. Thermal impedance as a fuction of temperature

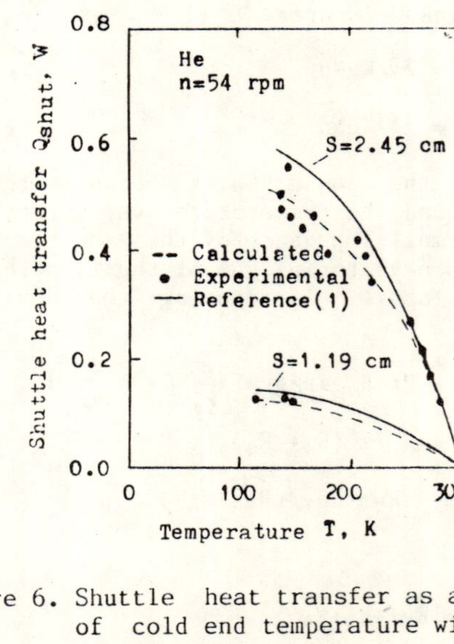

Figure 6. Shuttle heat transfer as a function of cold end temperature with the hot end at 300K

LIST OF SYMBOLS

a = thermal diffusivity (m2/s)
Λ_T= thermal penetration depth (m)

$\dot{n}$ = movement frequency of displacer (rpm)
k = thermal conductivity (w/m.K)
c = specific heat at constant volume (J/m3.K)
s = stroke of displacer (m)
R = thermal impedance (m2.K/w)
δ = gap between displacer and cylinder (m)
Nu= Nusselt number(dimensionless)
Q = heat transfer (W)
D = diameter of displacer (m)
L = length of displacer (m)
T = temperature (K)

SUBCRIPTS

v = volume
h = hot
c = cold
N = nylon
G = G--10
T = total
shu = shuttle
gas = gas

REFRENCES.

1. Ray Radebaugh and J. E. Jimmerman, "Shuttle heat transfer in a plastic displacer at low speeds", NBS TECHNICAL NOTE 1049, PP 46-51.

2. Chen Guobang, Sun Guohua et al, "The influence of non-ideal effects of the activating medium on the performance of a plastic cryocooler", ICEC-12, (1988).

3. N. Lambert, "Plastic low power cryocooler", 1986. (Doctorial theme, Horland)

4. F. J. Zimmerman and R. C. Longsworth, "Shuttle heat transfer", Advances in Cryogenic Engineering, Vol.16, PP 342-351.

5. Bian Shaoxiong, "Miniature cryocoolers", Machinery Industry Press, 1983, (in Chinese).

6. Ray Radebaugh, "Analysis of regenerator inefficiency for Stirling-cycle refrigerator with plastic displacer", NBS TECHNICAL NOTE 1049, PP 52-57.

7. H. S. Carslaw, J. C. Jeager, "Conduction of heat in solids", Oxford Univ. Press (1959).

CONCEPTUAL DESIGN OF MODIFIED PULSE TUBE REFRIGERATOR

Y. Matsubara, M. Kaneko* and A. Miyake**
Atomic Energy Research Institute, Nihon University, Tokyo 101, Japan
*Cryogenic section, Suzuki Shokan Co.,Ltd., Saitama 350, Japan
**College of Science and Engineering, Nihon University, Tokyo 101, Japan

ABSTRACT

The refrigeration temperature below 80 K has been obtained by the modified single stage pulse tube refrigerator such as orifice type, moving plug type and check valve type. These types of refrigerator may become the alternatives of Stirling or G-M cycle cryocoolers, if the reliability and the cost performance are improved.

This paper described the experimental results of the modified pulse tube refrigerator and the conceptual design of the pressure wave generator for the pulse tube refrigerator which may have a potential to improve the reliability and the cost performance.

INTRODUCTION

The development of reliable and low cost cryocoolers which meet the requirement of the application fields is coming to be an important subject. The unreliability of the cryocooler is primary originated by the frictional sealing components of the reciprocating piston. The pulse tube refrigerator has a potential as the reliable and low cost cryocooler because it has no mechanical expander.

The lowest attainable temperature of the pulse tube refrigerator is greatly improved by use of orifice and buffer volume [1,2]. However the refrigeration power per unit mass flow is still small in comparison to that of Stirling or G-M coolers.

With these considerations, we developed the alternative methods of the orifice pulse tube and the thermally actuated gas compression system.

ALTERNATIVE METHODS OF ORIFICE PULSE TUBE

According to the works developed by Radebaugh and coworkers, it is known that the refrigeration capacity of the orifice pulse tube refrigerator is increased as compared with that of basic pulse tube [3].

As one of the reason of this effect, we considered the pressure drop caused by the mass flow across the orifice. The gas compression work must increase when the mass flow occur at the orifice. This increased work is dissipated at the hot end of the pulse tube, which may induce the increment of refrigeration capacity at the cold end of the pulse tube.

As an alternative method which may also dissipate the gas compression work at the hot end of the pulse tube, we introduced the moving plug instead of the orifice as shown in figure 1(b). The difference between this moving plug and the expander or displacer of Stirling or G-M cycle is that the moving plug is operated at the room temperature. The motion of the plug should be controlled to obtain the optimum phase shift between the pressure and the volume change.

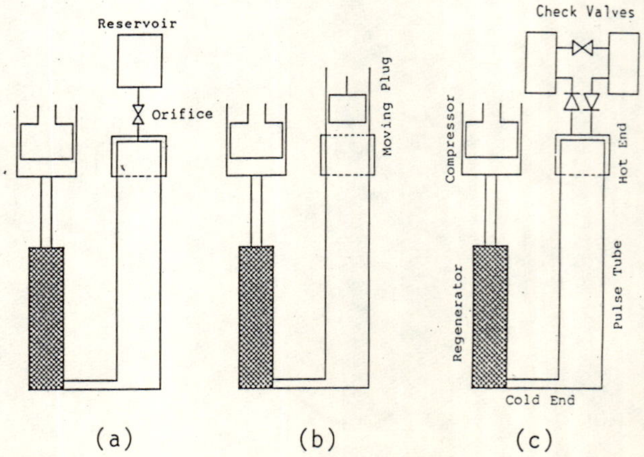

Fig.1 Schematic of pulse tube.
(a) Orifice type
(b) Moving plug type
(c) Check valve type

A pair of check valves instead of moving plug is also applicable as shown in figure 1(c). The gas compression work may be dissipated by the circulating gas at the hot end of the pulse tube. That gas flow is controlled by a valve installed between these check valves.

EXPERIMENT OF ALTERNATIVES

Figure 2 shows the schematic of test apparatus. The oscillating pressure was supplied by the switching valves and the compressor for such uses as G-M cycle cryocoolers.

The pulse tube is a stainless steel tube of 11 mm I.D., 0.5 mm thickness and 228 mm length. The copper isothermal region of 11 mm I.D. and 40 mm length was attached to the hot end of the pulse tube. Inside of the isothermal region, no heat transfer materials such as copper screen are installed. The water cooling channel was soldered at the isothermal region. A throttle valve as the orifice or a volume for the moving plug was connected by the copper tube of 2 mm I.D. and 180 mm length. The regenerator is a stacked 200 mesh copper screen of 36 mm diameter and 125 mm length.

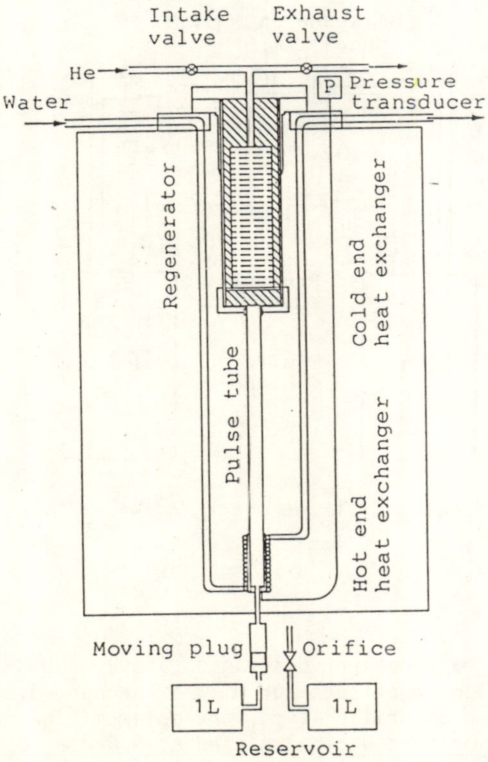

Fig.2 Schematic of test apparatus.

Figure 3 shows the experimental results obtained by a same pulse tube with different hot end condition. In case of basic pulse tube, minimum temperature of 163 K was obtained at the cycle frequency of 3.5 Hz. The optimum orifice opening decrease the minimum temperature down to 78 K with increasing frequency of 6.5 Hz. On the other hand, the minimum temperature of the pulse tube with the moving plug was 73 K at 10 Hz.

This plug was made of the phenolic resin formed in length of 20 mm with two cap seals and was installed in the stainless steel tube of 14 mm I.D.. The maximum displacement volume is 4.6 cm^3 which is 17 % of the pulse tube volume including the isothermal region. The displacement transducer was attached to the plug end. The pressure of the reservoir is maintained at the mean pressure of the oscillating pressure in the pulse tube. The friction force of the cap seals was used as the phase shifter of the plug. The actual displacement volume of the plug at 10 Hz was not obtained at the full stroke condition and was decreased to 3.4 cm^3.

During the experiment of the moving plug pulse tube, the temperature of the isothermal region has been maintained near room temperature without cooling water. This result indicates the isothermal region may not required for the case of moving plug pulse tube.

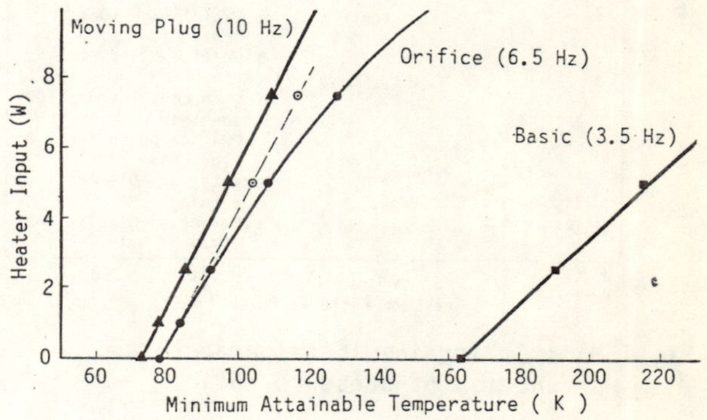

Fig.3 Refrigeration performance.

Figure 4 shows the relation between the pressure ratio and the minimum attainable temperature. This result indicates that the orifice type pulse tube refrigerator can be operated under the condition of smaller pressure ratio as compared with the basic type and further small pressure ratio will be abailable for the moving plug type one.

The ability of a cryogenic temperature refrigeration has been explained by Gifford and Longsworth using the mechanism of the surface heat pumping effect through the pulse tube wall [4]. It is supposed that the refrigeration mechanism of the orifice pulse tube is the phase effect of gas flow as used in a Stirling cycle. However the "orifice" is not an active device. Therefore there will be a limit to obtain the reasonable phase shift between the pressure and the temperature of the total mass in the pulse tube.

The moving plug instead of orifice is an active device, therefore it can be change the phase and the speed of the movement independently. The moving plug we used for this experiment is a simple plug only having the constant seal friction. Even though we have obtained the lower minimum temperature than the orifice type with same pulse tube. Further experiments using the cotrollable moving plug and the exact calculations should be performed to obtained the performance of the moving plug pulse tube refrigerator.

The merit of the simplicity of the pulse tube is decreased by adding the another moving parts, however, it may not serious problem because the moving components are still at room temperature.

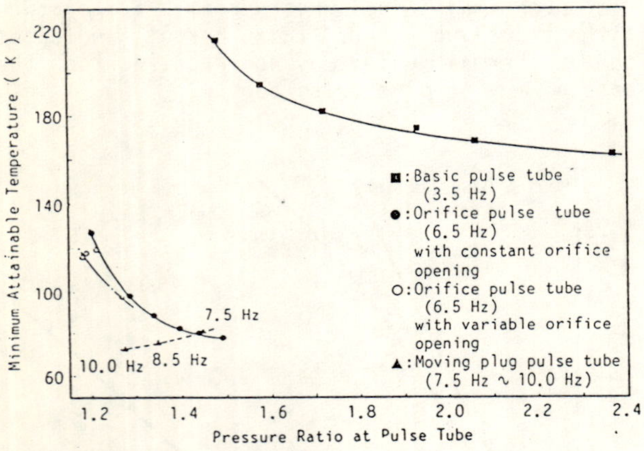

Fig.4 Minimum attainable temperature as a function of pressure ratio.

Figure 5 shows the experimental results of the check valve type of figure 1(c) in comparison to the orifice and the moving plug type.

The check valve used here is made of silicone sheet which has 0.8 mm thickness and 5 mm · in diameter and installed in a blass chamber having 1 mm port as shown in figure 6. Flow rate was measured using the mass flow meter installed between two check valves in series with a flow controll valve. The minimum temperature obtained by the check valve type pulse tube refrigerator was 74.5 K at 7.5 Hz. The measured flow rate at this condition was 0.012 g/cycle. It is noted that the data was obtained by the pulse tube without the copper

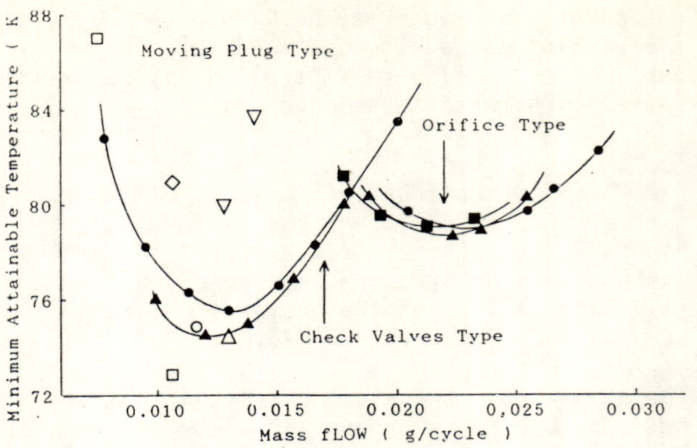

Fig.5 Performance of three types of pulse tube.
Operating frequency
Orifice; ● 7.0 Hz, ▲ 7.5 Hz, ■ 8.0 Hz
Check valves; ● 7.0 Hz, ▲ 7.5 Hz
Moving plug; ▽ 7.0 Hz, ◇ 7.5 Hz, ○ 8.5 Hz, △ 9.0 Hz, □ 10.0 Hz

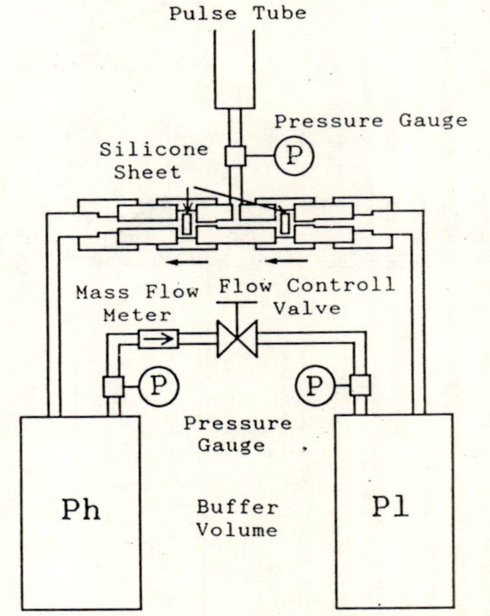

Fig.6 Schematic of the check valves unit.

isothermal region describ̶ed above, but the total length of tube 268 mm was unchanged. In the case of orifice type, the optimum flow rate at the orifice was calculated as 0.022 - 0.023 g/cycle. On the other hand, the flow rate of moving plug type was obtained almost same as the value of check valve type. This means the check valve type may act as the similar phase shifter as the moving plug type.

TYPE OF THERMAL COMPRESSOR

Thermal compressors can be classified into three types shown in the figure 7. Type (a) is abailable for such uses as a thermal compression part of the Vuilleumier cryocooler.

In the case of type (b), no external force is used to operate the system. Therefore the system is a kind of heat engine. This system has the drive shaft with a shaft seal. The displacer is actuated by the force which is generated by the pressure difference between two chambers separated by the shaft seal. The type (c) is also operated by only heat energy. This self actuation mechanism is almost the same as the type (b) except using spring and having no fixed stroke. This type of thermal compressor would not be unstable because the system of a spring and mass will acts such as a flywheel used in the type (b).

In order to operate the system, the work which is produced by using the drive shaft should be a positive value. The cyclic motion is also realized by replacing the expander to the throttle valve and reservoir because the pressure change of working gas does not trace on the reverse process.

If the work produced by the expander may not be used as the driving energy, the throttle valve method is the most simple system. Therefore the thermally actuated compression system having a valve has been selected for the pulse tube refrigerator.

EXPERIMENT OF THERMALLY ACTUATED PULSE TUBE REFRIGERATOR

Figure 8 shows the test machine of the thermally actuated pulse tube refrigerator. In the compression part, a gap regenerator is used and the displacer is made of 304 stainless steel. It is 63.4 mm in diameter and about 230 mm in length. The cylinder is also made of 304 stainless steel. The components have been welded. The radial clearance of the displacer section is about 0.3 mm. The heat injection part, which has the heat transfer area of about 0.01 m^2, is covered by electrical heater. The heat rejection part is covered by cooling water tube. The mass of the moving part except the springs is 0.87 kg. The charactristic of the spring is linear and its constant is 4.1×10^3 N/m. Opening the valve C and D, the pressure of spring chamber has become equall to the pressure of reservoir which has about 2 litters (2×10^{-3} m^3) of volume. The pressure variation of the reservoir has been small as compared with that of working gas. The ratio is about 1 to 2 % . The shaft which is sealed by cap seal has a diameter of 16 mm. The possible maximum stroke is about 14.2 mm.

The matrix of the cold regenerator are made of layers of phosphor-bronze mesh which is about 200 mesh. The regenerator has a diameter of 13 mm and a length of 50 mm. The pulse tube which is made of stainless steel has an inner diameter of 7 mm, a length of 140 mm and a thickness of 0.5 mm. The hot end of pulse tube which is made of copper has been cooled by water at room temperature. It has a length of about 30 mm and heat transfer area of 6.6×10^{-4} m^2.

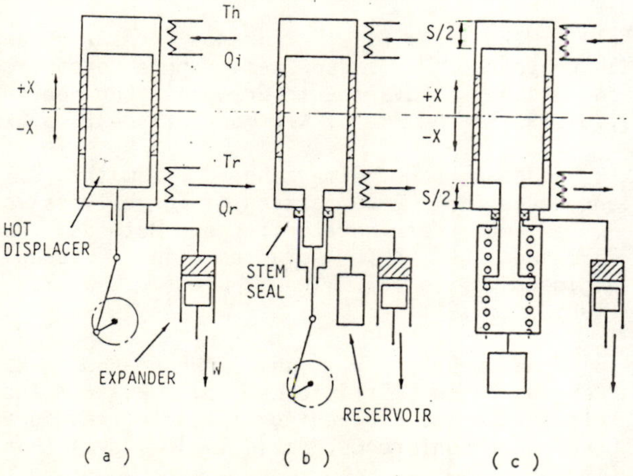

Fig.7 Three fandamental types of thermal compressor.
(a) Actuated by force with crank mechanism.
(b) Thermally actuated with crank mechanism.
(c) Thermally actuated without fixed stroke.

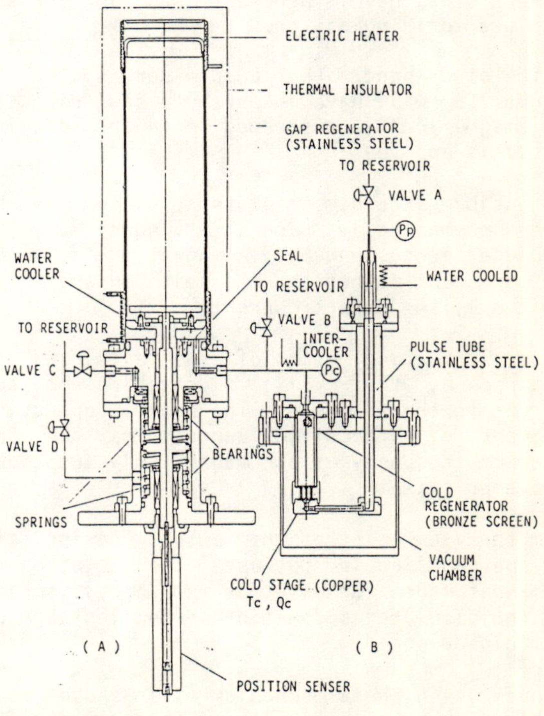

Fig.8 Thermally actuated pulse tube refrigerator.
(A) Thermal compression part.
(B) Pulse tube refrigerator part.

The temperatures of the electric heater portion and the water cooling portion were about 700 K and 300 K, respectively. The pressure of the reservoir was about 2.9 MPa. To keep the frequency at about 1.9 Hz the valve B has been adjusted. For the orifice pulse tube the valve A was slightly opened.

The thermally actuated pulse tube refrigerator has been operated at the temperature of about 200 K. The possibility of reaching 100 K has been presented under the condition of a low pressure ratio (about 1.4) and high frequency (about 10 Hz). The possibility of the high frequency operation has been suggested by the calculated results as well as the experimental results. Cryogenic refrigerators are required to be a reliable and a low cost. This particular refrigerator has a potential reliability with a low cost.

CONCEPT OF MODIFIED PULSE TUBE REFRIGERATOR

Our experimental study described above indicates that the combination of the thermal compressor and the check valve type pulse tube refrigerator is one of the interest feature for the low cost and reliable cryocooler.

The self actuated thermal compressor systems which are shown in figure 7(b,c) or figure 8 may have a posibility to be combined with the check valve type pulse tube. However, these systems still have a pressure seal ring for the displacer driving piston.

On the other hand, the compressor system in figure 7(a) will have no pressure seal ring if the magnetically suspended linear driving method is applide.

The posible arrengement of those system may be given schematically such as figure 9. The displacer is suspended by a magnetic field of the permanent magnet and the solenoid which is excited by the direct current.

The reciprocating motion of the displacer is obtained by the cyclic fluctuation of the current to the solenoid. The optimum operating frequency of alternating current may be the resonant frequency of the magneticaly suspended displacer system.

From the view point of the actual design, it may be applicable to use the stationary permanent magnet or a set of mechanical spring for the displacer suspention system instead of DC field solenoid.

However using those mechanism will cause some problems such as increasing of void volume for the working gas or increasing of friction force caused by the radial force of the suspension system.

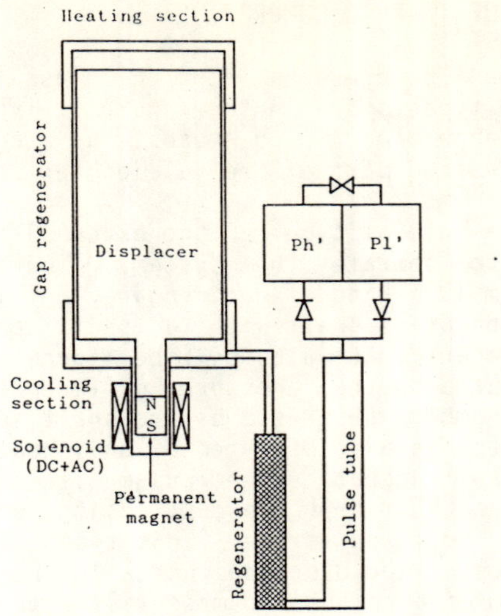

Fig.9 Conceptual design of simple pulse tube refrigerator.

SUMMARY

From the view point of the cost and the reliability, the pulse tube refrigerator has a good potential to be applied to several requirements. From our experiments, it has been proposed that the mechanism of the check valve type pulse tube has been a useful one for the sake of simplisity. The test operation of the refrigerator which is a combination of a orifice pulse tube and a self actuated thermal compressor has been done. Finaly we have proposed the conceptual design of the modified pulse tube refrigerator in order to meet the demand of high reliability and cost performance.

REFERENCES

[1] Mikulin, E.I., Tarasov, A.A. and Shkrebyonock, M.P., "Low temperature expansion pulse tubes", Advances in Cryogenic Engineering ,Vol.29, Plenum Press, New York, (1984), p.629

[2] Radebaugh, R., Zimermann, J., Smith, D.R. and Louie, B., "A Comparison of Three Types of Pulse Tube Refrigerators: New Methods for Reaching 60 K", Advances in Cryogenic engineering, Vol.31, Plenum Press, New York, (1986), p.779

[3] Radebaugh, R. and Herrmann, S., "Refrigeration Efficiency of Pulse Tube Refrigerators", Proceeding of 4th International Cryocooler Conference, David Taylor Naval Ship, (1987), p.119

[4] Gifford, W.E. and Longsworth,R.C.,"Surface heat pumping", Advances in Cryogenic Engineering, Vol.11, Plenum Press, New York, (1966), p.171

THE TEMPERATURE EFFECT OF THE DIFFERENT WORKING GASES OF A PHILIPS REFRIGERATOR

Jin Xin, Xu Xiaonong
Dapartment of Physics, Nanjing University,
Nanjing, China

A. L. Thomson
Physics Department, University of Sussex,
Brighton, England, BN1 9QH

ABSTRACT

In this paper, we used a strict three parameter equation to calculate the dependence on refrigeration temperature of the coefficients of performance and the amounts of refrigeration of the Philips refrigerator with different working gases. The results show that the coefficients of performance decrease in the order $\varepsilon_{Ne} > \varepsilon_{H_2} > \varepsilon_{He}$ for neon, hydrogen, helium as working gases respectively over the whole temperature range. The amounts of refrigeration are rather complex for different ranges of refrigeration temperature of the Philips refrigerator.

INTRODUCTION

It is well known that the coefficient of performance of a Philips refrigerator is the same as that of a Carnot refrigerator, and the amount of refrigeration is a linear function of the refrigeration temperature, T_1, if we use the ideal gas equation of state to calculate it. One of the authers[1] of this paper proved that the results of this are the same even though the Van der Waals equation of state of the gas is used. For dealing in detail with the real gas effect of a Philips refrigerator, it is necessary to find an accurate equation of state for the working gas of a Philips refrigerator. In Ref. 1, an accurate three parameter equation of state for low temperature gas was used to calculate the non-ideal gas effect of the coefficients of performance, ε, and the amounts of refrigeration, Q, of the Philips refrigerator dependent on the expansion ratio γ. The equation is

$$p = \frac{RT}{v-b} + \frac{a}{T^n v(v+b)} \quad (1)$$

where v is molar volume of gas, R = 8.31441 J mol^{-1}K^{-1} is ideal gas constant, a and b can be calculated from the following equations.

$$a = 0.42768 \frac{R^2 T_c^{2+n}}{p_c} \quad (2)$$

$$b = 0.08664 \frac{RT_c}{p_c} \quad (3)$$

where n = -0.064 for helium, n=0.28 for hydrogen and n = 0.48 for neon, T_c and p_c are the critical temperature and pressure of the gas respectively. From equation (3) and (4) a and b were calculated as shown in Table 1.

Table 1. Values for a and b from equation (1) for helium, hydrogen and neon

Gas	a($\times10^{-5}$ Pam^{6}K^n/mol^2)	b($\times10^{-5}$m^3/mol)
He	313.88	1.6354
H_2	6614.11	1.8372
Ne	13562.82	1.2055

For designing a real Philips refrigerator, the values of the coefficients of performance and the amounts of refrigeration and their dependence on the expansion ratio, γ, for different low temperature gases may rather important, but we think that for both designing and operating a Philips refrigerator, information on the coefficients of performance and the amounts of refrigeration and their dependence on the refrigeration temperature T_1 are also useful. In this paper, we will use equation (1) to calculate these results.

DERIVATION ABOUT EXPRESSIONS

The process cycle of a Philips refrigerator is a reverse Stirling cycle for which the PV diagram is shown in Fig. 1. The working gas is compressed isothermally at room temperature, T_h, from a state a to state b with the evolution of heat, and the working gas is cooled in the regenerator at a constant volume, V_b, to state c. Isothermal expansion takes place at the refrigeration temperature, T_1, until it returns to its original volume V_a at d. In this process the working gas absorbs heat which is called the amount of refrigeration of this cycle of the Philips refrigerator. The working gas is then transferred through the same regenerator and at constant volume V_a, to its original state at a.

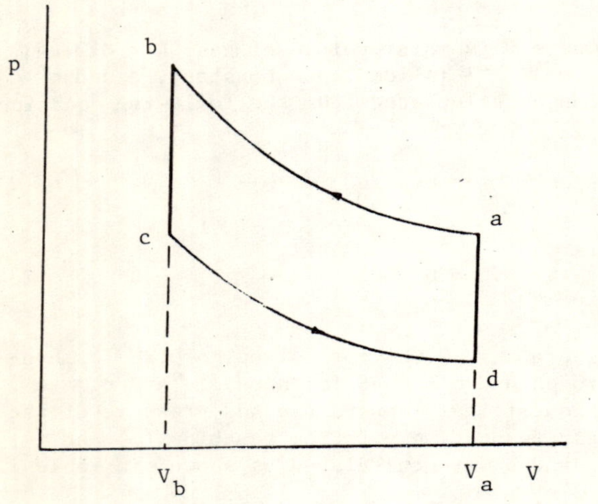

Fig. 1 The PV diagram of idealized refrigeration cycle of Philips refrigerator

From equation (1) and the Tds equation of thermodynamics[2], we can obtain the amount of refrigeration for one mole.

$$Q = \int_c^d T ds$$

$$= T_1 \int_c^d \left(\frac{\partial p}{\partial T}\right)_v dT \qquad (4)$$

Because

$$\left(\frac{\partial p}{\partial T}\right)_v = \frac{R}{v-b} + \frac{na}{T^{n+1} v(v+b)} \qquad (5)$$

So that

$$Q = T_1 \int_{V_b}^{V_a} \left[\frac{R}{v-b} + \frac{na}{T^{n+1} v(v+b)}\right] dv$$

$$= RT_1 \text{Ln} \frac{V_a-b}{V_b-b} + \frac{na}{T_1^n b} \text{Ln}\left(\frac{V_b+b}{V_a+b} \frac{V_a}{V_b}\right) \qquad (6)$$

The work done on the working gas by one cycle for one mole:

$$W = -\oint p dv$$

From Fig. 1 we can obtain

$$W = -\int_{V_a}^{V_b} \left[\frac{RT_h}{v-b} - \frac{a}{T_h^n v(v+b)}\right] dv$$

$$+ \int_{V_b}^{V_a} \left[\frac{RT_1}{v-b} - \frac{a}{T_1^n v(v+b)}\right] dv$$

$$= R(T_h - T_1) \text{Ln} \frac{V_a-b}{V_b-b}$$

$$+ \frac{a}{b}\left(\frac{1}{T_1^n} - \frac{1}{T_h^n}\right) \text{Ln}\left(\frac{V_b+b}{V_a+b} \frac{V_a}{V_b}\right) \qquad (7)$$

From (6) and (7), the coefficient of performance of Philips refrigerator can be obtained

$$\varepsilon = \left[RT_1 \text{Ln} \frac{V_a-b}{V_b-b} + \frac{na}{T_1^n b} \text{Ln}\left(\frac{V_b+b}{V_a+b} \frac{V_a}{V_b}\right)\right]$$

$$\div \left[R(T_h - T_1) \text{Ln} \frac{V_a-b}{V_b-b} + \frac{a}{b}\left(\frac{1}{T_1^n} - \frac{1}{T_h^n}\right) \text{Ln}\left(\frac{V_b+b}{V_a+b} \frac{V_a}{V_b}\right)\right] \qquad (8)$$

DISCUSSION

If we choose $V_a = 0.001 m^3/mol$, $T_h = 300K$, $\gamma = V_a/V_b = 3$, we can calculate the dependence of ε and Q on the refrigeration temperature T_1 for the different gases. Fig.2 shows the dependence of the performance coefficient ε_{He} for helium on the refrigeration temperature. The performance coefficients of hydrogen and neon have the relationship $\varepsilon_{Ne} > \varepsilon_{H_2} > \varepsilon_{He}$ over the whole temperature range, but they are not very different from helium. Table 2 shows the differences.

The amounts of refrigeration are rather complex as shown in Fig. 3. It shows the differences between the amounts of refrigeration of real gas, Q, of helium, hydrogen and neon for one mole and the amounts of refrigeration for an ideal gas Q_i, for one mole. In Fig. 3 we can see that $Q_{Ne} > Q_{H_2} > Q_{He}$ when T_1 is in the range of 30K to 60K, $Q_{H_2} > Q_{Ne} > Q_{He}$ when it is in the range of 60 K to 158K and $Q_{H_2} > Q_{He} > Q_{Ne}$ when it is higher than 158K.

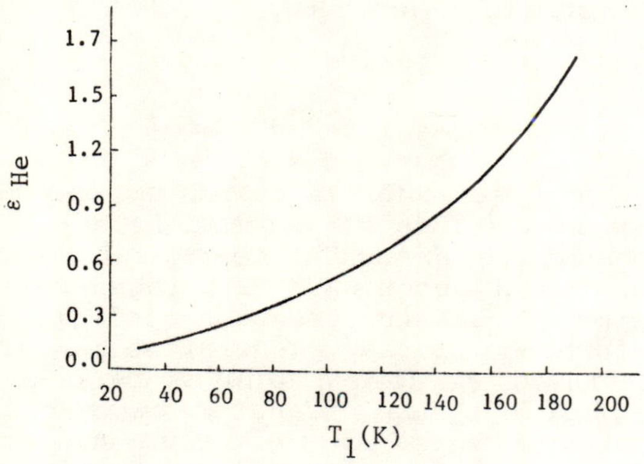

Fig. 2 The coefficient of performance of helium, ε_{He}, is dependent on the refrigeration temperature T_1.
Data: V_a=0.001m^3/mol, T_h=300K, $\gamma = V_a/V_b$ = 3.

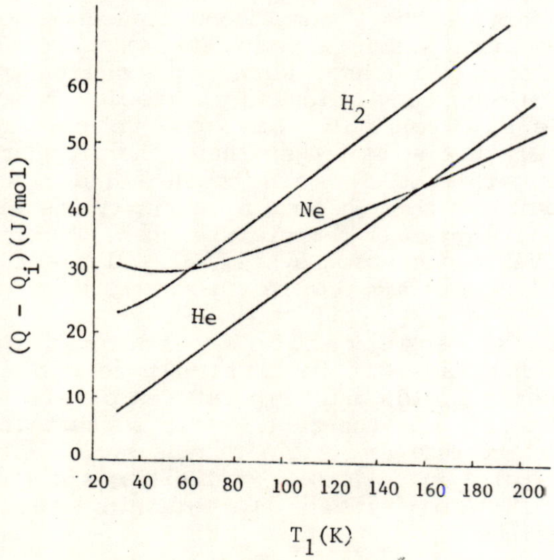

Fig. 3 The difference between the amounts of refrigeration, Q, of helium, hydrogen and neon for one mole and of ideal gas for one mole.
Data: V_a=0.001m^3/mol, T_h=300K, $\gamma = V_a/V_b$ = 3.

Table 2 The differences between the coefficients of performance between hydrogen and helium, and, neon and helium as a function of refrigeration temperature T_1.
Data: V_a=0.001m^3/mol, T_h=300K, $\gamma = V_a/V_b$ = 3.

T_1(K)	$\varepsilon_{Ne} - \varepsilon_{He}$	$\varepsilon_{H_2} - \varepsilon_{He}$
30	0.008	0.004
40	0.007	0.004
50	0.008	0.004
60	0.006	0.003
70	0.006	0.004
80	0.004	0.002
90	0.005	0.003
100	0.004	0.003
110	0.004	0.003
120	0.003	0.002
130	0.003	0.002
140	0.003	0.002
150	0.003	0.002
160	0.003	0.002
170	0.003	0.002
180	0.003	0.002

REFERENCES

1. Jin, X. and Chen, J. Q., 'thermodynamic analysis of Philips refrigerator', ICEC 12,(to be published).

2. Zemansky, M. W., 'Heat and Thermodynamics', Chapt. 11, Fifth Edition, McGraw-Hill Book Company, (1968).

THERMODYNAMIC ANALYSIS OF MAGNETICALLY ACTIVE REGENERATOR

K. Matsumoto and T. Hashimoto
Department of Applied Physics, Tokyo Institute of Technology,
Meguro-ku, Tokyo 152, Japan

ABSTRACT

Magnetically active regenerator produces refrigeration without gas expansion because it makes use of magnetocaloric effect. The model active regenerator had the Ericsson-like operation which is the simplest thermal cycle of active regenerators. $DyAl_{2.2}$ sintered compound which has the Curie temperature at 50 K was adopted as the magnetic regenerator matrix in order to study performance of active regenerator in the temperature range of about 30 to 70 K. Numerical analysis of thermal cycle gave the lowest temperature of this system. In the present analysis, the model active regenerator could expand operation temperature span over 30 K under the conditions of no refrigeration power and magnetic field of 3 T.

INTRODUCTION

Magnetically active regenerator is a thermal regeneration system combined with magnetic refrigeration. Ordinary thermal regenerator operates as a thermal sponge; thus it is passive. Contrary to this, magnetically active regenerator is active, because it produces refrigeration by itself.

Magnetic refrigeration is based on the magnetocaloric effect, where magnetic material changes its temperature and exchanges heats with heat reservoirs with variations of magnetic field. Magnetic refrigerators have the following advantages compared with gas compression-expansion system: ability of compactness due to high entropy density of magnetic material and possibility of reversible cycle in principle. The latter results in high thermal efficiency. Magnetic refrigerator may have high reliability because of simple mechanical parts. In magnetic refrigerator, thermal cycles are mainly composed of electrical control of magnetic fields.

A few magnetic refrigerators have been built and tested at various temperatures[1,2,3,4]. The authors have studied magnetic refrigeration at high temperatures to expand the operation temperature[5]. We have reported on a primary experiment on a magnetic refrigerator which operates at about 60 K with regenerative thermal cycle, the Ericsson cycle[6]. The performance of the regenerative magnetic refrigerator was limited mainly by two origins: insufficient heat transfer between the regenerator matrix and the working substance, heat leak due to thermal conduction of regenerator matrix.

Magnetically active regenerator is stuffed with magnetic material. Then, the regenerator matrix operates refrigeration cycle by periodically changing magnetic fields and gas flows.

Magnetically active regenerator has several advantages over the regenerative magnetic refrigerators previously reported. This system can improve the thermal conductance between regenerator matrix and transfer gas, because working material have large heat transfer area and forced convection is available. Active regeneration may be able to produce temperature span larger than other magnetic refrigerators. This is because individual elements of the regenerator matrix cycles with relatively small temperature span. Thus, an active regeneration system is equivalent to many cascaded magnetic refrigerators.

Active regeneration system has one disadvantage. It is difficult to control magnetic fields and temperatures of all the sections along the regenerator so that they cannot operate reversible thermal cycle. This difficulty in delicately controlling field and gas flow may yield decrease in thermal efficiency.

Magnetically active regenerator has been studied by some groups in the temperature range of 4.2 to 15 K. Claude et al.[7] studied active regeneration system with small spherical GGG (gadolinium gallium garnet) as regenerator matrix. Cogswell et al.[8] investigated the performance of active regeneration system with

GGG by computer simulation and they also made a testing regenerator.

In the present paper, we reported our analysis of magnetically active regenerator. The active regenerator had the Ericsson-like operation cycle which is the simplest operation of active regenerator. RAl_2 (R: rare earth atom) was adopted as regenerator matrix, the magnetic properties of which we showed to be suitable for magnetic refrigeration at high temperatures[5]. The lowest temperature of the model magnetically active regenerator was evaluated by computer simulation.

ANALYSIS OF THE MODEL ACTIVE REGENERATOR

(a) Basic operation of active regenerator

We explain the principle of magnetically active regenerator with the help of illustration, Fig.1. In this system, magnetic material is stacked in the regenerator. The magnet system gives time-varying magnetic fields which accomplish magnetization and demagnetization of regenerator matrix. Then, the displacer forces transfer gas through regenerator.

It is essential to cycle operation that the displacer movement should be synchronized with magnetization and demagnetization of the regenerator matrix. When the regenerator matrix is magnetized, the displacer moves down, forcing the heat transfer gas from the low heat exchanger to the high one. In this process, the gas absorbs heat from the regenerator matrix. As the regenerator matrix is demagnetized, the displacer moves up, returning the gas to the low heat exchanger. Regenerator matrix cools the gas in this process.

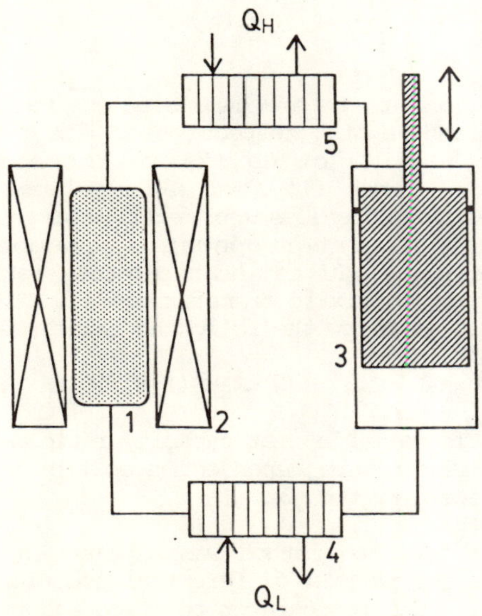

Fig. 1 Schematic of an active regenerator (1:regenerator matrix, 2:magnet, 3:displacer, 4:low heat exchanger, 5:high heat exchanger)

(b) The Ericsson-like cycle of active regenerator

The Ericsson-like operation is the simplest cycle of active regenerator. We show the operation cycle using Fig.2. Figure 2-a represents the temperature profiles of the regenerator matrix and the gas temperatures at the ends of the regenerator, when the active regenerator is in the stationary operation.

Time chart of the operation is shown in Fig.2-b where the magnetic fields, temperature profiles in the regenerator and the directions of gas flow are represented. The signs denote gas flow directions; the positive sign implies the flow direction from the hot side to the cold side. According to this definition, zero gas flow (represented by 0) indicates adiabatic change because of no heat transfer occurs between the regenerator matrix and the gas. As shown in Fig.2-b, the Ericsson-like operation is divided into the following four processes:

(1) ADIABATIC DEMAGNETIZATION PROCESS
Magnetic field decreases from the strength B_H to B_L with no gas flow. This field change yields adiabatic demagnetization of regenerator matrix, if the mass of gas contained within the regenerator is negligibly small. Consequently, the temperature profile changes from T_R^1 to T_R^2 with the adiabatic decrease in temperature of the regenerator matrix.

(2) COOLING PROCESS IN LOW FIELD
Heat transfer gas is forced to flow from the hot end to the cold end. Regenerator matrix cools the gas from hot inlet temperature T_{gH}^{in} to cold outlet temperature T_{gL}^{out}. At the same time the temperature profile of the regenerator matrix changes from T_R^2 to T_R^3 because of heat absorption from the gas.

(3) ADIABATIC MAGNETIZATION PROCESS
Magnetic field changing from B_L to B_H causes adiabatic variation in temperature profile of the regenerator matrix from T_R^3 to T_R^4.

(4) HEATING PROCESS IN HIGH FIELD
The regenerator matrix exhausts heat to the gas flowing from the cold end to the hot end. The temperature profile then returns to T_R^1 which is the initial temperature of the regenerator. The gas with temperature T_{gL}^{in} at the cold inlet is heated to T_{gH}^{out}, while it flows to the hot end.

Individual elements of the regenerator matrix operates the Ericsson cycle which consists of two adiabatic and two iso-field processes. By this reason, we call this active regenerator operation the Ericsson like-operation.

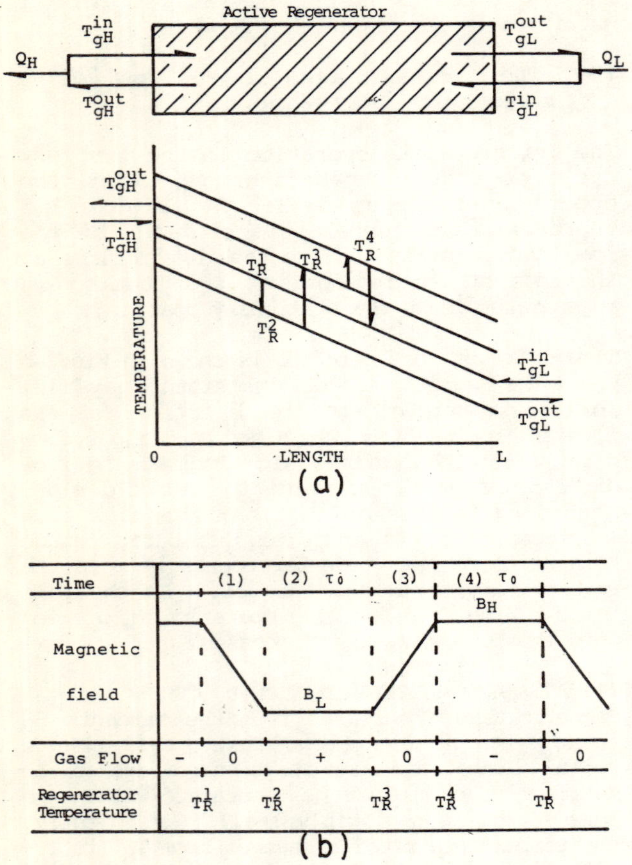

Fig. 2 Schematic drawing of the Ericsson-like operation of active regenerator.
(a) represents temperature profiles in the matrix and the gas temperatures.
(b) represents the time chart of the Ericsson-like operation.

(c) Computer simulation of the Ericsson-like operation

The most important point of the analysis is to introduce thermodynamic properties of the regenerator matrix which depend on temperature and magnetic field strength. The governing equations for the Ericsson type active regenerator may be derived on the following assumptions:

For simplification, heat capacity of the gas was regarded as constant. In the range of high temperatures, this assumption is not so bad. Beside this, the convective heat transfer coefficient between the regenerator matrix and transfer gas was constant throughout the regenerator.

In general, the mass of the gas within void of the regenerator is negligible compared to that of the regenerator matrix. Accordingly, the temperature change of the matrix in adiabatic process was calculated only from its entropy diagram.

The thermal conductivity of matrix was zero in the direction of flow and infinite in the direction perpendicular to flow.

The time necessary for magnetization and demagnetization was assumed to be zero. We neglected losses of the magnet system which depend on the operation speed. The cooling and heating process spent a half of the cycle period t_0, respectively.

The entropy of matrix is indispensable for the analysis. We used the entropy of regenerator matrix, $S_R(T,B)$ which was calculated with integration of specific heats in various magnetic fields.

The temperature change in the adiabatic process was obtained by solving the equation,

$$S_R(T_H,B_H)=S_R(T_L,B_L). \quad (1)$$

For iso-field processes, we applied a well-known method[9,10] used for the analysis of usual regenerator. For our model active regenerator, the energy conservation law yields the governing equations represented as follows with respect to gas and regenerator matrix:

$$\frac{\partial T_g}{\partial x} = \pm NTU\,(T_R - T_g) \quad ,(2)$$

$$\frac{\partial T_R}{\partial t} = \frac{NTU}{\Gamma'}\,(T_g - T_R) \quad ,(3)$$

where, x and t are dimensionless variables of time and length, respectively. The sign on the right side of eq.(2) represents the direction of gas flow. These equations show that two dimensionless parameters characterize active regenerator operation: number of transfer unit, NTU and heat capacity ratio of regenerator matrix to transfer gas, Γ'. The NTU and Γ' is given by the following relations:

$$NTU=hAL/mC_g \; , \quad \Gamma'=M_RC_R(T,B)/mC_gt_0 \quad .(4)$$

The Γ' is variable heat capacity ratio because of the temperature dependent specific heat of the regenerator matrix.

Finite difference method was applied in order to solve the partial differential equations (2) and (3). The regenerator position and process time were divided into 50 or 100 grid points. Since the temperature dependent specific heat of the regenerator matrix made eq.(4) non-linear, the grid temperatures could not be obtained explicitly. Calculation of the temperatures at the next grid point was iterated until the temperatures converged.

Two boundary conditions were imposed on the regenerator. First condition was that inlet gas temperature at the hot side T_{gH}^{in} was equals to high reservoir temperature and was constant during the cooling process. Second was concerned with inlet gas temperature at the cold end. The inlet gas temperature T_{gL}^{in} was also constant over the heating process and it equaled to cold reservoir temperature.

If refrigeration power equaled to zero, T_{gL}^{in} was given by the average of T_{gL}^{out}. In the present paper, the refrigeration power was fixed to zero, since we studied the lowest temperature attained by the active regenerator. We must, in general, evaluate the heats exchanged with external reservoirs from enthalpy of the gas, since the outlet gas temperatures are time dependent.

At the beginning of the simulation, the regenerator matrix and gas had high reservoir temperature throughout the regenerator. The four processes, demagnetization, cooling, magnetization and heating process were repeated in turn until the variation of average gas temperature T_{gL}^{out} became less than 10^{-3} K. Then, we obtained the temperature profiles of both the matrix and the gas as well as the temperatures of the gas at the both outlets of the regenerator for stationary operation.

RESULTS AND DISCUSSION

Active regenerator with use of $DyAl_{2.2}$ sintered compound was analyzed in order to study the potential ability of magnetically active regenerator in the temperature range about 50 K. Figure 3 shows the entropy diagram of $DyAl_{2.2}$, and it is evident that the material has maximum entropy change at the Curie temperature of 50 K.

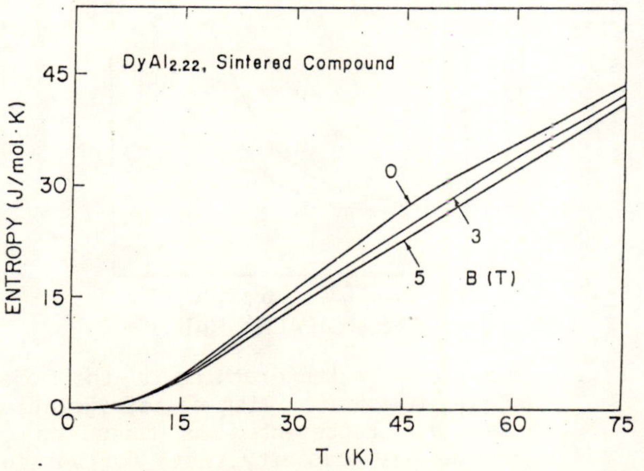

Fig. 3 The entropy-temperature diagram of $DyAl_{2.2}$ sintered compound

The dimensionless parameters, NTU and Γ are convenient to characterize the heat transfer processes. We used the specific heat of $DyAl_{2.2}$ at high reservoir temperature in zero magnetic field to evaluate heat capacity ratio Γ ($=M_RC_R(T=T_H,B=0)/mC_gt_0$), when the simulation condition was fixed.

Figure 4 shows an example of temperature profiles of regenerator matrix obtained by a calculation with the condition of NTU=50, $1/\Gamma$ =0.2, B_H=5 T, B_L=0 T and high reservoir temperature=70 K. The four temperature profiles respectively correspond to those represented in Fig.2-a (T_R^1: after the heating process, T_R^2: after demagnetization, T_R^3: after the cooling process, T_R^4: after magnetization).

The temperature variation of matrix becomes larger, as the operation range decreases in temperature. This is because small heat capacity of the matrix at low temperatures causes large temperature change during heat exchange process with gas. Thus, the temperature variation of the outlet gas T_{gL}^{out} is larger than that of T_{gH}^{out}. We defined the lowest temperature attained in this operation as the average temperature of T_{gL}^{out} over the cooling process.

The results of calculation were arranged as the lowest temperature attained by the active regenerator. Figure 5 and 6 represent the lowest temperatures as functions of 1/(heat capacity ratio) for various NTU values. The magnetic field strengths are 3 and 5 T in Fig.5 and 6, respectively, while both the high reservoir temperatures are 70 K. In both cases, the lowest temperatures decrease with increasing NTU value, and also increase with inverse heat capacity ratio.

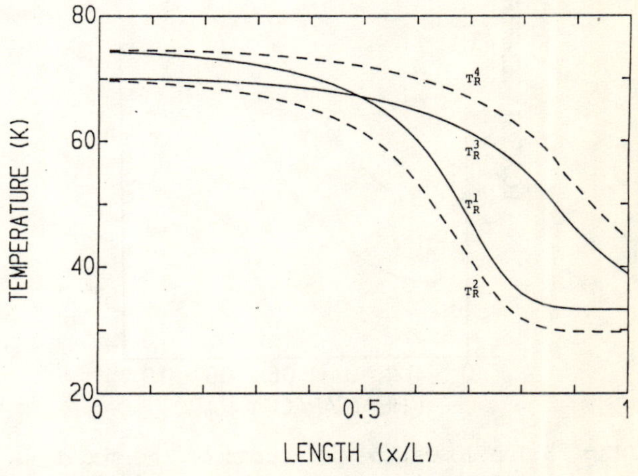

Fig. 4 Temperature profiles of the regenerator matrix at the end of each process in the condition of NTU=50, $1/\Gamma$=.2, B_H=5 T, T_H=70 K. The profiles correspond to those in Fig.2-(a).

We can readily understand these properties of active regenerator. The increase of the NTU value implies the improvement of heat transfer between regenerator matrix and gas. Large NTU value could, therefore, help to lower the gas temperature T_{gL}^{out}. The heat capacity ratio affected the change of temperature profiles in regenerator. Large thermal load to regenerator brought about large temperature variation of the matrix. Then, the outlet gas temperature rose because the matrix didn't have enough heat capacity to cool the gas.

The strengthening of magnetic field lowers the refrigeration temperature, since it gives rise to large entropy change. The temperature span of about 30 K was obtained with magnetic field of 3 T and that about 37 K was obtained with 5 T field. The temperature span was expanded only by a factor of 1.2 with a 67% increase in magnetic field.

In Fig.6, the lowest temperature is seen to be limited to about 40 K, even in the case of large NTU value and small inverse heat capacity ratio. The temperature dependent entropy change of $DyAl_{2.2}$ is considered the origin of this limitation of performance. As shown in Fig.3, the entropy change of $DyAl_{2.2}$ is large only around the Curie temperature of 50 K and it decreases to about the half of the maximum value at 30 K. $DyAl_{2.2}$, therefore, has no ability to lower the operation temperature below 30 K.

One method to lower the operation temperature is to make complex type regenerator, i.e. to stuff regenerator with several magnetic materials whose Curie temperatures vary from the hot side to the cold side. Thus, we can optimize the entropy change of the matrix throughout the regenerator for wide range operation. A series of ferromagnetic material RAl_2 is one of the candidate materials, because the Curie temperature can be controlled by exchange of the rare earth element, R. Since the optimization of the material combination needs full analysis of cycle operation, it is beyond the scope of the present paper.

It is obvious that refrigeration power remarkably depends on the heat capacity ratio of regenerator matrix to the gas. If heat capacity of the gas is closely smaller than that with which active regenerator can attain the low reservoir temperature, active regenerator can produce little refrigeration power. On the other hand, small amount of gas can absorb little heat from low reservoir. There is, consequently, the optimum heat

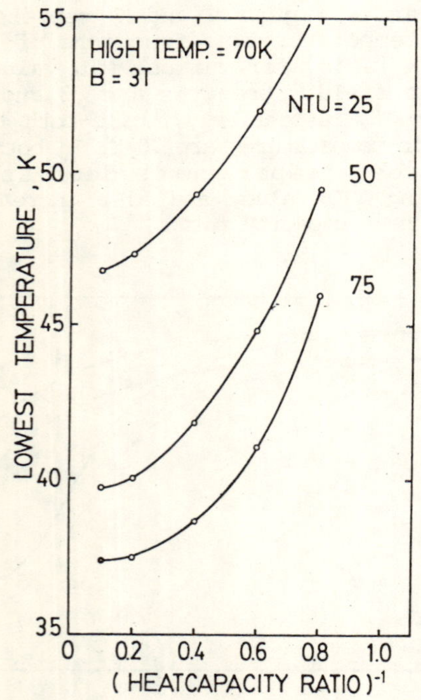

Fig. 5 The lowest temperatures of the model active regenerator with no refrigeration power are represented as functions of inverse heat capacity ratio for various NTU. The strength of magnetic field is 3 T. The high reservoir temperature is 70 K.

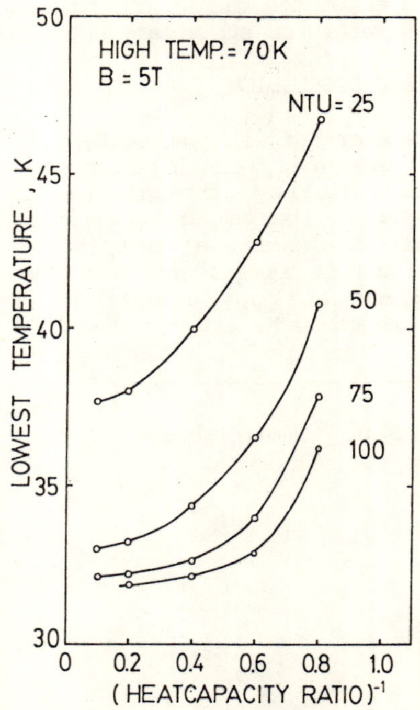

Fig. 6 The lowest temperatures of the model active regenerator with no refrigeration power are represented as functions of inverse heat capacity ratio for various NTU. The strength of magnetic field is 5 T. The high reservoir temperature is 70 K.

capacity ratio which gives a maximum refrigeration power. This problem will be analyzed elsewhere.

CONCLUSIONS

Magnetically active regenerator was analyzed numerically in the temperature range of about 30 to 70 K. The Ericsson-like operation and ferromagnetic material, $DyAl_{2.2}$ were adopted for the model regenerator. It was shown that the model active regenerator had a possibility of attainment the lowest temperature about 30 K from high reservoir temperature of 70 K. A regenerator matrix which have large entropy change over wide temperature range is required for wide operation. Combination of several magnetic materials enables to control the entropy change.

LIST OF SYMBOLS

A	= heat transfer area per unit length of regenerator
$B_{H,L}$	= magnetic field (high, low)
C_g	= specific heat of gas
$C_R(T,B)$	= specific heat of regenerator matrix
h	= heat transfer coefficient
L	= regenerator length
m	= mass flow rate
$S_R(T,B)$	= entropy of regenerator matrix
T_g	= temperature of gas
T_R	= temperature of regenerator matrix
$T_R^{1,2,3,4}$	= regenerator temperature profiles defined in the text
T_{gL}^{in}	= inlet gas temperature at cold end
T_{gL}^{out}	= outlet gas temperature at cold end
T_{gH}^{in}	= inlet gas temperature at hot end
T_{gH}^{out}	= outlet gas temperature at hot end
t	= dimensionless time
t_0	= period of time in heat transfer process
x	= dimensionless distance along regenerator
NTU	= number of transfer unit $= hAL/mC_g$
Γ'	= variable heat capacity ratio $= M_RC_R(T,B)/mC_gt_0$
Γ	= heat capacity ratio $= M_RC_R(T=T_H,B=0)/mC_gt_0$

REFERENCES

1 Brown, G.V. J.Appl.Phys., vol.47 (1976), pp 3673
2 Lacaze, A.F. et al. Adv.Cryog.Eng., vol.27 (1981), pp 703
3 Hakuraku, Y. et al. Cryogenics, vol.26 (1986), pp 171
4 Numazawa, T. et al. Adv.Cryog.Eng., vol.29 (1984), pp 589
5 Hashimoto, T. et al. Adv.Cryog.Eng., vol.32 (1986), pp 279
6 Matsumoto, K. et al. Adv.Cryog.Eng., vol.33 (1988), pp 743
7 Claudet, G.M. Adv.Cryo.Eng., vol.31 (1986) pp 733
8 Cogswell, F.J. et al. Proc.of Int.Cryocooler Conf. (1988)
9 Willmott, A.J. Int.J.Heat Mass Transfer vol.7 (1964), pp 1291
10 Rios, P.A. et al. Adv.Cryo.Eng., vol.13 pp 566

AN INVESTIGATION OF CYCLIC FLOW PRESSURE DROP
IN STIRLING CYCLE CRYOCOOLER REGENERATOR

Zhang Tiehan, Deng Xiaohui, Yi Nan, Wang Yuelan, Guo Fangzhong

Huazhong University of Science and Technology
Laboratory of Cryogenic Engineering
Wubei, China

BACKGROUND

For a split stirling cryocooler with pneumatically driven free displacer, the cyclic flow pressure drop in regenerator acts not only as a factor reducing the refrigeration capacity, but also as a factor determining the proper phasing of displacer motion in relating to the pressure wave from compression space. Recently, only few of the experimental pressure drop data for steady unidirectional flow through matrix have been published, while the experimental data in case of cyclic flow can scarcely be found in literature. These facts highlight the need for a comprehensive investigation of cyclic flow pressure drop both from the experimental and from the theoretical side.

Kim has published his experimental results of pressure drop on a cyclic flow regenerator test rig[1]. He came to the conclusion that cyclic flow pressure drop can also be correlated by dimensionless relationship of the form in which

$$f = \phi(Re)$$

definition of f is the same with that for the steady unidirectional flow, i. e.

$$f = -\partial p/\partial x \Big/ \left(\frac{1}{d_h}\frac{\rho u^2}{2}\right) \quad (1)$$

In his work, the comparison of experimental data of pressure drop amplitude with the calculated ones based on the Kays and London correlation[2] shows an averaged enhancement of friction factor cyclic flow under the given operating conditions. (frequency of cycling varied from 2.8CPS to 13.3CPS). D.R.Taylor in[3] has used the same correlation and comparison methods and has found the enhancement is much higher than in[1] in case of cycling frequency of 0.16CPS. A.J.Organ has pointed out in[4] properly that the local instantaneous pressure drop in the stirling cryocooler regenerator is a function of porosity but no experimental results have been given.

Based on the comparison of the performance prediction of a stirling engine with its experimental date[5], R.C.Tew has found that "the predicted indicated power agrees best with the test data if four times the reference regenerator friction factor is used", the later has been shown in the refernece as:

$$f = \frac{18.43}{Re^{0.6635}}, \text{ for } Re < 80 \quad (2)$$

$$f = 1.8e^{-(0.0172Re)} + 0.611, \text{for } Re > 80 \quad (3)$$

and almost the same suggestion has been proposed by Martini in[6], who used a multplier of 3.3 to 3.6 on the calculated pressure drop based on the steady flow correlation to bring the computed engine results in line with the measurements.

Chen in[7] analysed this problem from a basic understanding of the cyclic flow phenomena and has derived the friction factor from theoretical principles. Basically, the kinetic Re, defined as

$$Re_\omega = \omega r_0^2/\nu$$

is a unigue parameter that characterizes cyclic flow, and the ratio of friction factor between cyclic and unidirectional steady flow has been shown to vary with the kinetic Re number only. (See Fig 1). For regenerators, the author pointed out properly, because of the smallness of matrix Re_ω (order of 10^{-5}), the friction pressure drops for these two cases must not differ significantly. The comparison based on the amplitude attenuation phenomenon of the cyclic flow pressure drop in matrix from[2][3][5][7] is shown in Fig 2.

In [8], the authors have reported their investigations on the flow charateristics in cyclic flow regenerators. The phenomena of pressure wave phase shift between two section and the appearance of extremum of regenerator pressure

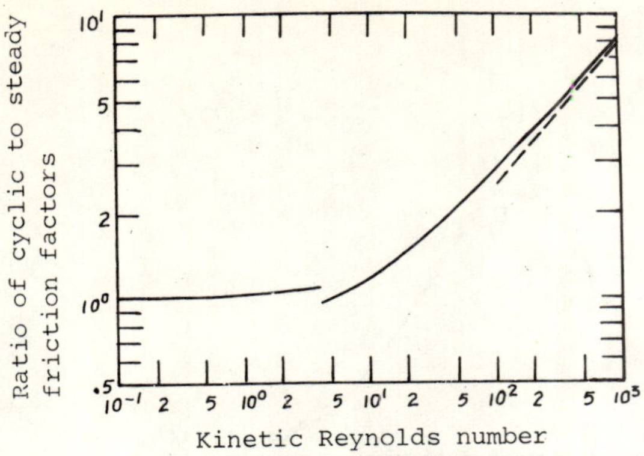

Fig.1 Fluid friction comparison between cyclic and steady unidirectional flow.

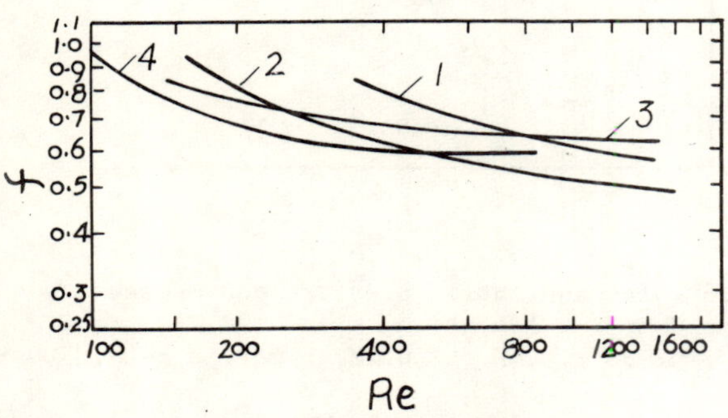

Fig.2 Friction factor correlations
1. Kim (cyclic flow 168 ∿ 800 r.P.m);
2. Kays and London; 3. Ergun; 4. NASA Lewis.

drop have been revealed and a mathematical model has been proposed to describe the phenomena. This reminds that the central issue of investigation on pressure drop in regenerator should not only be their amplitude-frequency characteristics, but also their phase-frequency side.

A step further, regenerator determines the performance of a free piston cryocooler not only by its heat transfer aspect, but also by its phase shift characteristics. The FPST Cryocooler may be considered as a system being composed of components shown in Fig 3.

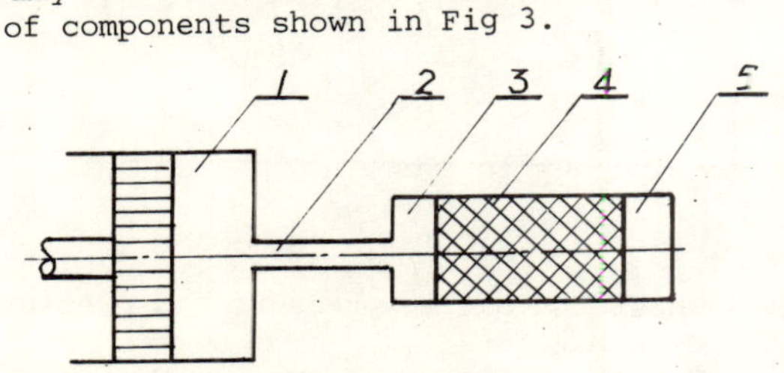

Fig.3 Schmeatic diagram of a FPST cryocooler.
1. compression space; 2. linking conduit; 3. Warm space; 4. regenerator; 5.cold space.

Theoretically from[8], we have had already for cyclic flow regenerator:

$$\begin{bmatrix} p(l,t) \\ \dot{m}(l,t) \end{bmatrix} = \begin{bmatrix} ch\lambda l & -Z_c sh\lambda l \\ -\frac{1}{Z_c} sh\lambda l & ch\lambda l \end{bmatrix} \times \begin{bmatrix} p(0,t) \\ \dot{m}(0,t) \end{bmatrix} \quad (4)$$

Which can easily be deduced to be (See [9],[10]) the reansfer function of pressure drop for regenerator:

$$W_{p_1 p_0} = \frac{p_1(j\omega)}{p_0(j\omega)} = \frac{1}{\frac{Z_2'(j\omega)}{Z_2'(j\omega)} sh\lambda l + ch\lambda l} \quad (5)$$

Z_2' is the load impedance of regenerator 4 in Fig. 3, i.e. of the cold space 5, for which

$$Z_2' = R_2 + jx_2$$

The transfer function $W_{p_1 p_0}$ indicates:

(1) $W_{p_1 p_0}$ depends strongly on the load Z_2' with which the regenerator connect;

(2) Resonanse phenomena appear clearly on certain fixed values of freguen cy independence on the end load reactance. Fig 4 is an example for a tube line connected with a void space as end load which can be described as

$$x_2 = \frac{B}{\omega V_0}; \qquad R_2 = 0 \quad (6)$$

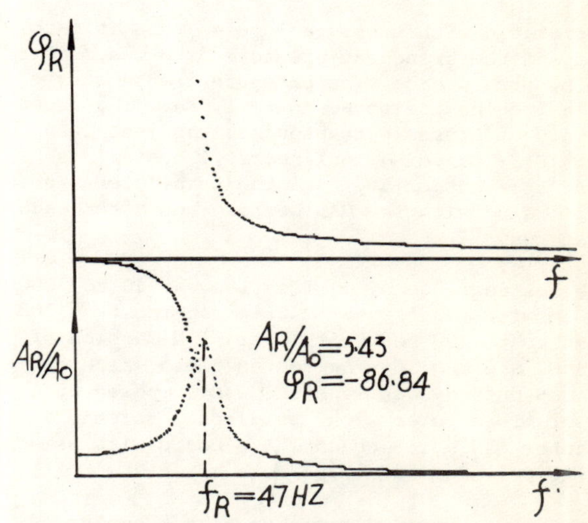

Fig.4 Resonance phenomenon for tube line connected with void space.

(3) In case of

$$Z_2' = Z_\omega'$$

there will be

$$W_{p_1 p_0} = e^{-\lambda l} \quad (7)$$

which means that the pressure wave attenuates continuously along the regenerator without reflection from the end. This implies that the warm space of FPST cryocooler should be considered as a void space for the linking conduit, while the cold space should be considered as the load reactance of regenerator. Their dimensions must be in matching with connecting components in order to achieve proper phasing pressure wave in cold space.

EXPERIMENTS AND RESULTS

A test rig has been constructed and instrumented with high speed transient data acquisition system (DAS), see Fig 5. The regenerator matrix

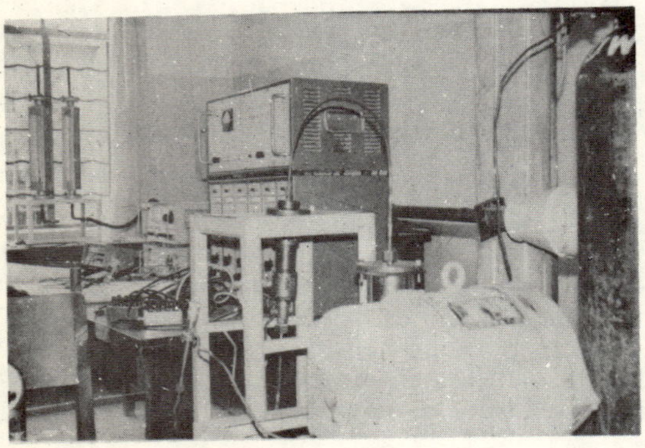

Fig.5 Cyclic flow regenerator pressure drop test rig.

was static with pressure impulse tube in each end and the transient pressure drop was measured by strain gage type pressure transducers with freqeney response up to 1.6KHz. The uncertainty of pressure transducers was less than 0.5% F.S(including nonlinearity, hysteresis and nonrepeatability) and the transducers were calibrated with the DAS be fore and after each test run.

The tested range of cyclic flow was 40 to 2000 r.p.m with regenerator matrix ranging from 200 mesh up to 400 mesh with overall dimension of $\phi 23\times 62$ and $\phi 8\times 50$. The tested fluid was high graded pure N_2 gas. The DAS was composed of a Super-AT computer with its display screen. a printer and a 16- channal A/D-board with samlping period down to 2μs/dot.

Fig.6 through Fig.9 give out four examples of the pressure wave at inlet and outlet ends of regenerator together with the pressure drop curves between then measured at 80 to 1200 r.p.m, the matrix was of 300 mesh, the outlet void volume of regenerator was $3cm^3$.

Table 1 demostrates the measured phase-shifts between inlet and outlet pressure of the regenerator the amplitudes of the pressure drop wave of the regenerator, and one can find from the table that:

(1) The phase-shift increased with the cyclic-- flow frequency from a value closing to zero

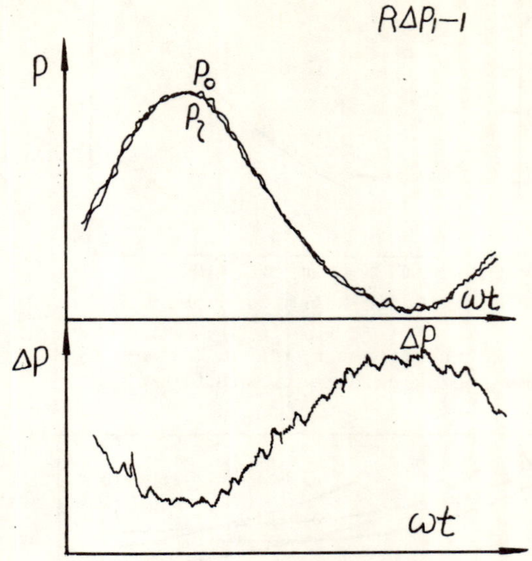

Fig.6 Inlet and outlet pressure and pressure drop ∿ ωt lot at:
n: 80 r.p.m; p_m: 0.2MPa; matrix mesh: 300.

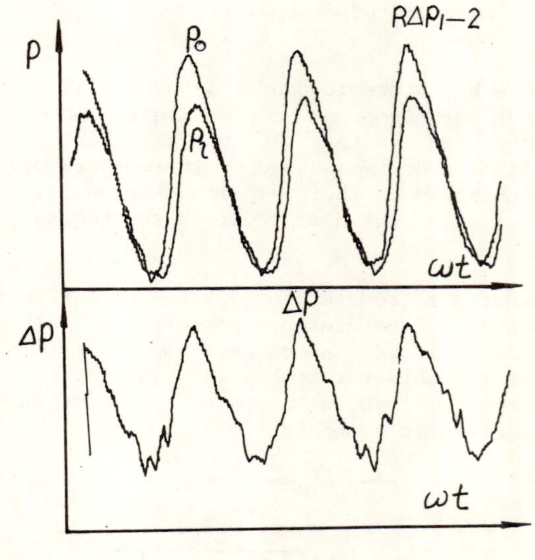

Fig.7 Inlet and outlet pressure and pressure drop ∿ ωt lot at:
n: 530 r.p.m; p_m: 0.2MPa;

matrix mesh: 300.

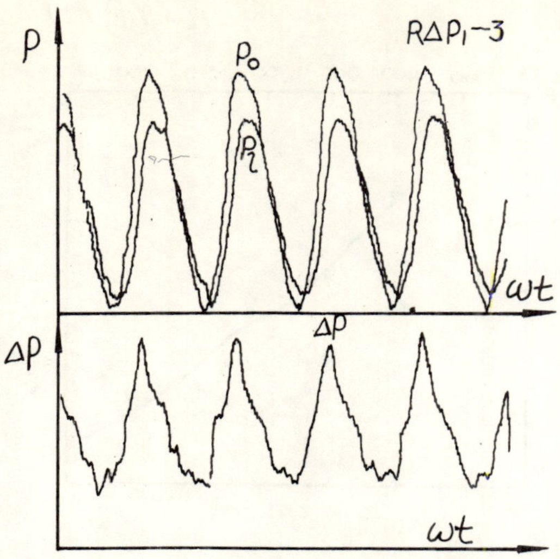

Fig.8 Inlet and outlet pressure and pressure drop∿ωt lot at:
n: 670 r.p.m; p_m: 0.2MPa;
matrix mesh: 300.

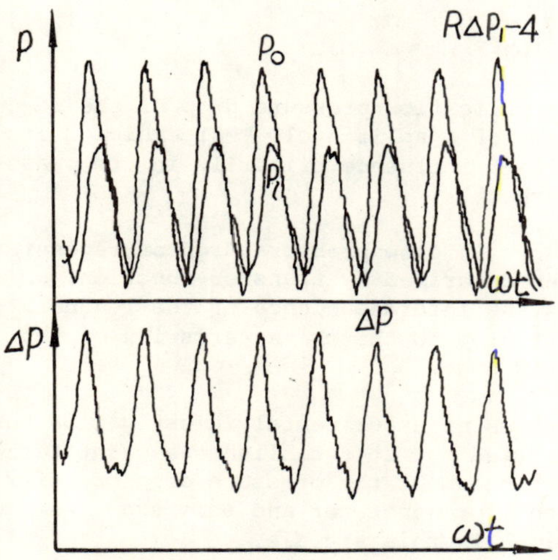

Fig.9 Inlet and outlet pressure and pressure drop∿ωt lot at:
n: 1200 r.p.m; p_m: 0.2MPa;
matrix mesh: 300.

(under 80 r.p.m) up to $\pi/4$ (under 1200 r.p.m).

(2) The amplitude of pressure drop wave of the regenerator had an obvious entremum at certain cyclic flow frequency (1200 r.p.m).

One can fine also from the figures that the pressure wave (especially for the pressure drop wave) dispays an evident non-harmonic character, which certainly, will effect the motion of displacer significantly. The sources leading to this non-harmonicity of pressure wave from the pressure wave generator (about which Mr D.Z.Zheng has reported his investigation on this conference) and from the nolinearity of wave propagation in the conduit and void space

Table. 1. Amplitude and phase shift characteristics of ΔP_{reg} (300mesh, L = 65mm, D = 23mm, inlet void = 3.2cm³, outlet void = 3cm³).

n (r.p.m)	$\phi_{\Delta p}$ (Degree)	$A_{\Delta p}$ (0.1MPa)
80	0	0.15
110	- 4	0.14
230	- 9	0.155
370	- 13	0.154
530	- 17	0.17
670	- 28	0.18
780	- 27	0.175
970	- 36	0.15
1120	- 40	0.16
1200	- 45	0.14

(including that of the regenerator). In order to makethis more quantitatively, a result of the, frequency-spectrum analysis of the pressure and pressure drop wave of the regenerator has been shown in Fig 10, which demonstrates

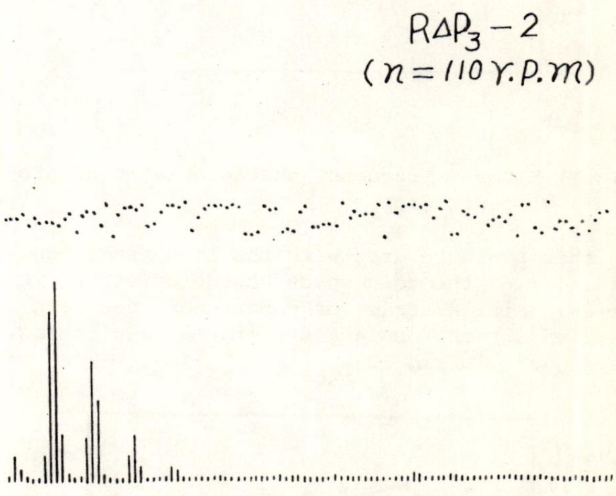

Fig.10 The amplitude-frequency spectrum of Δp-curve at n = 110 r.p.m, matrix of 300 mesh. $\Delta\omega = 10\pi$.

the nonlinearity involved in the pressure waves for the FPST cryocooler.

The data in Table 1 can be described by the amplitude-frequency and the phase-frequency characteristics of pressure drop as showm in Fig.11, these are easily to be transformed to the form of transfer function, i.e. the from of (5).

As has already been shown before, this transfer function describes not only the relationship between the outlet and inlert pressure drops of regenerator, but also the relationship

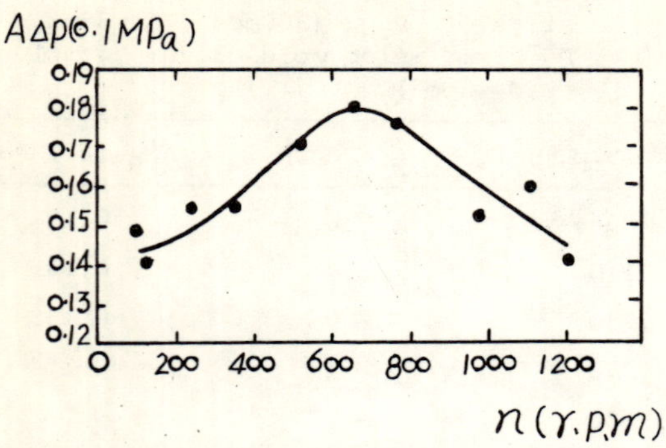

Fig.11 a) Amplitude-frequency characteristic of pressure drop.

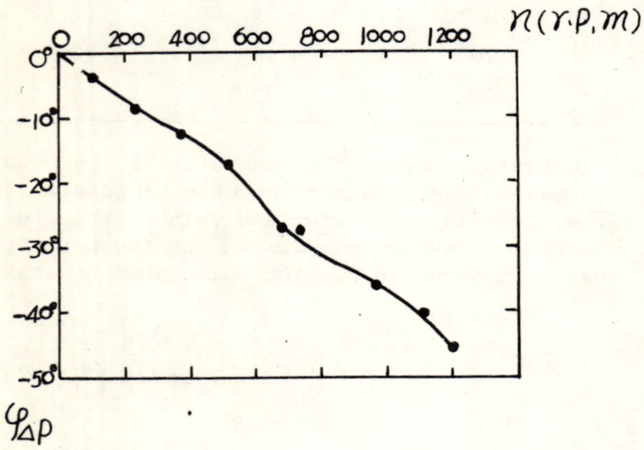

Fig.11 b) Phase-frequency characteristic of pressure drop.

of this pressure drop with the load reactance of it, i.e. the cold space characteristics. To reveal this, a series of experiments has been barried out and one example of the results has been given in Fig 12.

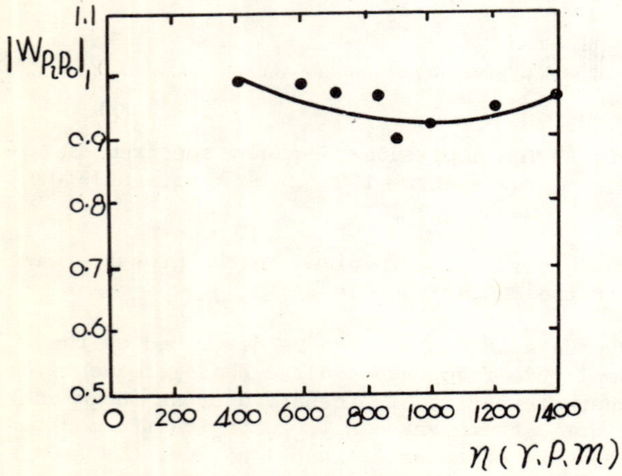

Fig.12 a) Amplitude-frequency characteristic of transfer function of pressure in regenerator.
(inlet void = 3.2cm³; outlet void = 1.8cm³).

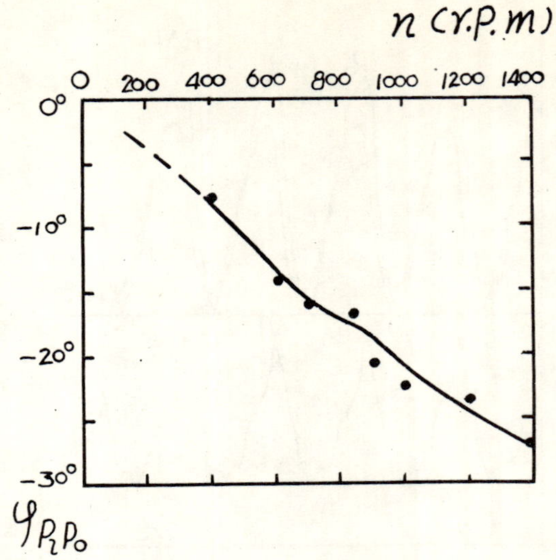

Fig.12 b) Phase-frequency characteristic of transfer function of pressure in regenerator.
(inlet void = 3.2cm³; outlet void = 1.8cm³).

CONCLUSION

1) The cyclic flow pressure drop of the regenerator of a split cycle FPST exhibits its significant influence primarily in its phase-shife aspect.

2) The cyclic flow pressure drop of regenerator has been described by transfer function, which reveals the interdependence of the regenerator pressure drop to the characteristics of the connected space.

3) A series of experimental investigation has been carried out to establish the quantitative information about the pressure drop transfer function of regenerator and some examples have been given in this article.

4) The nonlinearity involved in the pressure wave propagation and generation processes attracts certain interests for further investigations.

LIST OF SYMBOLS

A = cross section

B = module number of fluid $(B = \rho \frac{dp}{d\rho})$

R = dissipative resistance for wave
W = transfer function
Z = reactance
λ = propagation coefficient
0 — start
1 — end.

ACKNOWLEDGEMENT

The authors are grateful to Mr. C.Lee, Mr. D.

H.Yie and Me. Y.Xiang for their cordial cooperation in the process of mastering the DAS.

REFERENCES

1 Kim, J.C. 'An analytical and experimental study of flow friction characteristics for periodically reversing flow'. Paper No. 73-WA/FE-13. ASME. (1973).

2 Kays, W.M. and London, A.L. 'Compact Heat Exchangers'. Second ed. (1964).

3 Taylor, D.R. and Aghili, H. 'An investigation of oscillating flow in tubes'. Proc. 19th IECEC, (1984), Paper No. 849176.

4 Organ, A.J. 'An enquiry into the mechanism of pressure drop in the regenerator of the Stirling cycle machine'. Proc. 19th IECEC, (1984), Paper No. 849006.

5 Tew, R.C, Thieme, L.G. and Miao, D. 'Initial comparison of singe cylinder Stirling engine computer model predictions with test results'. DOE/NASA/1040-78/30, NASA TM-79044.

6 Martini, W.R. 'A nonproprietary, nonsecret program for calculating Stirling cryocoolers'. NBS S.P 698 (1984).

7 Chen, N.C.J, Griffin, F.P. 'Effects of pressure-drop correlations on Stirling engine predicted performance'. Proc. 18th IECEC, (1983), Paper No. 839114.

8 Guo, F.Z, Chou, Y.M, Lee, S.Z, Wang, Z.S and Mao, W. 'Flow characteristics of a cyclic flow regenerator'. Cryogenics (1987), 27 NO 3.

9 Popov, D.N. 'Transient Hydromechanical Process' Moscow (1982) (in Russian).

10 Guo, F.Z. 'Theory of Free Piston Stirling Cryocooler'. Textbook of HUST Wuhan China. (1988).

11 Doebelin, E.O. 'System modeling and respons: theoretical and experimental approaches'. N.Y. (1980).

A NEW 4K REFRIGERATING METHOD USING CYCLIC ADSORPTION/DESORPTION

Yan Xue and Zhou Yuan
Dept. of Urbanization Tech. & Municipal Eng., Beijing Inst.
of Civil Eng. and Architecture, Beijing, China
*Cryogenic Lab, Academia Sinica, Beijing, China

ABSTRACT

This article describes a new refrigerating method using the adsorption/desorption of an adsorbent (activated carbons) for helium gas to provide continuous cooling at liquid helium temperature. The principles of our new method, the designing of the experimental device and the experimental results are all presented here. The method stress simplicity and reliability through the absence of small orifices or moving parts at cryogenic temperature stage. With the experimental device, we had gained some amount of liquid helium by using this new method.

INTRODUCTION

Many developed or developing high techniques such as SQUID can only work at 8K or even at liquid helium temperature, but the amount of cooling power they require is very small (only several milliwatts). The novelty of our design is that it is able to provide continuous cooling at 4.2K by recycling an adsorption/desorption of activated carbon. For the initial stages of cooling we use a two-stage or three-stage regenerative cryocooler.

Because of definite condition, we can only build a thermostate to simulate the cool end of a regenerative cryocooler to realize our tentative idea. we have built an experimental device which can produce small amount of liquid helium using the method described below.

PRINCIPLES OF OUR METHOD

Adiabatic desorption

In our device, helium-carbon adsorption system in which helium is maintained at 1 atm is precooled to 8K by the thermostate. When an electric solenoid valve is opened to allow a pump to decrease the pressure of the system, the activated carbon will desorb helium gas. This process is very nearly adiabatic because at these low temperatures the specific heat of the metal vessel is negligible compared to the specific heat of helium gas. We know that if an adsorbent is heated it will evolve certain amount of gas. So in this adiabatic process the temperature of the helium-carbon system falls to 4.2K. Refrigerating capacity is produced with the carbon desorbing helium at 4.2K continuously. This refrigerating capacity liquifies another stream of helium gas holding 1 atm partly(1).

Principle of cycled desorption

In order to recycled the desorption of the activated carbon the system must be refilled with 1atm helium gas which is warm. This adiabatic adsorption process produces heating. The helium-carbon adsorption system called adsorption stage must be cooled again to 8K by the thermostate before the next desorption and liquefaction cycle may commence. When the temperature of adsorption stage falls, the carbon adsorbs helium gas at 1atm constantly with evolving heat to the thermostate. In this manner the desorption of the activated carbon may be cycled but the temperature of the adsorption stage will oscillate between 4.2K and about 11K.

To obtain a liquid helium bath with constant temperature, a seperate, well insulated, container called the liquid helium reservoir is connected with a long insulation tube to a heat exchanger(E3) at the bottom of the adsorption stage. A stream of incoming helium gas in the heat exchanger(E3) is condensated. The liquid is collected at the bottom of the reservoir. During the adiabatic adsorption process the adsorption stage heats up to 11K. This also heats the gas in the upper part of the insulation tube. However, due to the low conductivity of helium and due to the natural stratification of the gas, with the warm gas on top, littel heat is conducted down to the liquid reservoir. For the same principle, adsorption stage is connected to the thermostate above by another long insulation tube. The function of the long insulation tube is analogous to an electric diod, so we call it thermal diod. When the top end is cold, it conducts heat by convection currents. When the top end is hot, it becomes thermal insulator. The helium gas stratifies and the convection current cease(2).

Operating cycle

The operating cycle is shown in Fig.1. There are four phases in the cycle. At (a) the helium-carbon adsorption system (i.e. adsorption stage) at 1atm is cooled to 8K by the thermostate. At (b) the activated carbon desorbs helium gas adiabaticly with a valve being opened at room temperature and the temperature of the adsorption system falls to 4.2K. At (c) a stream of helium gas holding a pressure of 1atm is piped through a counterflow heat exchanger(E2) and the heat exchanger(E3) before entering the liquid reservoir. Good thermal contact with the adsorption stage causes condensation of the helium gas in heat exchanger(E3) with the carbon desorbing continuously. The final step in the cycle, (d), is the helium gas at 1atm fills back into the carbon chamber(i.e. helium-carbon system), that is activated carbon adsorbs helium gas adiabaticly.

DESIGN OF THE EXPERIMENTAL DEVICE

A schematic of the device is shown in Fig. 2. The adsorption stage and the final stage(LHe reservoir) is thermally connected to the thermostate. Helium gas lines goes from room temp. to the adsorption stage carbon chamber and to the final stage liquid helium reservoir. The two gas lines are seperated although they have the same helium gas supply. For the simplicity and ease of construction, the pressures are cycled by opening and closing electric solenoid valves. The gas line leading to the carbon chamber is alternately charged from a gas cylinder maintaining constant pressure or pumped by a pump. The gas line leading to liquid reservoir is alternately charged from or vented to the gas cylinder with condensation of the gas or vaporation of the liquid respectively.

The cycle time is the sum of adsorption time and desorption time. Adsorption time mainly determined by how long it takes to cool the adsorption stage to a temperature of the thermostate. Desorption time mainly determined by the structure of the device.

Part of vaccum room(2) holding high vacuity is immersed in liquid helium bath contained in the experimental Dewar(1). Cold current is transferred to the thermostate through a conduct temperature bar. The temperature of the thermostate is constant with controlling electric current of a heater in it.

The heat exchanger(E1) makes two streams of helium gas flowing into the carbon chamber and the reservoir respectively to be cooled.

The counterflow heat exchanger(E2) is very important. This is essential to liquify helium gas. Helium gas entering the liquid reservoir must be cooled to just above its boiling point before passing through the copper tube exchanger(E3). If it is not done the cooling power of adsorption stage will be wasted in irreversible heat transfer to hot helium gas and no liquid will be collected in the reservoir.

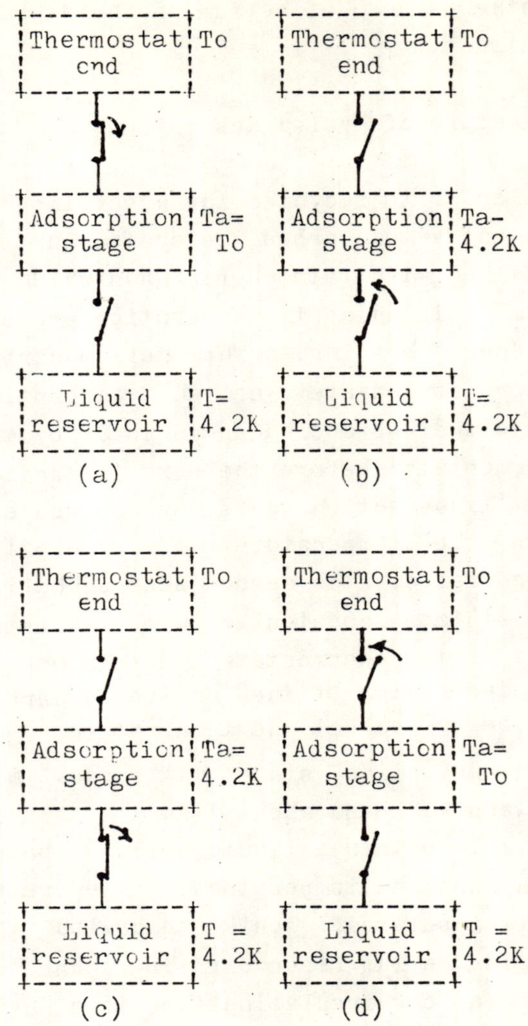

Fig. 1 Schematic of cyclic adsorption/desorption.
(a)→(b) adiabatic desorption.
(b)→(c) isotherm desorption, part of helium gas liquified.
(c)→(d) adiabatic adsorption.
(d)→(a) cool the carbon chamber.

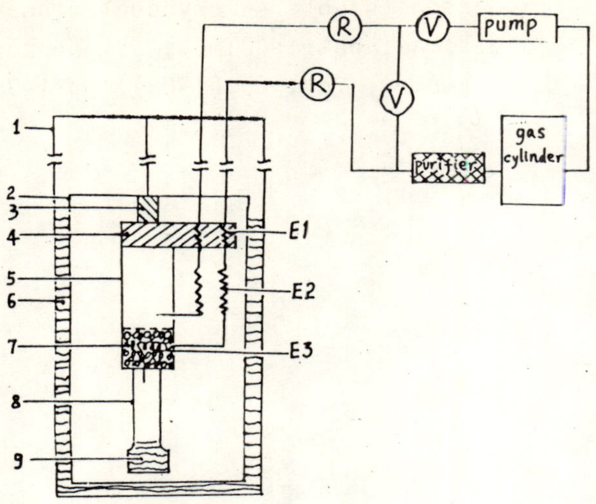

Fig. 2 Schematic diagram of device.

1. experimental Dowar.
2. vaccum room.
3. conduct temp. bar.
4. thermostat.
5. carbon chamber(i.e. insulation tube).
6. liquid helium used for thermostat.
7. activated carbon.
8. insulation tube.
9. LHe reservoir.

E. heat exchanger.
V. electric solenoid valves.
R. pressure regulator.

EXPERIMENTAL RESULT

The temperature of the thermostat is the refrigerating temperature of the cryocooler, i.e. the temperature previously cooled of the adsorption stage. Fig. 3 shows that the effects of increasing adsorption pressure is much better than decreasing desorption pressure for raising the precooled temperature.

Fig.4 shows that the temperature stability of the liquid reservoir is very good and cyclic temperature variations at the bottom of the carbon chamber. The observed temperature fluctuation at the bottom of the liquid reservoir was only 10mK, in spite of the 6K variation in the temperature of adsorption stage.

It was not possible to directly measure the amount of liquid helium in the vessel with our apparatus. We have estimated the amount of liquid helium produced by measuring the time required to boil it off by means of electric heater.

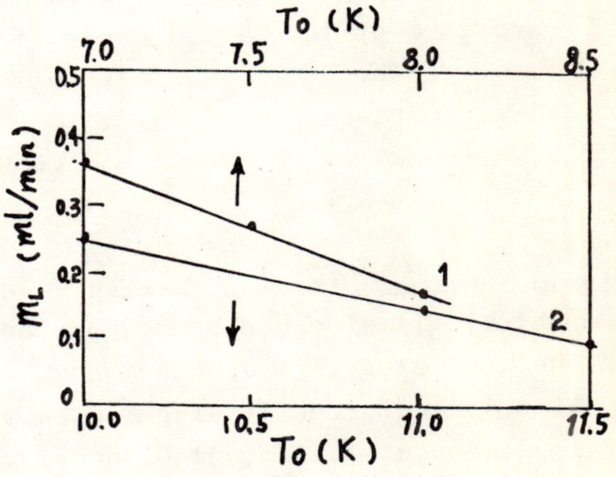

Fig. 3 The relation between the amount of liquid helium produced and the precooled temp. at different adsorption pressure.

1 -- adsorption pressure = 1atm
2 -- adsorption pressure = 5atm

CONCLUSION

A cycled adsorption/desorption experimental device which provides continuous cooling at liquid helium temperature has been demonstrated. Temperature stability of the liquid reservoir is very good because it is controlled by the boiling point of

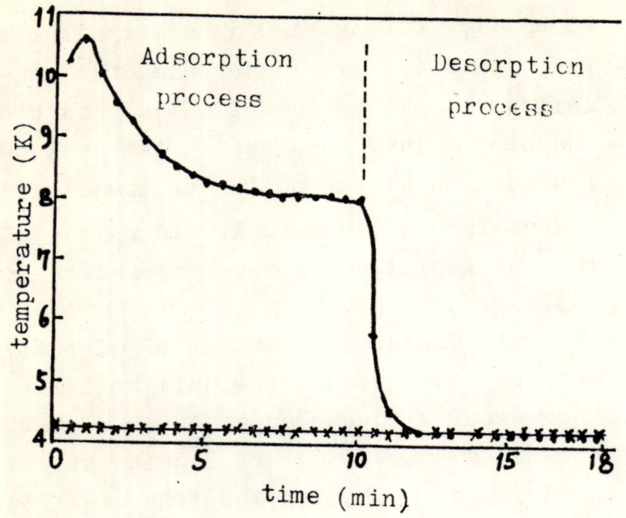

Fig. 4 Cyclic temperature variations at the adsorption stage and at the bottom of the liquid reservoir.
. --- Temp. of the adsorption stage.
x --- Temp. of the liquid reservoir.

liquid helium. It is a closed cycle completely. It also has no moving parts in the last stage. It does not have any small orifices prone to plugging too.

Although the amount of cooling power provided by the device is very little, it does have its own domain of application where the temperature stability and reliability make it more advantageous.

ACKNOWLEDGEMENT

The authers thank technician Zhang pin-sheng for his help in building up the experimental device.

REFERENCES

1 S.J.Gregg, K.S.W.Sing 'adsorption, surface area and porosity', Second edition, New York 1982.

2 C. Winter, S. Gygax, K. Myrtle and R.Barton 'stable 4K cryocooler using cycled adiabatic (Simon) expansion' Cryogenics, Vol. 25 (1985), pp71-74.

A STUDY OF THE LAST STAGE FEAT EXCHANGER OF A 4.2K CRYOCOOLER

Jin Sumin, Ji Siliang,
Zhao Kaitao, Yang Siwen.
(South-east University, Nanjing, China)

ABSTRACT

This paper studies the last stage heat exchanger of a 4.2K miniature cryocooler. A large number of experiments are made contrasting a resistance type last stage heat exchanger to a non-resistance type one. It is confirmed that a method with the aid of the negative Joule-Thomson effect of helium at low temperature is effective in improving and increasing the refrigeration capacity at 4.2K. There is an advantage in utilizing a resistance type heat exchanger as the last stage heat exchanger in a J-T loop.

INTRODUCTION

The main purpose of this paper is to study the last stage heat exchanger applied to a 4.2K miniature cryocooler composed of the J-T loop and a two stage G-M cryo-refrigerator. Fig.1 is the flow of the G-M/J-T miniature cryocooler designed by South-est University.
The working performance of the last stage heat exchanger directly influences the cooling capacity of the cryocooler. When the inlet pressure Ph and outlet pressure Pm of the G-M cryocooler are kept constant, the lower the temperature T5 and the smaller the temperature difference at the warm end of the last stage heat exchanger , the larger is the capacity of the cryocooler. Therefore, there are two ways to increase cooling capacity of the cryocooler. The first is to increasing the precooling temperature T5 which depends on the working performance of the G-M cryo-refrigerator, the high pressure Ph and the flow rate of the helium in the J-T loop G. The second is to decrease the temperature difference at the warm end of the last stage heat exchanger which depends on the thermodynamic property of helium at a low temperature. In this paper, we shall carry out a theoretical analysis and experiments by utilizing properties of the helium at low temperature to improve the operation condition of the last stage heat exchanger and to increase the cooling capacity of the cryocooler.

THEORETICAL ANALYSIS

The thermodynamic properties of helium are unusual at low temperature. The Joule-Thomson coefficient α_h depends completely on its pressure and temperature. α_h may have a negative value at low temperatures. Therefore, the temperature of the helium will rise even through it is expanded with a miniature prssure drop. For example, when the prssure of the helium is 1.862MPa, and its temperature is less than 9K, its Joule-Thomson coefficient will be negative. In the last stage heat exchanger of the J-T loop, the high pressure of the inlet helium is about 1.862MPa and its temperature is in the region of 5 to 14K. Therefore, the Joule-Thomson coefficient of the inlet helium in the last stage heat exchanger are negative at the temperature region below 9K. According to these analysis results, if the resistance of the last stage heat exchanger (known as the resistance heat exchanger) at the temperature region below 9K is increased, the working condition of the last stage heat exchanger would be improved , the temperature difference of the heat transfer would be raised and the cooling capacity of the cryocooler would be increased.
The pressure, temperature and entropy changes along the resistance tube are as follow:

$$\frac{\partial P}{\partial x} = \left[\frac{ug\beta W}{ta(C_p+\beta W^2)} + \frac{f\rho W^2}{2D}\right] / \left[W^2\rho K - \rho W^2\beta b - 1\right] \quad (1)$$

$$\frac{\partial T}{\partial x} = \left[\frac{ug\beta W^2}{ta\rho W(C_p+W^2\beta)} + \frac{f\rho W^2 b}{2D}\right] / \left[W^2\rho K - \rho W^2\beta b - 1\right] \quad (2)$$

$$\frac{\partial S}{\partial x} = \left\{\left[\frac{ug(W\rho K-1)}{ta\rho W(C_p+W^2\beta)} + \frac{f\rho W^2 b}{2D}\right]\frac{C_p}{T} - \frac{\beta}{\rho}\left[\frac{ug\beta W}{ta(C_p+W^2\beta)} + \frac{f\rho W^2}{2D}\right]\right\} / \left[W^2\rho K\right.$$

$$-\rho W^2 \beta b - 1]$$ (3)

where $b = \frac{KW + C_p \alpha_h}{C_p + \beta W^2}$, the Joule-Thomson effect may be written as:

$$\frac{\partial T}{\partial P} = \left[\frac{u g (W^2 \rho K - 1)}{a \rho W (C_p + \beta W^2)} + \frac{f \rho W^2 b}{2D}\right] / \left[\frac{u g \beta W}{a (C_p + \beta W^2)} + \frac{f \rho W^2}{2D}\right] \quad (4)$$

as the expansion process is adiabatic, relation (4)may be written as:

$$\frac{\partial T}{\partial P} = \alpha_h$$

On the other hand, the length and the construction of the resistance heat exchanger are important factors in enhancing its advantage. If the resistance tube is overlong or the reistance is greater or less in the tube, all of these will weaken the advantage of the resistance heat exchanger.

EXPERIMENT AND RESULTS

A resistance and a non-resistance type heat exchanger with the same heat transfer surface area have been designed to examine the superiority of the cryocooler.
The coaxial tube counterflow heat exchanger are designed as the last stage heat exchanger of the cryocooler. They are made of stainless steel. The diameter of the inner tube is 2mm and the thickness is 0.1mm. The diameter of the outer tube is 5mm and the thickness is 0.5mm. In the case of resistance tube, the inner tube is rolled into a hollow belt. Its cross section is approximately rectangular with a width of 3mm and a height of 0.4mm. The actual sizes of the two types of heat exchangers are shown Table 1.
The characteristics of the resistance and the non-resistance type heat exchanger are shown in Table2. with the following condition:
$Ph=1.921MPa$, $G=2.9NM^3/H$, and $T5=13.8K$. From Table 2. it is can be seen that though the exergy loss in the resistance heat exchanger is larger than that in the non-resistance one, the total exergy loss of the system with the resistance heat exchanger is less. Its cooling capacity is up to 0.14W.
The cooling capacity of the cryocooler with resistance or non-resistance heat exchanger vary with the flow rate in J-T loop at the different pressures Ph and Pm are shown in Fig 2.(a),(b),(c) and (d). From Fig 2. it is found that the cooling capacities of the cryocooler with the resistance heat exchanger are greater than those of the non-resistance system. At the same time, the higher the pressure Ph is, the better the effect of the resistance type heat exchanger is.

CONCLUSION

By theoretical analysis and the experimental results presented above, it is confirmed that the method which make use of the negative Joule-Thomson effect of the helium at low temperature is effective. The system with a resistance type heat exchanger as the last stage in a J-T loop can improve the working performance of the last stage heat exchanger and then increase the cooling capacity of the crocooler.

Table 1. The Actual Sizes of The Last Stage Heat Exchangers

Type of Heat Exchanger	Resistance	Non-resistance
Outer Tube Length (mm)	2500	2500
Inner Tube Length (mm)	2540	2550
Resistance Tube Length (mm)	480	0
Coil Diameter (mm)	130	130

Table 2. The Characteristics of The Last Stage Heat Exchanger

Type of Heat Exchanger	Resistance	Non-resistance
Temperature Difference At Warm End (K)	1.08	1.12
Cooling Capacity (W)	1.23	1.09
Exergy Loss In Last Stage EX E1 (W)	125.03	88.78
Exergy Loss In The Throttle Process E2 (W)	77.063	137.45
E1+E2 (W)	203.66	226.23

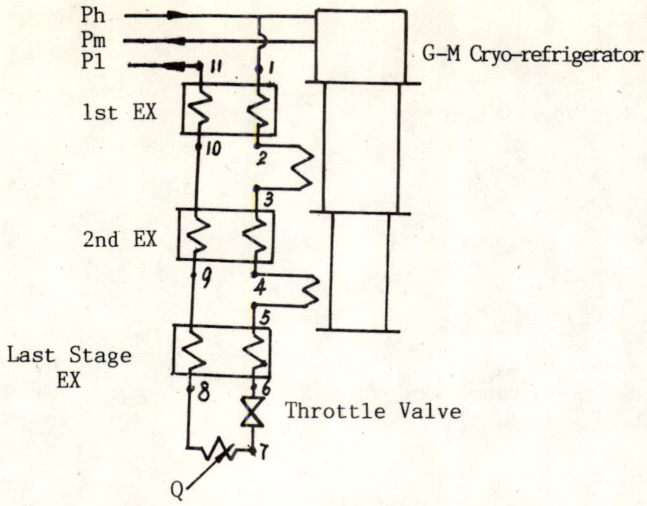

Fig 1. Flow Diagram of G-M/J-T Cryocooler

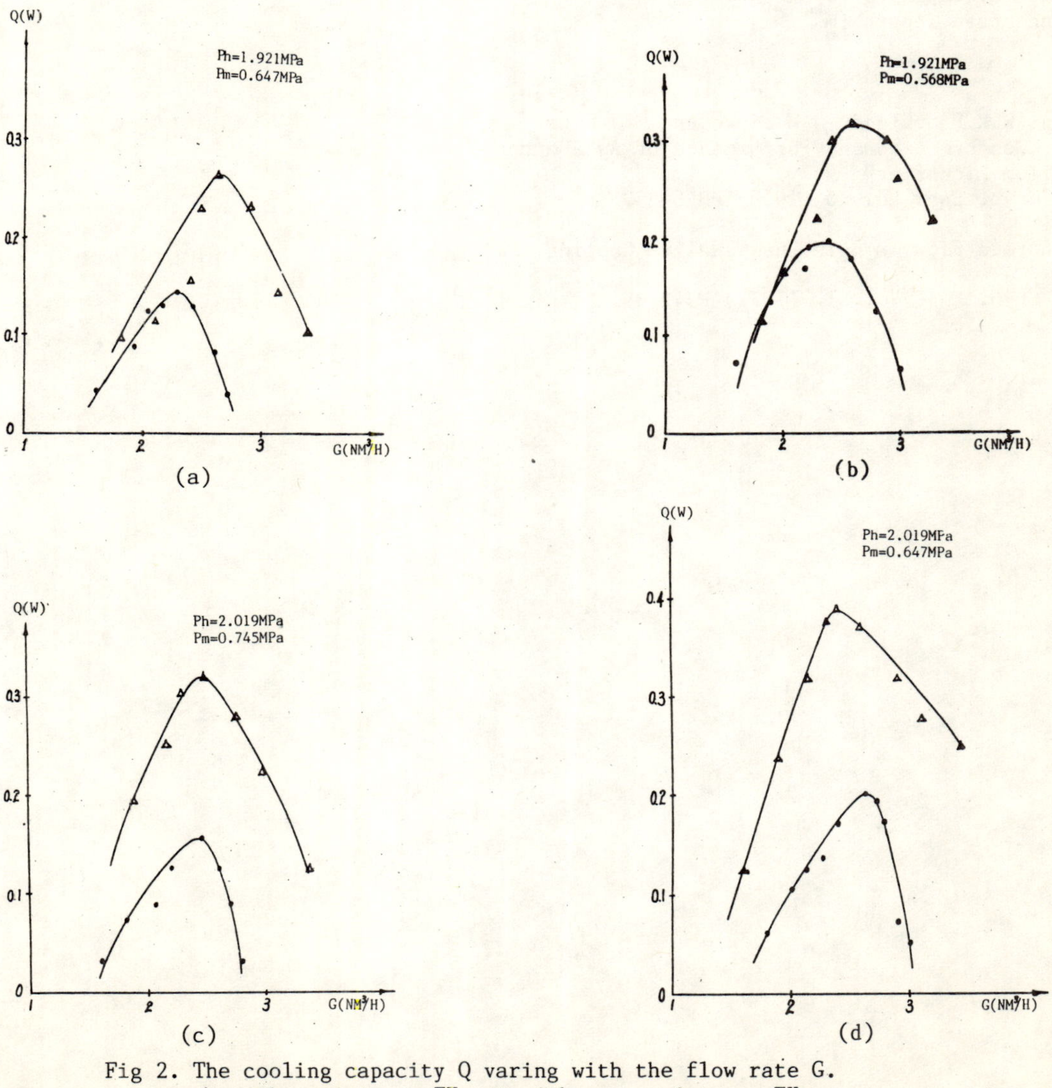

Fig 2. The cooling capacity Q varing with the flow rate G.
Δ-with resistance EX, ●-with non-resistance EX

LIST OF SYMBOLS

P =pressure (Pa)
Ph=inlet pressure (Pa)
Pm=the outlet pressure of the G-M cryo-refrigerator (Pa)
T =temperature (K)
q =heat flux (W/m^2)
a =cross sectional area (m^2)
u =length in girth (m)
ρ =density (kg/m^3)
W =velocity (m/s)
β =expansion coefficient (1/K)
K =compressibility coefficient (1/Pa)
D =diameter (m)
f =friction factor
t =the time of the helium flow dx length tube (s)
h =enthalpy (J/Kg)
s =entropy (J/Kg·K)
G =flow rate in J-T loop (NM^3/H)
Q =cooling capacity (W)
E =exergy (W)
α_h=Joule-Thomson coefficient (K/Pa)
Cp=specific heat capacity (J/Kg·K)
x =resistance tube length (m)

REFERENCES

1 D.B.Mann, W.R.Bjorklund, J.Macinko,and M.J.Hiza "Design, Construction and Performance of A Labortory-Size Helium Liquefier" Adv. in Cryo. Eng. Vol. 5 (1958) P346-353

2 V.Arp "Forced Flow, single-Phase Helium Cooling Systems" Adv. in Cryo. Eng. Vol. 17 (1971) P342-351

REGENERATIVE MATERIAL R_3T SYSTEM WITH EXTREMELY LARGE HEAT CAPACITIES BETWEEN 4 K AND ROOM TEMPERATURE

M. Sahashi, and Y. Tokai
Research and Development Center, TOSHIBA Corporation,
Kawasaki, Japan

R. Li, M. Ogawa, and T. Hashimoto
Department of Applied Physics, Tokyo Institute of Technology,
Tokyo, Japan

ABSTRACT

A new regenerative material R_3T system was developed for use in a conventional two-stage Gifford-McMahon type cryocooler. The specific heat of lead (P_b) as a conventional regenerator material is not large enough in the temperature range between 4 and 15 K for maintaining the refrigeration power and efficiency, because in most non-magnetic substance, that is Pb, only the lattice vibrations and the conduction electrons make a contribution to the specific heat.

In this paper we present a new magnetic material system (R; rare earth metal, T; transition metal) with the extremely large heat capacities between 4 K and room temperature. These compounds, formed between rare earth metals and 3d-transition metals, are ferromagnets or antiferromagnets with complicated spin configurations, and show a broad peak in specific heat at each transition temperature, for instance 7 K for Er_3Ni. In addition, this type compounds (Fe_3C) show the large lattice heat capacities even in the range above the transition temperature. Debye temperature for Er_3Ni is about 150 K, which is comparable to that for lead.

As the results of our refrigeration tests, it is clarified that by using this new material R_3T system, the no-load lowest temperature of 4 K level can be obtained in a conventional two-stage GM refrigerator.

INTRODUCTION

The needs for the new materials with much higher heat capacity values per unit volume in the temperature range below 15 K have been pointed out by several investigators (1,2). In a cryogenic engineering, a high heat capacity is the most important property of the materials, mainly used as heat regenerators. The specific heat of lead as a conventional regenerator material, however, decreases with decreasing temperature T, according to $(T/\Theta_D)^3$, where Θ_D is Debye temperature, and is not efficient in the temperature range below 15 K for use in a cryocooler regenerator, because in most non-magnetic material, including lead with the lowest Debye temperature, about 100 K, only the lattice vibrations (phonons) and the conduction electrons make a contribution to the specific heat.

On the basis of these backgrounds, several new materials have been proposed, as shown in Fig.1, which undergo magnetic transition or Schottky anomalies, at low temperature (3,4,5). Daniels and du Pré have reported the prototype test results for the refrigerator with the regenerator made from EuS (Europium Sulfide)(6). Remarkable improvement of the performance,

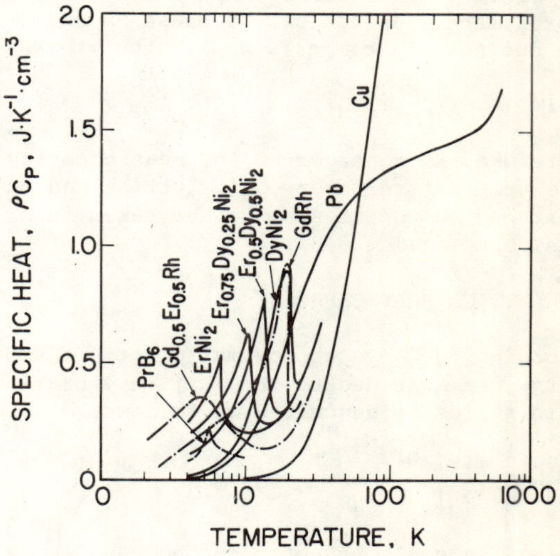

Fig. 1. Specific heat per unit volume of some new magnetic materials, proposed by several authors (3,4,5), lead and copper. Data for $Er_{1-x}Dy_xNi_2$ is from Yayama (3), GdRh, $Gd_{0.5}Er_{0.5}Rh$, Pb are from Buschow (4) and PrB_6, Cu are from Barbisch (5).

however, was not able to be achieved in this type refrigerator, where the lowest temperature of only 7.8 K was achieved even in the three stage type. Recently the successful results have been exhibited by Nakashima et al. (7), in which the compounds, GdRh (Gadolinium Rhodium) and (GdEr)Rh (Gadolinium Erbium Rhodium), as the regenerator materials were employed, based on the investigation by Buschow et al. (4). But these materials, including EuS, are too expensive to put to a commercial use.

In this paper we proposed the new type magnetic material system with large heat capacities between 4 K and room temperature, useful for a conventional two-stage GM refrigerator, which is commonly spread for many cryogenic systems, especially MRI magnets and cryopumps, and the magnetic and lattice specific heats for the rare earth - 3d transition metal intermetallics rich in rare earth metals are discussed.

HEAT CAPACITIES FOR MAGNETIC SUBSTANCES

The temperature variation of the specific heat, C(T), is related to the total entropy, S(T), by the following equation, in general.

$$S(T) = \int_0^T C(T)/T dT \qquad (1)$$

The total entropy of magnetic substances is mainly composed of three kinds of entropies, that is magnetic entropy, S_M, lattice entropy, S_L, due to the lattice vibrations, and the entropy due to the free electrons. In the temperature range above 4 K, we can write it approximately as follows, because S_M and S_L are dominant in comparison with the others.

$$S(T) = S_L(T) + S_M(T) \qquad (2)$$

Therefore, in considering the heat capacity in this temperature range, only lattice and magnetic heat capacities should be taken into consideration.

(1) LATTICE HEAT CAPACITY

According to Debye approximation, the thermal energy per mole due to the lattice vibrations, U, in solids is described as follows,

$$U = 9R\frac{T^4}{\Theta_D^3}\int_0^{\Theta_D/T} \frac{\xi^3}{e^{\xi}-1} d\xi \quad . \qquad (3)$$

where R is gas constant. In the case where only lattice vibration mainly contributes to the specific heat, for non-magnetic substances, we can write the specific heat per unit volume, ρC_p, in low temperature limit, approximately as follows,

$$\rho C_p = 1943 \cdot \frac{d_A}{N_A}\left(\frac{T}{\Theta_D}\right)^3 \qquad J \cdot K^{-1} \cdot cm^{-3} \quad , \qquad (4)$$

where ρ is the density of solids, $g \cdot cm^{-3}$, Cp is the isopiestic specific heat, $J \cdot K^{-1} \cdot g^{-1}$, d_A is the number of atoms per unit volume, cm^{-3}, and N_A is Avogadro's number. Thereby, in order to obtain the larger lattice heat capacity, we shall systematically search for magnetic substances with the lower Debye temperature and the larger number of atoms per unit volume.

Debye temperatures for the typical low melting temperature metals, for instance In (Indium), are shown in Table 1, as reference. Figure 2 shows the equilibrium phase diagram for the Er-Ni (Nickel) system (8). This diagram

Table 1 Debye temperatures for some low melting temperature metals.

Element	M.P.* (K)	Θ_D (K)
In	430	108
Hg	234	79.1
Tl	576	78.5
Pb	601	105
Bi	544	119

*M.P. ; MELTING POINT

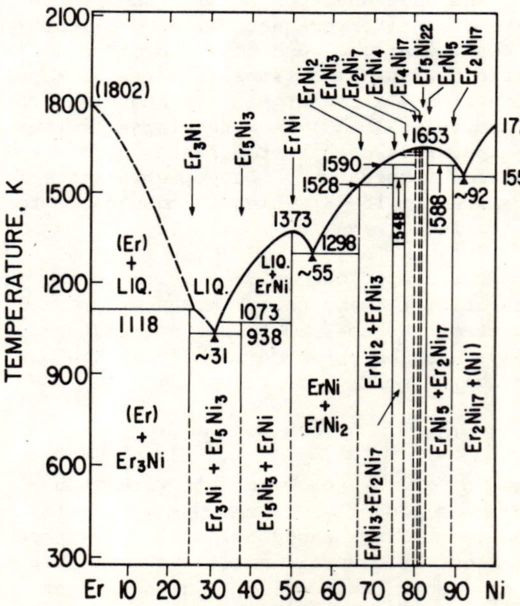

Fig. 2. Equilibrium phase diagram for the Er-Ni system, from Buschow (8).

represents the typical character as observed in every rare earth-transition metal system, where there is a eutectic reaction over the composition range rich in rare earth metal. For example, the eutectic point exists near the composition of about 30 atomic percent nickel for Er-Ni system, in which the lowest melting temperature, about 1000 K can be achieved. Therewith, every compounds formed in these systems are magnetic ones. Thereupon, from the view point of realizing the low Debye temperature, intermetallics rich in rare earths are hopeful magnetic substances.

Figure 3 shows the transition metal, M, dependence of the number of atoms per unit volume for rare earth Laves phase compounds, only in which all combinations, shown in Fig. 3, crystallizes in the same structure. As shown in Fig. 3, with the number of 3d electrons increased, the number of atoms per unit volume is inclined to increase. Therefrom, the larger value in the number of atoms per unit volume can be obtained in rare earth-cobalt (Co) and rare earth-nickel compounds, which are desired ones from the view point of the volumetric heat capacity.

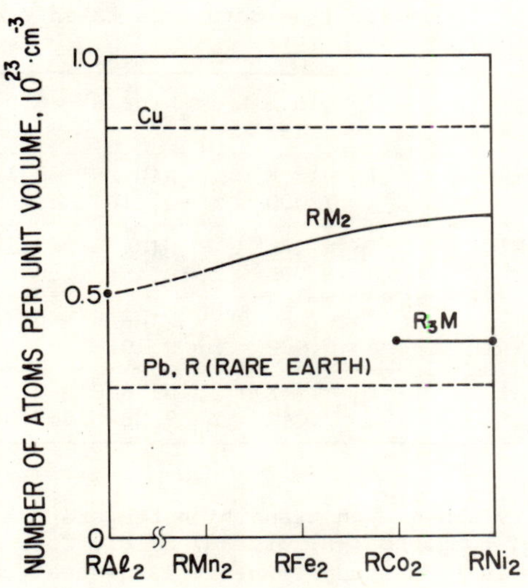

Fig. 3. Transition metal, M, dependence of the number of atoms per unit volume for rare earth Laves phase compounds.

(2) MAGNETIC HEAT CAPACITY

In the simple ferromagnets or antiferromagnets, where there is only one kind of magnetic interaction between magnetic spins, the energy due to magnetic spins per unit volume, E, is described as follows, in molecular field approximation,

$$E = -dm \cdot Z \cdot J<S_z>^2 \qquad (5)$$

where dm is the number of magnetic atoms, Z is the number of nearest magnetic neighbors, J is the exchange interaction integral, and $<S_z>$ is the expectation for the magnetization direction component of magnetic spin. By differentiating the above equation, we can obtain the isovolumic specific heat per unit volume, C_V, as follows,

$$C_V = \left(\frac{\partial E}{\partial T}\right)_V = -2dmZJ<S_z>\frac{\partial <S_z>}{\partial T} , \qquad (6)$$

$$\lim_{T \to T_c - 0} C_V = 5R\frac{dm}{N_A}\frac{S(S+1)}{S^2 + (S+1)^2} , \qquad (7)$$

where T_c is Curie temperature and S is quantum number for magnetic spin. It is well-known that large heat capacities occur in connexion with magnetic phase transition, characterized by T_c. According to the equation (7), the value of specific heat at the transition point is proportional to dm and $\frac{S(S+1)}{S^2 + (S+1)^2}$. Thereupon, suppose T_c exists in the low temperature range, large heat capacity can be obtained near T_c and in some cases its value may get to be about ten times as large as that of lead. Figure 4 shows the rare earth metal element, R, dependence of Curie temperature for some rare earth-transition metal Laves phase compounds (R; La to Tm). As shown in Fig. 4, praseodymium (Pr), neodymium (Nd), and erbium compounds have relatively low Curie temperature.

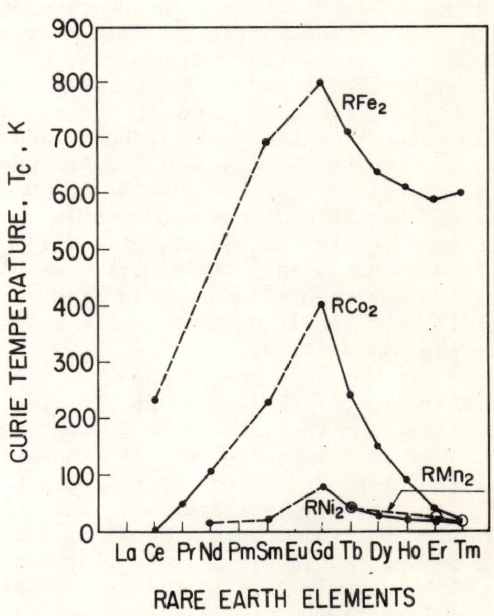

Fig. 4. Rare earth metal element, R, dependence of Curie temperature for some rare earth-transition metal Laves phase compounds.

(3) Selection of rare earth-transition metal system

Based on the investigation of the lattice and magnetic heat capacities, we selected the intermetallic compounds rich in rare earth metal, RM_2, RM, and R_3M (R; Nd or Er, M; Ni or Co). The results of our experiments on these compounds are reported in the next section.

EXPERIMENTAL RESULTS AND DISCUSSION

The heat capacities were measured in an adiabatic calorimeter using the pulse heating method. The temperature of the sample was measured by a carbon glass resistance thermometer. The samples were prepared by cutting after arc melting in an atmosphere of purified argon gas. The purity of the rare earth metals used in this study was 99.9 %. The samples were analyzed by powder X-ray diffraction technique. By the results of the diffraction patterns, it was ceritified that $ErNi_2$ crystallises in the $MgCu_2$ type cubic structure, ErNi in FeB type orthorhombic structure and Er_3Ni, Er_3Co and Nd_3Ni in Fe_3C type orthorhombic structure.

Figure 5 shows the temperature dependence of the specific heat for three kinds of erbium-nickel compounds, lead, and copper. The maxima in the specific heat, ρCp, versus temperature curves were observed, which are at 6.5 K in $ErNi_2$, 10.5 K in ErNi, and 7 K in Er_3Ni, respectively. Except for $ErNi_2$, the ordering temperatures almost agree with the values derived from magnetic measurement in previous paper (Table 2)(9). In the case of $ErNi_2$, the sharp peak in the specific heat was observed at the transition temperature, but, compared to the value for T_c in the previous paper (Table 2), there is remarkable disagreement. In respect of the magnetic heat capacity, with increasing Er content, the broadening of the specific heat anomaly is inclined to be more remarkable. This tendency is considered to be related to the magnetic spin structure and the magnetic interaction mainly between the localized 4f moments in Er sites (Fig. 6)(10,11), resulting in a double-peaked specific anomaly in Er_3Ni.

In the other aspect, there are the striking characteristics in the lattice heat capacity in the temperature range above each transition temperature. With Er content increased, the lattice heat capacity increases and in Er_3Ni its value becomes comparable to that of lead. As our experimental results, it is clarified that in the intermetallics rich in rare earths there is a expectable possibility of new magnetic materials with large magnetic and lattice heat capacities existing.

Figure 7 shows the temperature dependence of the specific heat for Er_3Ni, Er_3Co, Nd_3Ni and ErRh (4). In the aspect of the lattice heat capacity above each transition temperature, all R_3T compounds (R; Er, Nd, T; Ni, Co), employed in this study, exhibit large heat capacity values in the range above 15 K, as expected. For all three compounds, shown in Fig. 7, the specific heat reaches the two times as large value as that of ErRh (4) at 30 K. On the other hand, there is a remarkable difference between three compounds in respect of the specific heat in the range below 15 K. For Nd_3Ni a very sharp peak is observed at the transition temperature, 12.4 K, and this specific heat versus temperature curve indicates a typical antiferromagnetic behavior same as ErRh, while for Er_3Co exhibits a considerable broadening of the specific heat anomaly as observed in Er_3Ni. It is very interesting to consider the reason why the specific heat anomaly due to the magnetic ordering depends on the rare earth element for the compounds with the same crystal structure.

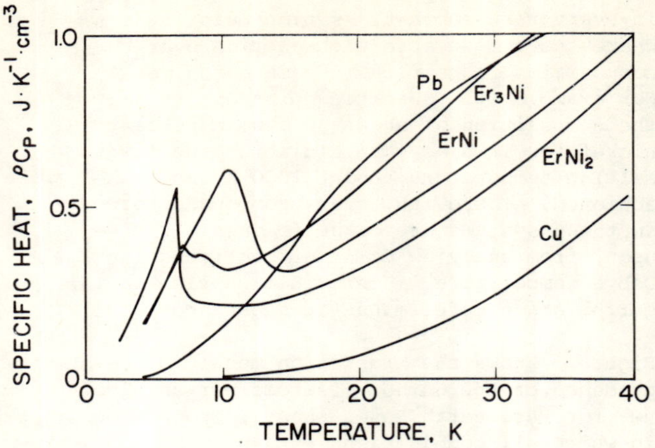

Fig. 5. Temperature dependence of the specific heat for three kinds of erbium-nickel compounds, lead and copper.

Table 2 Magnetic ordering temperatures (T_c or T_N), saturation moments (μ) and lattice constants of some rare earth-nickel and rare earth-cobalt compounds. (The data were taken from reference (9).)

$ErNi_2$	T_c = 18.5 K a = 7.125	$\mu(\mu_B/R)$ = 6.9	
ErNi	T_c = 12 K a = 7.000	$\mu(\mu_B/R)$ = 8.1 b = 4.118	c = 5.414
Er_3Ni	T_N = 7.7 K a = 6.79	$\mu(\mu_B/R)$ = 7.1 b = 9.45	c = 6.23
Er_3Co	T_c = 7 K a = 6.902	$\mu(\mu_B/R)$ = 6.9 b = 9.191	c = 6.189
Nd_3Ni	T_N = 15 K a = 7.04	$\mu(\mu_B/R)$ = 0.7 b = 9.86	c = 6.43

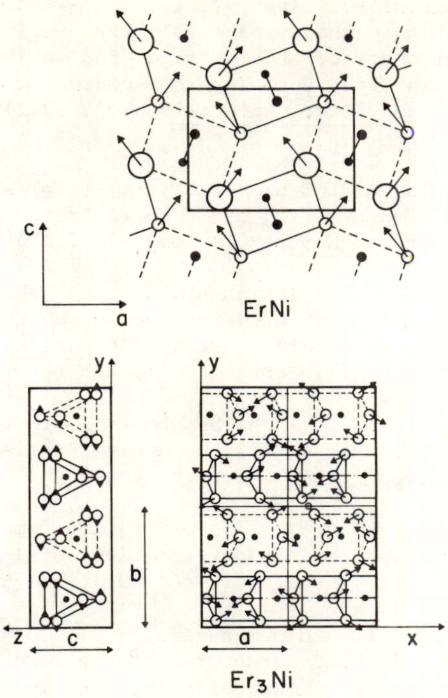

Fig. 6. Spin configurations for ErNi and Er_3Ni.

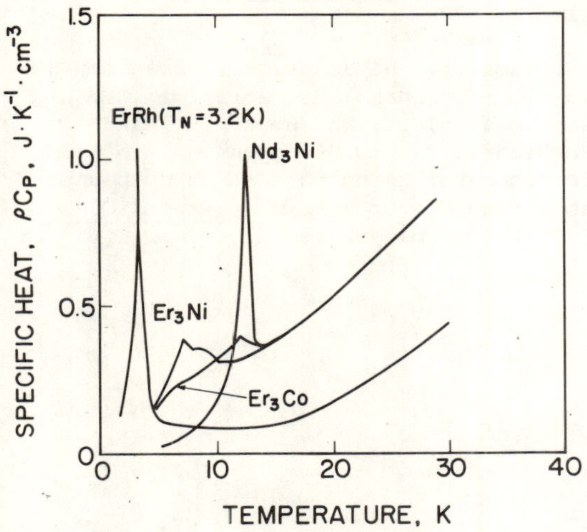

Fig. 7. Temperature dependence of the specific heat for Er_3Ni, Er_3Co, Nd_3Ni and ErRh (4).

As the results of our investigation for a series of compounds rich in rare earth metals, we developed a new magnetic material Er_3Ni for use in a conventional cryocoolers, in which the specific heat exhibits a constant character in the range between 7 and 15 K and the large value, 0.39 $J \cdot K^{-1} \cdot cm^{-3}$, which is equal to that for lead at 15 K.

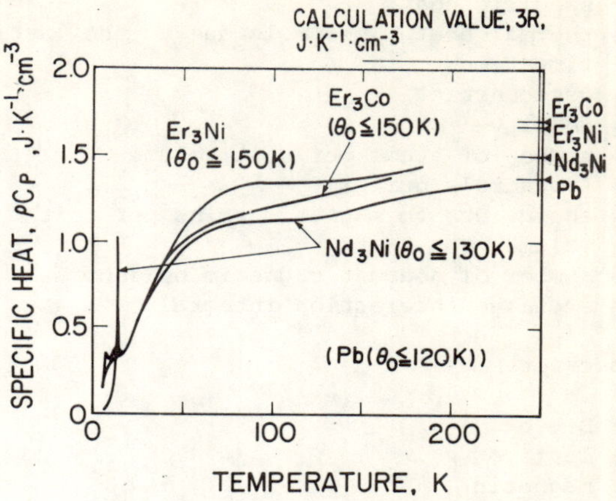

Fig. 8. Temperature dependence of the specific heat for E_3Ni, Er_3Co and Nd_3Ni in the high temperature range

Finally, for the purpose of clarifying the lattice heat capacity, the specific heat measurements were performed in the high temperature range above 40 K. Figure 8 shows the temperature dependence of the specific heat for the above-mentioned three R_3T compounds. The Debye temperatures are 150 K for Er_3Ni and Er_3Co, and 130 K for Nd_3Ni, obtained by using the relation that $Cv/3R = 0.503$ at the temperature of $\Theta_D/4$. All these values are nearly equal to that of lead.

We have used only the newly developed compound Er_3Ni, instead of lead, with great success in powdered form for the second stage regenerator matrix of a conventional two-stage GM refrigerator (12). As our test result, the no-load lowest temperature of 4 K level can be achieved even in a conventional two-stage GM type. Thereto, the price of this material, that is Er_3Ni, is less expensive enough for commercial use, in which the price is no problem.

CONCLUSION

Newly developed magnetic material R_3T system is presented, which exhibits the extremely large heat capacity between 4 K and room temperature, especially suitable for a conventional two-stage GM refrigerator.

Thereto, this material system composed of erbium and nickel is less expensive enough for commercial use.

LIST OF SYMBOLS

T = temperature
Θ = characterization temperature
S = entropy or quantum number for magnetic spin

C = specific heat
U = thermal energy per mole due to the lattice vibrations
R = gas constant
ρ = density
d = number of atoms per unit volume
N = Avogadro's number
E = energy due to magnetic spins per unit volume
Z = number of nearest magnetic neighbors
J = exchange interaction integral

Subscripts

D ; Debye
L ; lattice
M ; magnetic
A ; atom or Avogadro
m ; magnetic atom
z ; magnetization direction component
<>; expectation
c ; Curie temperature or transition
<>; expectation

REFERENCES

1. Walker, G. "Cryocoolers", Part 2, P.33, Plenum Press, New York (1984).

2. Smith, Jr. J. L., Robinson, Jr. G. Y., and Iwasa, Y. 'Survey of the state of the art of miniature cryocoolers for superconductive devices'. Superconductive Technology in Review, No.84-3, 1 (1972).

3. Yayama, H., Tomokiyo, A., Hashimoto, T., Kuzuhara, T., Li, R., Sahashi, M., and Inomata, K. 'Specific heat of regenerative material RNi_2 system'. IEEE Trans. Mag., Vol. Mag-23 (1987), 2850.

4. Buschow, K. H. J., Olijhoek, J. F., and Miedma, A. R. 'Extremely large heat capacities between 4 and 10 K'. Cryogenics, Vol. 15 (1975), 261.

5. Barbisch, B., Olsen, J. L., and Kwasnitza, K. 'High enthalpy materials for use in superconductor stabilization and in low temperature cryocooler regenerators'. Adv. Cryo. Engr. Materials, Vol. 32 (1986), 271.

6. Daniels, A. and du Pré, F. K. 'Triple-expansion stirling cycle refrigerator'. Adv. Cryog. Eng., Vol. 16 (1971), 178.

7. Nakashima, H., Ishibashi, K., and Ishizaki, Y. 'Development of 4-5 K cooling stirling cycle refrigerator. Proc. 4th Int. Cryocoolers Conf., Sept. (1986), 263

8. Moffatt, W. G., "The handbook of binary phase diagrams", Vol. 3, Genium Publishing Corporation, New York (1984).

9. Buschow, K. H. J., 'Intermetallic compounds of rare-earth and 3d transition metals'., Rep. Prog. Phys., Vol. 40 (1977), 1179

10. Gignoux, D. and Gomez-sal, J. C., 'Magnetic properties and structures of the rare earth compounds $RNi_{1-x}Cu_x$ with the FeB-type structure'., J. Magn. Magn. Mat., Vol. 1 (1976), 203

11. Gignoux, D., Lemaire, R., and Paccard, D., 'Etude des structures magnetiques des composes Er_3Co et Er_3Ni par diffraction neutronique', Sol. St. Commun., Vol. 11 (1972), 1709

12. Kuriyama, T., Hakamada, R., Nakagome, H., Tokai, Y., Sahashi,M., Horigami, O., Li, R., Yoshida, O., Matsumoto, K., and Hashimoto, T., 'Development of a 5 K GM refrigerator using rare earth compounds as a regenerator matrix'. Proc. ICCR (1989), (to be published).

THE DEMAND OF A CLOSED CYCLE REFRIGERATOR FOR A MAGNETIC RESONANCE IMAGING (MRI) MAGNET

Lin Minhe, Wang Kueiwu, Li Yuanren, Ma Yongxiang
Kejia Corporation Ltd. Academy Sinica

ABSTRACT

A superconducting magnet equipped with a closed cycle refrigerator can make superconducting system that is easy to operate and maintain. Such a system will be welcomed by hospitals and factories.
How much capacity of refrigerator that can meet magnet need, and what kind type of closed cycle refrigerator suited for Magnetic Resonance Imaging (MRI) are discussed. The work situation of three types of closed cycle refrigerator, which were made by CVI, APD, KCCH, and operate in China, are given.

1. INTRODUCTION

Superconducting magnets have been developed and applied in Physics Laboratories since the 1950's. After appearing Nuclear Magnetic Resonance(NMR), superconducting magnets have come out the Laboratory and have been applied in chemistry and biochemistry field. In the 1980's Magnetic Resonance Imaging(MRI) has appeared, and superconducting magnets have been applied and developed in the medical area. Besides the High Gradient Magnetic Separation (HGMS), and other applications are being tested in industry and will be applied soon. But the problem is that expensive liquid helium (LHe) and difficult to maintain cryogenic techniques are needed and they are not easy to apply in industry and other non-physics areas. For this reason, scientists and engineers are doing their best to look for a high critical temperature superconducting materials, and to carry out research on a closed cycle refrigerator for making superconducting system easy to operate and maintain.
Since MRI possesses a special value in medical applications, it is being developed repidly. But at present, its market is in America and Europe, where LHe is not expensive and easy to supply. Recently as MRI markets expand, the supply of cheap LHe is a big problem. So more attention is being paied on the closed cycle refrigerators and begin the subject of research by many companies.

2. THE CLOSED CYCLE REFRIGERATOR SYSTEM

The closed cycle refrigerator was applied first to use a two stage Gifford-Mcmahon (G-M) refrigerator, which can supply 77K and 20K of cryogen to cool LN and 20k shield, for reducing LHe evaporation. For example, an Oxford NMR magnet equipped with a two stage G-M refrigerator can delay the time for refilling LHe for several months. Recently two closed cycle systems have bee developed which can recondense helium fully with no need to refill LHe any more.
1) The first is a three stage closed cycle refrigerator which incorporates a two stage G-M refrigerator and a Joule-Thomsom(J-T) valve Its first and second stages can supply 77K and 20K of cryogen to the cool shield. The third stage (J-T) can supply of 4.2k cryogen, its cold head is plugged into magnetic cryostate neck and make evaporating helium into LHe, Figure 1 is a typical G-M refrigerator interface.
2) The second is klaud cycle liquefier system. From the end of the first heat exchanger, 77k of cryogen is supplied to a liquid nitrogen (LN) bath to recondense the boil-off nitrogen (N). From a J-T valve, to accept an external throttle directly supply LHe into the magnet.

3. THE OPERATING SITUATION OF THREE TYPE CLOSED CYCLE SYSTEMS

Now there are eight MRI superconducting magnets working in China. But helium resources are poor in Asia, helium is expensive in China. For this reason a magnet equipped with a clsoed cycle refrigerator is used. Six of these magnets are equipped with the LHe recondenser, and other are applied with the LN recondenser only. Their operating detail are introduced as follows:

a. CGR 511-4.5 Refrigerator (CVI)
This includes a 2 stage G-M refrigerator and a J-T valve.(see Figure 2.) Its 4.5k recondenser, whose dimentions are 41mm diameter and 50mm long, passes through a 1.5m long transfer line from the expender and then is plugged into a magnetic cryostate neck. It is necessary to avoid ferromagnetic material and vibration to effect homogeneity and image quality. The performance is as follows:

power at refrigerator,		power at recondenser,
4.5k	2.7 watts	2.0 watts
20k	5.0 watts	3.0 watts

The compressor is a CBST 10.0-12KW Rotary compressor system. It utilizes oil flooded machines, whose most prevalent non-cryogenic use is in standard Freon air condition system. So this machine is reliable.
The CGB 511-4.5 Refrigerator system is equipped with a MRI magnet which was built by Wang NMR Inc. and Kejian Corporation Ltd. It has run since February 1987. and continuously run for nine months, at Shekou Shenzhen, China. In this period, it was stopped several times for trouble by power supply or cooling water, and its most long stop time is 36 hours, but its working situation is satisfaction and it does not need refilling with LHe. There are no vibrations and noise as the machine is runing. If we cut down the transfer line and then put the expender on cryostate directly it can supply more cryogen.

b. HS4 Refrigerator (APD)
This includes a 2 stage G-M refrigerator and a J-T valve. The expender is put on to the magnet directly and the recondenser is plugged into the cryostat neck.(see Figure 3.) The performance is follows;

Temperature	cryogen power
16k	3.5 watts
4.5k	0.65 watts
input power	4.5 kw

The compresser is HC-4MK, a rotary vane compressor. This is taken from a standard production line and has a history of long reliable routine operation experience. Four compressors are needed; three units are in parallel to supply high pressure helium to the G-M refrigerator, another supplies high pressure helium to a J-T valve. Each unit of power is 3KW, and the total input power is 12 KW.
Two ADP cycle systems equipped with IGC magnets have being worked since March 1988. Up to now their working situation are satisfactory, even they stopped work about more 24 hours,for trouble by power supply or cooling water, they do not need LHe refilling. There is no vibration and noice. Another APD cycle system equipped with IGC magnet has being worked since October 1988. These cycle system are installed at the Long-march hospital, the Hua-shan hospital in Shanghai, and the Xian medical college in Xian.

c. Module 1200 helium liquefier (KOCH)
This is a klaud cycle helium liquefier. From the first heat exchanger, take 65-70k cool steam helium gas to the external LN condenser supplying 99 watts of cryogen,(equivalent to 2 liter/hr). After J-T valve, accept an external throttle directly to supply LHe into the magnet a about 1 liter/hr.The M 1200 liquefier diagram is shown in figure 4.This cycle system is equipped with the IGC magnet at the Beijing cancer hospital and has been adjusted by KOCH technicans, since July 1988. When the technican left China, this machine stopped working for trouble by compressor oil leaking out. So operating experience has not been obtained.

4.DEMAND OF THE CLOSED CYCLE REFRIGERATOR

The typical cryogen consumption. of MRI magnet is shown as follows;

	IGC		Wang-KJ	Oxford	Siemens
central field (T)	0.5	1.5	0.6	0.5-1.0	1.0
LHe boil-off rate (liter/hr)	0.3	0.5	0.5	0.5	0.4
LN boil-off rate (liter/hr)	1.2	1.2	1.5	1.5	0.7

*without the refrigerator equipped with

usually for an MRI magnet, the cryogen consumption must be kept less than 0.5 liter/hr of LHe and 1.0 liter/hr of LN. According to cryophysics data, LHe latent heat (at 4.2k) is 0.72 Whr/litre, and LN latent heat (at 77k) IS 45 Whr/litre. But when the magnet is chargging and the leads are inserted in the magnet, more LHe will be evaporated. After retracting the leads, it should take 2-3 days that rate of evaporation can be steady. So the recondenser power should be desired more than the maximum evaporation rate about 2-3 times. The correct cryogen power is difficult to calculate. It depend on matching structure of the cryostat and the refrigerator. A rough estimate can be made as follows; 0.3-0.5 liter/hr rate of LHe evaporation needs a recondenser power about 0.6-1.8 watts (at 4.2k). As for the LN recondenser, 100 watts (at 77k) is needed for 1 liter/hr rate of LN evaporation. Other demand of the closed cycle system are light vibration, non-ferromagnetic material, a small volume and a light weight.

5. DISCUSSION

a. In the G-M closed cycle system, the helium stream is individual, not mixed with the helium vapor as in the magnet cryostat, so as to eliminate potential comtamination.
b. Perhaps the most advanced concept at this time* is the G-M cycle system. A large capacity system of this type be installed on the magnet

directly which can provide 4.2k, 20k, 77k of cryogen to inside and cooling shields, and enable the equipment to run without cryogen service.

c. Operating experience at several sites indicates no image degration from magnet, acoustic or R.F. interference from refrigerators.

d. For a 0.5-1.0 T magnet, the G-M cycle system is convenient, economic and suitable. A Claud cycle liquefier system, need a high power consumption is not economical. It will be economical if it operate in a multi-magnet hospital where one refrigerator could serve several magnets, and serve more than a 1.5 T magnet that will consume cryogen.

REFERENCE

1. R.E.Schwall. "MRI-Superconductivity in the Marketplace" IEEE Trans. Magn. MAG-23.No.2 March (1987), pp 1287-1297.
2. Garry R. Morrow "Superconducting Magnets For MRI Applications" IEEE Trans. Magn. Mag-23. No.2 March(1987) pp1294-1298.

EXPERIMENTAL INVESTIGATION ON THE PERFORMANCE
OF PRESSURE WAVE GENERATOR FOR SPLIT STIRING CRYOCOOLER

Zheng Dazhen
Laboratory of Cryogenic Engineering
Huazhong University of Science and Technology
Wuhan, Hubei, China

ABSTRACT

For the split stirling cryocoolers, pressure wave generator plays an important role in determining the performance and cold capacity of the complete cryocooler set. Nevertheless, the published experimental results can scarcely be found.

This paper presents a comparison between the thermodynamic aspect of a theoretical process and that of an actunl process of pressure wave generator. The pressure wave of actual process has been determined and discussed.

Experimental investigation has been carried out, including the effects of a series factors such as revolution speed, temperature rise, dead volume and dimensions of connecting pipeline, ets. They all influence the wave form significantly. From the experimental results the main approaches for obtaining suitable pressure wave form for split stirling cryocooler have been deduced.

INTRODUCTION

The split stirling cryocooler consists of a pressure wave generator and a split cold finger. The pressure wave generator produces pressure waves in working spaces, and causes refrigeration effects in the cold space with the displacer/regenerator shuttling back and forth in the split cold finger. The pressure wave generator plays an important role in determining the performance and cold capacity of the complete crcryocooler set. The reliability of it always determines that of cryocooler. Nevertheless the relevent published experimental results and analysis can scarcely be found.

The pressure wave generator is conventionally called a compressor. However it is distinct from the conventional compressor, as its action is not only to raise the gas pressure and there are not entrance and exit vaves and so isobaric processes of adimition and exhaust can not be maintained. Its piston is driven by crank case shaft or by linear electromagnetic motor. So that the volume of cylinder is changed periodiclly, and the gas in working space of the cold finger connected with the cylinder is compressed and expanded alternatively to generate some type of pressure wave. Therefore, it would be more reasonable to name it as the gas pressure wave generator.

Generally, for the split stirling cryocooler great difference exisits between the predicted and the experimented performances. One of the reasons is that the pressure wave and displacer motion have a direct bearing on the performance and technical problems, e.g friction and leakage, of the generator. Hence it is necessary to determine the actual working parameters and find out their pattern of propagation in the generator, otherwise we'll be unable to find out the correct design procedure.

REVIEW ON ACTUAL PROCESSES

Fig.1 shows a schematic diagram of the experimental set-up for the pressure wave generator. During compression, the piston moves from the bottom dead point (BDP) towards the top dead point (TDP) and the pressure is increased as the

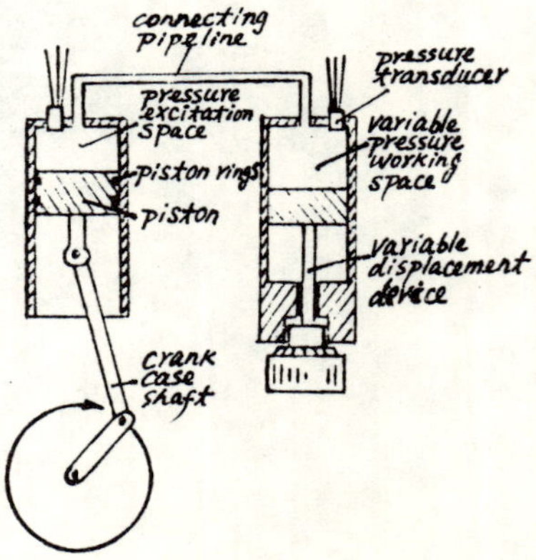

Fig.1 Schematic diagram of the experimental set-up.

total volume is decreased. During expansion, the

pistoń moves to the BDP and the pressure is ćecreased as the total volume is increased.

Under following assumption (1) the piston moves sinusoidally; (2) the working fluid obeys the perfect gas equation, $PV = RT$; (3) the instantneous pressure keeps the same value throughout the system; (4) the piston rings are in perfect sealing; (5) the position of piston is at the BDP during charging, and under the charging pressure P_{ch}; than the pressure wave equation for the working space will be (see fig.2)

$$P = \frac{P_{ch}(V_d + V_0)^n}{[V_d + \frac{1}{2}V_0(1+\cos\omega t)]^n} = \frac{B}{(A+\tau\cos\omega t)^n}.$$

I which

$$B = P_{ch}(\frac{V_d + V_0}{V_d})^n = P_{ch}(1+2\tau)^n; \qquad \tau = \frac{V_0}{2V_d};$$

$$A = 1+\tau.$$

provided the values of n during the compression process and expansion process are the same, these two processes are theoretically reversiable.

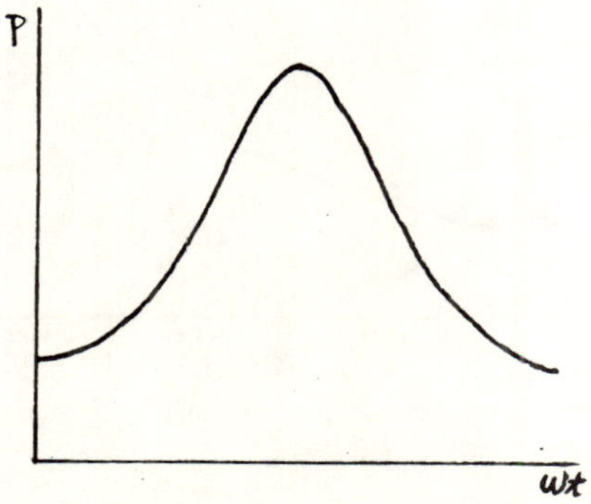

Fig.2 p-wt curve of the theoretical process.

Disregarding a series of factors such as friction, viscocity and structure damping, ets, and whether the process is isothermal or adiabatic, the work expanded in the former half period shall be equal to the work released in the later half period. So that once the first urge has been put, the piston can move continually anu does not expend any work.

The actual process of pressure wave generator is more compliated than the theoritical. The complication may be shown as below:

1) The actual compression and expansion processes are diverse from each other and the values of n at different points on each curve are different (Fig,3). This is caused by the temperature difference between the fluid and the wall of cylinder.

When the frequcy is high, and the thermal capacity of the cylinder, in which the temperature is at average value over the compression and

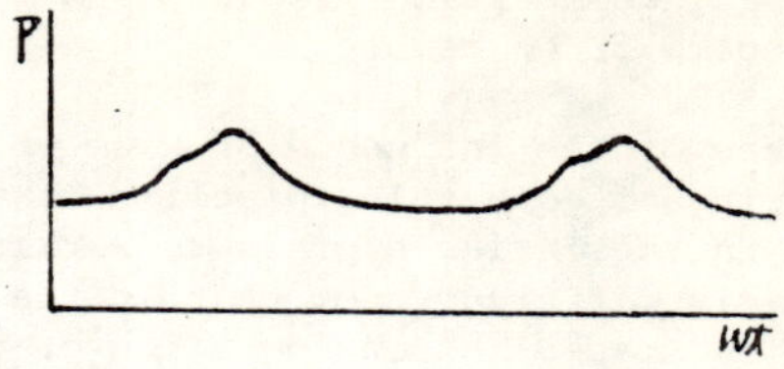

Fig.3 p-wt plot of practical process.

expansion process under steady state, is large, there may be heat continuously transferred from the wall of cylinder to the fluid at the initial stage of compression process. So that the endothermic compression is resulted $(n > k)$. Later on, the temperature of fluid is raised and the compression becomes exothermic $(n < k)$. At the initial stage of expansion process, the expansion is exothermic, as the fluid temperature is higher than that of the wall and there is heat continuously transferred from the fluid to the wall. Later on, along with reduction of temperature, there is heat absorbed from the wall and endothermic expansion takes place.

2) In addition to the factor of volume, the friction, charging pressure, revolution speed and resistance of connecting pipeline are the primary factors affecting the work consumption.

3) The effects of friction may be shown as below:

(1) Friction causes frictional heat.

The frictional work consumed between the piston rings and cylinder wall in one turn may be shown as:

$$W_m = 2\mu\pi DhS \sum_{i=1}^{Z} P_i$$

which is released in the form of heat. The most part of frictional heat is carried away by the cooling medium, and the others take part in the working processes to affect the value of n and the fluid temperature. It is quite evident that the frictional work consumtion relates to the factors such as P_{ch}, N and ε_1. In the practical cycles there are other parts, in which frictional work is consumed, such as crank pin, cross head, neck of shaft and bearing, etc.

(2) Friction can bring about departures from perfect motion of piston.

(3) Friction can cause leaking and thus influences the stability of pressure waveform.

(4) Friction can cause contamination of working fluid and thus the regenerator may be blocked.

4) The sealing performance of piston rings has effects on working processes. The piston rings can't seal up absolutely, and certainly the fluid leaks from the higher pressure side to the lower. On this condition, the mass of fluid during expansion process is slightly larger than

that during compression process, and the processes became irreversible.

5) Because of the influence of wave propagation and damping of connecting pipeline, the pressure wave in the variable pressure working space is entirely different from that in the pressure excitation space. In consideration of the propagation speed of the pressure wave and the fluid blocking which may occur when the fluid passes through the connecting pipeline, the amplitudes, pressure ratio and phase relation of the pressure waves in the variable pressure working space are changed. As the speed of revolution, resistance of pipeline and the ratio of volumes are raised, the changes reach to a greater degree.

EXPERIMENTAL RESULTS AND ANALYSES

In Fig.1 the driving motor is suplied by d.c, and its revolution speed may be controlled by a voltage requlator. The piston has a diameter of 20mm and two values of strokes of 14mm and 18mm, with two sealing rings. The volume of the variable pressure working space is adjustable. There are two groups of linkage pipeline: copper pipes, d_i = 1.5mm, L = 200, 400, 600 and 800mm; stainless steel pipes, d_i = 0.6mm, L = 200 and 400 400mm. The major instruments for measurement are strain gage type pressure transducer dynamic strainometer and light oscilloscope. The temperaticre is measured with a copper-constantan thermocouple and a intelligent digital multimeter of Model 1905_a.

1) Determination of the sealing performance of piston rings.

The effects of static and dynamic sealings for two groups of rings which have different degrees of wearness were measured and compared as shown in Fig.4 and Fig.5, respectively. From Fig.4 and Fig.5, it is obvious that the pressure wave is stable and smooth for the rings having good sealing performance, while the wave is distorted for the rings having bad sealing performance.

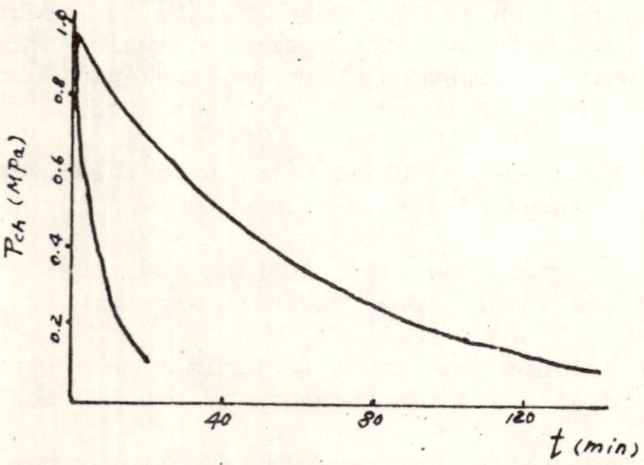

Fig.4 Static sealing effects of two groups of piston rings:
A: with good sealing performance
B: with bad sealing performance

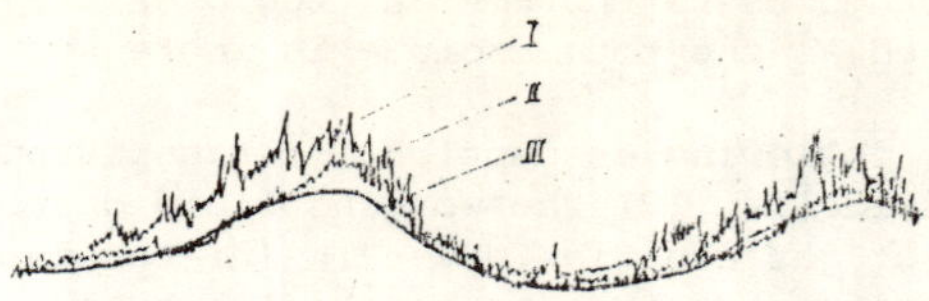

Fig.5 Comparison of pressure wave forms for two groups of piston rings:
I: group B at starting. II: group B in static state. III: group A in static state.

It is noticeable that at the start of the process or with the high values of P_{ch} and N, the wave is severly distorted.

2) Temperature rise of pressure excitation space.

Fig.6 shows the measured temperature of pressure excitation space vs N, under naural cooling condition and P_{ch} = 0.5MPa (gauge pressure). Increasing the revolution speed, the temperature is raised rapidly. When N = 1680rpm, the temperature may rise to 117°C for N_2. If the effect

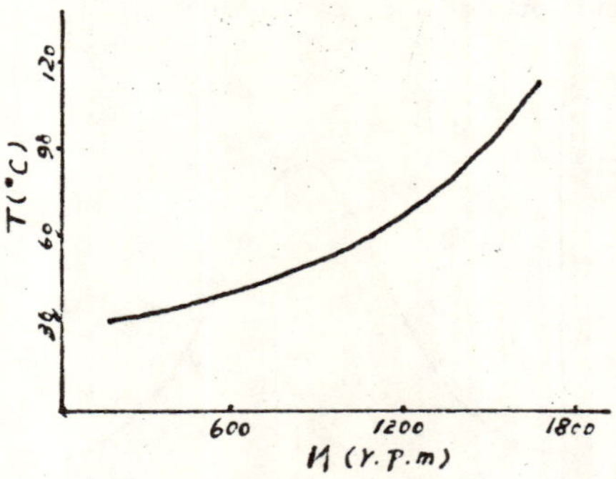

Fig.6 The temperature rise curve of pressure excitation space.

of sealing is relatively poor, the temperature rise is comparatively low. In view of this, the major factor for designing cooling device shoud be the sealing effact of piston rings.

3) Influence of revolution speed.

(1) The revolution speed affects the temperature rise. This can be seen clearly from Fig 6.

(2) The revolution speed affects the pressure waveform.

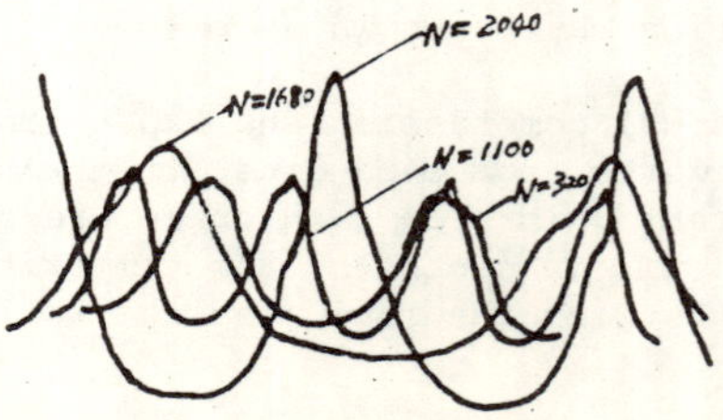

Fig.7 The pressure waveforms in pressure excitation space.

Fig 7 shows the measured waveform of pressure excitation space with a stroke of 18mm. For different revolution speeds, the waveform are clclearly distinct. As compared to the theoritical curve, the slope of the curve during compression process is greater than that expansion process, because the work expended during compression is more than the work released during expansion. Increasing the revolution speed, there is obviously a step change in the compression process. This may be caused by the imperfectness of crank case shaft.

(3) The revolution speed affects the pressure ratio.

From the point of view of thermodynamics the revolution speed has nothing to do with the pressure ratio. But experiments have indicated that they are dependent on each other closely. Fig 8 shows the change of amplitude of pressure wave, which was recorded with quick dropping of the revolution speed from N = 2000rpm to N = 300 rpm. Along with the dropping of N, ε_1 is reduced obviously. It can be shown elsewhere through dynamic anlysis, that the reactance of connecting pipline plays a deteminant role for ε_1, in addition to the effect of the frictional heat.

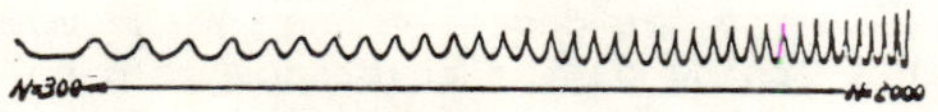

Fig.8 Variation of amplitude of pressure wave vs N.

4) Pressure waveforms in pressure excitation and variable pressure space. Under the conditions of P_{ch} = 0.5MPa, S = 14mm, V_{od} = 164mm³, N = 1600rpm, the pressure waveforms in the pressure excitation space and the variable pressure space were recorded at the same time as shows in Fig 9, from which it is obvious that the pressure ratios, waveforms and phase relations in these two spaces are different. This is caused by the substantial distinction between the thermodynamic processes in these two spaces. The

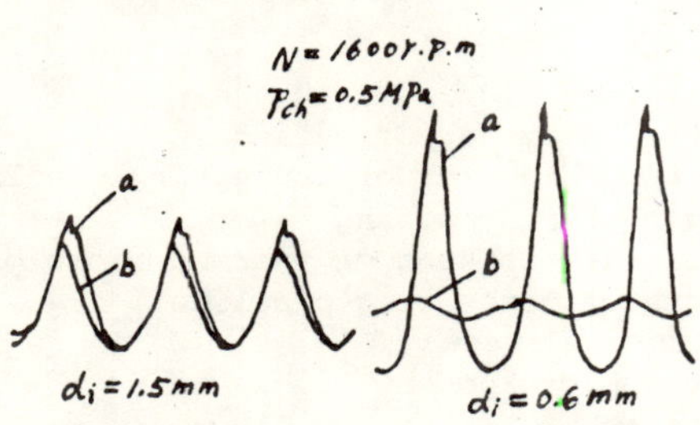

Fig.9 Comparison of Pressure waveforms in two spaces.
a- In pressure excitation space
b- In Variable pressure working space.

processes of gas in pressure excitation space may be approximately taken as compression and expansion of variable fluid mass, while the processes in the variable pressure space are charging and releasing under variable pressures. The fluid dynamic relationship between these two spaces is controlled by the propagation law of pressure wave and mass flow wave in connecting pipeline. Only by matching on the characteristics of the pressure wave in the variable pressure space can we design a properly acting split stirling cryocooler.

5) The influence of the reactance of connecting pipeline.

The fundamental factor deciding the waveform in the two spaces is the reactance of pipeline. The waves in these two spaces have been measured for a series of pipelines having different values of di and L as shown in Fig.10, The measured results demonstrated that:

(1) For the connecting pipeline with same inside diameter and finite lengths (200 - 800mm) and working under the same frequency, the influences of the pipeline length on the waveform, phase shift and pressure ratio in the two spaces are quite small. As the frequency is changed, the pressure ratios in two spaces are changed as well, but the change of pressure ratio in pressure excitation space is much larger than that in variable pressure space, ε_1 is increased while ε_2 decreased with the increasing of ω. However, the influence of ω on phase shift between the pressure waves in the two spaces is relatively small.

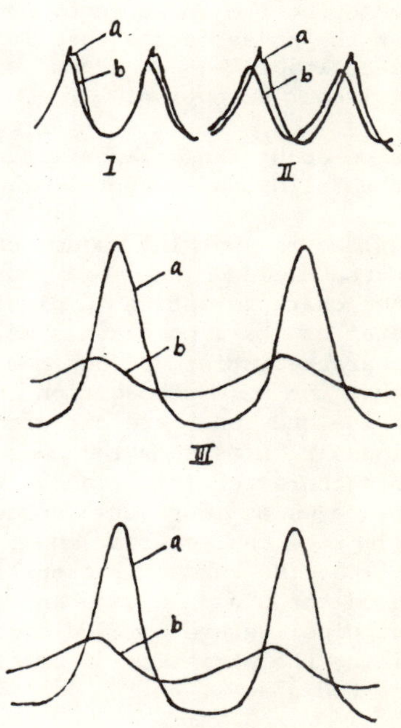

Fig.10 Pressure waveforms(1)
a-In pressure excitation space. b-In variable pressure working space.
I-d_i = 1.5mm L = 200mm N = 1600rpm. II-d_i = 1.5mm L = 600mm N = 1600rpm. III-d_i = 0.6mm L = 200mm N = 320rpm. IV-d_i = 0.6mm L = 400mm N = 320rpm.

(2) The influence of the inside diameter of the pipeline on the pressure ratio and phase shift of the pressure waves in the two spaces is significant. The smaller the diameter of pipeline is, the greater the value of ε_1 and the smaller the value of ε_2, as well as the clearer the phase shift will be.

(3) The influence of local resistance on the pressure wave may be shown as Fig 11 which shows the waves in two spaces connected with the pipeline ($d_i = 1.5$mm $L = 400$mm) having a certain

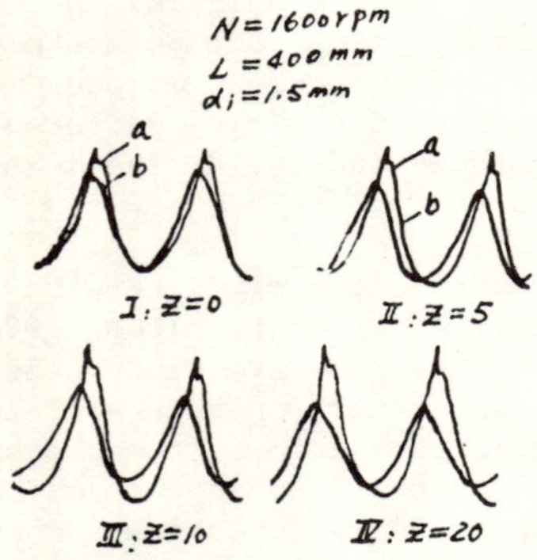

Fig.11 Pressure waveforms (2)
z- Numbers of throttle holes

number of throttling holes. From Fig 11 we can find that a finite increace of local resistance has a smell influence on waves in the two spaces.

(4) After checting the above mentioned figures we find that the phase of the wave in variable pressure space leads that in pressure excitation space without exception.

The discussion about these experimental results led us to the following concepts:

For cyclic flow, the working volume characteristics should be divided into void characteristics and tube characteristics. A pipeline of large diameter can be a compatible volume of these two characteristics. It has been shown elsewhere that the wave propagation characteristics in voids and tubes are different. The above-mentioned figures show the measured results of void characteristics in the two spaces. As the pressure waves in the two spaces are tied together by tube characteristics of connecting pipeline, So the tube characteristics of propagation of the pressure wave and mass flow in different pipeline are the decisive factor for determining the variation of pressure waveforms in the two spaces.

6) Influence of dead volume. Experiments have proved that when di and s are constant, the amount of dead volume can affect the stability of the pressure waveform, in addition to the effect on values of ε_1 and ε_2. The comparison of pressure waveforms varied with the change of dead volume in exciting space under the same revolution speed is show in Fig 12. When the dead volume is relatively large, the wave-

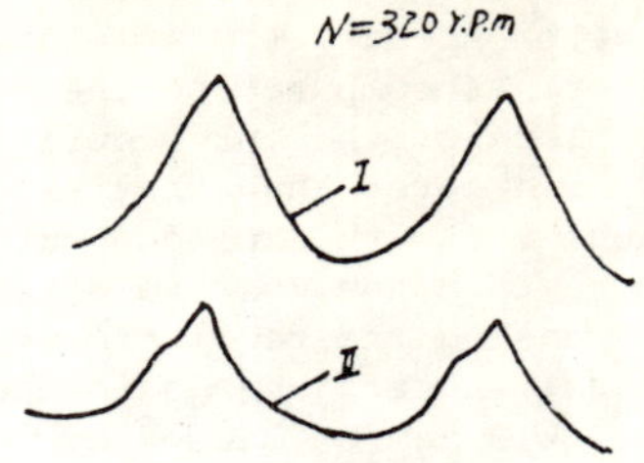

Fig.12 Pressure waveforms (3)
I- $V_{od} = 164mm^3$ II-$V_{od} = 7.85mm^3$

form is not apparently distorted.

CONCLUTION

The pressure wave generator is one of the critical elements for split stirling cryocooler. Its working mechanism is distinct from the conventional compressors and its performance is influenced by many factors. The investigation has indicated that except for the characteristic dimensions, the main factors affecting performance are the sealing of piston rings, revolution speed and reactance of connecting pipeline. Experimantal results have proved that the inside diameter of pipeline is the decisive one of the main factors influencing on the reactance of pipeline. There are some other less important factors having influence on the pressure waveform, (pressure ratio and phase shift), such as the pipelength, local resistance and dead volume, etc.

These factors constitute an important aspect of detailed investigation in improving the performances of split strling cryocooler.

ACKNOWLEDGEMENT

I am very grateful to prof F.Z.Guo for his suppport and suggestion, and also to Mr Z.Q.Zheng and Mr D.G.Zhang for their works during the preparatory stage of experiments.

NOMECLATURE

D = Inside diameter of cylinder
d_i = Inside diameter of connecting pipeline
h = Height of piston ring
k = Adiabatic exponent of thermodynamic process
L = Length of connecting pipeline
N = revolution speed
n = Polytropic exponent
p = Pressure
P_{ch} = Charging pressure
S = Stroke
T_m = Mean temperature of fluid in working volumes.
t = Time
V_a = Variable pressure working volume

$V_d = V_a + V_{od} + \frac{\pi d_i^2}{4} L$ — Total dead volume

V_0 = Swept volume of piston
V_{od} = Clearance volume

ω = Angular frequency
Z = Numbers of rings
z = Numbers of throttle hole
μ = Friction coefficient

$\tau = \frac{V_0}{2V_d}$

ε_1 = Pressure ratio in pressure excitation space
ε_2 = Pressure ratio in variable pressure working space

REFERENCES

1 M.Lin and S.Y.Sun.'The priciple of piston type compressor'. Xian Jiaotong University.

2 Graham Walker.'Cryocoolers'. Plenum Press (1983).

3 W.Mao.'The investigation on regenerator-cryocooler System'. Thesis of Master Degree HUST Wuhan China (1986).

4. CRYOGENIC INSULATION AND DEWAR

BLOWDOWN ANALYSIS OF A SOLID CO_2 TANK IN ORBIT

CHANG-KENG LIU
Lockheed Research Laboratory
Palo Alto, California, USA

ABSTRACT

The cryogenic limb array etalon spectrometer (CLAES) to be launched on board the NASA Upper Atomophere Research Satellite (UARS) in late 1991 must be cooled to temperatures lower than those of the gases under observation. The CLAES cryogenic system consists of two components: a solid neon (SNe) tank at 14.1 K to cool the detectors to minimize detector noise, and a solid carbon dioxide (SCO_2) tank at 128 K to reduce the background emissions from the optical system. These stored cryogen coolers represent an attractively simple alternative to other active refrigeration systems.

Previous works have been conducted on the safety analysis of the cryogen dewars before launch in the catastrophic case of air leaking into the vacuum insulation space. As the sudden heat load causes the pressure to rise in the cryogen dewar, a burst disc in the emergency vent lines will rupture to allow the escape of the vaporized cryogenic fluid. When the satellite is launched to a 600 km orbit, one major concern was that a higher than nominal vent rates for a time after the pyro vent valve is opened. The resulting vent thrusts and torques may have an adverse impact on the spacecraft altitude control system (ACS). An analysis on orbital venting from the SCO_2 tank is presented here.

INTRODUCTION

The cryogenic limb array etalon spectrometer (CLAES) to be launched on board the NASA Upper Atmosphere Research Satellite (UARS) in late 1991 must be cooled to temperatures lower than those if the gases under observation. The CLAES cryogenic system consists of two components: a solid neon (SNe) tank at 14.1 K to cool the detectors to minimize detector noise, and a solid carbon dioxide (SCO_2) tank at 128 K to reduce the background emissions from the optical system (Fig. 1). These stored cryogen coolers represent an attractively simple alternative to other active refrigeration svstems.

Previous works have been done on the safety analysis of these cryogen dewars before launch, in the case of air leaking into the vacuum insulation space (Ref. 1, 2). As the sudden heat load causes the pressure to rise in the dewars, a burst disc in the emergency vent lines (Fig. 2) will rupture to allow the escape of vaporized cryogen. These previous works investigated the catastrophic heat loads, sizing of the emergency vent lines, and the temporal history of the tank pressure after the burst disc rupture.

When the satellite is launched to a 600 km orbit, one major concern was that a higher than nominal case tank temperature may cause higher than nominal vent rates for a time after pyro valve is opened. The resulting vent thrust and torque at the exit of the vent lines may have an adverse impact on the spacecraft altitude control system (ACS).

For present analysis, it was assumed that the vent pyro valve opened at SCO_2 temperatures reaching 150 and 194.4 K (with CO_2 saturation pressures of 6.34 and 760 torrs respectively). The CO_2 mass flow rates and the exit velocities in the orbital vent lines for both cases are determined after the pyro valve is opened and pumping on the cryogen is initiated. The CO_2 vent exit momentum is given by the product of the mass flow rate and the exit velocity.

CLAES SCO_2 TANK AND ORBITAL VENT LINES

The SCO_2 guard tank has a surface area of 31,500 sq cm and net volume of 152 litres. It is to be filled with 85% of SCO_2 during the last servicing, and subcooled with liquid helium (LHe) to about 25 K before launch. The vent lines are closed during launch and deployment. Normally, the pyro valve opens before the SCO_2 warms up to 128 K.

For present analysis, the CO_2 orbital vent line system can be divided into sixteen sections as follows.

1) .83"x2.19" well end fitting inside CO_2 tank.
2) 1.0"x1.44" Teflon bellows.
3) .92"x1.48" aluminum sleeve
4) 1.0"x1.44" Teflon bellows
5) .92"x2.34" clamshell/bulknead fittings
6) 1.0"x1.50" bulkhead fittings
7) 1.0"x2.06" CO_2 orbit vent flange
8) 1.58"x2" CO_2 vent manifold
9) .92"x2.54" vent line
10) .92"x.92" EL
11) .92"x2.04" line to pyro valve
12) 1.0"x2.36" line from pyro valve
13) 1.25"x5" vent line
14) 1.25"x6.45" vent line
15) 1.25"x4.75" metal bellows
16) 1.19"x30" vent line on spacecraft

The temperature distribution along the CO_2 orbital vent lines is assumed as follows. Section 1 is at the temperature of the cryogen, varying from a high of 194.43 & 150 K to a low of 120 K. Sections 5 through 16 have the constant wall temperature at 207 K, same as the vacuum shell. The temperature gradients between 207 K and the cryogen temperature are equally and linearly distributed in the bellows Sections 2 and 4. The temperature of Section 3 is assumed to be constant at the mean of 207 K and the temperature of the cryogen.

ANALYSIS AND COMPUTATIONS

The end of the CO_2 orbital vent lines discharges into the space vacuum of 2.6×10^{-9} torr at orbit of 600 km. Hence it can be assumed that the vent CO_2 gas (GCO_2) will be choked at the end of the vent lines. These critical flow rates in the orbital vent lines are computed by the program CO281.S as a function of the SCO_2 temperatures in the tank. The procedure is similar to that in Ref. 2, except the new subroutines PROPCO & PGCO which were generated for CO_2 properties being used in CO281.S. The thermodynamic and transport properties of CO_2 used are obtained from open literatures (see APPENDIX). The inlet section of the vent line protrudes about 1.4" into the SCO_2 tank. Hence, the entry loss is assumed to be one velocity head. The stagnation temperature in the SCO_2 tank is assumed to range from a high of 194.4 K down to the equilibrium temperature of the SCO_2 tank. This later is defined as when the SCO_2 sublimation rate equals to the CO_2 vapor vent rate while the tank temperature remains constant for tank heat load of 2.13 W. The critical flow rates so determined are given in Figs. 3a,b, and were used in the program CO285.S to determine the temperature and pressure history in the SCO_2 tank. The history of the CO_2 vent exit momentum were then obtained from the tank temperature history.

As shown in Fig. 3, the critical vent rates of CO_2 range from a high of 95.09 g/s at tank temperature of 194.4 K (760 torrs), to .7106 g/s at 150 K (6.34 torrs) and .006 g/s at 128 K (.173 torr). At equilibrium temperature of 126.6 K, the critical CO_2 vent rate is found to be .0035 g/s.

The steady state equilibrium temperature of the SCO_2 tank can be determined as follows. If the CO_2 tank is taken as the control volume, from the first law of thermodynamics,

$$\dot{Q}_t = (C_s \cdot M_s + C_v \cdot M_v) \cdot dT/dt + H_v \cdot \dot{m}_v \qquad (1)$$

where $\dot{Q}_t$ is the tank heat load, C and M are the specific heats and masses of SCO_2 and GCO_2, dT/dt is the rate of temperature change, H_v the enthalpy of the GCO_2 vented at the critical flow rate $\dot{m}_v$. At steady state, the first two terms on the right hand side of Eq. 1 disappeared. When the sublimation rate $\dot{m}_s$ equals to the vent rate m_v, Eq. 1 becomes

$$\dot{Q}_t = H_v \cdot \dot{m}_v = Q_s \cdot \dot{m}_v \qquad (2)$$

In addition to the critical vent rates (Fig. 3), the enthalpy H_v, which can be approximated by the heat of sublimation Q_s below the triple point, is also a function of temperatures (Fig. 4). At any given temperature, the values of Q_s and $\dot{m}_v$ were used to calculate a value of $\dot{Q}_t$. The temperature at which the value of $\dot{Q}_t$ becomes 2.13 W has been found to be 126.6 K (Fig. 5).

The CO_2 temperature in the guard tank and the CO_2 vent rate are plotted in Figs. 6 as a function of time after the pyro valve is opened. It is seen that the SCO_2 temperature decreases to 128 K in about 61.6 and 67.9 hours for pyro valve opened at 194.4 and 150 K. The difference is due to the much higher initial vent rate (95.1 g/s versus .711 g/s) and a smaller SCO_2 mass if the pyro valve opened at 194.4 K. For the tank heat load at 2.13 W, the sublimation rate equals to the critical vent rate when the cryogen temperature reaches 126.6 K. In theory, this point is reached when the time approaches infinity. The SCO_2 tank temperature will then remain constant.

The exit momentum of the vent CO_2 are shown in Figs. 7a,b for the cases of pyro valve opening at 195.4 K and at 150 K. In the first case, the initial momentum is 21.014 N (4.724 lbf) when the pyro valve opened at 194.4 K. It then decreases to 1.9 N (0.427 lbf) at .1 hour, .147 N (.003 lbf) at 1 hour and .00127 N (.0002855 lbf) when the SCO_2 tank temperature approaches to 128 K at 61.6 hour.

Fig. 7b shows that when the pyro valve opened at 150 K, the initial vent CO_2 exit momentum is .147 N (.033 lbf). It decreases to .072 N (.0162 lbf) when the SCO_2 tank temperature approaches to 128 K at 67.9 hour. A summary of the results of computation is given in Table 1.

CONCLUSIONS

It is seen that all precautions should be taken to make sure that the pyro valve be function properly, and open at the right moments. Otherwise, the temperature of the SCO_2 tank may be too high for the guard, and the vent exit momentum could have an adverse impact on the spacecraft ACS.

ACKNOWLEDGEMENT

This work was sponsored by the NASA Goddard Space Flight Center under contract NAS5-27752.

REFERENCES

(1) Liu, C. K., Naes, L. G. and Manikowski, Jr., A. F., 'Catastrophic Heat Load on a Solid-Hydrogen Dewar'. J. of Spacecraft Rockets, Vol. 23, No. 4 (1986), pp 428-435.

(2) Liu, C. K., Naes, L. G. and Manikowski, Jr, A. F., 'Sizing an Emergency Venting System for a Cryogenic Dewar'. Advances in Cryogenic Engineering, Vol. 31(1986), pp 973-981.

(3) Reynolds, W. C., 'Thermodynamic Properties in SI'. Department of Mechanical Engineering, Stanford University, 1979.

(4) Newitt, D. M., Pai, M. U., N. R. Kuloor, N. R. and Huggill, J. A. W., "Carbon Dioxide", in 'Thermodynamic Functions of Gases'. Vol. 1, Din F. (ed), Butterworths Scientific Publications, 1956.

(5) Quinn, E. L. and Jones, C. L.,'Carbon Dioxide'. American Chemical Society Monograph Series, Reinhold Publishing Corporation, 1936.

(6) Egan, C. J., 'The Heat Capacity, Vapor Pressure, Heat of Sublimation, and Entropy of Solid Carbon Dioxide'. Ph.D. Dissertation, University of California, 1936. Also Giauque, W. F. and Egan, C. J., Jour. Chem. Physics, 5, 45, 1937.

(7) Ibel, W. "Thermophysical Properties", Section 2, in 'Handbook of Heat Transfer', Rohsenow, W. M. and Hartnett, J. P. (eds), McGraw-Hill Book Company, 1973.

TABLE 1 Blowdown Analysis of SCO_2 Guard Tank

Pyro Valve Opens at CO_2 Tank Temperature of	194.4 K	150 K
At t = 0		
Tank pressure, torr	760	6.34
Mass of SCO_2, kg	202.32	209.15
Mass of GCO_2, g	62.9	0.68
Critical vent rate, g/s	95.09	0.71
Exit velocity, m/s	221	207
Exit momentum, newton	21.014	.147
(lbf)	(4.724)	(.003)
SCO_2 at 128 K (.1734 torr)		
t, hour	61.6	67.9
Mass of SCO_2, kg	178.24	200.44
Critical vent rate, g/s	.006	
Exit velocity, m/s	212.2	
Exit momentum, newton	.00127	
(lbf)	(.0002855)	
SCO_2 at 126.7 K (.1348 torr)		
t, hour	202	227
Mass of SCO_2, kg	176.07	198
Critical vent rate	.00367	
Exit velocity,m/s	218	
Exit momentum, newton	.0008	
(lbf)	(.00018)	

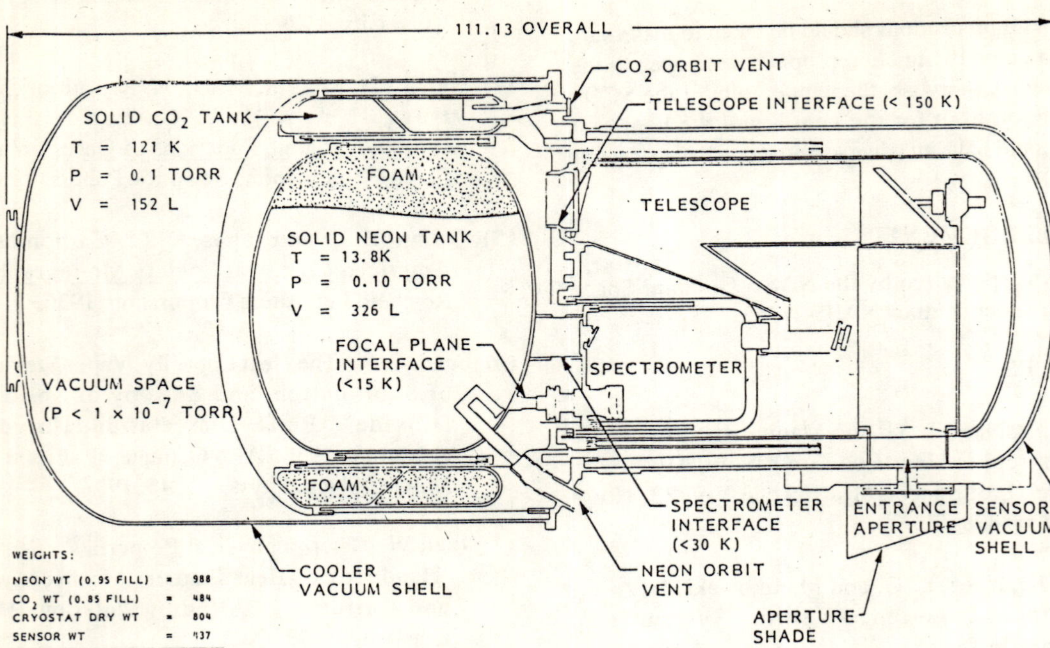

Fig. 1 CLAES Cooler/Sensor Schematic

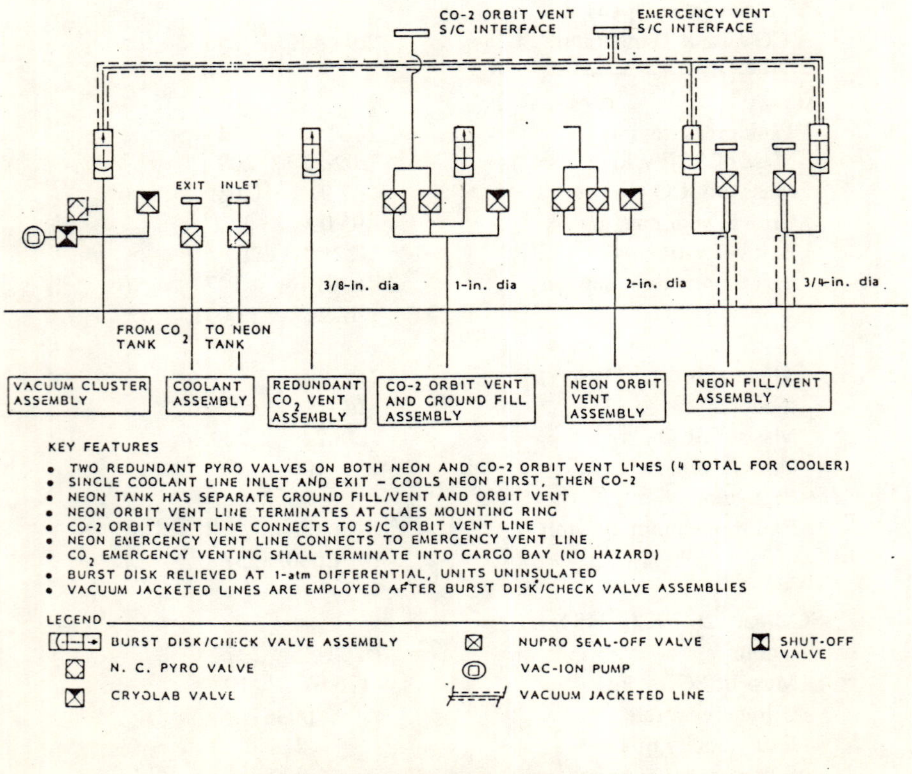

Fig. 2 External Plumbing Schematic

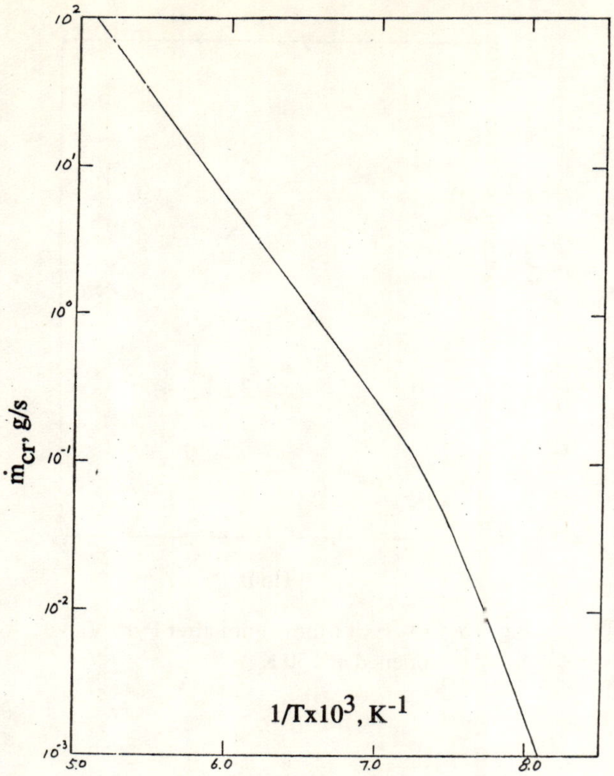

Fig. 3 CO_2 Critical Flow Rate

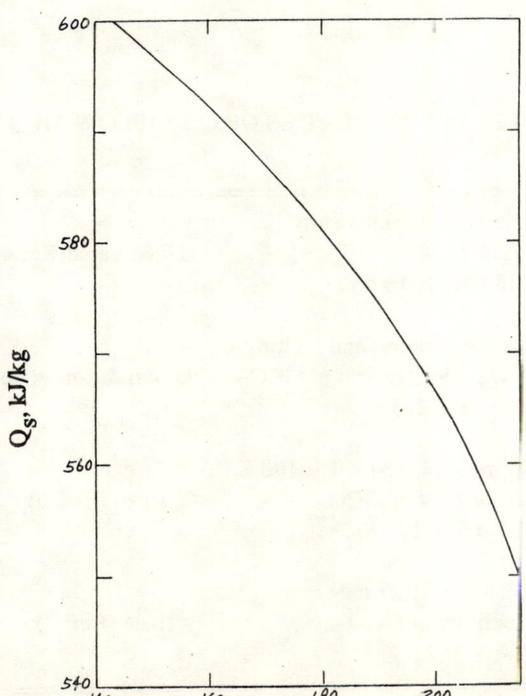

Fig. 4 SCO_2 Heat of Sublimation

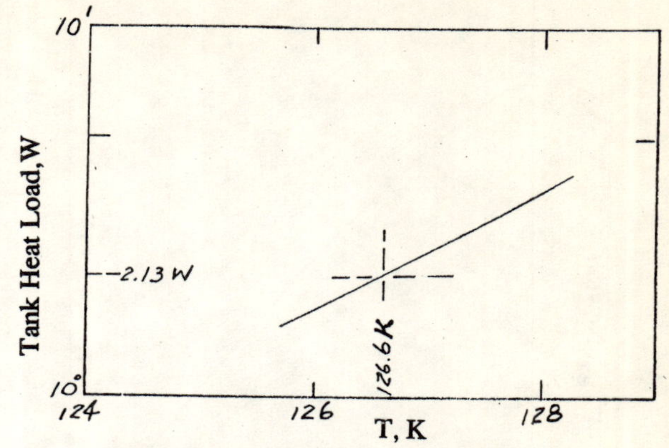

Fig. 5 Equilibrium Temperature of SCO_2 after Pyro Valve opened

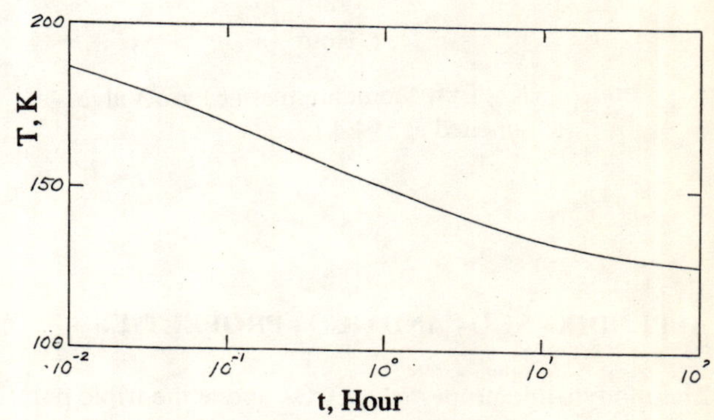

Fig. 6a CLAES Guard Tank CO_2 Temperatures after Pyro Valve opened at 194.4K

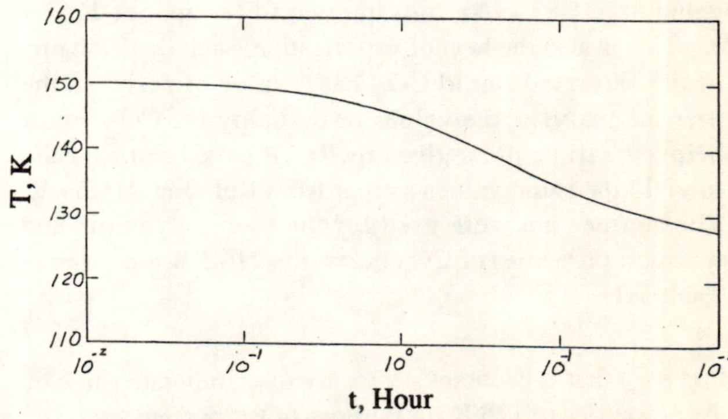

Fig. 6b CLAES Guard Tank CO_2 Tempertures after Pyro Valve oened at 150 K

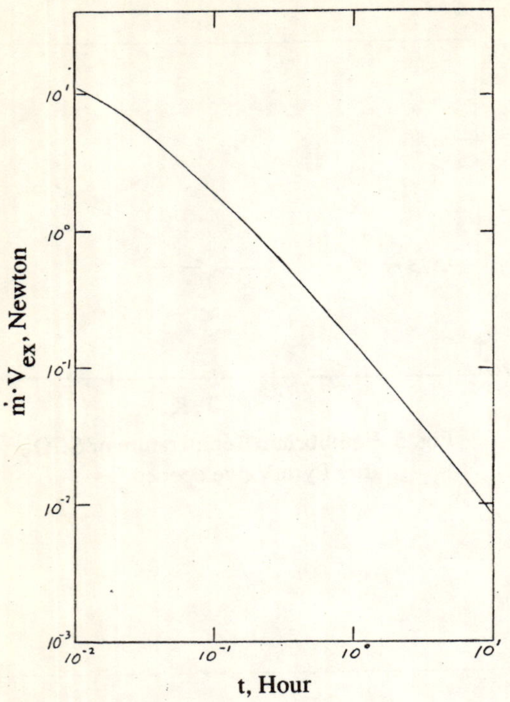

Fig. 7a CO_2 Exit Momentum after Pyro Valve opened at 194.4 K

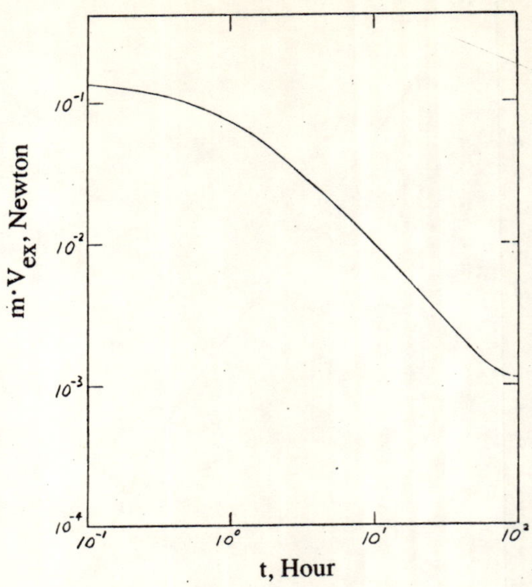

Fig. 7b CO_2 Exit Momentum after Pyro Valve opened at 150 K

APPENDIX - SCO_2 AND GCO_2 PROPERTIES

Thermodynamic properties of CO_2 above the triple point of 216.54 K are obtained from Ref. 3. Thermodynamic and transport properties of solid and vapor CO_2 below the triple point are also obtained from the open literatures. The enthalpy function is a composite extensive property defined as the sum of the internal energy and the product of pressure and the specific volume. It is a point function measured above any convenient datum. In Ref. 3, it has a value of 351.87 kJ/kg for saturated CO_2 vapor at 216.54 K. This is also the heat of vaporization, such that enthalpy of the saturated liquid CO_2 has a value of zero. In the present analysis, the values of enthalpy for CO_2 vapor below the triple point given by Ref. 4 have been adjusted to yield the same values as that from Ref. 3 at 216.54 K. The sources hat were used for the thermodynamic and transport properties of CO_2 below the 216.5 K are given in Table A1.

It is seen that it becomes necessary to extrapolate some of the properties to 125 K for purpose of the present analysis. This is true in the cases of most of the thermodynamic and transport properties.

TABLE A1 PROPERTIES OF CO_2 BELOW 216.5 K

Saturated SCO_2 and Vapor [B T, P, S, H, V $143.11 < T < 216.5$ K	Newitt et al(Ref.4)
Heat of sublimation and melting of SCO_2 specific heat of GCO_2 $173.1 < T < 216.5$	Quinn & Jones(Ref. 5)
Vapor pressure, $154 < T < 196$ K Specific heat of SCO_2 $15.5 < T < 190$ K	Egan (Ref. 6)
Viscosity and thermal conductivity of GCO_2 $187 < T$ 252.6 K	Ibele (Ref. 7)

A PROTOTYPE TEST ON A SPECIALLY SHAPED DEWAR CONTAINER USED FOR THE 300 KW SUPERCONDUCTING UNIPOLAR MOTOR

Hu Lianfang, Li Ying, Jin Xingyue
Refrigeration and Cryogenics Lab, Zhejiang University
Hangzhou, China

ABSTRACT

This paper describes the characteristics of the structure and manufacturing technique of a liquid helium Dewar container for the superconducting unipolar motor. The temperature distribution of the different insulating parts in the vacuum space is analysed by means of the tests on liquid helium and liquid nitrogen. According to the results of the tests and analysis, an improved design of the insulation structure of the specially shaped Dewar container is proposed.

INTRODUCTION

Research on a superconducting unipolar motor is an important subject under the National Science and Technology Commission. In this motor, there are a pair of superconducting coils and corresponding liquid helium Dewar contairers. To meet the specificiations of the container structure, special requirements are developed for the support and the insulation, including the inner container support and the narrow vacuum space insulation. The support will have to stand 330 kilograms weight and another 15 tons of axial force when the motor is running. A shield and multi-layer insulation were arranged to limit the radiation heat transfer from the surroundings. It was rather difficult to wrap the multi-layers, position the shield and make the support. We made a prototype the same sige as the practical one. Analysis and experiments on this prototype with liquid nitrogen and liquid helium provide necessary parameters for the design of the practical container.

STRUCTURE OF THE CONTAINER

The structure and size of the prototype are shown in Fig.1. The inner container neck tube 1 is 50 mm in diameter and 400 mm long, with a corrugated tube on the top to compensate it for the temperature variation. The lower end of the outer neck tube 2 which is made of carbon steel is connected with the outer container 3 by a flange. The shield 4 is made of 1 mm thick copper plate. The winding tube 5 is a ϕ8x1 stainless steel one. It ducts the evaporated storage from the lower end of the inner neck tube 50 mm away from the inner shell to the center of the fibreglass epoxy support 230mm away from the hot end. The tube is tied to the support with stainless steel straps. The winding tube ducted evaporation first cools the support, then goes to the shield and finally to the neck tube and the open air. The superconducting magnet 6 is placed in the inner container. The central cylinder 7 is made of 3 mm thick carbon steel plate. The fibreglass epoxy support 8 is 401 mm long, 288 mm in diameter, 10 mm thick. The inner container 9, which is made of 1.5 mm thick stainless steel, has a volume of 80 liters. Between the inner container and the shield there are ϕ8 ceramic balls. The sealing circular plate on the larger end of the outside shell is 20 mm thick, and those on the smaller end and in the middle 8 mm. The container weighs about 200 kg in all.

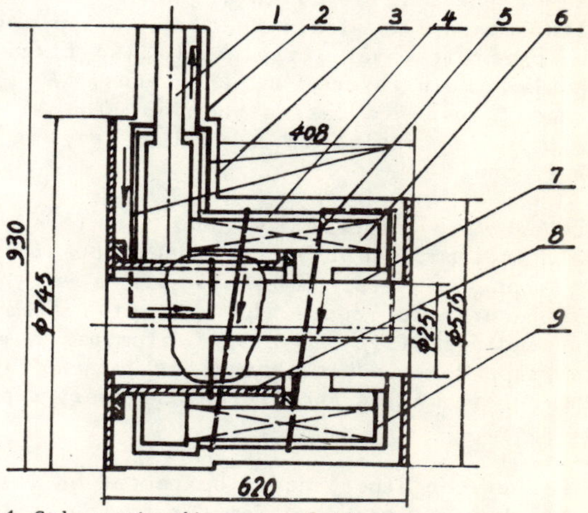

Fig.1 Schematic diagram of the container
1-inner neck tube; 2-outer neck tube;
3-outer container; 4-shield;
5-winding tube; 6-magnet;
7-central cylinder; 8-support;

MANUFACTURING TECHNIQUE

Argon-arc-welding had been adopted throughout to make the container, except for the welding between the outer neck tube and the flange. A thick and short fibreglass epoxy supporting cylinder is designed to stand 330 kg of weight and 15 tons of axial tension stress, thereby introducing considerable conductive heat transfer. The winding tube, which ducts the evaporated helium or nitrogen, is very effective in lessening the conduction through the support by reducing the heat transfer temperature difference, as shown in Fig.2

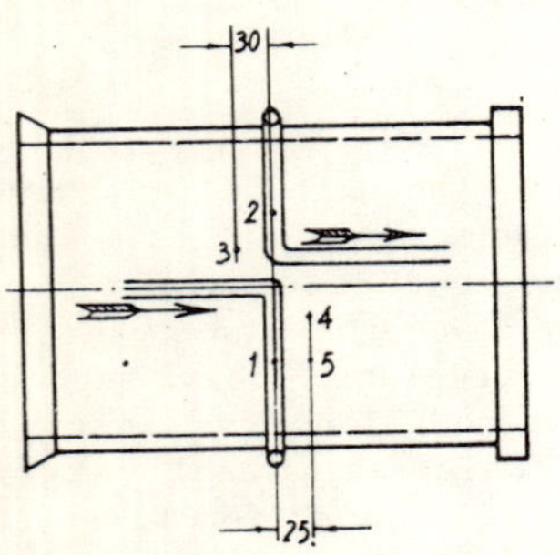

Fig.2 Temperature measuring points of the support and winding tube
1,2 - points on the wall of the tube;
3,4 - points on the support surface;
5 - point inside the support at a deepth of 9 mm

Aluminized mylar of four siges is wrapped around five different parts of the inner container. The multilayers on the ends, which are circular in shape, are cut in the radial direction and folded in the center to meet the size of the central circular hole. The larger cylinder and the middle head are wrapped with one long multilayer strap, which also fixes the circular multilayers on the ends. A nylon thread is used to fasten them up. The smaller cylinder is handled in a similar way to the larger one.

Although the central cylinder of the outer shell is separated from the inner container by the copper shield, it still has a very low temperature. To reduce the radiation from the surroundings, fifty layers of aluminized mylar are wrapped around it. Epoxy 618 is used as an adherent in making the fibreglass epoxy support cylinder.

The inner container should be tested on its own by the helium mass-spectrometer to see if there are any leaks. 500 kg of activated carbon are placed at the ends of the inner container. After finishing the assembly the container was given an overall leak detection test. The outgassing rate became less than 7×10^{-5} Torrl/sec after three days of evacuation by a machine pump, and less than 3×10^{-5} Torrl/sec when the container was filled with liquid nitrogen.

RESULTS OF TESTS

The main purpose of the tests was to examine the structure of the container for its quality. Therefore, thermocouples of gold-ferro and nichrome were attached to different positions of the support and the shield, as shown in Fig.2 and Fig.3.

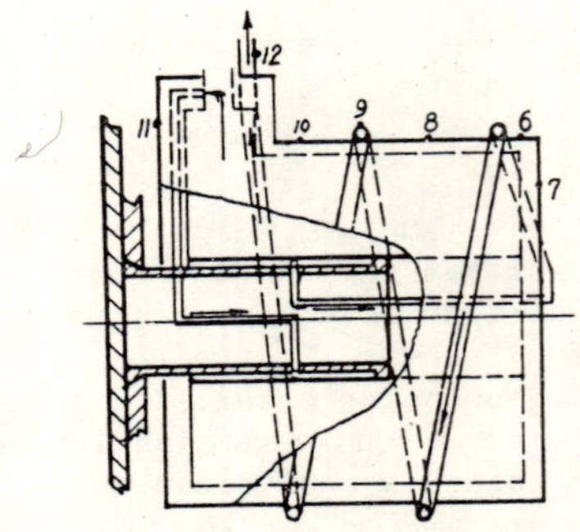

Fig.3 Temperature measuring points on the shield
9,12 - points on the wall of the tube;
6,7,8,10,11 - points on the outer suface of the shield

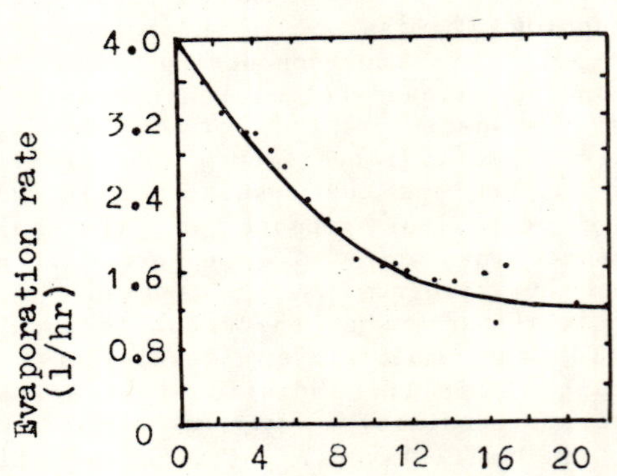

Fig.4 Evaporation rate of liquid helium

Fig.5 and Fig. 6 show the temparature distribution of the support and shield of the tests on liquid nitrogen and helium respectively.

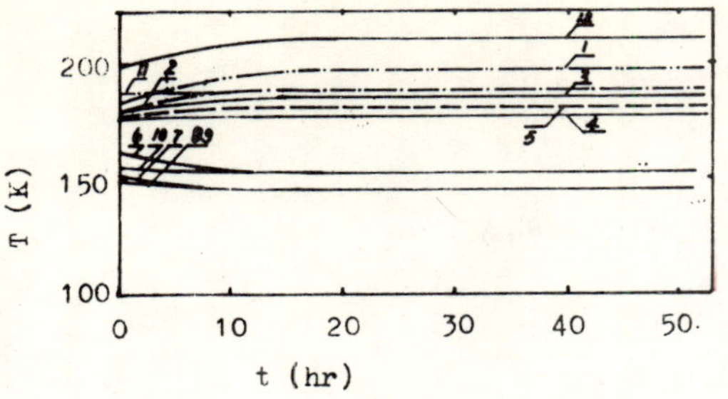

Fig.5 Temperature of the points when the container is filled with liquid nitrogen

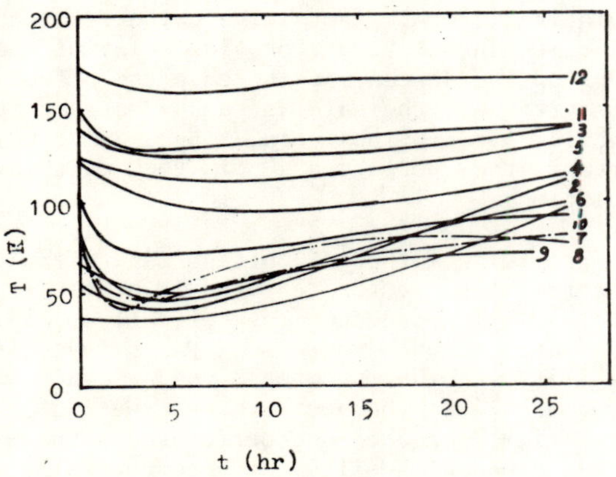

Fig.6 Temperature of the points when the container is filled with liquid helium.

ANALYSIS OF TESTS

Table 1 gives the results of computation and measurement. Obviously, the measured evaporation rates are higher than the computed ones. The comparison of the measured and computed data will give us some hints on improving the insulation structure. Firstly, in the test on liquid nitrogen, the tested temperature values of the support are higher than the calculated ones, and the recovery heat of the winding tube around the support is only 1/4 of the calculated one. The deviation is even larger for the support. Thus we can conclude that the support should be better cooled.

Secondly, tests show a close agreement with the calculation on the temperature distribution and the recovery heat of the shield. The evaporated nitrogen flowing to the shield has a higher temperature than the support, and this results in a negative recovery heat. The temperature of point 11 is higher than the others on the shield. This suggests that this part of the shield should be better insulated and cooled.

Table 1. A comparison of calculations and measurements

		Liquid nitrogen		Liquid helium	
		Calculated	Measured	Calculated	Measured
Support	Mean temperature of cooling point (k)	148	185	19	111
	Entry and exit temperature of winding tube (K)	77–220	182–198	4–34	52.5–62.5
	Recovery heat by winding tube(Kcal/h)	2.36	0.55	3.59	2.88
	Conducted heat on hot end (Kcal/h)	2.89	1.96	3.72	3.63
	Conducted heat on cold end (Kcal/h)	0.46	1.26	0.17	0.56
Shield	Mean temperature of shield (K)	226	160	66	76.7
	Entry and exit temperature of winding tube (K)	220–232	198–187	34–98	62.5–94.5
	Recovery heat (Kcal/h)	0.2	–0.38	7.7	7.148
	Emission heat (Kcal/h)	3.55	1.92	0.10	0.067
Conducted heat by multi-layer (Kcal/h)		3.77	5.94	7.89	7.64
Conducted heat by neck and others(Kcal/h)		0.229	3.37	0.251	0.294
Evaporation rate (1/h)		0.11	0.17	0.86	1.50

CONCLUSION

This prototype for the 300 kw superconducting unipolar motor has an evaporation rate of 1.5 litres per hour, which is acceptable for an initial design of the container. However, the design of the shield and support should be improved in order to achieve a better insulation.

REFERENCES

1 Henning, C.D., Markiewicz, W.D. et al, Advances in Cryogenic Engineering, Vol.21, Plenum Press, New York (1976), P.9

2 Henning, C.D., Rhodeniger, R.L. et al, Advances in Cryoenic Engineering, Vol.21, Plenum Press, New York (1976), P.7

3 Aulonov, Yu.F., Unipolor Super-conducting machine for shipbord, Power Engineering (USSR), Vol.21, No.1(1983), P.69-73

THE EFFECT OF PERFORATION ON MULTILAYER INSULATION

Chen Hong, Zheng Jianyao, Chen Guobang
Cryogenics Laboratory, Zhejiang University
Hangzhou, China

ABSTRACT

Since the pressure distribution between layers of the multilayer insulation (MLI) is actually existed, the more number of layers used, the greater the evacuation resistance. For the sack of reducing heat leak of residual gases and evacuation time, the perforated radiation shields are used. Experimental results have shown that the radiation heat leak may increase and the character of MLI may decrease if any improper holed shields are used. In order either to keep effective radiation area or to reduce the evacuation resistance, one-dimensional slits with width about zero cut on screen are developed. The performance of three types of MLI (i.e. flat screen, one perforated with small holes and one with slits) are compared at theoretical and experimental consideration, and the practical method to improve the performance of MLI is discussed.

INTRODUCTION

Vacuum multilayer insulation (MLI) which consists of low-emissivity radiation shields separated by low-conductivity spacers is a well known super-insulation and it has been widely used in cryogenic technology. To minimize gas conductance, the interstitial pressure must be less than 0.01 Pa. In practice, since the edging evacuating capability is very small, insterstitial gas and outgassing from insulating materials is very difficult to remove, and the pressure of internal layers is much higher than that of the outside (1,7). Obviously, the greater the thickness and the layer density, the more difficult the evacuation.

Many authors have studied methods for improving the evacuation process of multilayer insulation and reducing the interstitial residual gas pressure, Examples are the evacuation process of heating and carbon dioxide replacement (4,5), adapting carburized paper as a spacer (6), and perforating the radiation screen (1,7,8). However, heating and carbon dioxide replacement may result in the solid conduction increasing due to the compressing of the insulating layers. Installing carburized paper will raise the cost. perforating a lot of small holes on the screen may increase the effective emissivity of the shields and decrease the effective radiation area, and so finally decrease the performance of the multilayer insulation.

The promising method of cutting many slits on the screen (its perforated area is almost equal to zero) will not reduce the effective radiation area, but increase the evacuation capability. The thermal and evacuation performance of the above three types of MLI (i.e. flat screen, one perforated with small holes and one with slits) are theoretically and experimentally studied in this paper. The results have shown that the method of cutting slits on the screen is superior in improving the evacuation and thermal performance. It can be used in engineering applications.

THEORETICAL ANALYSIS

1. Heat Conduction of the Interstitial Residual Gas

In order to describe the interstitial pressure distribution across MLI blankets, the amount of the outgassing rate inside the blankets must be known. The free outgassing rate or the desorption rate of molecules from a surface is both time and temperature dependent. An approximate theoretical outgassing formula can be expressed by (8), which is derived as follows:

$$V = V_0 T \exp\left(-\frac{H}{RT}\right) \tag{1}$$

where v_0 is the outgassing parameter; $1.47*10^{12}$ molecules/cm^2-sec-k, for 20 to 30 layer MLI; $2.94*10^{12}$ molecules/cm^2-sec-k, for 40 to 50 layer MLI;
H is the activation energy of desorption of Van der Waals molecules, 0.92 KJ/mol for hydrogen; and 5.61 KJ/mol for nitrogen;
R is the universal gas constant;
T is the temperature, K.

The equation for the free-molecular-conduction Heat transfer rate in an MLI based on energy equilibrium is

$$q_g = \frac{5}{8}\alpha \rho \bar{c} k (T_2 - T_1) \tag{2}$$

where α is the accommodation coefficient;
ρ is the gas density ;
c is the mean gas molecular velocity;
k is the Boltzmann's constant;
T_1 , T_2 is the lower and the higher temperature respectively.

Let t be the perforation rate of the insulation sheet, m_n , m_{n-1} be the net free molecular flux across the two adjacent radiating shields, then

$$t m_n = t m_{n-1} + 2(1-t)\,v(T)$$

$$m_n - m_{n-1} = \frac{2(1-t)}{t}\,v(T) \tag{3}$$

When multishields are used and the distance between the shields is small compared with the total multilayer thickness, m can be expressed as

$$m = \frac{1}{4}\bar{c}\frac{d\rho}{dn}$$

Therefore:

$$m_n - m_{n-1} = \frac{d}{dn}\left(-\frac{1}{4}\bar{c}\frac{d\rho}{dn}\right)$$

$$\frac{d^2\rho}{dn^2} = -\frac{8(1-t)}{\bar{c}t}\,v(T) \tag{4}$$

The boundary conditions are:

$$\frac{d\rho}{dn} = 0 \qquad \text{at } n = 0$$

$$\rho = \rho_0 \qquad \text{at } n = N$$

the maximum molecular density can be obtained as:

$$\rho_{max} = \frac{4v(1-t)n^2}{\bar{c}t} \tag{5}$$

Combining equations (5) and (2), the gas conduction heat of an interstitial gap is

$$q_g = \frac{5}{2}\,\frac{v(1-t)n^2}{t}\,k(T_2 - T_1) \tag{6}$$

Finally, an equation for the overall gaps of the insulation thickness is obtained

$$q_g = \frac{5}{2}\,\frac{v(1-t)n}{t}\,k(T_N - T_0) \tag{7}$$

where T_0 , T_N are temperatures of the first layer and the Nth layer respectively.

2. Radiation Heat Transfer on a Perforated Shield

The radiation heat flow in the multilayer insulation with perforated shields can be estimated from a one-dimensional infinite parallel plate model (Fig. 1).

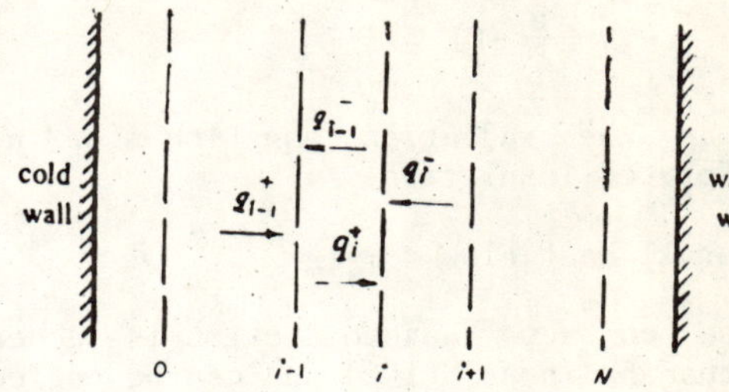

Fig. 1 One-dimensional infinite parallel plate model

According to radiation heat transfer theory, equations of radiant flow in the ith interspace (i=0,1,.....,N) can be written as:

$$q_{i-1}^- = t_i q_i^- + (1-t_i)[\varepsilon_i \sigma T_i^4 + (1-\varepsilon_i) q_i^+]$$

$$q_i^+ = t_{i-1} q_{i-1}^+ + (1-t_{i-1})[\varepsilon_{i-1} \sigma T_{i-1}^4 + (1-\varepsilon_{i-1}) q_{i-1}^-] \tag{8}$$

where ε_i is the emissivity of the ith screen;
σ is the Stefan-Boltzmann radiation constant.

The first term on the right side of equation (8), $t_i q_i$ is a portion of q_i transmitted through the perforation area. The second term represents the effective radiation emitted by the unperforated area, which is the sum of the radiation itself $\varepsilon_i \sigma T_i^4$ and the reflected radiation $(1-\varepsilon_i) q_i$.

At the steady state, the radiation heat flow between adjacent perforated radiating shields is equal. If the perforation rate and the emissivity are both constant, the radiation heat flow across the insulation blankets can be obtained

$$q_r = \frac{t(q_N^- + q_N^+ - q_0^- - q_0^+) + (1-t)\varepsilon\sigma(T_N^4 - T_0^4)}{n(1-t)(1-\varepsilon)} \tag{9}$$

on the other hand, the value of each screen

$$(1-t)\varepsilon(q_i^- + q_i^+ - 2\sigma T_i^4)$$

is equal, equation (9) is therefore replaced by the following

$$q_r = \frac{\varepsilon(1-t)+2t}{n(1-t)(1-\varepsilon)}\sigma(T_N^4 - T_0^4) = \frac{\sigma(T_N^4 - T_0^4)}{(\frac{2}{\varepsilon_t} - 1)n} \tag{10}$$

where $\varepsilon_t = \varepsilon + (2-\varepsilon)t$ (11)

is the effective emissivity of perforated foils.

This shows that the greater the value of t, the larger the value of ε_t and q_r, when t=0, equation (10) gives

$$q_{ro} = \frac{\sigma(T_N^4 - T_0^4)}{(\frac{2}{\varepsilon} - 1)\, n} \qquad (12)$$

This is the radiation equation for normal unperforated insulation.

3. Overall Heat Flow

Because contact heat conduction is generally less than 5% in an MLI(2) and can be neglected, the equation for the total heat flow through the perforated insulation becomes

$$q = q_r + q_g$$

$$= \frac{\sigma(T_N^4 - T_0^4)}{(\frac{2}{\varepsilon_t} - 1)\, n} + \frac{5}{2}\frac{v(1-t)n}{t} k(T_N - T_0) \qquad (13)$$

This means that, q_r will increase with t while q_g will decrease. Thus, an optimum perforation rate which corresponds to the mininum overall heat flow should exist

$$\frac{\partial q}{\partial t} = 0$$

The optimum value of t is then given by

$$\frac{1-t}{t} = \frac{1}{n}\left[\frac{(\frac{2}{2-\varepsilon})\sigma(T_N^4 - T_0^4)}{\frac{5}{2} v k (T_N - T_0)}\right]^{1/2} \qquad (14)$$

As a result, a corresponding minimum value of q_{min} is given by

$$q_{min} = \frac{\sigma(T_N^4 - T_0^4)}{(\frac{2}{\varepsilon} - 1)\, n} + \left[10 v k (T_N - T_0)\sigma(T_N^4 - T_0^4)(\frac{2}{2-\varepsilon})\right]^{1/2} \qquad (15)$$

Equation (15) shows that, no matter how great n may be, the heat flow q can never be less than that of the second term.

4. Radiation Heat Transfer with Slits on the Screen

Consider an MLI blanket with slots on the screen (Fig. 2). When radiation enters a slot, it is absorbed or reflected by the edges of the slot walls. The flux reflected by the colder layer is still distributed in all directions due to the diffusion and specular reflection, and then into the next slot. The ensuing re-reflection among layers provides additional opportunities for energy absorption. The slot in an MLI blanket behaves like a black cavity which will eventually absorb all the radiation entering it from external sources. Therefore the energy absorbed by a slot in an MLI blanket depends on the view factor between the slot and the warmer wall as well as the temperature and the emissivity of the warmer wall.

As shown in Fig. 2, the view factor of an element area dA_1, to a finite area A_2 is (10)

$$F_{dA_1 - A_2} = \frac{1}{\pi X}\left[\sqrt{1+X^2}\tan^{-1}\left(\frac{Y}{\sqrt{1+X^2}}\right) - \tan^{-1} Y + \frac{XY}{\sqrt{1+Y^2}}\tan^{-1}\left(\frac{X}{\sqrt{1+Y^2}}\right)\right] \qquad (16)$$

where X=b/c, Y=a/c

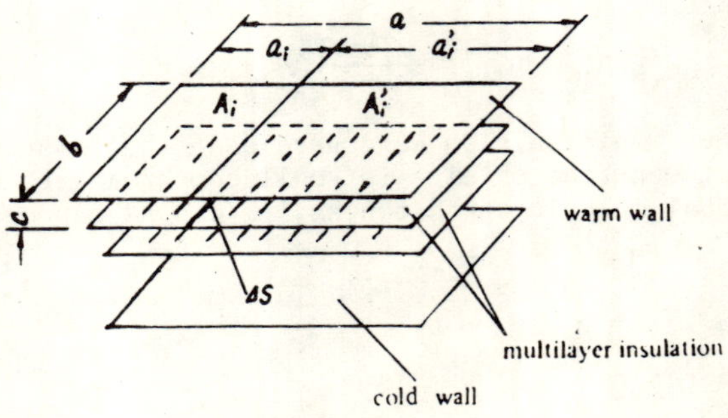

Fig. 2 MLI blanket with slot

The radiation energy emitted from the warmer wall directly into the slots is (11)

$$Q_{\Delta S} = \sigma \varepsilon_2 T_2^4 F_{\Delta S - A} S \qquad (17)$$

where S is the area of slot;
ε_2 is the emissivity of warmer wall.

The overall heat flow through MLI blanket with a slot is:

$$q_{r,slot} = \frac{\sigma(T_N^4 - T_0^4)}{(\frac{2}{\varepsilon} - 1)\, n} + \sigma\varepsilon_2 T_2^4 F_{\Delta S - A}\frac{S}{A} \qquad (18)$$

for a one-dimension slit S=0, then $q_{r,slot} = q_{ro}$. We can see from equation (18) that perforation with slits on screen does not increase radiation flux.

5. Option of Perforation Type

It is known that both the evacuation capability and the radiation heat of a hole are greater than that of a slot. Furthermore, in the case of the same venting capability (same area) the performance of insulation with slots is better than that with square holes. Due to the effect of the Blank Cavity (12), the increment of heat flux per unit slot area is strongly dependant on the slot width. The effect of the Black Cavity vanishes as the width tends to zero; therefore one-dimensional slits do not increase the radiation flux.

EXPERIMENTAL RESULTS

The insulation performance of three types of MLI (i.e. flat, perforated with small holes and with slits) were compared with experimental measurements.

A standard cylindrical cryostat with two guard chambers was used to measure the heat flow and the apparent thermal conductivity of the MLI (13).

Three types of sample have been tested. The materials used, the processing parameters and the design of the sample are shown in Table 1 and Figure 3.

Table 1 MLI test sample

Type	(A) Unperforated	(B) Perforated with holes	(C) Perforated with slits
Radiation shield	Al-foil (0.016mm)	Al-foil (0.016mm) t=0.03 D=3mm	Al-foil (0.016mm) L=80-100mm W=0
Spacer	Fiber-glass clothe (0.025mm)		
Layer number	20		
Layer density	10 (shield/cm)		

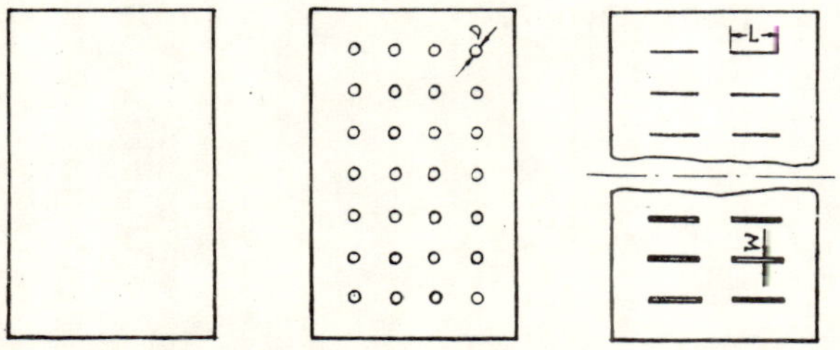

(a)unperforated (b)perforated with holes (c)perforated with slits

Fig. 3 Three types of MLI sample

The experimental results are summerized in Table 2.

From Table 2 we can see that the gas heat flow due to residual gas conduction is as large as 43% of the overall flow for the unperforated insulation (A), 3.4% for holes (B), and 18% for one-dimension slits (C). Perforating shields can therefore improve the evacuation performance of an MLI and sample (B) is better than (C).

The overall heat transfer of the MLI blanket with perforated shields is 1.63-fold and 1.13-fold greater than that of the unperforated blanket (A) for samples (B) and (C) respectively. This is because perforation (especially for holes), causes the effective emissivity and the radiation heat flow to increase. It is thus clear that the effect of perforation with slits on the increment of radiation is 4-5 fold less than that with holes.

Table 2. Expereimental results

	Type	(A)	(B)	(C)
Experimental data	Vacuum (Pa)	$1.0\cdot10^{-2}$	$0.76\cdot10^{-2}$	$1.9\cdot10^{-2}$
	K_{app} (W/m·K)	$6.91\cdot10^{-5}$	$8.65\cdot10^{-5}$	$7.94\cdot10^{-5}$
	q (W/m^2)	0.735	1.195	0.833
	q_i/q_A	100%	163%	113%
Calculated data	q_r (W/m^2)	0.419	1.156	0.679
	q_g (W/m^2)	0.316	0.039	0.154
	q_g/q	43%	3.4%	18%
Evecuation time	t (h)	72	40	48
	t_i/t_A	100%	56%	67%

Comparing the evacuation time of the three types of MLI sample, the evacuation time can be reduced to 56% and 67% for holes (B) and slits (C) respectively. We therefore find that perforations are very effective in accelerating the evacuation.

However, the data from Table 2 show that the total heat flow across a blanket with a slit is still larger than that without perforations. It seems to be in contradiction with the fact that perforation with a slit on the screen can reduce the gas conduction. The reason may be that the slit crack has been enlarged due to the wrapping tension, causing the radiation flow to increase. The experiment has shown that the temperature distribution associated with slits is different from that with an unperforated screen. This is due to the effect of radiation-absorption of the slots and it also tends to make heat transfer larger. It can also be predicted that a slit will have a significant effect on reducing the overall heat transfer if many more layers and a higher layer density are used.

In summary, the performance of an MLI with a slit is superior for both the insulation property and the evacuation process.

In addition, two experiments with different slit lengths have shown that the insulation performance of short slits is higher than that of long slits. The former can avoid the cracking tendency of the slits under the wrapping tension. More slits with a shorter length should be used in practice.

CONCLUSIONS

The perfomance of insulation and evacuation by cutting slits on the MLI screen is superior to that obtained by perforating holes. The advantages of one-dimension slits are not only in making the evacuation easier but also not in increasing the radiation flow through an MLI blanket. The results will be better if slits with a shorter length are used.

ACKNOWLEDEMENTS

The authors would like to express their gratitude to Senior Engineer Hu Xianwen for his kind instruction and helpful discussion and to Sun Guohua, Cui Peihong as well as Zhuang Tanyun for the assistance in experiments.

REFERENCES

1. Price J. W., "Measuring the gas pressure within a high-performance insulation blanket", Adv. Cryo. Eng., Vol.13, (1967) PP662

2. Verkin B. I., Mikhalchenko, R. S., "Effect of moltilayer insulation thickness on apparent thermal conductivity", Second International Engineering Conference (1968) PP 123-127.

3. Mikhalchenko R. S. "Heat and mass transfer in multilayer insulations", Sixth International Cryogenic Engineering Conference (1976).

4. Van Gundy D. A. et al, "A study of condensing vacuum insulation", Adv. Cryo. Eng., Vol.5 ,(1960).

5. Xiao J. C. et al, "A study of new evacuation technology of CO2 replacement and heating for cryogenic container", Cryogenic engineering , No.1, (in Chinese) (1982).

6. Scurlock R. G and Saull, "Development of multilayer insulation with thermal conductivities below 0.1 $*10^{-6}$ W/(cm.K)", Cryogenics, Vol.16, No.5, PP 303-311.

7. Crawford R. F., " Space evacuation of NRC-2 insulation for liquid hydrogen tankage", J. Spacecraft Vol.4, No. 12, (1967) PP 1598

8. Chau H. and Moy H. C., "Thermal characteristics of multilayer insulation", AIAA 70850.

9. Verkin B. I., "Multilayer insulation with diffraction shields", Seventh International Cryogenic Engineering Conference, (1978), PP586-592.

10. Sparrow E. M., Cess R. D., "Radiation heat transfer", (1978), PP 364

11. Shu Q. S., "Systematic study to reduce the effect of cracks in multilayer insulation I", Cryogenics, Vol.27, (1987), PP249-256

12. Shu Q. S., "Systematic study to reduce the effect of cracks in multilayer insulation II", Cryogenics, Vol.27, (1987), PP298-311

13. Kropschot R. H. et al, "Multiple-layer insulation", Adv. Cryo. Eng., Vol.5, (1960), PP 197.

ESTIMATION AND MEASUREMENT OF THE INTERSTITIAL PRESSURE IN MULTILAYER INSULATION

S. L. Bapat, K. G. Narayankhedkar and T. P. Lukose
Indian Institute of Technology, M. A. College of Engineering
Bombay - 400076 INDIA, Kothamangalam - 686666 INDIA

ABSTRACT

An analysis is carried out for MLI using fabric spacer. The interstitial pressure is predicted for different layers. The predicted values are compared with experimental values for same layer densities and number of layers. It is observed that the interstitial pressure increases with number of layers and as a consequence contribution by gas conduction increases with number of layers.

INTRODUCTION

Many investigators have tried to develop models which are capable of predicting the thermal performance of multilayer insulation (MLI). However, a satisfactory theoretical model is not yet achieved. The difficulties arise from the following major areas:

1) the complex structure of the materials
2) the non-availability of the properties of materials in the temperature range of application of MLI
3) the unpredictable variations of the parameters like interstitial pressure, contact pressure etc.

In the present work, the interstitial pressure is predicted and compared with the experimental values for the same layer densities and number of layers. The model considers the exact arrangement of MLI using fabric spacer. The overall heat transfer through MLI occurs with the interaction of different modes of heat transfer such as solid conduction through the materials and the points of contact, radiation through the fibrous spacer and gas conduction through the residual gas in the clearance space between the layers and also in the voids in the spacer. A few analytical or empirical equations developed by various investigators are discussed by Narayankhedkar et al. (1).

HEAT TRANSFER BY GAS CONDUCTION

Most of the investigators neglected the gas conduction with the assumption that the pressure in between the layers will be less than 10-4 Torr. However, a few investigators have considered the heat transfer due to gas conduction. Leohard and Tatra (2) introduced an empirical coefficient COEFF. Barron (3) and Folkman and Lee (4) developed equations using the kinetic theory of gases. Streed (5) used an equation developed by Verschoor and Greebler (6) to account for the gas conduction through voids of spacer materials. Bell et al. (7) used Knudsen equation for computing gas conduction. Price and Lee (8) and Keller et al. (9) used empirical correlations. Bell et al. (7) also observed that the ratio of effective thermal conductivity measured and predicted using empirical relations lies between 1.3 to 4.1 which indicates the need for improvement in prediction equations. Price (10), from his experimental investigations observed that the interstitial pressure is orders of magnitude higher than the chamber pressure. The theoretical as well as experimental study of Lin (11) also proved that, even with broad-side pumping (perforated shields), the pressure in between the layers is much higher than the chamber pressure. Verkin (12) concluded that major part of heat flow through MLI is due to gas conduction. Scurlock (13) observed considerable improvement in the performance of MLI when loaded with getter materials. These studies show that the gas conduction which is a function of interstitial pressure plays an important role in the performance of MLI. Liu (14) from his study on the outgassing rate of MLI,

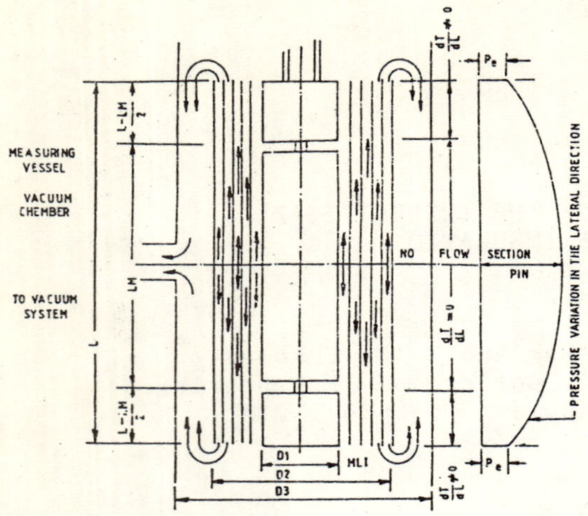

Fig. 1 Model for interstitial pressure analysis

concluded that the residual pressure in MLI is controlled by the outgassing rate of the spacer materials.

MODEL FOR INTERSTITIAL PRESSURE ANALYSIS

The model for analysis is shown in Fig. 1. The molecules within the layers will take a parallel path between the layers of MLI and escape through the edges. This mechanism is called 'edge-pumping'. The gas flow will be in the opposite directions from the middle section of MLI. This section is termed as 'no-flow section' and the pressure at this section will be maximum.

ASSUMPTIONS

1) The flow is one dimensional i.e. the gas molecules move in the lateral direction.
2) The flow is isothermal throughout the length of the measuring vessel.
3) The spacing of layers of MLI is uniform.
4) The residual gas is nitrogen.
5) The gas has constant fluid properties.
6) The gas flow is free molecular as pressure is below 0.01 Torr.

CONDUCTANCE OF PASSAGES

The path of the molecules during pumping consists of the annular passages between

1) the layers of MLI, where dT/dL=0, the length of which is LM/2,
2) the layers of MLI, where dT/dL≠0, the length of which is equal to the length of the guard vessel.
3) Outer surface of insulation and the inner surface of vacuum chamber.

The general equation for conductance through an annular passage in molecular flow regime (15) is given by

$$CO = (\pi g_c \bar{R} T)^{1/2} (D_o - D_{in})^2 (D_o + D_{in}) * KO/[(18M)^{1/2} L] \tag{1}$$

Table 1 gives magitude of KO as a function of ratio of diameters.

Table 1 Correction factor KO

D/Din	0.0	0.254	0.500	0.707	0.866	0.966
KO	1.0	1.672	1.154	1.254	1.430	1.675

The above equation has been extended to compute the conductance of different paths mentioned below:

a) Conductance between the Layers of MLI in dT/dL=0 Region

Substituting for T, L1, Do and Din in Eq.(1), the conductance in the ith passage can be expressed as

$$CO1'_i = (\pi g_c \bar{R} \bar{T}_{gi} / 18M)^{1/2} * \{4 \delta_c^2 \, 2 (DL_i - \delta_c) KO \cdot 2/LM\} \tag{2}$$

The paths between the layers are parallel and hence, total conductance in the dT/dL=0 region

$$CO1' = \sum_{i=1}^{N} CO1'_i = \sum_{i=1}^{N} \{(\pi g_c \bar{R}/18M)^{1/2} * [16 T_{gi}^{1/2} \delta_c^2 (DL_i - \delta_c) KO/LM]\} \tag{3}$$

b) Conductance between Layers of MLI in dT/dL≠0 Region

By extending the same procedure,

$$CO1'' = \sum_{i=1}^{N} \{(\pi g_c \bar{R}/18M)^{1/2} [16 T_{gagi}^{1/2} \delta_c^2 * (DL_i - \delta_c) KO]/(L - LM)\} \tag{4}$$

c) Total Conductance in between the Layers of MLI

The flow path in between the layers of MLI at dT/dL=0 and dT/dL≠0 region are in series. Hence total conductance in flow path in between the layers,

$$1/CO1 = (1/CO1' + 1/CO1'')$$

$$CO1 = (CO1' \cdot CO1'')/(CO1' + CO1'')$$

The overall conductance is due to the conductance of the flow path in between the layers, CO1 and the conductance of the annular passage between the surface of MLI and vacuum chamber, CO2. These paths are in series and hence, overall conductance can be calculated as follows

$$1/CO = 1/CO1 + 1/CO2$$

$$CO = (CO1 \cdot CO2)/(CO1 + CO2) \quad (5)$$

INTERSTITIAL PRESSURE

For analysis MLI can be assumed to be consisting of concentric tubes, closed at one end (no flow section) and evacuated from the other end (edge).

For an element at a distance x from the edge, the perimeter of MLi considering both sides,

$$B = \sum_{i=1}^{N} 2\pi Di \quad (6)$$

The gas load (outgassing rate) on the element of length dx,

$$-dQ = qo\ B\ dx \quad (7)$$

From the equation for throughput through the element,

$$Q = dc/dp = CO(l/dx)\,dp \quad (8)$$

where, CO = conductance
qo = specific outgassing rate, i.e. outgassing rate per unite area

On differentiating,

$$dQ = CO\ [l \cdot dx](d^2p/dx^2) \quad (9)$$

Equating eqs. (9) and (7)

$$-q_0 B\ dx = CO \cdot l \cdot (d^2p/dx^2)dx$$

$$(d^2p/dx^2) = -q_0 B/(CO \cdot l) \quad (10)$$

On integrating Eq.(10)

$$(dp/dx) = -q\ (Bx/CO) + C_1 \quad (11)$$

On further integrating Eq.(11)

$$p = -q\ (Bx^2/2CO) + C_1 x + C_2 \quad (12)$$

Applying the boudary conditions,

At $x = 1$; $(dp/dx) = 0$
At $x = 0$; $P = Pe$, the pressure at the edge.
$C_1 = q_0 B l/CO$ and $C_2 = Pe$

Substituting for C1 and C2 in Eq.(12), and simplifying the interstitial pressure at any section can be expressed as

$$Px = (qoB/2COx)\ (2l-x) + Pe \quad (13)$$

PRESSURE DISTRIBUTION IN THE LATERAL DIRECTION

The pressure at various sections of MLi, in the lateral direction, can be calculated by using Eq.(13). This is done dividing the length into different sections and calculating conductance between the edge and the section considered.

a) Pressure Distribution in dT/dL ≠ 0 Region

Dividing the length of guard vessel, LGV into 'n' equal divisions, the conductance of jth section can be computed using the equation

$$CO1j'' = \sum_{i=1}^{N} \{(\pi g_c \bar{R}/18M)^{1/2} [16\, T_{gag\,i}^{1/2} * \delta^2 (DL_i - \delta) K0]/(LGV_j)\} \quad (14)$$

Thus from the Eq.(13), Pressure at any section j,

$$Pinj = (q_0 B/2CO1_j'')(2l - j) + Pe \quad (15)$$

where, j = 1,-------, n

b) Pressure Distribution in dT/dL = 0 Region

Dividing the length of the measuring vessel into n number of equal divisions, the conductance upto jth section can be calculated by knowing the conductance in the dT/dL ≠ 0 region and conductance upto the section in dT/dL = 0 region. The conductance upto jth section in dT/dL = 0 region can be calculated using the Eq.(7) by changing LM to LMj, where j=1,2,---,n.

$$CO1j' = (\pi g_c \bar{R}/18M)^{1/2} \sum_{i=1}^{N} \{[T_{gi}^{1/2} \cdot 16\, \delta_c^2 * (DL_i - \delta_c) K0]/(LM_j)\} \quad (16)$$

Conductance in dT/dL = 0 region is given by Eq.(8)

As total conductance upto jth section,

$$COj = (CO1_j' \cdot CO1_j'')/(CO1_j' + CO1_j'') \quad (17)$$

Thus the pressure at various sections can be calculated by using the equation

$$Pinj = (q_0 B/2CO_j)[2l - (L - LM + j)/2] + Pe \quad (18)$$

Using the equations (15) and (18), the interstitial pressure at different sections in the lateral direction can be

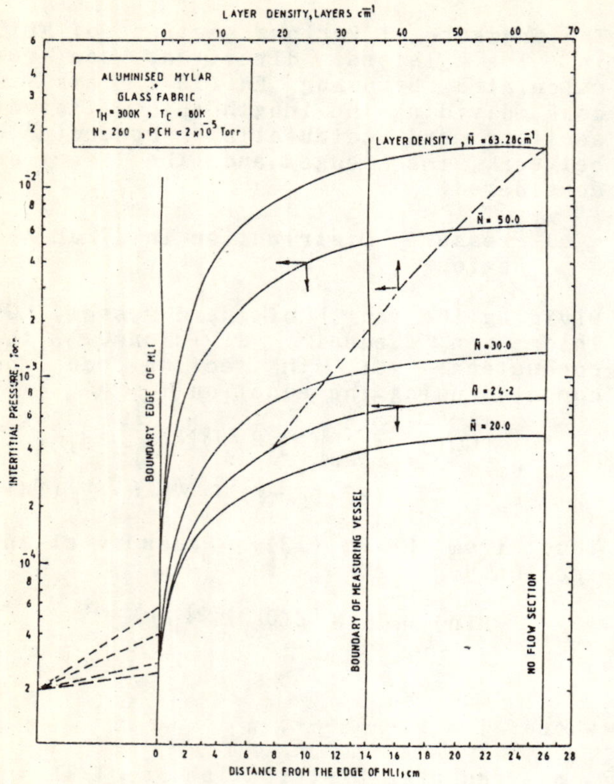

Fig. 2 Profile of interstitial pressure in the lateral direction and variation of pressure at no flow section with layer density

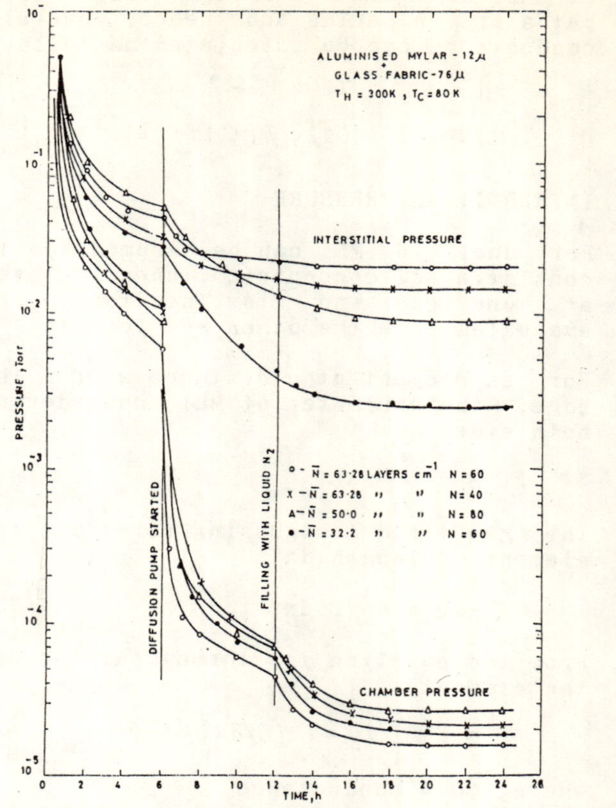

Fig. 3 Variation of chamber pressure and interstitial pressure with time for the Aluminised Mylar and Glass Fabric combination

calculated.

MEASUREMENT OF INTERSTITIAL PRESSURE

For the present investigation a system was developed for the measurement of interstitial pressure with a view to reduce the disturbance to the MLI system. This system can be used for the measurement in cyclindrical arrangements.

The device developed for the interstitial pressure measurement consists of a heating filament and a thermocouple. Thermocouple is spot welded to the mid-point of heating element. A variable resistance to regulate the current through the filament, a millivoltmeter, a milliameter, and a power source are necessary. The filament and the thermocouple assembly are mounted on a flexible disc to facilitate its mounting on the MLI. It is designed in such a way that a standard thermocouple vacuum gauge element can be used for the measurements.

The elements are calibrated with reference to the gauge fitted to the vacuum chamber of the present set-up. The calibration is done after simulating similar conditions at which they are used.

The elements are mounted at the selected position of MLI and the measurements are taken using the thermocouple vacuum gauge. Since thermocouple sensing element is used, its pressure measuring range is limited to 10-3 Torr.

CHAMBER PRESSURE

THe chamber pressure is measured using thermocouple gauge (upto 10-3 Torr) and Penning gauge (below 10-3 Torr and upto 10-6 Torr). They are located at the outlet of the vacuum chamber.

RESULTS AND DISCUSSION

MLI under study comprises of ALuminised Mylar (12μ thick, coated on one side only) as shield and glass fabric (76μ thick and mesh density 16 threads/cm as the spacer. The variation of interstitial pressure in the lateral direction and its variation with layer density are computed from the prediction equations (15) and (18). The results are shown in Fig.2. It can be observed that as the layer density increases, the interstitial pressure increases. It is also observed that the variation in number of layers for constant layer density has negligible effect on the interatitial pressure. Eventhough the gas load increases as number of layer increases, the conductance also increases consequently the change in interstitial pressure is only marginal.

PRESSURE HISTORY

The variation of chamber pressure and the corresponding variation in the interstitial pressure with time is shown in Fig.3. It is observed that as the number of layers increases, the interstitial pressure also increases for the same layer density. It is also observed that for the same number of layers, the interstitial pressure increases as layer density increases.

Table 2 Variation of interstitial pressure with number of layers

Layer density, N = 50 layers/cm

Number of layers	20	40	60	80	100
Interstital pressure Pin(Torr*10^2)	.674	.641	.630	.625	.622

Table 3 Influence of layer density and number of layers

$\bar{N}$	63.2	63.2	63.2	50	30
N	80	60	40	80	60
PinM	22.0	20.0	14.0	9.0	2.50
PinT	15.8	16.3	16.3	6.3	1.343
A* %	28.2	18.5	16.4	30.0	46.4

* A=(PinM-PinT)/PinM
PinM -- Interstitial pressure measured at 25th layer from the surface of the cold boundary
PinT -- The predicted mean value of interstitial pressure

This because of the change in the dimensions of the gas flow path.

It is observed that there is appreciable change in the interstitial pressure (Table 2) with number of layers for the same layer density. Even though outgassing rate increases with numberof layers, the conductance also increases. Hence the prediction of interstitial pressure by using throughput equation will not change appreciably. Consequently, the MLI system becomes homogeneous with respect to structure contact pressure and gas pressure etc. In the homogeneous material, the conductivity independent of thickness.

During the present experimentation it is observed that the interstitial pressure increases with the number of layers with the same layer density (Table 3). But these experiments are limited to a few readings and the same have not been repeated to come to definite conclusion.

CONCLUSIONS

It may be concluded from the present investigations that the interstital pressure increases with number of layers and as consequence the contribution by gas condition increases with the increase in number of layers. Thus the increase in effective thermal conductivity with number of layers, may be attributed to gas conduction.

LIST OF SYMBOLS

A - Area
a - Area, radius of contact patch
C - Circumference
CL - Characteristic length of clearance
CO - Conductance
COEFF - Expirical coefficient
D,d - Diameter
KO - Correction factor
k* - Thermal conductivity
kg - Ideal thermal conductivity of gas
LGV - Length of guard vessel
LM - Length of measuring vessel
LS - Characteristic length
l - Length of element
M - Molecular weight
N - Number of layers
$\bar{N}$ - Layer density
n - Number of layers
P - Intersitial gas pressure; contact pressure between layers
Qo - Outgassing rate
q - Heat flux
R - Gas constant; thermal resistance
$\bar{R}$ - Universal gas constant
T - Absolute temperature
t - Thickness
α - Accomodationb coefficient, constant

γ - Ratio of specific heats
δ - Solid fraction
δ_c - Clearance between layers
λ - Mean free path

SUSCRIPTS

c - cold boundary
d - diameter
eff - effective
f - fibres
g,gc - gas conduction
gi - gas conduction at ith layer
H - hot boundary
i - ith layer
in - inside; insulation
p - padding/spacer
1,2 - surface 1,2

REFERENCES

1. Narayankhedkar, K. G., Bapat, S. L., Lukose, T. P., "Theoretical investigations on multilayer insulation". Second International Conference on Cryogenics, Calcutta, Dec.6-10 (1988).

2. Leohard, K. E. and Tatra, R. F., "Superinsulation for cryogenic space tankage". Proc. ICEC Vol.3 (1970), PP490-496.

3. Barron, R. F., "Principles of evacuated cryogenic insulations". AIChE Symp. Series, Vol.68 (1972), PP40-49.

4. Folkman, N.R. and Lee, T.G., "Thermodynamic design fundamentals of high performance insulation". J. Spacecraft and Rockets, Vol.5 (1968), PP954-959.

5. Street, E. R., Cunnington, G. R. and Zierman, C.A. "Performance of multilayer insulation system for 300-800K temperature range". Prog. in Astro. and Aero., Vol.18 (1965), PP735-772.

6. Verschoor, J. D. and Greebler, P., "Heat transfer by radiation and gas conduction in fibrous insulations". Trans. ASME Mechanical Engineering Vol.74 (1952), PP961-968.

7. Bell, G. A., Nassat, T. C. and Wedel, R. K. "Thermal performance of multilayer insulations applied to small cryogenic tankage". Adv. in Cryo. Vol.22 (1975), PP272-282.

8. Price, J. W. and Lee, T. G., "Analysis design and testing of heat insulation components for high performance insulation systems". Adv. in Cryo., Vol.22 (1975), PP265-273.

9. Keller, C. W., Cunnington, G. R. and Glassford, A. P., "Final report--Thermal performance of multilayer insulations". NASA CR-134477 (1974).

10. Price, J. W., "Measuring the gas pressure within a high performance insulation blanket".

11. Lin, J. T., "Analysis of gas flow through multilayer insulation system". AIAA Journal, Vol.11 (1973), PP995-999.

12. Verkin, B. I., "Effect of multilayer insulation thickness on apperent thermal conductivity". Proc. ICEC-2 (1968), PP123-127.

13. Scurlock, R. G. and Saull, B., "Development of multilayer insulation with thermal conductivity below 0.1 μW/cm.K". Cryogenics, Vol.13 (1976), PP303-311.

14. Liu, C. K. "Outgassing rate of MLI materials at ambient temperature". J. Vac. Science Technol., Vol.17 (1980), PP696-700.

15. Roth, A., "Vacuum Technology". North Holland Publishing Co., Amsterdam, PP60-102, (1976).

A STUDY OF COOL-DOWN PROCESS ON LOADING LH_2 TO LAUNCH VEHICLE

Fu Xili
Beijing Institute of Special Mechanism
P.O.Box 9200-71, Beijing

ABSTRACT

In this paper, the conservation of energy during cool-down process is analysed, calculating formulaes for cool-down time are develped and discussed. At last, a controlling method for cool-down process, which is based on pressure of LH_2 tank on board is presented.

THEORETICAL ANALYSIS ON COOL-DOWN PROCESS

The simplified LH_2 loading system is shown in fig. 1, In which 1-1 denote inlet, 2-2 denote outlet. Generally, as LH_2 has been transferred into the LH_2 tank on board and the liquid level has reached to a certain high, we say that the cool-down process is ended. Take the loading system as a subject studied, analyse the energy transferred into the system and out from the system.

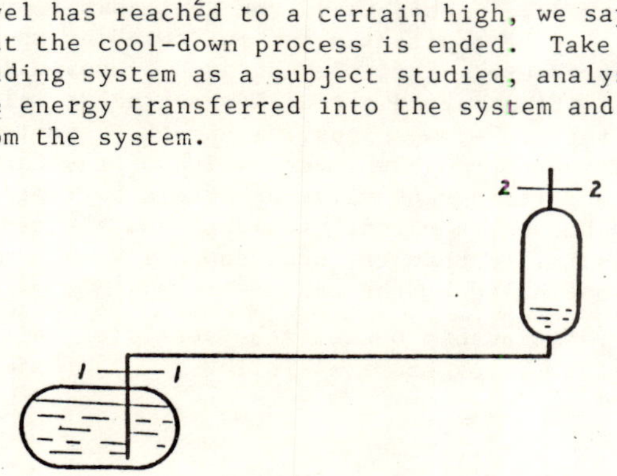

Fig.1

Energy transferred into the system as follows:
Heat transferred from ambient to the system

$$\int_0^{t_2} \dot{Q}dt = \bar{\dot{Q}}\Delta t$$

where $\dot{Q}$ — heat transfer rate

Δt — cool-down time

$\bar{\dot{Q}}$ — average heat transfer rate

Heat released from the system cooled

$$-\Sigma\int_{T_1}^{T_2} m_i c_i dT = \Sigma m_i \bar{c}_i \Delta T$$

where m_i — mass of cooled portions

c_i —specific heat of cooled portions

$\bar{c}_i$ — average specific heat

ΔT — temperature drop

Heat carried by LH_2 into the system

$$\int_0^{t_2} \dot{m}_1 h_1 dt = \bar{\dot{m}}_1 h_1 \Delta t$$

where $\dot{m}_1$ — mass flow rate of LH_2 for cool-down

$\bar{\dot{m}}_1$—average mass flow rate

h_1 — enthalpy of LH_2 entered to the system

Energy out from the system as follows:

Heat carried by the medium out from the system

$$\int_0^{t_2} \dot{m}_2 h_2 dt = \bar{\dot{m}}_2 \bar{h}_2 \Delta t$$

where $\dot{m}_2$ — mass flow rate of the medium out from the system

$\bar{\dot{m}}$ — average mass flow rate

h_2 — enthalpy of the medium out from the system

$\bar{h}_2$— average enthalpy

The change of internal energy of medium within the system is

$$M_2U_2-M_oU_o$$

where M_2, U_2 — mass and specific internal energy of the medium within the system at the end of cool-down process respectively;

M_o, U_o — mass and specific internal energy of the medium within the system at the initial of cool-down process respectively.

Applying the principle of conservation of energy and mass

$$M_2U_2-M_0U_0=\dot{Q}\Delta t+\Sigma m_i\bar{c}_i\Delta T+\bar{\dot{m}}_1h_1\Delta t-\bar{\dot{m}}_2\bar{h}_2\Delta t \qquad (1)$$

$$M_2-M_0=(\bar{\dot{m}}_1-\bar{\dot{m}}_2)\Delta t \qquad (2)$$

from (1),(2), solving for cool-down time

$$\Delta t=\frac{\Sigma m_i\bar{c}_i\Delta T-M_2(U_2-\bar{h}_2)+M_0(U_0-\bar{h}_2)}{(\bar{h}_2-h_1)\bar{\dot{m}}_1-\dot{Q}} \qquad (3)$$

or

$$\Delta t=\frac{\Sigma m_i\bar{c}_i\Delta T-M_2(U_2-h_1)+M_0(U_0-H_1)}{(\bar{h}_2-h_1)\bar{\dot{m}}_2-\dot{Q}} \qquad (4)$$

From equation(3), we note that the characteristic of cool-down process as follows:

(1) The cool-down time decreases, as the LH_2 flow rate increases. In order to decreasing cool-down time, the LH_2 flow rate ought to be larger. However, as the flow rate increases, the gas pressure within LH_2 tank on board and the stress of structure increase. Because LH_2 tank is the weakest part of the system, the maximum LH_2 flow rate for cool-down is limited by the pressure within the tank.

(2) As the heat capacity of piping system cooled increases, the cool-down time and the quantity of LH_2 needed for cool-down increase. Therefore the piping system ought to be light in mass as possible. Multilayer insulation has mary advantageous, such as small in mass, excellent in performance, so that this type insulation has been usually used for LH_2 lines.

(3) If the insulation performance of piping system is improved, the heat leaked to the system and the cool-down time will be decreased. While $(\bar{h}_2-h_1)\bar{\dot{m}}_1-\dot{Q}\leqslant 0$, according to equation (3), the cool-down time will be infinite or negative in value. In this case, performing cool-down for the system is impossible. Because $(\bar{h}_2-h_1)\bar{\dot{m}}_1$ is the heat absobed by cooling medium, $\dot{Q}$ is the heat transferred to the system from ambient surroundings.

As $(\bar{h}_2-h_1)\bar{\dot{m}}_1=\dot{Q}$, the heat transfer is in equibrium, the temperature is no longer reduced and the cool-down time will be limitless. LH_2 tank on board is a part cooled. It is well known that heat transferred to LH_2 tank is much larger than that to the piping system. During the initial period of cool-down process, the gas temperature within LH_2 tank is higher, the heat transferred to the LH_2 tank is lesser. The tank temperature reduces, the heat transferred to the tank increases, so that the cool-down flow rate ought to be increased correspondingly. According to this rule, the optimal cool-down flow rate for loading LH_2 ought to be increased gradually.

(4) As the cool-down flow rate increases, the cool-down time decreases, the time for heat exchange between cold gas vaporized from LH_2 and piping system shortens. In this case, the LH_2 used for cool-down increases, because the utilizing of senisible heat of the cold vapor is not perfect.

(5) From equation (4) we note that the cool-down time is also dependant on exhaust flow rate of vapor $\bar{\dot{m}}_2$. At the initial period of cool-down process, While LH_2 enters the line, it quickly boils and changes into vapor perfectly. At the same time, the flow resistance increases, that will lead to decreasing cool-down flow rate automaticly. In order to decreasing cool-down time for a long line,some venting valves are setted along the transfer line, those valves are used for venting cold vapor boiled from LH_2 during cool-down process. But that will lead to complicating the piping system.

CONTROLLING FOR COOL-DOWN PROCESS

As mentioned above, cool-down flow rate is the most important factor, which influences the cool-down process is that controlling for cool-down process is that controlling for cool-down flow rate. Because a great portion of LH_2 entered the line vaporises, the actual flow rate of LH_2 can not be measured by the turbine type flowmeter. In this case, the cool-down flow rate may be controlled and monitored by the gas pressure within the LH_2 tank on board. At the initial period of cool-down process, all LH_2 for cool-down vaporizes and the vapor(gas) will vent through the vent valve of the tank. For a certain vent valve, there is a relationship between the vent flow rate and the gas pressure, so that the cool-down flow rate may be controlled by the gas pressure.

The relationship between the vent flow rate and the gas pressure as follows:

when

$$\frac{P_1}{P_2}\leqslant\left(\frac{2}{k+1}\right)^{\frac{k}{k-1}}$$

$$\dot{m}_2=AP_2\left(\frac{2}{k+1}\right)^{\frac{1}{k-1}}\sqrt{\frac{2kM}{RT_2(k+1)}} \qquad (5)$$

where P_1, P_2 — gas pressure after vent valve and gas pressure within LH_2 tank respectively (Pa)

$\bar{\dot{m}}_2$ — mass flow rate of gas vented (kg/s)

A — venting area of the vent valve (m^2)

$K=\frac{C_p}{C_v}$, $K\doteq 1.4$ for GH_2

M — molecular weight of gas

T_2 — gas temperature within tank (K)

R — universal gas constant (N.m/kg.mol.K)

Supposing that the LH_2 for cool-down are all vaporizes during cool-down process and the LH_2 mass flow rate for cool-down equals to that of vapor vented out of the tank. For exemple, as A = 18 cm^2, T_2 = 100K and P_2=0.25MPa (absolute pressure), the correspounding LH_2 cool-down flow rate equals to 0.478kg/s (410 l/min LH_2). The controlling of cool-down flow rate by using the LH_2 tank pressure is based on the supposition that the LH_2 for cool-down all vaporizes to GH_2 and vents out from the tank. Therefore, those controlling method mentioned above is only used for controlling at the initial period of cool-down process. As the cool-down runs, the LH_2 tank temperature drops gradually. If the LH_2 tank pressure is constant, the corresponding LH_2 cool-down flow rate increases gradually, that is expectant. As mentioned above, if P =0.25MPa =constant, the relationship between cool-down flow rate and GH_2 temperature within tank can be obtain from equation (5), that as shown is fig. 2.

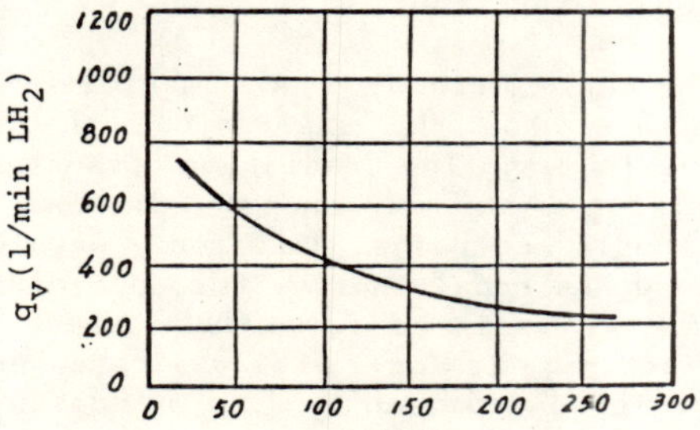

Fig. 2 The relationship between cool-down flow rate and GH_2 temperature

LH_2 tank pressure P =0.25MPa =constant

q_v — cool-down flow rate of LH_2(L/min)

T_2 — temperature (K)

As the cool-down process runs, the LH_2 for cool-down has only partly varporized to GH_2, that the GH_2 mass flow rate is less than that of LH_2. At this time, the cool-down flow rate may be controlled and monitored by using turbine type flow meter which runs in single phase flow of LH_2. In order to measure the actual LH_2 flow rate, the flow meter ought to be setted near the inport of the line. The throttling valve which is used to adjusting the LH_2 flow rate ought to be placed down the streamof the line.

REFERENCES

1. Randall F. Barron Cryogenic Systems (Second Edition) OXFORD UNIVERSITY PRESS, NEW YORK CLARENDON PRESS, OXFORD 1985.

COMPUTATION OF THE TEMPERATURE DISTRIBUTION OF THE MULTI-LAYER INSULATED SHIELDS BY A MONTE CARLO METHOD

Shen Luochan, Li Ying, Chen Guobang
Zhejiang University
Hangzhou, China

ABSTRACT

This paper uses a Monte Carlo method to caluculate the temperature distribution of multi-layer insulated shields and compares it with the general simplified calculation method.

INTRODUCTION

One of the thermal calculation tasks used for a multi-layer insulated shield is the determination of the shield temperature distribution.

In technological calculations it has been simplified by assuming
(1) a shape factor F=1 between a series of shield cylinders with an identical axis, as if it were a series of infinite plates,
(2) neglecting the influence of the shield emissivity on the "shape factor",
(3) the shield emissivity is taken as a constant and each shield as a gray diffuse surface.

According to these facts it is considered in this paper that
(1) the shields consist of different diameter cylinders,
(2) the shield emissivity is a function of temperature $\epsilon=f(T)$,
(3) the influence of the shield emissivity on the "shape factor" (so we have called it "transfer factor", and we use a Monte Carlo method to calculate it).

After obtaining all the transfer factors from one shield to another, the temperature distribution of the insulated multi-layer shields can be obtained by a network method.

Because it is assumed that the surface emissivity is close to 1 in integration calculation of the shape factor, the shape factor is a function of the geometry parameters. But with the Monte Carlo method it is easier to handle problems in which the surface shapes are complicated, and the surface emissivities are variable. This method is also accurate. According to Bian [1], a calculation result of the integration method when compares with the Monte Carlo method for a standard shape surface with emissivity of 1 is correct within an error of 0.2%.

THE CALCULATION METHOD

The Monte Carlo method is a numerical method which is based on the statistical theory of random variables. The radiation transfer is simulated by a set of energy bundles. Each energy bundle contains different wavelength photons and has equal energy. For determination of the energy radiated from the surface i to the surface j, it only follows the energy bundles emitted from surface i and calculates how many energy bundles are absorbed by the surface j, and the number of energy bundles multiplies the energy of each bundle is equal to the radiation energy. The emitting direction of the energy bundle from the surface is decided by the cone angle θ and the circumferential angle φ. They are all random numbers. For the diffuse surface this relation is

$$\cos\theta = 1 - R_\theta \qquad (1)$$

and

$$\varphi = 2\pi R_\varphi \qquad (2)$$

where R and R are random numbers from 0.0 to 1.0.

When the energy bundle arrives at the other surface, it is absorbed or reflected randomly by the surface absorptivity. For the gray diffuse surface, if the random number

$$R_\alpha \leqslant \epsilon \qquad (3)$$

the energy bundle is absorbed, otherwise it is reflected (ϵ is the surface emissivity of the arrival point). For the same reason the reflecting direction can be decided by equations (1) and (2). The above processes should be repeated until the bundle is absorbed by a surface.

Finally the number of all energy bundles absorbed by the surface j from the surface i is Nij. If it is assumed that Qi and Ni are the radiation energy per unit time and the number of the energy bundles from the emitting surface i respectively, then each energy bundle has the same energy

$$W=Qi/Ni \tag{4}$$

and the radiation energy from the surface i to surface j is

$$Qij=Nij \cdot W \tag{5}$$

The transfer factor from surface i to surface j then is

$$\mathcal{F}ij=Nij/Ni \tag{6}$$

Since the surface shape is a cylinder, the energy bundles emitted from the inside surface can all be at the outside surface. However, for a bundle from the outside surface to the inside one, only when

$$\theta \leq \theta_{cr} \tag{7}$$

can it reach the inside surface, as shown in fig. 1

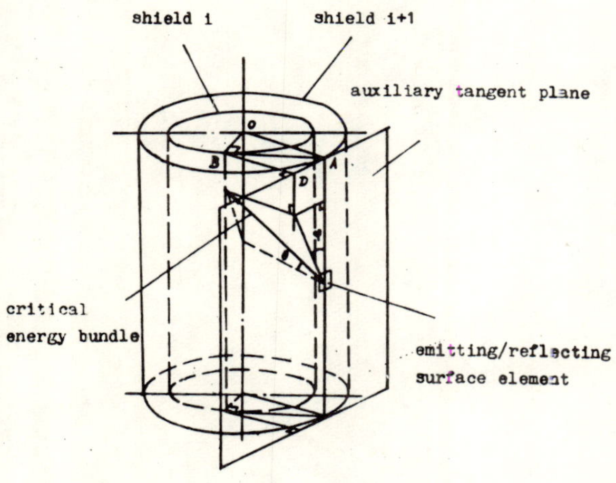

Fig. 1 Geometric relation of radiation transfer factor

At the steady state the radiation heat transfer is a constant Q. Each shield temperature can be decided within the network in reference [3] by equations

$$Ti=(Ti+1 - Q\cdot\sigma\cdot Ri)^{\frac{1}{4}} \tag{8}$$

$$Ri=\frac{1-\epsilon_i}{\epsilon_i \cdot A_i}+\frac{1-\epsilon_{i+1}}{\epsilon_{i+1}\cdot A_{i+1}}+\frac{1}{\mathcal{F}_{ij}A_{i+1}} \tag{9}$$

where σ, $\mathcal{F}_{i+1}$, Ri are the Stephan-Boltzmann constant, the transfer factor and the radiation resistance, respectively. In general the surface emissivity ϵ depends on the kinds of the surface and temperature. so it should be interated among the distribution of shield emissivities, transfer factors and temperature in the computation.

COMPUTATION RESULTS AND ANALYSIS

The sample calculation is for a multi-layer insulated shield of a liquid hydrogen container, as follows
the inside diemeter Di=450mm
the outside diameter D_o =510mm
the shield number N=14
the distance between shields=2mm
the shield emissivity ϵ =6.13x10^{-4}T[2]
the inside wall temperature Twi=20.38K
the outside wall temperature Two=293.0K

In the computation only the heat transfer due to radiation is calculated. The gas and shield conduction are neglected, the end effect is not considered, and the shields are assumed to be isothermal diffuse surfaces. There is radiation heat transfer only in the radial direction.

The computation involving energy bundles Ni=100000 takes about 13 hours in the IBM-PC computer. For comparison, in the simplified calculation we take ϵ =0.03, F=1 and the same boundary conditions. The distribution of the shield temperatures and emissivities are shown in fig. 2 and fig. 3

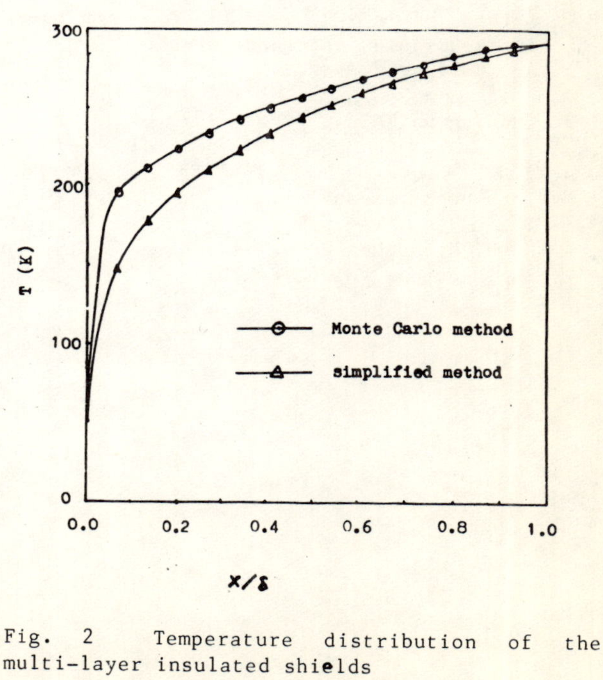

Fig. 2 Temperature distribution of the multi-layer insulated shields

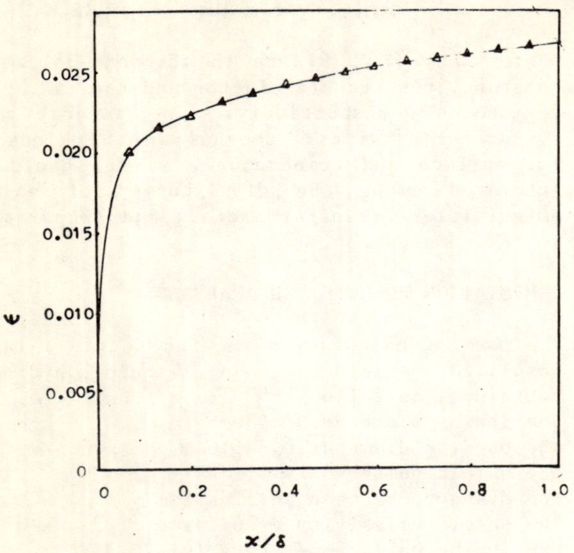

Fig. 3 The emissivity distribution of the shields

It is shown that the nearer one is to the inside wall the larger shield temperature gradient becomes because of the increase of the shield emissivity gradient. There is some developmenc rather than the simplified calculation. Of course, considering the heat transfer the shield temperature distribution will have some change, it should be studied furhter.

REFERENCES

1 Bian Baihui, The Computation and Analysis of the Radiation Transfer Factor by Monte Carlo Method, in the first Engineering Thermoscience Conference of the Universities and Colleges, 1983.9.

2 Cunnington,NASA CR72605, 1971.

3 J. P. Holman, HEAT TRANSFER, MacGraw-Hill Int. book company, 1981.

4 R. H. Kropschot et al, Advances in Cryogenic Engineering, Vol.5, P. 189.

5. HEAT AND MASS TRANSFER AT LOW TEMPERATURE

HEAT TRANSFER IN PRESSURIZED SUPERFLUID HELIUM

Mieko Ohtsuka, Takashi Suzuki, Hiroyuki Tetsuka
Experimental Center for Very Low Temperature and Energy Technique,
Tokyo Institute of Technology,
Oh-okayama, Meguro, Tokyo 152, Japan

and
Akio Sato
Toshiba Research and Development Center
Ukishima, Kawasaki, Kanagawa 210, Japan

ABSTRACT

The steady state heat transfer characteristics in a channel of pressurized superfluid helium (He II) have been investigated. The test channel is made, and a foil heater of stainless steel is installed in the channel. The channel is maintained in He II bath pressurized to 1 atmospheric pressure. The temperature distribution of He II along the channel and the temperatures at the heater are measured with respect to continuous heat flux applied to the heater. The critical heat flux, where the temperature of liquid helium in the channel reaches the λ - transition temperature of liquid helium, is determined with different He II bath temperatures. These results are compared with numerically calculated results. Also, the characteristic helium boiling behavior dependent on channel orientation is discussed.

INTRODUCTION

Superfluid helium (He II) is known to have advantages for cooling superconducting magnet because of its excellent thermal properties. In connection with the research on superconductive magnetic energy storage, it is required to cool superconducting magnet without any trouble and efficiently. To obtain the optimum efficient conditions, however, it is necessary to understand more about the heat transfer behavior of He II as coolant in detail.
Several works have been done on the liquid helium heat transfer in channels, but those were, for example, with a heater at the end of the channel [1] or at the middle point of the channel [2]. One experiment with foil heater has been done in the case of normal He I [3].
In the present work, we have studied the steady state heat transfer in a channel of He II pressurized to 1 atmospheric pressure. Pressurized He II has more advantages than saturated He II from the viewpoints of cooling capacity and the technical handling to keep the He II state. Some results previously obtained have been reported [4]. The purpose of this work is to understand the heat removal characteristics of He II and the helium boiling phenomena in a channel. The test channel is made simulated to a cooling channel of superconducting magnet windings or a superconductive conductor channel. Also, numerical calculation has been made, and the results are compared with the experimental results.

EXPERIMENTAL SYSTEM AND PROCEDURES

The test channel is made of glass-fibre-reinforced plastics (GFRP). The schematic view of the test channel is shown in Fig.1. Two kinds of test channel, A and B, have been made, the dimensions of which are shown in Table 1.
For each channel, the surface of one of the walls along the channel is covered with a foil of 10 μm thick stainless steel as heater. To measure the temperatures of the liquid helium along the channel, three carbon resistors are used, which are located at the center and near the both ends of the channel. The carbon resistors used were fabricated by ourselves from commercially available carbon resistors. The temperatures at the heater are measured with three AuFe-Chromel thermocouples, the reference junctions of which are immersed in He II bath (Fig.1). The He II bath temperature is measured with Germanium thermometer.
The whole channel is supported horizontally or vertically in He II bath pressurized to 1 atmospheric pressure. The cryostat is shown

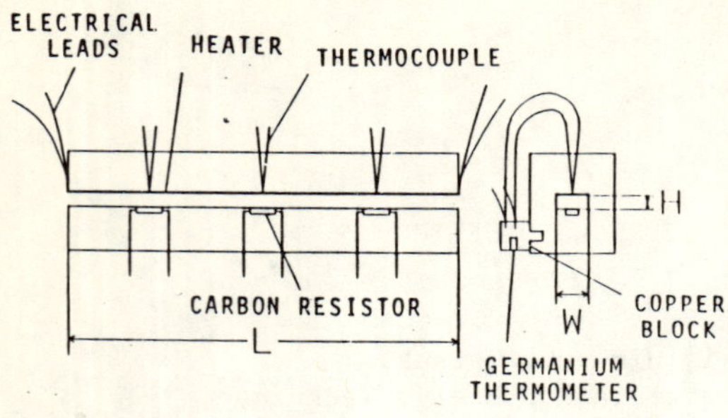

Fig. 1 Test channel

	L(mm)	H(mm)	W(mm)
A	100	2	10
B	160	2	6

Table 1 Channel dimensions

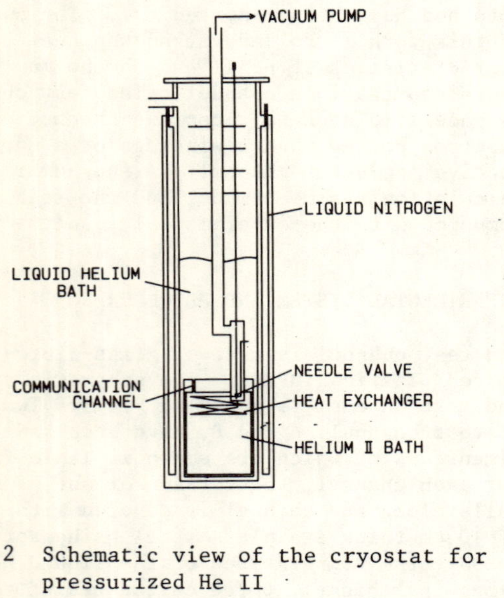

Fig. 2 Schematic view of the cryostat for pressurized He II

in a schematic form in Fig.2. The type is of Claudet [5]. The diameter of He II bath is 280 mm and the depth is 500 mm. He II is obtained by using heat exchanger made of copper. The liquid helium inside the heat exchanger is being pumped throughout the experiment.
Continuous heat input is applied to the foil heater, and the liquid helium temperatures in the channel and the temperatures at the heater are measured with data acquisition system. The system consists of scanner, digital multimeter, personal computer, and so on.

RESULTS AND DISCUSSION

The measurements were made at various He II bath temperatures between 2.2K and 1.9K. Typical temperature variations with respect to the heat flux applied to the foil heater are shown in Fig.3(a) and Fig.3(b). Figure 3(a) shows the temperature variations at the center of the channel A. T_c is the temperature of liquid helium in the channel and ΔT_h is the temperature difference between the heater and the He II bath temperature, T_h and T_b respectively. In this case, the bath temperature T_b = 2.105K. Heat flux Q is expressed for unit area of the foil heater. With the heat flux of about Q = 0.09 W/cm^2, the temperature of the liquid helium reaches the T_λ (the transition temperature from superfluid He II to normal He I), and with a little higher Q, the temperature at the heater starts to rise up to more than 10K, even while the liquid helium temperature stays at T_λ. On the other hand, as shown in Fig.3(b), the liquid helium temperature near the end of the channel rises up gradually with respect to Q up to about T_λ. The temperature at the heater starts to rise suddenly with Q $\sim$ 0.125 W/cm^2, although the liquid helium temperature is still below T_λ. It goes up with higher Q as same as at the center of the channel.
When we consider He II as coolant for superconductor, the value of heat flux where the temperature of liquid helium in the channel reaches T_λ is important; therefore, we call the heat flux thus defined as the critical heat flux Q_c.

Similar experiments were carried out with the channel, in the cases of supported horizontally or vertically, or in the cases of the foil heater faced downward to the channel or upward. Our attention was focused to the behavior of the onset of the temperature rise in He II and at the heater. In all cases of the experiments, the values of Q_c are about the same as long as with the same He II bath temperature. However, the liquid helium boiling phenomena seem to be a little different. The relations between the temperatures of liquid helium in the channel and at the heater are slightly different with different channel orientations. It suggests the co-existence of He II and He I in the channel, for example, as shown in Fig.4.

Since the value of Q_c is significant, we determined those values with different He II bath temperatures. Also we made calculation as describe below.
We proposed a one-dimensional heat transfer model as shown in Fig.5 [4], where Q is the heat flux from the heater and q(x) is the heat flux along the channel at the position x in

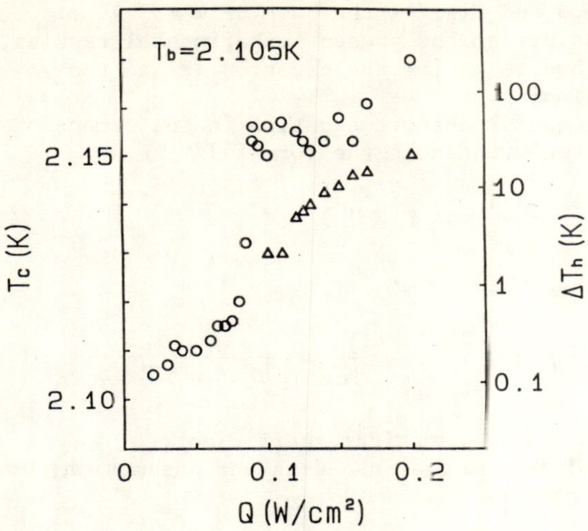

Fig. 3(a) Temperature variations at the center of channel A versus heat flux to the heater;
T_c (O) : temperature of helium
ΔT_h (Δ) : $T_h - T_b$

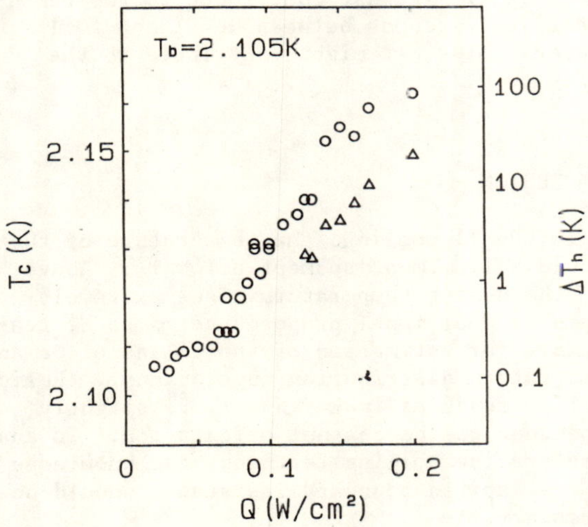

Fig. 3(b) Temperature variations near the end of channel A versus heat flux to the heater;
T_c (O) : temperature of helium
ΔT_h (Δ) : $T_h - T_b$

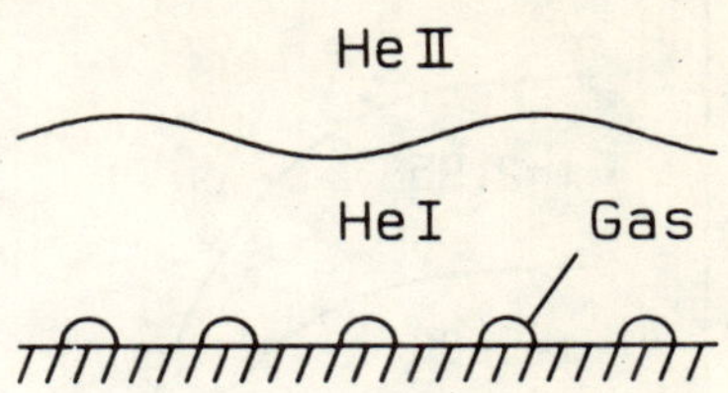

Fig. 4 Helium boiling phenomena near heater surface

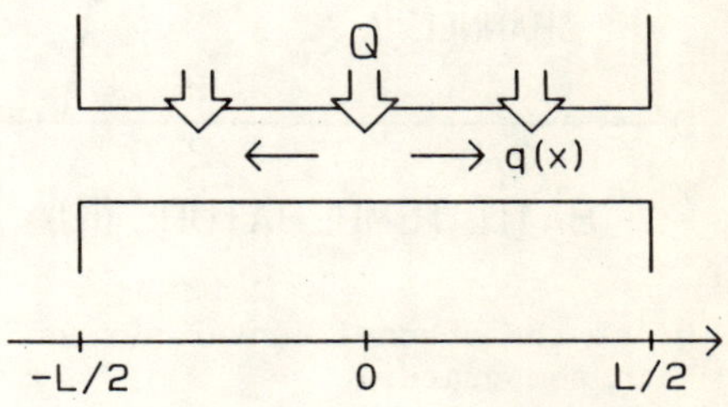

Fig. 5 One-dimensional heat transfer model for calculation [4]

the He II channel. In the case of Fig.5, we assumed

$$q(x) = Q\,x\,W/A\,, \tag{1}$$

where W is the width of the channel and A is the cross section of the channel. According to Soloski et al. [6], q(x) is expressed as

$$q^3(x) = -K_s \cdot \frac{\rho_s^4 s^4 T^3 \eta}{\rho_n \rho} \cdot \frac{dT}{dx} \tag{2}$$

(see LIST OF SYMBOLS).
Then, Q_c is obtained for our model as

$$Q_c = \frac{A}{W}\left[4\left(\frac{2}{L}\right)^4 K_s \cdot \int_{T_b}^{T_\lambda} \frac{\rho_s^4 s^4 T^3 \eta}{\rho_n \rho}\,dT\right]^{1/3} \tag{3}$$

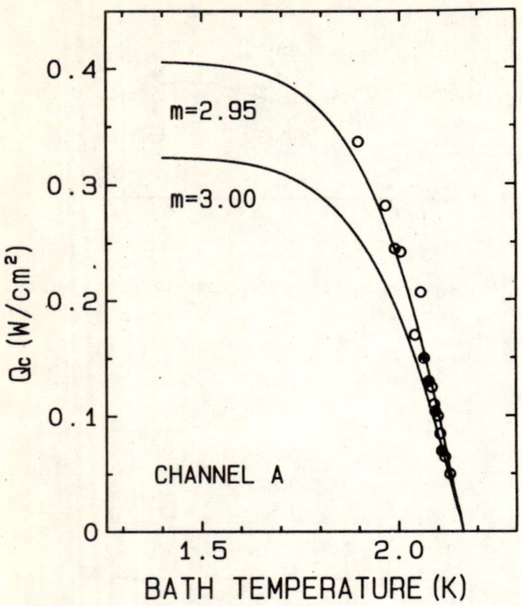

Fig. 6 Qc at the channel center versus He II bath temperature
— : calculated
o : experimental results with the heater faced downward
• : experimental results with the heater faced upward

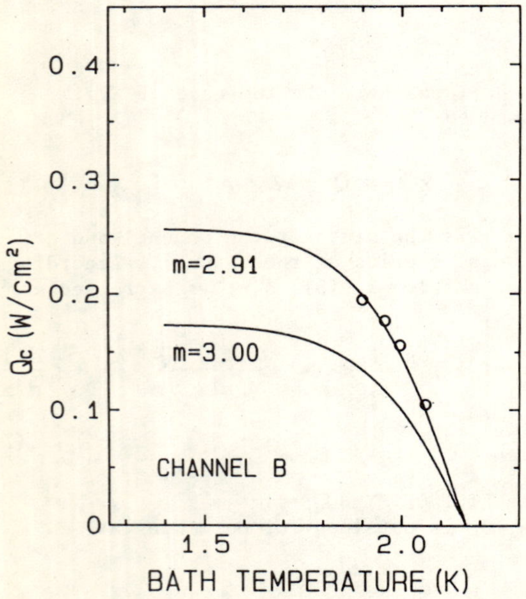

Fig. 7 Qc at the channel center versus He II bath temperature
— : calculated
o : experimental results

We calculated Qc numerically with different bath temperatures and obtained as shown in Fig.6 and Fig.7 (the case of m = 3). Considering the present experimental results, we had to modify the equation (2) as the following.
The heat transport equation in one dimension can be written in the form [7]

$$\frac{dT}{dx} = f(T)\, q^{m} \tag{4}$$

where

$$f(T) = -\frac{1}{K_s} \cdot \frac{\rho_n \rho}{\rho_s{}^4 s^4 T^3 \eta} \tag{5}$$

and m is a numerical coefficient.
Applying to the case with our assumption, Qc is obtained as

$$Q_c = \frac{A}{W}\left[(m+1)\left(\frac{2}{L}\right)^{m+1} K_s \cdot \int_{T_b}^{T_\lambda} \frac{\rho_s{}^4 s^4 T^3 \eta}{\rho_n \rho}\, dT \right]^{1/m}, \tag{6}$$

which corresponds to eq.(3). We obtained m = ?.95 to give the best fit to the experimental results for the channel A as shown in Fig.6, and m = 2.91 for the channel B as shown in Fig.7. Some other values of m have been reported [8], but the difference has not been interpreted physically. Interface between solid and He II creates the Kapitza boundary resistance. It must contribute to the temperature differences between He II and foil heater. Also, it might contribute to the value of m.

CONCLUSION

As for He II cooling, the temperature of the liquid helium must be kept below T_λ. Above T_λ, the heater temperature rises extremely rapidly. Our model proposed seems to be reasonable for estimation of the values of Qc and temperature distribution in a channel. Helium boiling features in a channel are slightly dependent on the channel orientation. To analyse precise boiling features, the knowledge of the Kapitza boundary resistance should be necessary.

LIST OF SYMBOLS

K_s = constant

ρ_s = superfluid density

ρ_n = normal fluid density

ρ = total density

S = entropy

η = viscosity

ACKNOWLEDGEMENTS

The authors would like to express our thanks to T.Yazawa of Toshiba Research and Development Center, K.Ishito and S.Okabe of Tokyo Institute of Technology, for their valuable contributions to the experiment.

REFERENCES

1 Kamioka, Y. 'The prediction of steady state heat flux and related temperature profiles in pressurized superfluid helium II'. Cryogenic Engineering (Journal of the Cryogenic Society of Japan, in Japanese), vol. 18 (1983), pp 119-125.

2 Gentile, D. and Francois, M.X. 'Heat transfer properties in a vertical channel filled with saturated and pressurized helium II'. Cryogenics, vol. 21 (1981), pp 234-237.

3 Chen, Z. and Van Sciver, S.W. 'Channel heat transfer in He I - Steady state orientation dependence'. Advances in Cryogenic Engineering, vol. 31 (1986), pp431-438.

4 Ohtsuka, M. 'Helium II heat transfer in a channel'. Proceedings of the 12th International Cryogenic Engineering Conference, (Southampton, 1988), to be published.

5 Claudet, G., Lacaze, A., Roubeau, P. and Verdier, J. 'The design and operation of a refrigerator system using superfluid helium'. Proc. ICEC 5, Kyoto, (1974), pp 265-267.

6 Soloski, S.C. and Frederking, T.H.K. 'Dimensional analysis and equation for axial heat flow of Gorter-Mellink convection (He II)'. Int. J. Heat Mass Transfer, vol. 23 (1980), pp 437-441.

7 Van Sciver, S.W. 'Helium Cryogenics' (1986), pp 143

8 Claudet, G. and Seyfert, P. 'Bath cooling with subcooled superfluid helium'. Advances in Cryogenic Engineering, vol. 27 (1982), pp 441-449.

CONDUCTIVE CONVECTION CONJUGATE HEAT TRANSFER OF THE PRESSURE PULSE TUBE AND EXPERIMENTAL INVESTIGATION OF GAS STREAM WAVE REFRIGERATION OF A ROTATING HEAT SEPARATOR

Shao Jian, Gao Jinling, Wang Jianfeng
Refrigeration & Cryogenics Laboratory, Zhejiang University,
Hangzhou, China.

ABSTRACT

A physical model of the pressure pulse tube (PPT) is described in this paper. Conductive convection conjugate heat transfer of the vertical hollow pin with variable heat fluxes based on the analysis of gas flow in the PPT. A numerical solution is made by means of finite difference technique. The computational results are agree satisfactorily with the experimental results. Comparable tests are made with Laval nozzle contributor and convergence nozzle contributor with expansion ratios (ε) from 1.5 to 7.0 in order to increase the isentropic efficiency of rotating heat separator (RHS)

KEYWORDS: rotating heat separator, isentropic efficiency, conjugate heat transfer, comparable test, gas stream wave refrigeration

INTRODUCTION

Rotating heat separator (RHS), a new type of refrigerator, accomplishes its low temperature refrigeration based on the gas stream wave expansion by directing a rapid periodical jet of compressible fluid towards the entrance of a long tube closed at the other end. RHS is different from the conventional reciprocating and turbine expanders and has been widely used for recovering light hydrocarbons from petroleum gas or natural gas and the treatment of chemical tail gas because of its simple construction, low rotation speed (2000--4000rpm) and high isentropic efficiency (>60%)(1-7). The pressure pulse tube (PPT) is the key component of momentum exchange and energy transmission of the RHS which is different from the resonant tube, Hartmann--Sprenger and pulse tube refrigerator in many aspects (8). The heat dissipating characteristics have important effect on the refrigeration of the RHS. The two types of nozzles have different effects on the isentropic efficiency of the RHS.

Gas Flow and Conjugate Heat Transfer

The important function of the residual gas (oscillating gas) in the PPT is in absorbing the kinetic energy of the fresh gas (expanding gas) and transmitting heat which is dissipated through the outer wall surface of the PPT by momentum exchange and the Hugniot heating effect (shock wave compressible thermal effect). The heat dissipating characteristics depend on the propagation of waves (shock waves, expanding waves, compressible waves). The wave diagram is as Fig.(1). The residual gas is heated by the two shocks. The heat is dissipated to atmosphere by the conductive convection conjugate heat transfer of the PPT. The high temperature section is formed near the middle of the PPT because of the interaction of the two shocks while the fresh gas subsequently furturely expands and the low temperature gas is

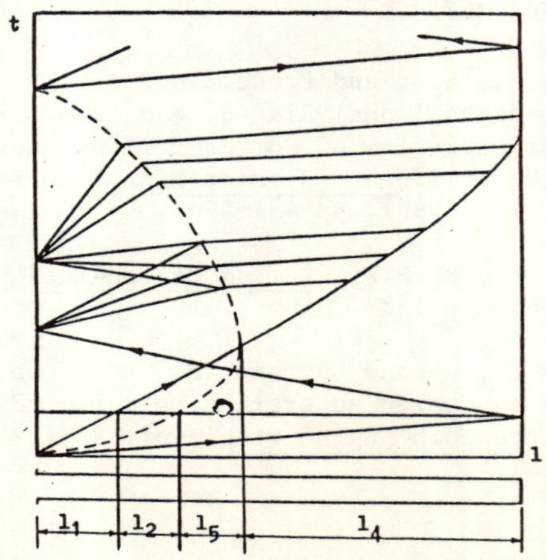

Fig.(1) Wave diagram in the PPT

obtained as result of expanding wave bounds.

The calculation method of the gas parameters (,t,p,v) in every zone of the diagram was given by the author (9). The segments 11, 12, 15, 14 and the corresponding heat fluxes are obtained by the following equations:

$$L1=(V1-W2'(L/W3 \qquad \varepsilon(x)=0 \tag{1}$$

$$L2=(V1-W2')\cdot\tau 1$$

$$\varepsilon(x)=\rho RT2\,\frac{k+1}{12k^2}\cdot\eta^3_{1-2}\cdot f_2 \tag{2}$$

$$L5=\Delta X$$

$$\varepsilon(\ddot{X})=\rho RT5\,\frac{k+1}{12k^2}\cdot\eta^3_{5-4}\cdot f_5 \tag{3}$$

$$L4=L-L1-L2-L$$

$$(X)=\rho RT4\,\frac{k+1}{12k^2}\cdot\eta^3_{5-4}\cdot f_4 \tag{4}$$

while x=(W3-V1+W2')/(V1/2+W1=W2')/W3

Because the heat capacity of the PPT is much greater than that of the fresh gas and residual gas, the wall temperatures of the PPT are almost constant when the RHS is on its working condition. The heat fluxes are the results of the Hugniot heating effects depending on the wave diagram. The time--averged values are calculated by the equations (1-4) which are approximately regarded as only the function of x. As the large temperature gradient of the PPT exists, the heat transter problem cannot be treated by the conventional Newton's cooling law or Fourier conduction law and must be solved with the conductive convection conjugate heat transfer (10,11,12).

We assume for convenience of our analysis:

(1). L>>D, L>>δ (δ is the thickness of the boundary layer);
(2). The external natural convection is steady;
(3). The heat specific Cp, Cv, the conductivities Ke, Kf, and the ratio of heat specific K are constant;
(4). The external radiant heat transfer between the PPT wall and atmosphere can be neglected.
(5). The heat dissipation on the closed end surface of the PPT;
(6). The PPT wall temperature is a function of x only;
(7). The inner heat fluxes are functions of x only; ....

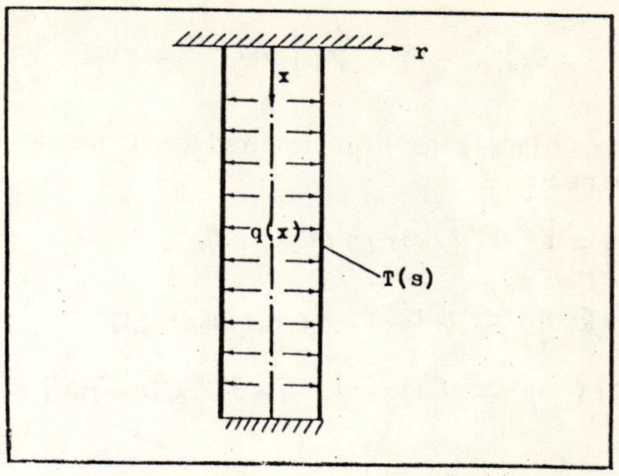

(Fig.(2) Hollow pin

Thus as in Fig. (2), the one dimension energy equation of the PPT is

$$\frac{d^2T_s}{dx^2}=\frac{2\alpha}{\lambda_e\cdot r_0}(T_s-T)+\frac{1}{\lambda_e}q_s(x) \tag{5}$$

The corresponding boundary conditions are

$$T_s\big|_{x=0}=T_0\,,\qquad \frac{dT_s}{dx}\Big|_{x=L}=0 \tag{6}$$

For the external boundary layer

$$\frac{\partial(ru)}{\partial x}+\frac{\partial(rv)}{\partial r}=0 \tag{7}$$

$$u\frac{\partial u}{\partial x}+v\frac{\partial u}{\partial r}=g\beta(T-T_\infty)+\frac{\nu}{r}\frac{\partial}{\partial r}\left(r\frac{\partial u}{\partial r}\right) \tag{8}$$

$$u\frac{\partial T}{\partial x}+v\frac{\partial T}{\partial r}=\frac{\alpha}{r}\frac{\partial}{\partial r}\left(r\frac{\partial T}{\partial r}\right) \tag{9}$$

The boundary conditions are:

$$u=v=0 \qquad T=Ts(x) \qquad \text{at} \qquad r=r0$$
$$u\to u_\infty \qquad T\to T_\infty \qquad r\to\infty$$
$$u=u_\infty\,, \qquad T=T_\infty \qquad \text{at } x=0, r\geqslant r0 \tag{10}$$

We introduce the nondimensional parameters:

$$\xi=x/L$$
$$\theta_s(\xi)=[T_s(x)-T_\infty]/(T-T_\infty] \tag{11}$$
$$h^*(\xi)=2L/(K_f\,Re\,y_0)$$
$$N_c=2K_f\,Re\,y_0/\lambda_e r_0$$

Equation (5-6) are rewritten as follows:

$$\frac{d^2\theta_s}{d\xi^2}=N_c h^*\theta_s \tag{12}$$

$$\theta_s|_{s=0} = 1, \quad \frac{d\theta_s}{d\xi}\Big|_{\xi=1.0} = 0 \tag{13}$$

Definiting the nondimensional parameters:

$$\eta = (r^2 - r_0^2)/(4r_0L)(\theta_s/\xi)\, y_2 \tag{14}$$

$$f(\xi, \eta) = \psi(x, r)/[r_0(rux)\, y_2] \tag{15}$$

$$\theta(\xi, \eta) = [T(x, r) - T_\infty]/[T_0 - T_\infty] \tag{16}$$

equations (7-8) are rewritten as:

$$(1 + \xi^{1/2}\lambda\eta) f''' + (f + \xi^{1/2}\lambda) f'' + 8\xi\, G_r Re^{-2}\, \theta = 2\xi\left(f'\frac{\partial f'}{\partial \xi} - f''\frac{\partial f}{\partial \xi}\right) \tag{17}$$

$$P_r^{-1}(1 + \xi^{1/2}\lambda\eta)\theta'' + (f + P_r^{-1}\xi^{1/2}\lambda)\theta' = 2\xi\left(f'\frac{\partial \theta}{\partial \xi} - \theta'\frac{\partial f}{\partial \xi}\right) \tag{18}$$

And the boundary conditions are:

$$f(\xi, 0) = f'(\xi, 0), \theta(\xi, 0) = \theta_s(\xi)$$
$$f'(\xi, \infty) = 2.0 \quad , \theta(\xi, \infty) = \theta'(\xi, \infty) = 0 \tag{19}$$

while f',f'',f''' stand for partial derivatives with respect to . And we have

$$T_s(x) = T_w(x),$$

$$h^*(\xi)(T_s - T_\infty) = -\lambda_e \frac{\partial T}{\partial r}\Big|_{r=0} \tag{20}$$

$$h^*(\xi) = \left[-\frac{\partial \theta}{\partial \eta}\Big/\theta|_{\eta=0}\right]\Big/(2\xi^{1/2}) \tag{21}$$

The computation conditions are:

L=1.5 m,	r0=0.004 m	
Ph=15 atm,	Pl=2.5 atm,	ε=6.0
Tin=280 °k,	T =293 °K,	T0=243 'K
Ks=110 w/m,'k	Kf=2.59 x 10^{-2}w/m, 'K	

The results calculated are shown in Fig.(3). The wall temperatures are high at x/L=0.3~0.4. In summary, the results are fairly good compared with the tested ones, although many simplifications are made. The calculation values are higher near the temperature peak and lower near the closed end. As we know the η_s values of the RHS will increase if the cold end is insulated and the heat transfer of the high temperature section is enhanced.

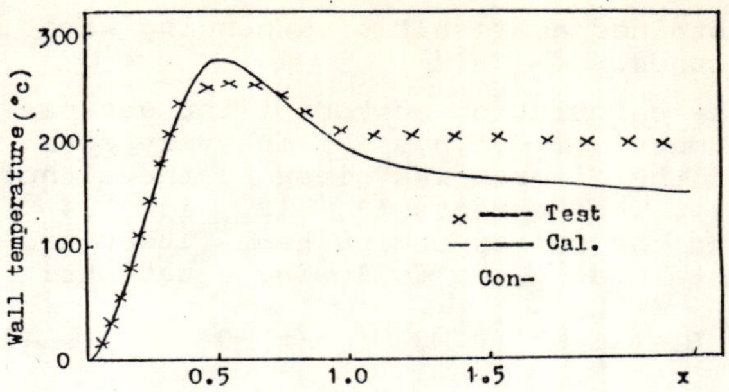

Fig.(3) Wall temperature distribution

Comparable Tests on the Laval Nozzle and Convergence Nozzle

The convergence nozzles are used in our previous experimental prototypes and RHS used in industrial engineering. As the continual jet of the rotating nozzle excites a periodical (not a continous) outer flow which is different from the static jet of a motionless nozzle, the outer flow cross-sectional area is variable with time and such a flow is very complicated. The experimental apparatus and testing system were described in detail in the reference (1).

The isentropic efficiency (η_s) is

$$\eta_s = (Hh - Hl)/(Hh - Hs) \tag{22}$$

for an idea gas

$$\eta_s = (1.0 - Tl/Th)/(1 - \varepsilon^{\frac{1-k}{k}}) \tag{23}$$

As the working gas is air, pressure are not very high and temperatures are very low (>200 °K). In our experiment the equation (23) is used to estimate the values of η_s. For air,K=1.4.

There exist optimal rotating speeds for certain operating conditions, therefore the tests at different rotating speeds are important

$$N = 60\ f/M \tag{24}$$

while N is the main shaft rotating speed (rpm),M is the number of the nozzle of main the shaft, f is the pressure oscillating frequency of the gas in the closed end (l/sec.).

We understand from the experiment that the rotating speeds have important effects on the values of η_s of the RHS, as shown in Fig. (4). For certain ranges of the rotating speeds, there exist high η_s values. In our testing range, the η_s values have two peaks. The first corresponds to the rotating speeds 2900-3400 rpm, the second to 5800-6800rpm. The second optimal speed is double the first for certain conditions and the η_s values can almost be kept unchanged when the rotating speeds change $\pm 5\%$ near an optimal one. The relative speed is defined as the ratio of the rotating speed to the first optimal speed in Fig.(4).

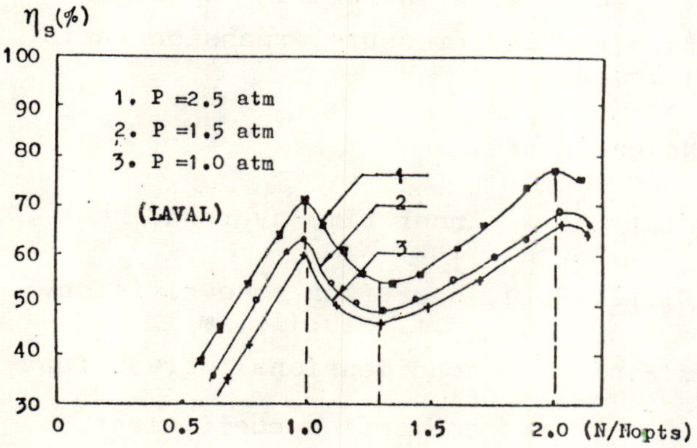

Fig.(4) Effect of rotating speeds on η_s

As shown in Fig.(5), the η_s is 55-70% for the Laval nozzle, but 50-65% for the convergence nozzle when the back pressure (Pl) keeps at Pl=1.5 atm.

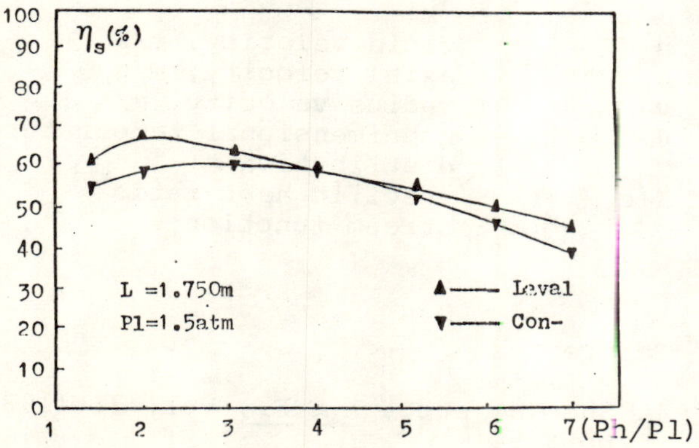

Fig.(5) Effect of expansion ratio on η_s with constant Pl

The back pressure has some effect on the value of η_s when ε/is constant, as shown in Fig.(6), ε=2.0. High η_s can be obtained while Pl=1.5-2.5 atm with a Laval nozzle. This is because every factor which leads to energy losses is

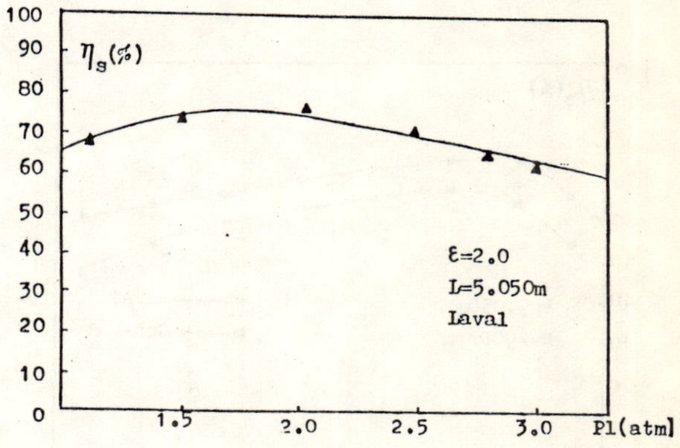

Fig.(6) Effect of Pl on η_s with constant expansion ratio

directly or indirectly related to the back pressure. A different factor has a different effect on the η_s values and the total energy losses have minimal values when the back pressures change from 1.5~2.5 atm.

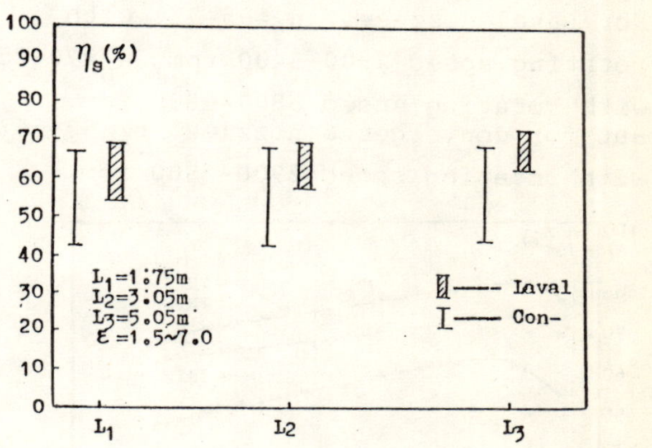

Fig.(7) Effect of the length of PPT on the η_s of RHS

As shown in Fig.(7) with PPT L=1.75, 3.05,5.05 m, ε=1.5-7.0, the longer the PPT, the higher the η_s and the smaller the η_s changing range compared with convergence nozzles. So we know Laval nozzles are able to increase the η_s and decrease the η_s changing range.

In general RHS works with constant rotating speed. But the speed transmision system would increase the size of the RHS's construction. The results are as Fig.(8) with rotating speed 2900 rpm. For Laval nozzles, η_s is 60-75%, but for convergence nozzles, η_s is 50-65%.

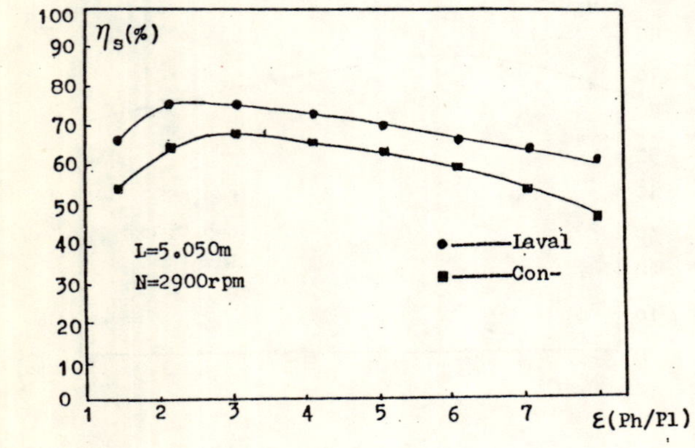

Fig.(8) Effect of expansion ratio on the η_s value with constant speed

On the optimal thermal conditions and with optimal frequency matching, the test results are plotted in Fig.(9). For Laval nozzles, η_s=65-75% with rotating speed 2900-3400 rpm; η_s=70-85% with rotating speed 5800-6800 rpm. But for convergence nozzles. η_s=50-65% with rotating speed 2900-3500 rpm.

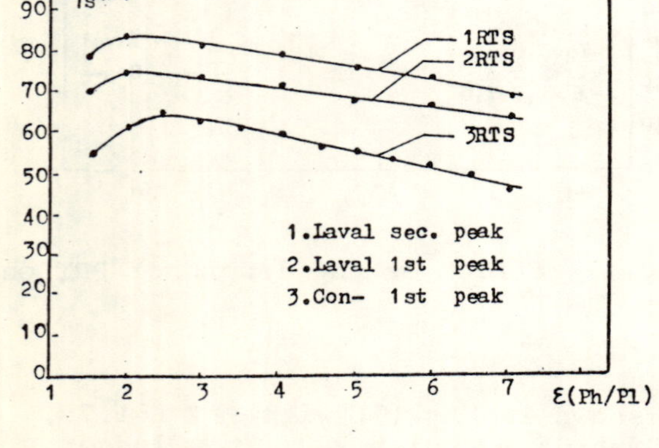

Fig.(9) The isentropic efficiency of RHS with optimal matching

Conclusions

1. The heat dissipating characteristics of the PPT can be approximately solved with a conductive convection conjugate heat transfer model of a hollow pin with variable inner heat fluxes. Computational results are in satisfactory agreement with the experimental ones.

2. Compared with convergence nozzles, Laval nozzles are able to increase the isentropic efficiency and decrease the changing range of the isentropic efficiency of RHS, high η_s can be obtained (η_s=65~75%) with rotating speed 2900-3400 rpm and pressure expansion ratio 1.5~7.0.

Nomenclature

$\varepsilon(x)$	inner time-averaged heat flux, W/m^2
L1,L2,L5,L4,	heat flux zone division, c.f. Fig.(1), m
$f(\xi,\eta)$	nondimensional stream function
α	convection coefficient, $W/m^2, ^oK$
$h^*(\xi)$	nondimensional heat transfer coeff. $W/m^2, ^oK$
Kf	conductivity of atmosphere (air), $W/m. ^oK$
λe	conductivity of PPT, $W/m. ^oK$
Nc	conductive convection conjugate parameter
Re	Reynolds number
Pr	Prandtl number
r0	outer radius, m.
T0	base temperature of the pin, ^ok
u_∞	fluid velocity, m/s
u	axial velocity, m/s
v	radius velocity, m/s
θ	nondimensional temperature distribution
$K(c_p/c_v)$	specific heat ratio
	stream function

References

1. Shao J. etc., A.C.E., Vol. 31 (1986) pp 685.
2. Shao, J. etc., Int. J.Cryogenics, Vol. 26, No. 11 (1986), pp 634.
3. Shao, J. etc., Proc. ICESR, Xi'an, China, (1986)
4. Shao, J. etc., Proc. ICEC, Berlin-West (1986)

5. Shao, J. etc., Proc. SCEC'85,B-12, Shanghai, China (1985), pp 93.
6. Shao, J. etc., J. Zhejiang University, Vol. 18, No. 3, (1984)
7. Shao, J. etc., J. Zhejiang University, Vol. 22, No. 5, (1988)
8. Wang, J.F., M. D. thesis of Zhejiang University, (1989)
9. Gao, J.L., M. D. thesis of Zhejiang University, (1988)
10. Sparrow, E.M., J.Heat Transfer, Vol. 104 (1982), pp 204.
11. Huang, M.J., Int. J.Heat Mass Transfer, Vol. 28, No. 3, (1985), pp 523.
12. Luikov, A.V., Heat and Mass Transfer, Mir-Publishers, Moscow, (1980), pp334.

HEAT EXCHANGE ON POROUS COATINGS IN SUPERFLUID HELIUM

N.M.Grigorenko, N.M.Volkova,
A.V.Kondrashkin, E.V.Zhuravleva
Scientific and Industrial Association of
Helium Engineering Moscow, USSR

The results of studies on He II heat exchange on porous surfaces have shown that an important role in this process is played by specific mechanisms of heat transfer, characteristic of He II only, which leads to a behaviour that conventional liquids do not demonstrate.

Some of these properties displayed by He II heat exchange on porous coatings were pointed out in reference /1/. That paper studied heat transfer through a thin-walled disk, one or both sides of which had a porous coating. The disk was squeezed between two sections with He II. The lower section was connected to a bath with saturated He II; its temperature was controlled by vacuum pumping. The upper section was connected to a vessel with pressurized He via a heat exchanger. A heater in the upper section maintained the He II temperature on a higher level than that in the lower section. The maximal temperature of He II in the upper section was 2.17K. The methods of the investigation made the liquid boiling impossible, so the studies were carried out only in the Kapitza regime.

The experiments demonstrated a proportional increase (up to 5 times) in transferred heat due to several coating layers of fraction powders (90μm-10μm). Conventional classic liquids do not possess such properties: under the conditions of natural or forced convection the effect of these coatings on the heat flow would be negligible.

This paper is an attempt to explain the phenomenon of a sharp increase in the heat exchange rate (confirmed by our experiments) by the intra-pore internal convection of the superfluid and normal components. This convection is described in terms of the presentday concepts of heat transfer in channels under these conditions.

The authors studied the integral effective coefficient of heat conductance (heat exchange) and critical heat flux (corresponding to the onset of boiling) in accordance to the geometries of coatings and physical experimental conditions.

To this end, our methods differed from those used in reference /1/. First, our experiments measured the temperature of the wall with a coating and, secondly, the heat flux was measured within a wide range including critical values. Our investigations were carried out on samples shown in Fig.1. The end face of cylinder 1 was sintered with a copper powder of various fraction: 20,60,80 and 100 μm. The coatings had a diameter of 10 mm and a depth of 3-10 layers. The purity of the sintered powder was measured by quantitative microanalysis. The coating structure was estimated with the Epignant electronic microscope. As a result, the pore size distribution, as well as the mean, maximal and minimal diameters in various coating cross-sections were obtained. The powder was being sintered in hydrogen at 1050°C during two hours. A bifilar heater 2 made of constantan wire was wound on the opposite end of the copper cylinder. Three resistance thermometers 3, resistance type, were mounted in line with the experimental sample at different heights. The distance between the heater and thermometers was large enogh to provide for the equalization of the temperature profile across the cylinder in the range of measurements. The copper cylinder was soldered into vacuum casing 4 through thin-walled stainless steel spacer 5. To prevent heat leaks from the copper cylinder to the He II bath the end face of spacer 5 was insulated with a layer of epoxy adhesive 6 that was renewed prior to each experiment. The vacuum casing was a long pipe that ended in pressuretight joint 7 for current leads. The sample under investigation was placed in helium dewar 8 with a nitrogen screen so that the upper end of the vacuum casing with the pressuretight joint was in the vapor region. In the experiments the He II temperature was kept constant with vacuum pumping. To eliminate radiation heat leaks to the copper cylinder copper baffle 9 was mounted in the upper portion of the vacuum casing.

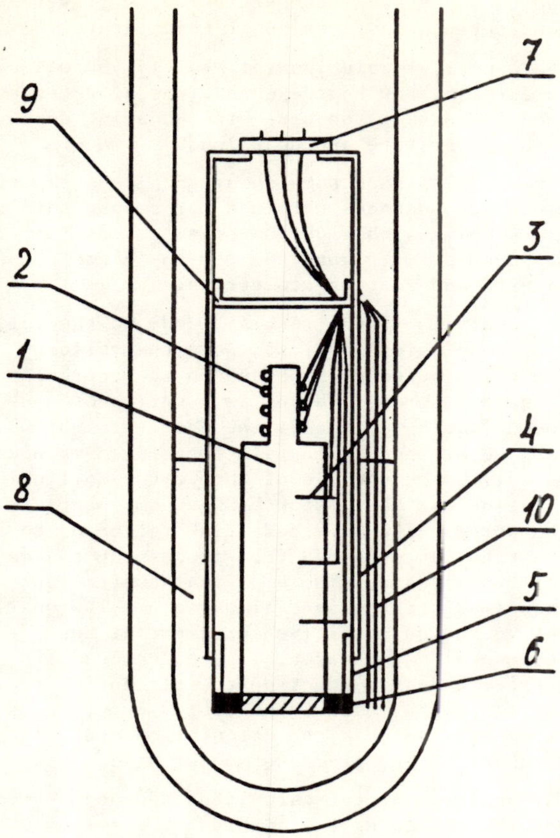

Fig. 1 Experimental set-up

To ensure operation at a low level of He II in the dewar, wire bundle 10 was soldered to the casing wall in the screen area, its free end reaching the bottom. The motion of the He II film along the wire and its evaporation caused the screen temperature to remain low. To minimize heat input due to the conductance of the current leads the latter were fixed to the cop per baffle.

Many experiments on metal properties and thermal physics preceeded the investigations of heat exchange on porous coatings. The preliminary tests included: 1) a study of the heat conduction coefficient of the copper cylinder needed to calculate the temperature of the heated wall; 2) the determination of the geometric characteristics of the porous coating; 3) the determination of the porous material pupurity and 4) a study of Kapitza conductance in materials used as powders. The heat conduction coefficient of the cylinder material was experimentally determined on the same sample before a porous coating was applied to it. The investigations were performed at various temperatures of He II in the dewar. From 1.6 to 10K they gave heat conduction coefficients of copper used for the samples.

The effectiveness of finning was estimated with a mathematical model where a porous coating was assumed to be a system of non-communicating channels. The channel configuration was determined by statistically processing the data on metal properties. We used samples with a clearly defined maximum of pore size distribution to prevent the pores from being filled with small particles and a denser packing with lower porosity and higher non-uniformity being formed (Fig. 2).

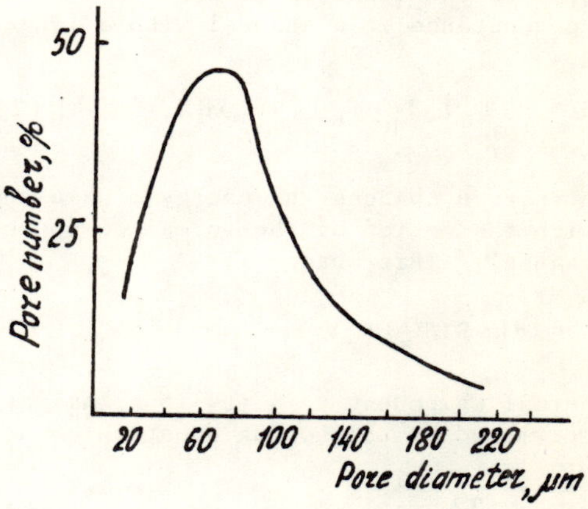

Fig. 2 Histogram of pore size distribution.

The processing of the obtained data made it possible to produce a geometric characteristic of a model for each type of sintered powders. Samples of powder materials were also tested for their Kapitza conductance. These data were used to calculate heat exchange in porous coatings. The samples were identical to the earlier ones.

The sample end surface under investigation was subjected to turning. Some samples were additionally annealed at 850°C with susequent water cooling and air storage. Conductance was studied at He II bath temperatures from 1.5 to 2.1K. The investigations demonstrated that annealing increases the conduction coefficient almost twofold; the effect of surface purity was not observed.

In a first approximation, heat exchange in a porous coating seems to be analogous to heat transfer on a finned surface. Heat transferred from the initial surface by conduction of a fin is removed by the liquid due to heat exchange. However, this process in He II has its peculiarities: 1) the heat exchange (heat conduction, in this case) is proportional to the third power of temperature and, consequently, rises with a decrease in the spacing between fins and with a load increase: 2) heat in the interfin spacing is transferred due to internal convection which is a very strong mechanism of heat transport smoothing out the temperature gradient in He II; 3) all the processes, i.e. heat conduction and internal convection greatly depend on the temperature of He II in which the heat exchanging surface is submerged; 4) heat conduction coefficients at He II temperatures are very low.

The process under discussion was analysed using the solution of the following equations:

A change of the heat flux due to the fin conduction determined as

$$dQ = -d[\lambda_\mu(T_\mu)\frac{dT_\mu}{dX}F_\mu(X)] \qquad (1)$$

is equal to the quantity of heat removed by heat conductance in a channel with a length of, dX

$$dQ = h(T_*)[T_\mu - T_*]\pi D(X)dX, \qquad (2)$$

which in turn changes the energy flux associated with the motion of the normal component in the channel. This change is

$$dQ = d[\rho S(T_*)T_* V_n \frac{\pi D^2(X)}{4}] \qquad (3)$$

The normal component velocity in a laminar flow is determined by the London equation:

$$\frac{dT_*}{dX} = 32\frac{\eta}{d^2}\cdot\frac{V_n}{\rho S} \qquad (4)$$

whereas in a turbulent flow it is obtained from the Gorter-Mellink equation:

$$\frac{dT_*}{dX} = \frac{A\rho_n\rho^3}{S^3\rho_s}\cdot V_n^3 \qquad (5)$$

Boundary conditions for Eq.(1)-(5) are determined as follows:

- at the coating base - by the equality of heat flows

$$\frac{\lambda_{cm} dT_{cm}}{dX} = h(T_*)[T_{MO}-T_*]$$

where the temperature of the fin and the disk outer wall

- at the outer face of the coating - by the equality of heat flux determined by the fin heat conduction, and heat flux removed by He II

$$\lambda_M(T_\mu)\frac{dT_\mu}{dX} = h(T_*)[T_\mu - T_*]$$

as well as by the equality of the He II temperature at the pore inlet and the He II temperature in the bath.

Here T_μ and T_* are current values of the metal and He II temperature, respectively; λ_M is the metal heat conduction; ρ, ρ_n, ρ_S - are densities of He II, normal and superfluid components, respectively; V_n is the normal component velocity; S is entropy; A is the Gorter-Mellink constant; Q it the heat quantity; F_M is the fin cross-section; D is the channel diameter; h is the Kapitza conductance coefficient; X is the coordinate. The subscripts are as follows: M is metal, $*$ is liquid, cm is the wall at the coating base.

The thermophysical properties of He II were considered as a function of temperatures.

The numerical solutions of Eq. (1)-(5) was used to estimate how heat exchange was affected by such factors as the pore size, coating depth, bath temperature and heat load.

Fig. 3 and 4 show some theoretical and experimental dependences obtained for smooth samples and samples with a porous coating. Similar dependences were obtained for any other temperatures and coating geometries.

The analysis of the results leads to the following conclusions. First, note the validity of the calculation model based on the representation of a porous coating as a series of non-communicating channels, the geometries of which were determined by the aforementioned methods, as well as on the use of the Gorter-Mellink equation for the motion of the components in the pores. The Gorter-Mellink factor up to a constant was obtained from processing the experimental data. The empirical coefficient cannot affect the general pattern. The qualitative and quantitative agreement of the experimental and theoretical dependences of the effective heat conduction coefficient on the coating depth, bath temperature, etc., leads us to believe that the calculation model is adequate to the processes under discussion.

The analysis of the calculated dependences gives the following explanation of the experimental data. The heat exchanging surface quickly develops with a decrease of the sintered powder diameter.

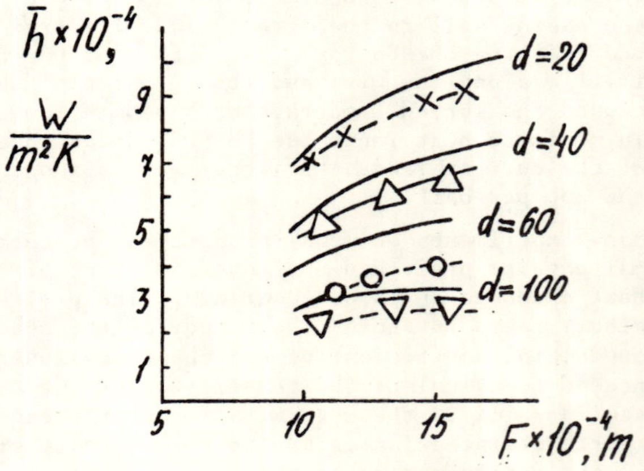

Fig. 3 The reduced coefficient

$$\bar{h} = \frac{Q}{\Delta T\cdot F_2\Lambda ag}$$ versus the calculated heat exchange surface with varous diameters d of the sintered power.

T = 1.6K; Q = $0.9.10^4$ W/m^2; theory: ————;
experiment: x - d = 20 μm, Δ - d = 40 μm, o - d = 60 μm, ∇ - d = 100 μm

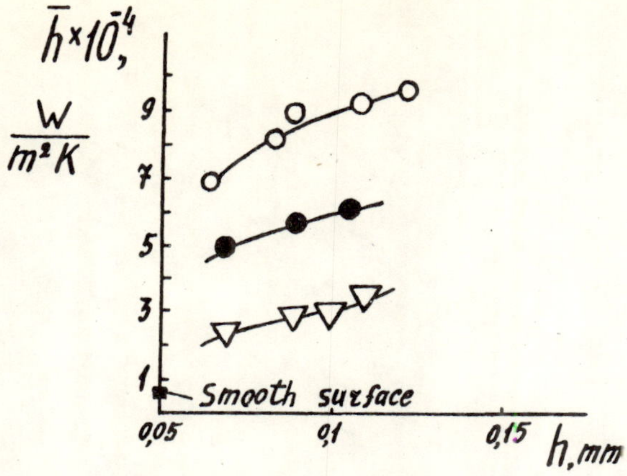

Fig. 4 The reduced coefficient of heat conduction $\bar{h}$ versus various bath temperatures T_b and the coating depth h. d = 20 µm; Q = $0.8 \cdot 10^4$ W/m^2; experiment: o - T_b = 1.6K, • - T_b = 1.66K, ∇ - T_b = 1.8K, ■ - T_b = 1.6K (smooth surface)

However, the gain in the heat exchange rate, as is mentioned in reference /1/ is not equal to the growth of the heat exchange surface although the proportionality of these values is evident (see Fig. 3). The authors of reference /1/ relate this fact to the coating imperfection, the existence of dead pores and the increase of the Kapitza resistance. The analysis of the calculated dependences shows, however, that this phenomenon, taking into account structural defects, should be attributed to the very nature of heat transfer in a porous coating. The internal convection of the components that provides for the heat transport in the pores causes the temperature of He II in the pores to increase. If the temperature rises the heat conduction coefficient increases whereas the temperature head drops. In the 1.6K-2.17K temperature range the added effect of these dependences results in the decrease of the quantity of the transferred heat. A greater effect on the heat exchange rate is caused by the inefficiency of the coating fins due to low heat conduction of the metal. The fin temperature approaches the He II temperature when the coating depth is 2-4 layers. The studies have shown that the lower the He II bath temperature and the greater the pore diameter, the greater the coating depth should be. (See Fig. 4). Indeed, in both cases the quantity of heat removed from the fins decreases while the temperature at the fin base remains high enough. Critical heat flux resulting in a vapor phase on smooth surfaces has been investigated by many researchers since these values are the most important characteristics used in the design and operation of various superconducting devices.

The values of critical heat flux obtained on samples without a coating at different temperatures and submersion depths agree with the data in the literature.

Data on critical heat flux on porous coatings have not been published. This paper suggests the following approach to the determination of these values. According to reference /2/, critical heat flux under the conditions of internal convection in saturated He II occurs when helium in the channel is superheated to the values presented in Table 1.

Table 1

He II temp. in channel	1.6	1.7	1.8	2	2.1
Max. He II temperature in channel	1.9	1.95	1.9	2.05	2.12

These data were used to numerically determine the critical heat flux in a porous coating. Eq.(1)-(5) were solved using the following boundary conditions and additional equations of energy balance.

The He II temperature at the fin base is equal to T_{max}; the critical heat flux is

$$Q_{KP} = h(T_{max}) \cdot (T_{cm} - T_{max}) \frac{\pi D^2}{4} + \lambda \frac{dT_{cm}}{dX} F_M$$

The obtained critical heat flux was compared to the experimental values (Fig.5). The analysis of this comparison shows that in a first approximation the calculated dependence $q_{kp}=f(T_b)$ agrees with the experimental data in the pattern and order of its values.

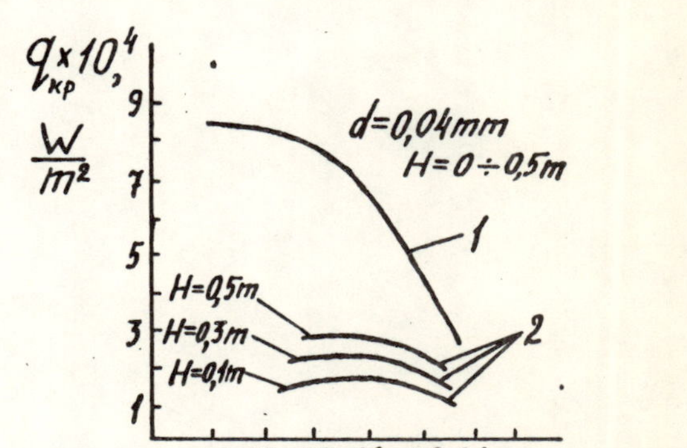

Fig. 5 Critical heat flux q_{kp} versus bath temperature T_b and submersion depth H; ——— - smoothed-out experimental curves: 1 - smooth surface; 2 - porous coatings

In contrast to smooth surfaces, on porous surfaces does not depend on the depth of sample submersion (see Fig.5). The result is identical to the data obtained from investigations of critical heat flux in channels. It can be explained by the high value of pressure loss during the normal component motion in the pores compared to the magnitude of the hydrostatic column.

The principal conclusion is that it is possible to intensify heat exchange and critical heat flux on a surface with porous coatings made of copper powders of fine uniform fraction with a depth of 3-5 layers.

REFERENCES

1 Khalil A. 'performance of sintered copper heat exchange surfaces in helium II'. Proc. 10th Int. Cryog. Eng. Conf., 1984, pp 483-486.

2 Krafft. Report KFK-1786, Karlsruhe NRS, ANL-TRANS-1032 (April 1973).

AN EXPERIMENTAL STUDY ON HEAT TRANSFER ENHANCEMENT OF AMMONIA FLOODED EVAPORATOR AT AIR-CONDITIONING CONDITIONS

Zhuang Lixian Huang Quanxing Lu Yingsheng Deng Songjiu

Chemical Engineering Research Institute
South China University of Technlolgy

ABSTRACT

This paper is an experimental report on heat transfer enhancement of ammonia flooded evaporator using a tube bundle of carbon steel T-shaped fin tube. The experimental results have shown that the T-shaped fin tube is very effective for ammonia flooded evaporator. Its overall heat transfer coefficient is about 120% higher than that of a plain tube at air-conditioning conditions of low heat flux, or its heat exchange area can be cutdown by over 50%.

The tube side and shell side heat transfer coefficient α_i and α_o of T-shaped fin tube evaporator is as much as 1.75 and 3.9 times of plain tube respectively. The designing correlations of plain tube and T-shaped tube flooded evaporator at air-conditioning conditions were obtained.

The experimental results and mechanisms of heat transfer enhancement of T-shaped fin tube were discussed.

INTRODUCTION

As an efficient element for enhancing pool boiling heat tranfer, T-shaped fin tube (Gewa-T) developed by Wieland Worke, has been the focus of the researchers' attention, since it was reported in U.S. patent (1) in 1979.

Stephan K. et al (2),(3), Marto J.P. et al (4),(5), and Yilmaz, S. et al (6) had studied the performance and mechanism, two-phase flow characteristics of liquid and vapor and critical heat flux of copper T-shaped fin tube respectively. Their studies have shown the superior heat transfer performance of T-shaped fin tube.

In our previous study (7), a transparent model of T-shaped tube was adopted to observe the two-phase flow partern in T-shaped tunnel, the heat transfer performance of T-tube was also studies. However, the copper T-tube can not been adopted in ammonia flooded evaporator and reboiler of oil-refining industry. Thereby, developing carbon steel T-tube and studing its heat transfer performance is significant for ammonia refrigeration, oil-refining and petrochemical industry. Simulating air-conditioning operation conditions, taking liquid ammonia as working medium, the experimental studies of tube bundle of 24 plain tubes and T-shaped fin tubes were carried out. Using Wilson' analytical method, the designing correlations for engineering purpose were obtained.

EXPERIMENTAL SET-UP AND PROCEDUER

The experimental studies on flooded evaporators were made with the apparatus shown in figure 1 and the configuration of tested tubes was shown as the photo in figure 2.

Three evaporators were fixed side by side. The refrigerant and refrigerated water can be supplied to indivitual evaporator by means of controlling valves. Evaporator P was made of carbon steel with shell diameter of 219mm and length 1661mm, 24 ϕ 20x2.5x1500mm plain tested tubes were arranged horizontally in tube bundle the total heat exchange area based on the inner tube diameter was 1.696m^2. Evaporator T was the same shell diameter as evaporator P and length of 1361mm, 24 ϕ 20x2.5x1200mm T-shaped fin tubes were arranged in tube bundle. The total inner heat exchange area was 1.356m . The structure parameter of T-shaped fin, the pitch b, the mean width of slit e, the fin height h and the channel width 2a was 2.15mm, 0.258mm, 1.008mm and 0.773mm respectively. Evaporator (T+TC) was a bilateral augmented one, on which the outer surface of tubes was T-shaped fin and the inner was transversally corrugated surface. Unfortunately, several tested tubes were broken by ice while the liquid ammonia was been filling into the set. Because the refrigerated water pump didont worked at that time. So, the data of evaporator (T+TC) had not been measured.

The expermental operation was carried out as

* The project supported by National Natural Science Foundation of China

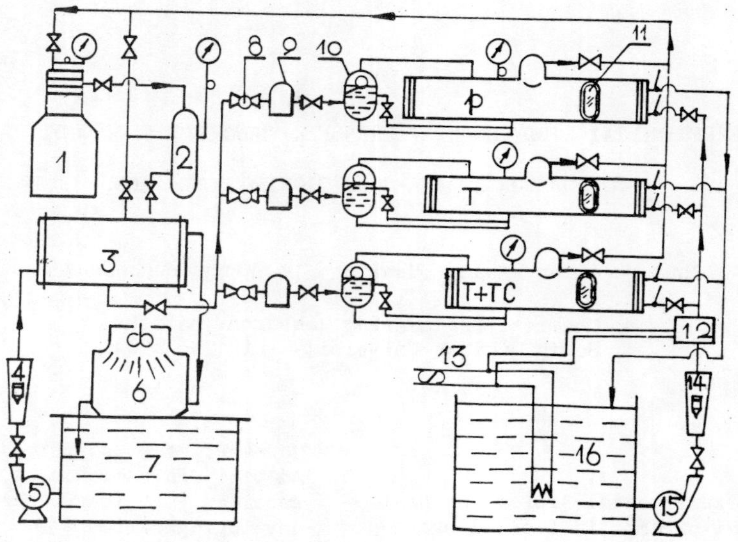

Figure 1. Schematic diagram of experimental set-up

1. compressor 2. oil separator 3. condenser 4,14. flowmeter 5,15. water pump 6. cooling tower 7,16. water tank 8. expansion valve 9. filter 10. float valve 11. glass 12. temperature controller 13. electric heater P-plain tube evaporator, T-T-tube evaporator, (T+TC)-(T+TC) evaporator.

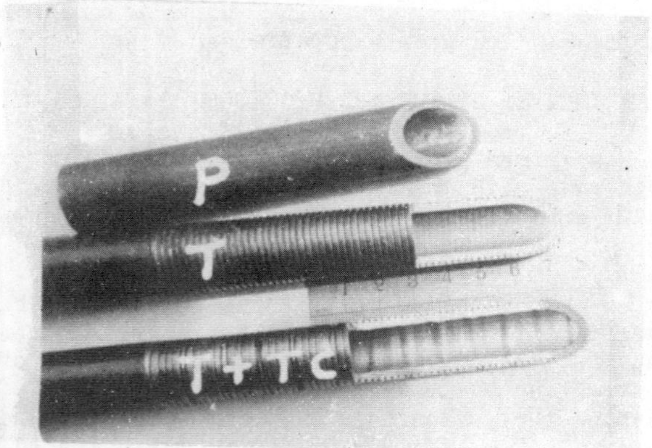

Figure 2. configuration of tested tubes

follows: After de-oil, high pressure ammonia vapor coming from compressor 1 was condensed in condenser 3. After expansion and filtration, the liquid ammonia flowed in-to the evaporator. The liquid level of evaporator could be contr-lled and observed by float valve 10 and glass 11, respectively. In the evaporator, heat required to evaporate ammonia was supplied by circulating water coming from water tank 16 and its flowrate was measured by rotameter 4. The ammonia vapor returned to compresser 1. The circulating water returned to water tand16 and was heated by electric heater 13. Maintaining the vaporating pressure P at the value of 5.16 bar (abs) and circulating water inlet temperature T_i within the range of 10-14°c, changing the flowrate of circulating water in proper order, different outlet temperature T_o of circulating water can be meassured. In addition the heat flux cf evaporator could be changed by means of regulating by-path valve of condenser 3 to returned some high pressure ammonia vapor to compressor 1. A series of overall heat transfer coefficient can be calculated from the experimental data obtained from different steady state operation conditions.

EXPERIMENTAL RESULTS AND DISCUSSION

1. Overall heat transfer coefficient K. Figure 3 showed the relation between overall heat tranfer coefficient K and tube side water velocity W. It can be found that the T-shaped tube have a significant effect of heat transfer enhancement. As an example with the data group of symbol "▼", "●", the overall heat transfer coefficient of T-shaped tube is about 120% higher than that of a plain tube at the same water velocity of 0.8 m/s.

2. Heat transfer coefficient of tube side. Following equation of convective heat transfer was used for water flowing and cooled in tube side.

$$Nu = C_i \, Re^{0.8} \, Pr^{0.3} \qquad (1)$$

According to experimental data of plain tube and T-shaped tube, a serial C_i value could be decomposed by wilson' schematic analytical method (refer to fig.4). Both the values of 0.023 and 0.056 were selected on which the variation of heat flux and scatter of data points were minimun. Thereby, convective heat transfer equations could be obtatned for plain tube and T-shaped tube evaporator,respectivbly.

$$Nu_{i,P} = 0.023 \, Re^{0.8} \, Pr^{0.3} \qquad (2)$$

$$Nu_{i,T} = 0.056 \, Re^{0.8} \, Pr^{0.3} \qquad (3)$$

For T-tube, taking account into the influence of water velocity increase on Reynolds number induced from reduced inner diameter after fabrication. A correcting factor of 0.795 was cited, following equation was obtained.

$$Nu_{i,T} = 0.0402\ Re^{0.8}\ Pr^{0.3} \qquad (4)$$

Thus, the enhancing rate of tube side was $Nu_{i,T}/Nu_{i,P} = 1.75$

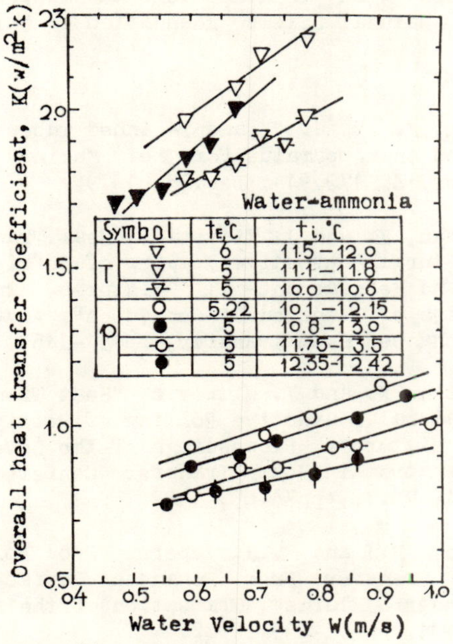

Fig.3.K Versous W

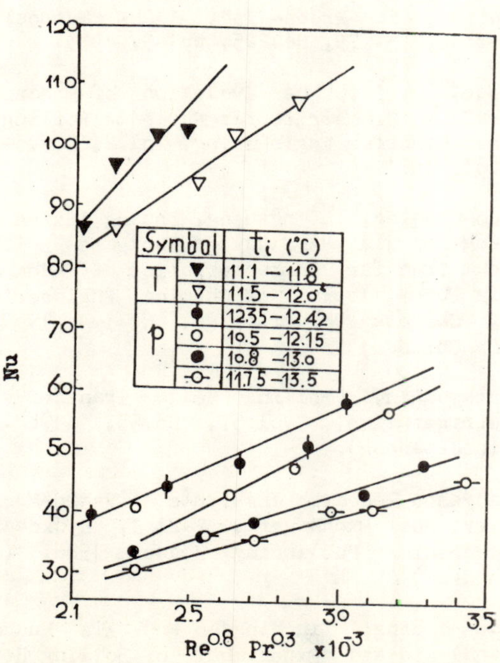

Fig. 4. Nu Versous $Re^{0.8}$ $Pr^{0.3}$ of tube side

3. Heat transfer coefficient α_o of shell side For given medium, reference (8) suggested following expresson of nucleate pool boiling.

$$\alpha_o = CG(R_p)\ F(P) q^n \qquad (5)$$

where, C, $G(R_p)$ and F(P) presents the experimental constant, the influence of surface state and the operating pressure respectively.

Mostinski (9) also has given a nucleate boiling heat transfer correlation and suggested:

$$n = 0.7$$

$$F(P) = 1.8\ P_T^{0.17} + 4\ P_T^{0.12} + 10\ P_T^{10} \qquad (6)$$

where, $P_T = P/P_C$. P_c is the crictical pressure. For ammonia, $P_C = 112.6$ bar. Let $CG(R_p) = C_M$, neglecting the term $10P_T^{10}$, equation (6) and (5) become :

$$F(P) = 0.804\ P^{0.17} + 0.013\ P^{1.2} \qquad (7)$$

$$\alpha_o = C_M(0.804\ P^{0.17} + 0.013\ P^{1.2})q^{0.7} \qquad (8)$$

Regressing the experimental data, equation (9) and (10) were obtained for plain tube and T-shaped tube respectively (see Fig.5)

$$\alpha_{o,P} = 8.14(0.804\ P^{0.17} + 0.013\ P^{1.2})q^{0.7} \qquad (9)$$

$$\alpha_{o,T} = 31.77(0.804\ P^{0.17} + 0.013\ P^{1.2})q^{0.7} \qquad (10)$$

For air-conditioning conditions of P = 5.16 bar, the designing correlations of ammonia

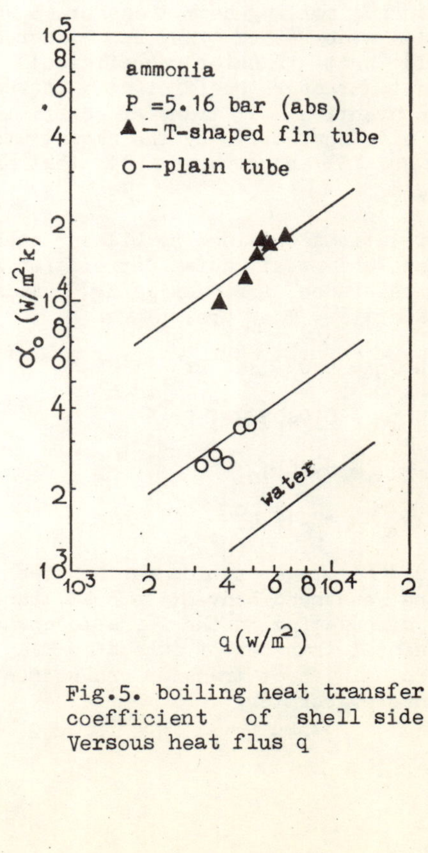

Fig.5. boiling heat transfer coefficient of shell side Versous heat flus q

flooded evaporator for plain tube and T-shaped tube were obtained.

$$\alpha_{o,P} = 9.46\ q^{0.7} \qquad (11)$$

$$\alpha_{o,T} = 37.12\ q^{0.7} \qquad (12)$$

The boiling heat transfer enhancement rate between T-shaped tube and plain tube is 37.12/9.46 = 3.9.

4. Analysis and discuss of heat transfer mechanism. For nucleate boiling of T-shaped tube, two kinds of hypothesis, or "vapor bubble contact length longering" and "couter vapor-liguid phase flowing model" had suggested by Stephan, K. et al (2) and Cao Yiding et al (10) respectively. However, our visual test (7) and Marto's photo experience (4) showed that Stephan's hypothesis was existent but not perfectly sure. Thereby, we suggested summarily that the enhancement of heat transfer of T-shaped fin tube may be resulted from:

(1) The high efficient evaporation of very thin liquid film in the T-shaped tunnel;

(2) High circulation rate of liquid in the T-shaped tunnel promote the convective heat transfer between the liquid on the metal surface and in the mainbody;

(3) The inner fine spiral ridges formed after the T-shaped fins were fabricated mean that the Roughness of inner wall was increased artifically. So, the inside forced convection heat transfer coefficient was also promoted.

CONCLUSION AND SUGGESTION

1. The results of tube tundle test showed, that carbon steel T-shaped fin tube is very effective for enhancing boiling heat transfer of ammonia flooded evaporator. They can increase the overall heat transfer coefficient K, boiling heat tranfer coefficient α_o in shell side and convective heat transfer coefficient α_i in tube side respective by the amount about 120%, 290% and 75% as compared with that of a plain tube;

2. The correlations obtained by Wilson' schematic method and test data correlation have supplied a reference for designing ammonia flooded evaporator. They are:

$$\begin{cases} Nu_{i,P} = 0.023\ Re^{0.8}\ Pr^{0.3} \\ Nu_{i,T} = 0.0402\ Re^{0.8}\ Pr^{0.3} \end{cases}$$

$$\begin{cases} \alpha_{o,P} = 9.46\ q^{0.7} \\ \alpha_{o,T} = 37.12\ q^{0.7} \end{cases}$$

3. The liquid film evaporation in the T-shaped channel, the enhanced convective heat transfer of liquid circulating in the T-shaped channel, the enhancedheat transfer of tube side play the important role in heat transfer enhancement of T-shaped tube evaporator;

4. Limited by the experimental conditions of present studies, the circulating water velocity was kept in the range of 0.5-1 m/s. Higher water velocity was ussally adopted in industrial ammonia flooded evaporator. So, it is better to do some experimental studies at the higher water velocities to simulate the actual operation conditions.

ACKNOWLEGEMENT

The authors wish to acknowlege the support of National Natural Science Foundation of China.

REFERENCE

(1) Saior, M. et al, "Y and T-Finned Tubes and Method and Apparatus for Their Making", U.S. Patent, 4, 179,911. Dec.25, 1979.

(2) Stephan, K. and J. Mitrovic, "Heat Transfer in Natural Convective Boiling of Refrigerants and Refrigerant-oil Mixtures in Bundles of T-shaped Finned Tubes", Advances in Enhanced Heat Transfer, pp131-145, 1981.

(3) Stephan, K. and J. Mitrovic, "Heat Transfer in Natural Convective Boiling of Refrigerant -oil Mixture", Proceedings of the Seventh International Heat Transfer Conference, Vol.4, PB.12, pp73-85, 1982.

(4) Marto, P.J. and V.J. Lepere, "Pool Boiling Heat Transfer From Enhanced Surface to Dielectric Fluids", Transation of the ASME, Vol.104, pp292-299, 1982.

(5) Marto, P. J. and Hernandez, B., "Nucleate Pool Boiling Characteristics of a Gewa-T Surface in Freon-113", AIChE Symposium Series, Vol.79, No.225, pp1-9, 1983.

(6) Webb, R. L., "The Evolution of Enhanced Srrface Geometries for Nucleate Boiling", Heat Transfer Engineering, Vol.2, Nos.3-4, P.63, 1981.

(7) Luo Guoqing, Lu Ingsheng, Zhuang Lixian and Deng Songjiu, "A Study on Nucleate Boiling Heat Transfer Characteristics of T-shaped Fin Tube", Journal of Chemical Engineering of Chinese Universities, No.1, 1989. (in Chinese)

(8) Totagushin, "Boiling Heat Transfer", Refrigeration, Vol.57, No.55, 1982. (in Japanese)

(9) Langzhou Research Institute of Petro-Machinery, Heat Exchangers, Part I, Hydrocarbon Processing Publishing House, 1986. (in Chinese)

(10) Cao Yiding, Xin Mingdao and Xie Huande, "Analysis and Experiment of Boiling Heat Transfer on T-shaped Finned Surface", Journal of Engineering Thermophysics, Vol.7, No.1, pp63-66, 1986. (in Chinese)

CORRECTLY DETERMINING THE WORKING STATE OF NON-AZEOTROPIC MIXTURES AT THE OUTLET OF CONDENSER

Xie Guozhen Wu Yezheng
Division of Refrigeration Department of Power Machinary Engineering
Xi'an Jiaotong University, Xi'an, China

ABSTRACT

Energetic improvements are of special interest. They are achieved when temperature of refrigerant flowing through condenser and evaporator varies considerably. The nonazeotropic mixtures can be used for this purpose. To achieve an improvement over the present level by using the non-azeotropic mixtures, it is necessary to study the thermodynamic and heat transfer process, as the conditions get more complex in the transfer from one-component to multi-component systems. One thing we have to do during the study is to determine the working state of non-azeotropic mixtures at the outlet of condenser, as the relation of pressure with temperature of a great quantity non-azeotropic mixtures has not been known.

The authors have discovered that by changing the heat load of condenser and by observing the pressure at the outlet of condenser in a refrigerating system one can define exactly the working state of mixtures at that place.

To explain the discovery, the view of heat and mass auto-feedback equilibrium has been raised. The view has been verified by test during the author's investigation on the performances of a window air conditioner KC-28S.

INTRODUCTION

In a refrigeration test equipment, it is very important to correctly determine the working state (supercooled or saturated) of refrigerant at the outlet of condenser in a refrigerating cycle.

It was well know that in an azeotropic refrigerating cycle employed single-refrigerant, shown in Fig.1 a-b-c-d-e-a. The relation of saturation pressure with saturation of temperature is often used to determine the working state of refrigerants at point d.

Since the relations of saturation pressure with temperature have not been acquired, we can not determine the working state of refrigerants at point d' in a non-azeotropic refrigerating system employed mixtures, shown in Fig.1 a-b-c-d'-e'-a, by using above method. In this paper, a method applied to determine the working state of mixture at point d' is presented.

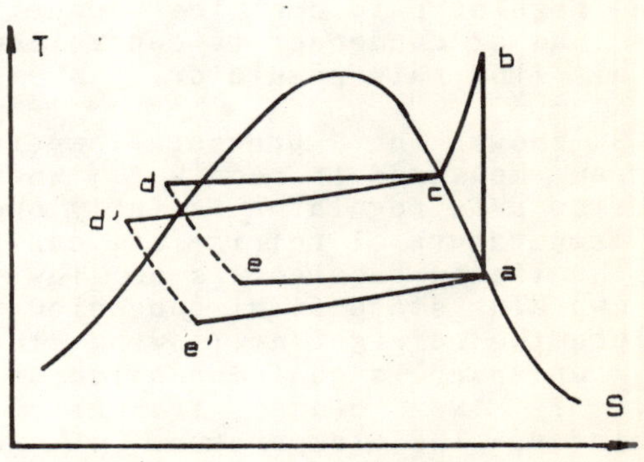

Figure 1

APPEARANCES AND ANALYSIS

During investigating the varying condition performances of a window conditioner, it has been discovered that by changing the pressure at the outlet of condenser we can correctly determine the working state at point d'.

Fig.2 shows the equipment in which the overheated refrigerant vapour coming from the compressor enters the air-cooled condenser from right to left. After being cooled refrigerant enters

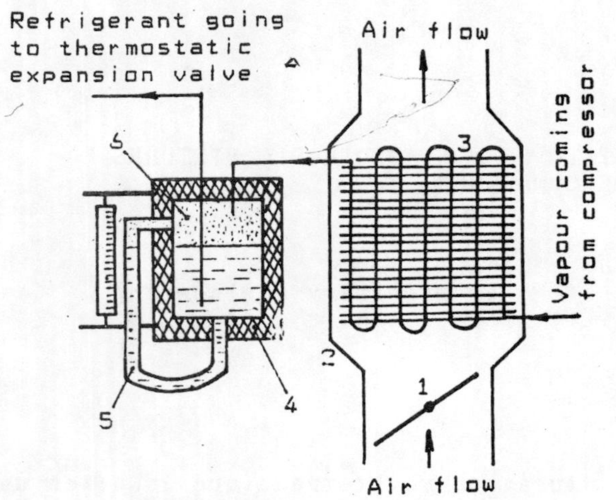

1--Air flow rate regulator
2--Air-cooled condenser
3--Fins
4--Heat-insulating material
5--Level gauge
6--Liquid receiver

Figure 2

the liquid receiver. The receiver is connected with a level gauge made of nylons and is wrapped up with heat-insulating material. Air flows through the condenser equipped with an air flow rate regulator to cool the vapour. The heat load of condenser is controlled by the air flow rate regulator.

Fig.3 shows the pressure-temperature diagram measured in receiver. When the air flow rate regulator is fully opened, the temperature of refrigerant contained in the liquid receiver is so low that the working state is at subcooled point 5 since the refrigerant flowing through the condenser is cooled considerably by air. If we close gradually the regulator, the temperature of refrigerant at outlet of condenser rises and approaches to the saturated state (point 4).

It seems that the working state goes into vapour-liquid zone (two phases zone) if the regulator was closed continuously. As a matter of fact, it is not so. We find that

1) During test the level of liquid receiver is very tall so the refrigerant flowing out the receiver is saturated liquid instead of vapour-liquid mixture.
2) The real pressure-temperature diagram shows that if the air flow rate continuously decreases the

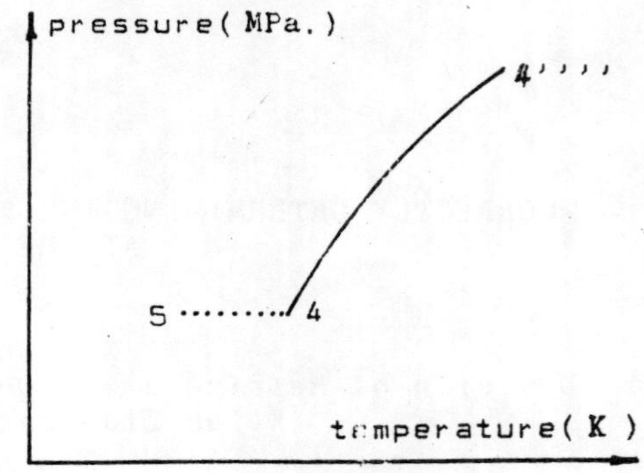

Figure 3 The pressure versus temperature in receiver

working state at point 4 will vary along the path from point 4 to 4''''. This can be explained as follows:

When working state approaches to the point 4, the heat flux released by the refrigerant vapour is equal to the one absorbed by air. If the air flow rate is regulated quite little, the refrigerating system will be in the heat and mass auto-feedback equilibrium stage. As the air flow rate decreases, a small amount of saturated vapour which is not condensed by air enters the liquid receiver. In consequence, the pressure rises, the volumetric efficiency, the mass flow rate M of compressor and the heat flux Qr released by the refrigerating vapour are reduced [1].

Simulaneously, as the temperature difference between the air and vapour increases, the heat flux Qa absorbed by air rises. Finally, the pressure of receiver arrives automatically at a value, at which the term Qr is equal to the term Qa.

The above description can be expressed in Fig.4. As the air flow rate decreases, the working state varying from point 5 (subcooled) to point 4 (saturated) and along the path of 4-4'-4''-4'''-4'''' results in a series of new equilibrium working states. The path 4-4'-4''-4'''-4'''' is saturation curve.

The procedure of auto-feedback equilibrium can be shown as follow:

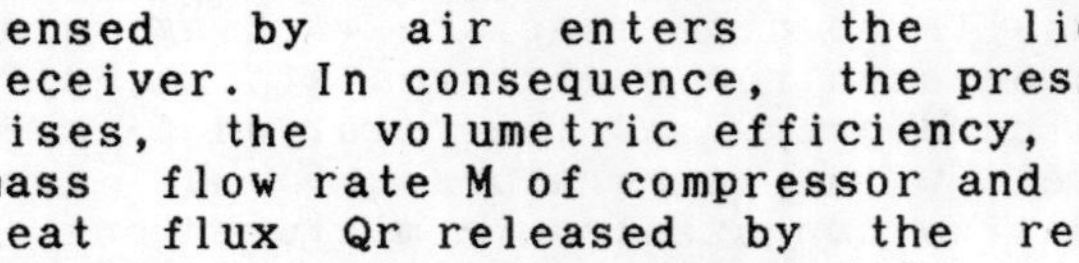

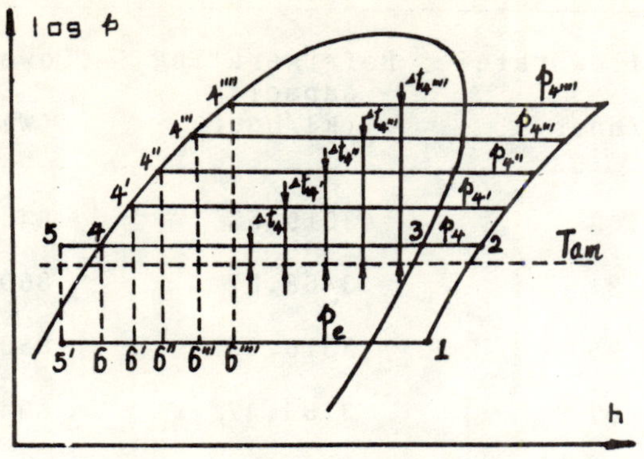

Figure 4

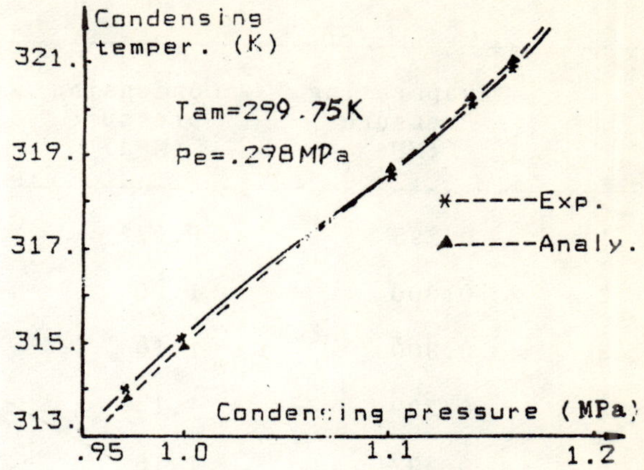

Figure 5 The saturation pressure versus saturation temperature

Table 1

	Ambient temp. (K)	Evapo. pressure (MPa)	Condensing press. tested (MPa)	Staurated temp. tested (K)	Calculating satur. temp. (K)	Relatively deviation of temper.
1	299.75	0.295	0.927	313.89	313.75	-0.044%
2	299.95	0.300	1.00	314.98	314.85	-0.041%
3	299.85	0.300	1.10	318.60	318.73	0.041%
4	2.9955	0.299	1.14	320.19	320.24	0.016%
5	299.65	0.297	1.16	320.91	320.95	0.013%

In short, as the heat load of condenser increases, the working states approach from subcooled zone to saturated zone, vice versa. The constant pressure part is the subcooled zone. The pressure turning point 4 is in the saturated working state.

EXPERIMENT

In order to verify the view raised above, a test equipment has been established by authors on a window conditioner KC-28S. Pressure gauges with high precision have been used to measure pressure. Copper-constantan thermoelectric couples of 0.1 mm diameter have employed to measure temperature. The potentiometer UJ35 has been used as recorder. The apparatuses above have been calibrated. Refrigerant R12 was used as working medium. The results were shown in Table 1 and Fig.5.

Fig.5 and Table 1 show that the relatively deviation of temperature is less than 0.045%. It is feasible to determine correctly the working state of non-azeotropic mixtures at the outlet of condenser by observing the condensing pressure.

Table 2 shows that as the air flow rate decreases the condensing pressure is raised, the power increases, the heat load of condenser, the mass flow rate and the refrigerating capacity reduce. These correspond to the law of heat and mass auto-feedback equilibrium (as Fig.4).

CONCLUSION

The varying law of working state of non-azeotropic mixtures at outlet of condenser has been discovered. The view of heat and mass auto-feedback equilibrium has been raised and explained. This view has been verified by experiment.

Because the relations of pressure with temperature of a lot of non-azeotropic

Table 2

	Evaporating Pressure (MPa)	Condensing pressure (MPa)	Mass flow rate (Kg/hour)	Refrigerating capacity (kJ/hour)	Power (W)
1	0.295	0.972	27.63	3619.53	844
2	0.300	1.00	27.31	3468.37	860
3	0.300	1.10	26.96	3316.08	888
4	0.299	1.14	26.07	3154.47	894
5	0.297	1.16	25.56	3067.20	898

mixtures have not been known, it is beneficial for investigating the performances of refrigerating cycle to determine the working state of non-azeotropic mixtures at the outlet of condenser by test. The method stated above can arrive at this purpose.

NOMENCLATURE

Q Heat flux, J/hour
M mass flow rate, kg/hour
P pressure, MPa
T temperature, K

SUBSCRIPTS

r releasing
s absorption
e evaporation
am anbient
c condensation

REFERENCE

1. Mao Daoping, Reciprocating Refrigerating Compressor, China Machine Press, 1983.

AN INVESTIGATION ON AN AIR-DUCT STRUCTURE FOR THE FREEZING ROOM OF A COLD STORE

Hu Xiongfei, Liu Chuyun, Huang Xiaosong
Refrigeration and Cryogenics Lab, ZheJiang University
Hangzhou, China

ABSTRACT

In this paper the flow-field in the empty freezing room of a cold store is obtained by solving the Eulerian Equation. The insignificant difference of the flow-field obtained by the calculation to the practical one shows that the structure of an air-duct can be optimized by a numerical calculation.

Several new criteria (substandard ratio, substandard degree of frozen food and complete freezing time), to assess air-duct structure are also suggested. From assessing four types of air-duct structure, it is found that a new structure of air-duct is better than that of the others.

INTRODUCTION

The freezing process is a very important link in the production chain of freezing and storing food. A great amount of power is consumed in the process. For example, about 80% of total power is consumed in a meat cold store. Obviously, it will lead to reduced energy consumption if an excellent air-flow in the freezing room can be obtained.

A type of air-duct structure with air-distribution plates of equal span (structure A and B in Table 1),and a structure with no plate (structure C in Table 1) are widely used in China. In order to investigate the effect of the air-duct structure on the flow-field in a freezing room, and to explore a new structure of the air-duct, the flow-field are responding to structures A, B and C were obtained by numerical calculation and experimental measurement respectively. Then a new structure for an air-duct, structure D, was suggested and experimentally measured. Several new criteria (substandard ratio, substandard degree of frozen food, complete freezing time,) to assess the air-duct structure of a freezing room are also presented. The comparison shows that the structure D is the best one in the four air-duct structures.

MATHEMATICAL MODEL OF THE AIR-FLOW AND ITS SOLUTION

In order to describe the distribution of air in an empty cold store and to obtain the best air-duct structure, it is assumed that:

1. The length of the freezing room is much longer than the width and the height; the velocity of air-flow is constant; and because the flow-field in the freezing room is symmetry about the section which is vertical to the predominant direction of cooling units,the three-dimensional air-flow can be simplified to a two-demensional one.

2. The thickness of the air-flow distribution plate can be reglected compared with the width.

3. The flow-field will not change with time.

The General Eulerian equation can be expreseed as following (1):

$$\underbrace{\text{div}(\rho u \phi)}_{\text{convection}} = \underbrace{\text{div}(\Gamma_\phi \text{grad}\,\phi)}_{\text{diffusion}} + \underbrace{S_\phi}_{\text{source}} \quad (1)$$

Where ϕ : all dependent variables;
Γ_ϕ : exchange coefficient related to ϕ ;
ρ : density of air.

The representative values of ϕ, $\Gamma\phi$ and $S\phi$ in Eq.(1) are listed in Table 1.

Table 1. Representative values of variables in Eq.(1)

Eq. of conservation	ϕ	$\Gamma\phi$	S
Mass (continuity)	1	0	0
Momentum in Y direction	u_y	μ	$-\frac{\partial P}{\partial y}$
Momentum in Z direction	u_z	μ	$-\frac{\partial P}{\partial z}$

The boundary conditions are:

(1) The air-flow velocity and the volume at the inlet are given;
(2) The dependent variable at the outlet is determined by calculation;
(3) The flow velocities at the surfaces of the plates and walls are zero.

The SIMPLE method presented by Patankar and Spalding (2) is used here, which involves the definition of a network, a related formula and a low slack control, expressed as follows:

1) The network: Fig 1 shows the net in a flow-field. The values of all the variables except the velocity $\vec{u}$ are stored in the points of intersection of the net. $\vec{u}$ is stored between the neighbouring points of the intersection.

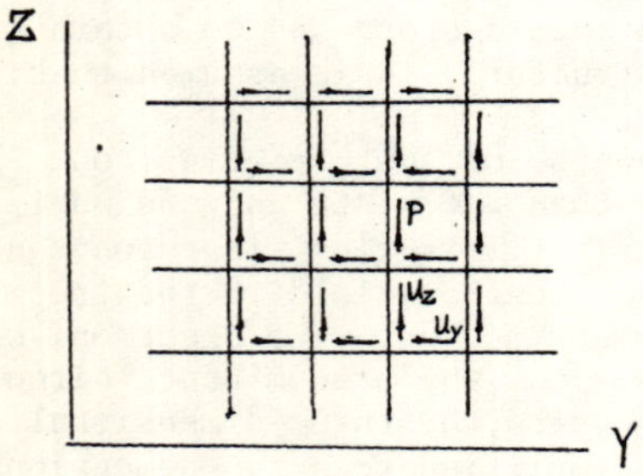

Fig.1 The network

2) Discretization of the equation (1). It is assumed that:
a) The flux properties at the surface of a control volume are homogeneous.
b) The ϕ value between any two neighbouring points changes linearly for the diffusion term.
c) The ϕ value at the surface of a control volume is obtained by using an up-wind method for the convection term.
d) The source term, S_ϕ, in a control volume is homogeneous and linearized as

$$S_\phi = S_u^\phi + S_p^\phi \cdot \phi_p \quad (2)$$

The standard form of the limited difference equation of the Eulerian Equation is expressed as

$$\phi_f = \frac{\sum_{I=E,W,N,S} A_I \phi_I + S_u^\phi}{\sum_{I=E,W,N,S} A_I - S_p^\phi} \quad (3)$$

where A is the convection coefficient and diffusion coefficient;
f is the foot-note, any points of the net;
I is the four directions(left, right, up and down) about point f.

Eq.(3) can be solved by the Gauss-Seidel method. In the process of calculation, continuity of mass is kept by a pressure correction equation, which is obtained by the use of the continuity equation. The form of the pressure correction equation is as same as Eq(3), and can be written as

$$P_f' = \frac{\sum_{I=E,W,N,S} A_I P_I' + S_u}{\sum_{I=E,W,N,S} A_I - S_p'} \quad (3')$$

The steps for solving Eq(1) by using a computer are as the follows:
1. Estimate the pressure-field p^*;
2. Solve the momentum equation, obtain u_y^* and u_z^*;
3. Solve the corrected equation of pressure p';
4. Calculate a new pressure-field, $p = p^* + p'$;
5. Calculate a new flow velocity, i.e. u_y' and u_z', then obtain u_y and u_z

$$u_y = u_y^* + u_y', \qquad u_z = u_z^* + u_z';$$

6. The calculation will be stopped when $|p-p^*| < E$;
7. Otherwise, repeat steps 1 to 6.

THE PRESENTATION OF A NEW TYPE AIR-DUCT STRUCTURE AND ITS ASSESSMENT

The freezing room being investigated is shown in Fig2. Its width, height and length is 5.70m, 4.70m and 18.60m respectively. The distance of the air-distribution plates from the floor is 2.80m. The measurements of the air-distribution plate and the air-duct are shown in Table 2 .

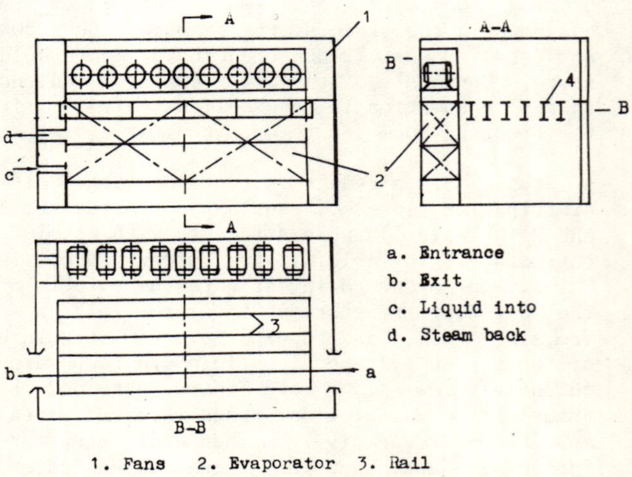

Fig 2 The structure of the investigated freezing room

The points which are investigated here are:
y= 1.38, 1.95, 2.64, 3.16, 4.06, 4.77, 5.46 (m)
z= 0.68, 0.94, 1.17, 1.45, 1.68, 1.93, 2.19 (m)
when x=3.00, 9.30 and 15.60m

Table 2 Specification for the air-distribution plate and the air-duct

structure	width of plate (mm)	width of air-duct (mm)
A	600	60
B	500	160
C	no plates	

The results of measurement and calculation (Fig.3 A,B,C) show that the flow-field obtained by numerical calculation is a close approximation to that of the experimental one. The relative error is within 8%, and the relative error of mean values about the y direction is 1%. It is shown that the air-duct structure could be optimized by the numerical calculation.

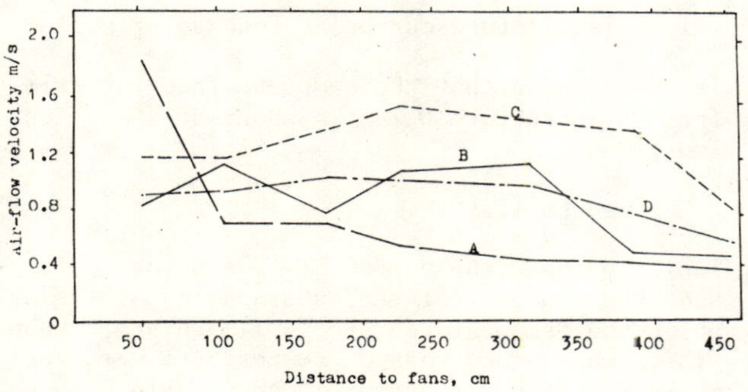

Fig 3 The results of measurement and calculaion for structures A,B,C,D

On the basis of these results,a structure D,(Fig.4) which has unequal width of air-duct has been suggested. The cycle volume of flow, the flow-field, the temperature-field and the weight loss of food corresponding to the structure have been measured.

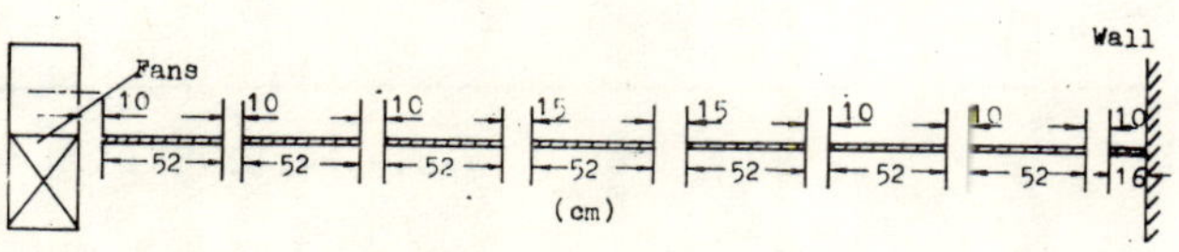

Fig. 4 Structure D

The conventional assessment on the air-duct structure is based only on the temperature and weight loss of the frozen food. It is not very reasonable. Therefore we introduce the follow concepts.

1. Substandard ratio Sr

The substandard ratio of a frozen food is defined as the follows: After a certain freezing time, the ratio between the food of which the temperature is still higher than that required and all required frozen food. This ratio may be expressed in weight,volume or quantity. Normally, an average temperature value of the center and the surface of the frozen food is the so-called food temperature.

2. Substandard degree Sd

A substandard coefficient H of the food is firstly defined as:

$$H=\begin{cases}0 & \text{when } (T-Ts)<=0 \\ 0.1\,[\text{Integer}(T-Ts)+1] & \text{when } (T-Ts)>0\end{cases} \quad (4)$$

where Ts: required freezing temperature;
T : freezing temperature.

The substandard degree then is defined by Eq(5)

$$S_d = \frac{\sum_{I=1}^{n} H_I \cdot W_I}{\sum_{I=1}^{n} W_I} \cdot 100\% \quad (5)$$

Where W_I: weight or volume of the food;
n : total quantity of food.

For the same reason,Eq(5) can be rewritten as

$$S_d = \frac{\sum_{I=1}^{n} H_I}{n} \cdot 100\% \quad (5')$$

3. Complete freezing time τ

The freezing time is the time required for the food to be cooled to a required temperature. For a certain batch of food, the freezing time should correspond to a certain substandard ratio, because the freezing time varies with different substandard ratios.

The complete freezing time τ is defined as the required freezing time when a zero substandard ratio is reached.

Obviously, for a given volume of flow and evaporation temperature of the refrigeration system of the cold store, the more homogeneous the flow-field, the shorter the complete freezing time.

EXPERIMENTAL RESULTS AND DISCUSSION

1. The cycle flow volume and the homogeneity of air-flow velocity.

The experimental data for the room under study which was occupied by food is shown in Table 3. It was found that for the cycle volume of air-flow, C D B A; for the homogeneity of air flow velocity, D C B A;

Table 3 Comparison of cycle flow volume and flow velocity

structure	cycle volume of flow (m3/h)*	average flow velocity (m/s)	homogeneity of flow velocity**
A	76147 (79017)	0.594	worst
B	78445 (85770)	0.713	good
C	119370 (129838)	0.940	bad
D	96422 (112995)	0.821	better

* the values in brackets are the cycle volume of flow of an empty room.
** the homogeneity of flow velocity is judged by the area which is surrounded by the curve of experimental flow velocity and the curve of experimental average flow veloeity i.e. u=const.

Obviously, comparing with structure C, the cycle volume of flow for the structure D is somewhat lower, due to the increase of flux resistence, but the homogeneity of the flow velocity is much better.

2. History of the cooled food

Fig 5 shows the temperature history of the cooled food. Table 4 shows the time needed for the center temperature of food from a value cooling to another, and also shows the complete freezing time. The initial temperature of the food center is adopted as 15 °c , and the required temperature is -15 °c.

Table 4 The freezing time, h

structure	15～-10 (°c)	4～-2 (°c)	Complete freezing τ
A	11.85	6.80	14.80
B	12.20	6.25	14.30
C	10.85	5.25	13.50
D	10.55	2.75	12.40

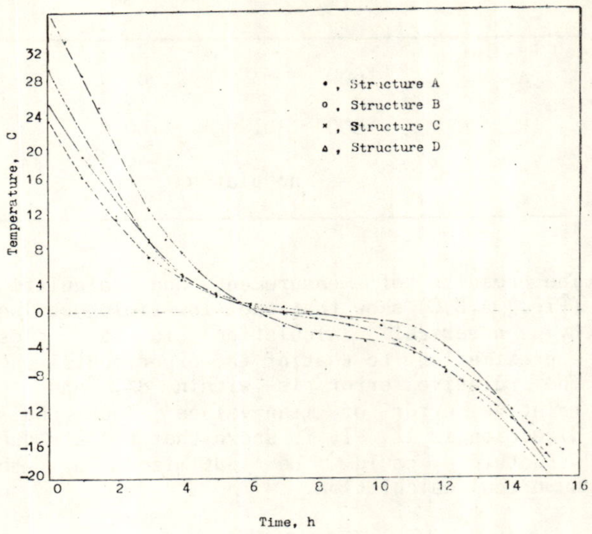

Fig 5 Temperature curves of food being cooled

It was shown that the complete freezing time for the structure D is the shortest among the four.

3. Frozen quality

Table 5 shows the number of measuring points and the number of the substandard points at which the temperatures are still higher than -15 °c when the average temperature of all measured points reaches -15 °c. Table 6 shows Sr and Sd of the frozen food under two conditions for various types of structure of air-duct.

Table 5 The number of points at which the center temperature of the food is still higher than -10 °c(corresponding to average temperature of -15 °c) and their values for the center temperature.

structure	number of measuring points	number of sub-standard points *
A	8	4(-9.7,-5.4,-4.9,-5.5 °c)
B	8	4(-5.4,-8.8,-7.6,-5.1 °c)
C	8	3(-2.6,-3.2,-9.1 °c)
D	13	5(-6.9,-9.6,-6.4,-4.5, -7.4 °c)

*:The values in the brackets are center temperatures of substandard points.

Table 6 A quality comparison of frozen food

condition	structure	Sr (%)	Sd (%)
I	A	50.0	21.3
	B	50.0	18.8
	C	37.5	20.0
	D	38.5	13.8
II	A	62.5	35.0
	B	50.0	26.3
	C	37.5	23.8
	D	0	0

* condition I:The average temperature -15 ℃ of all the measured points was reached; condition II: Sr for the structure D is zero;

Under condition II, the freezing time for the various structures is the same. Therefore, the comparison is more reliable than that under condition I. Results in table 5 and table 6 show that the quality of frozen food under the condition of the structure D is the best among the four.

4. Weight loss of food

Table 7 shows that the weight loss of the structure D is 73% of that of the structure A.

Table 7 value of weight loss

structure	value of weight loss (%)
A	2.06
D	1.50

5. Power consumption

Power consumption in a freezing process is approximately proportional to the freezing time as shown in Table 8.

Table 8 Relative value of power consumption for various air-duct structure, %

structure	when average temp. of all food reached the required temp.	according to the complete freezing time
A	100	100
B	103	97
C	92	91
D	89	84

CONCLUSION

1. The structure of an air-duct in a freezing room can be optimized through numerical calculation by using a computer;

2. Based on the new concepts of substandard ratio, substandard degree and complete freezing time, the effect of homogeneity of flow-field and the value of air-flow velocity on the food freezing can be accurately described, and the structure of an air-duct can be assessed more reasonable.

3. The structure D is the best one among structures A, B, C and D. It is recommended that in order to obtain a best structure of the air-duct, further studies are required.

REFERENCES

1 Patankar, S.V. and Spalding, D.B. 'A calculation procedure for heat, mass and momentum transfer in three dimensional parabolic flous'. Int. J. Heat Mass Transfer, Vol.15 (1972), pp 1787.

2 Patankar, Ś.V. 'Numerical Heat Transfer and Fluid Flow'. MCGRAW-HILL Book Company, New York(1979).

SIMULTANEOUS HEAT AND MASS TRANSFER IN FILM ABSORPTION ON HORIZONTAL TUBE BUNDLES

Tang Jingsong, Lu Zhen, Wang Jian, Liu Huanchen, Yu-chi Bin
Department of Power Machinery Engineering
Shanghai Jiao Tong University
Shanghai 200030, P. R. China

ABSTRACT

A theoretical analysis of simultaneous heat and mass transfer in falling film adsorption on horizontal tube bundles is presented. The distributions of temperature and concentration across the film, the effects of tube span and spraying density on heat and mass transfer are discussed. An experimental setup has been established to testify the theoretical model. The experimental results are satisfactorily in agree with the theoretical solution, which indicates that the presented model is well enough to describe the real falling absorption process.

INTRODUCTION

LiBr/H2O absorption refrigerators and heat pumps, which may be powered by low potential thermal energy, have been widely used in refrigeration, air conditioning and waste heat utilization. In order to improve the performance of these systems, it is necessary to understand the real process in absorber, because it is the foundation for optimization of absorber and system to save energy and material [Ref.1-2]. However, as absorption process includes simultaneous heat and mass transfer, it is much more complex than condensation and evaporation; and there are only a few papers [Ref.3-8] which deal with this field.

In this paper, a modified theoretical model on film absorption is presented; and the numerical solution is compared with experimental results.

MATHEMATICAL MODEL

The system being analyzed is schematically described in Fig.1:

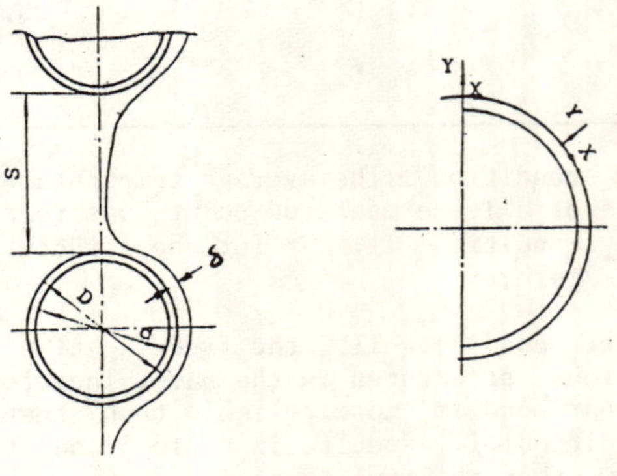

Fig.1 Physical and coordinate system

Ai represents the ith tube in the tube bundles. Cooling water flows inside the tubes. The space outside tube bundles is full of pure vapour of absorbate. The liquid solution falls freely from the bottom of Ai-1 to the top of Ai, then forms thin film; the film flows downwards along the tube surface, whilst absorbs vapour into the film in this process, the heat of absorption is generated. Finally the film flows off the bottom of Ai to the top of its next tube. This process is repeated in every tube, which forms the whole process in tube bundles.

Omitting the effects of strike and the surface tension of solution, the liquid flow is simply considered as a continuity and steady falling film around the outside surface of tube with an initial velocity. This initial velocity is simply assumed to equal the vecolity of the free falling body, that is, $U_o = \sqrt{2gs}$. As we have discussed in Ref.8, the value of Uo has little effect on film flow.

The following assumptions have also been

made in formulating the mathematical model:

1) The flow and heat transfer of the cooling water inside tubes is in fully developed region. The temperature of water inside each tube is approximatelly constant and equals the average value of inlet and outlet temperature inside that tube.
2) The heat transfer inside tube wall is one dimensional heat condition.
3) The vapour is pure; and its pressure is constant, that means there is no mass transfer resistance in vapour phase.
4) The equilibrium between the vapour and liquid exists at their interface.
5) The flow of film is laminar; and the diffusion terms along the direction of flow are negligible.
6) The physical properties of liquid on each tube are constant.
7) No shear forces are exerted on the film by the vapour.
8) All the heat of absorption is transfered into the film; and there is no other heat exchange between liquid- and vapour- phase, except the heat of absorption.
9) In the space between each two tubes, the absorption is negligible; and the distribution of temperature and concentration in the bottom of Ai-1 is the same as the distribution in the top of Ai respectively.

Under these assumptions, the simultaneous heat and mass transfer in the tube bundles at steady state may be described by the following partial differential equations in the coordination described in Fig.1, where X and Y direction is along and perpenticular to the tube periphery respectively.

$$\frac{\partial U}{\partial x} + \frac{\partial V}{\partial y} = 0$$

$$U\frac{\partial U}{\partial x} + V\frac{\partial U}{\partial y} = -\frac{\partial^2 U}{\partial y^2} + g\sin\left(\frac{2x}{D}\right)$$

$$U\frac{\partial T}{\partial x} + V\frac{\partial T}{\partial y} = a\frac{\partial^2 T}{\partial y^2}$$

$$U\frac{\partial C}{\partial x} + V\frac{\partial C}{\partial y} = Dm\frac{\partial^2 C}{\partial y^2}$$

$$\dot{m} = \int_0^{\delta} \rho U \, dy$$

The boundary conditions applying to these equations are as follows:

at $x=0$ $\quad U = 2gs; \qquad V=0$

$$Ti(y)|_{x=0} = Ti\text{-}1(y')|_{x=\pi r}$$

$$Ci(y)|_{x=0} = Ci\text{-}1(y')|_{x=\pi r}$$

$$\int_0^{\delta} Udy|_{i,x=0} = \int_0^{\delta} Udy|_{i-1,x=r}$$

here, y, y' apply to $\int_0^{y} Udy = \int_0^{y'} Udy$

at $y=0$, $\quad U=0 \;; \quad \frac{\partial U}{\partial y} = 0$

$$\frac{\partial C}{\partial y} = 0 \;; \lambda\frac{\partial T}{\partial y} = \frac{(Tw-Tc)}{(Rw+Rc)}$$

Nusselt number for the calculation of Rc can be calculated by using the following correlation:

$$Nu = \begin{cases} 1.86(RePr)^{0.33} & Re<2200 \\ 0.012(Re^{0.87}-280)Pr^{0.4} & 2200<Re<10000 \\ 0.023Re^{0.8}Pr^{0.4} & 10000<Re \end{cases}$$

at $y=\delta$; $\quad \dot{m} = -\rho V + \rho U\frac{d}{dx} \;; \quad \frac{\partial U}{\partial y} = 0$

$$\lambda\frac{\partial T}{\partial y} = \rho\frac{DmDH}{1-C}\frac{\partial C}{\partial y}$$

$$F(Ti, Ci, Pa) = 0$$

The equations are transfered into X-ω coordination and solved by using the numerical method of Partankar and Spalding. The numberical method is unconditionally stable and converged.

The physical properties of LiBr aqeous solution, are calculated from the computer programm, using the formulae in [Ref.9].

CALCULATION RESULTS AND ANALYSIS

In the analysis of absorption process, the parameters listed in Table 1 are used. These parameters are in agreement with the feature of the experimental setup.

Table 1 Parameters of experimental setup

D (m)	d (m)	L (m)	w (W/m.K)	S(m)	N
0.018	0.014	0.4	16.6	.022	10

Fig.2 shows the distributions of temperature and concentration acroos the film. As Dm is much less than a, the mass transfer rate is very slow, so the absorbed vapour is accumalated near the interface. As a result, Ci is much less than Cx, and the mass transfer driving force decreases rapidly. In addition, as Ci is lower than Cx; Ti is higher than Tx; and equilibrium exists between Ci and Ti, therefore, the liquid film is

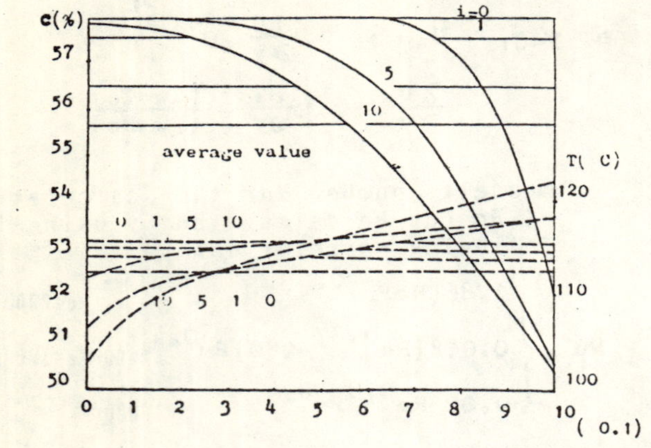

Fig.2 temperature and concentration distribution
--- temperature; —— concentration
To=112.7 ℃ Co=57.95% Pa=455mmHg
Tco=81.9 ℃ Gc=121.9kg/hr F=372.0Kg/hr.m

always subcooled. Furthermore, in the situation under our consideration, the subcooled degree is relatively large. These phenomena direct us the promising way to enhance absorption by using tube with structured surface or adding surfacetant into solution to disturb or mix it On the other hand, since the subcooled degree is relatively large its effect on absorption should be carefully considered on the designing of an absorber.

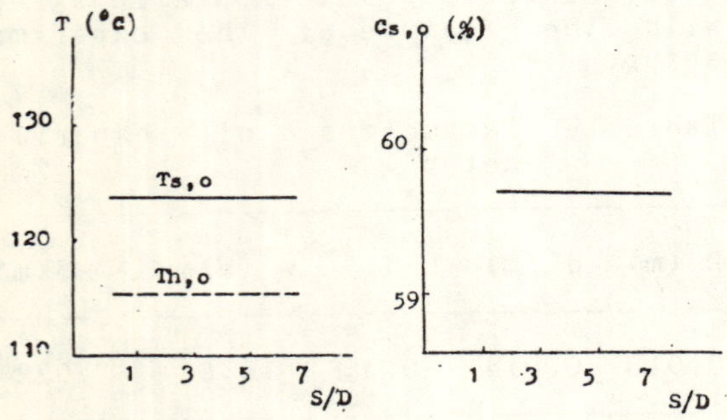

Fig.3 Effect of tube span
To = 120 Co = 61.5% Pa = 400mmHg
Tc,o=120 Gc=120kg/hr F=400Kg/hr.m

Fig.3 describes the effect of tube span S on absorption. When S increases, Ts,o, and Cs,o vary slightly. Consequently, on designing of an absorber, the tube span could be as small as possible to reduce the size of absorber, which would not deteriorate the absorption process.

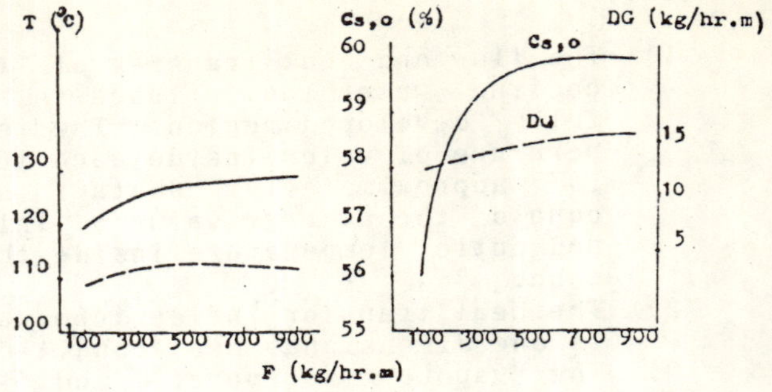

Fig.4 Effect of spraying density
To = 120 ℃ Co = 61.5% Pa = 400mmHg
Tc,o=120 ℃ S/D= 1.22 Gc = 120Kg/hr

Fig.4 describes the effect of spraying density F on absorption. When F, increases, Ts,o, Cs,o and the amount of absorbed vapour all increase. But Tc,o has a maximum value. This phenomeon is owing to the effects between heat and mass transfer. For a larger spraying density, the variation of concentration becomes smaller, consequently, the mass transfer driving forces and the amount of absorbed vapour increase. However, a larger spraying density corresponds to a larger heat transfer resistance because the film thickness is increased. In the end, there is an optimum spraying density.

There is anthor factor which will affect the value of optimum spraying density. This factor is the power of solution pump. So, the exact value of optimum spraying density should be calculated from the optimization of the whole system, which is beyond the present paper. According to Fig.4, the optimum spraying density should be less than 500Kg/hr.m.

EXPERIMENTAL SETUP

an experimental setup has been established to testify the present theoretical model. The construction of the setup is briefly shown in Fig.5.

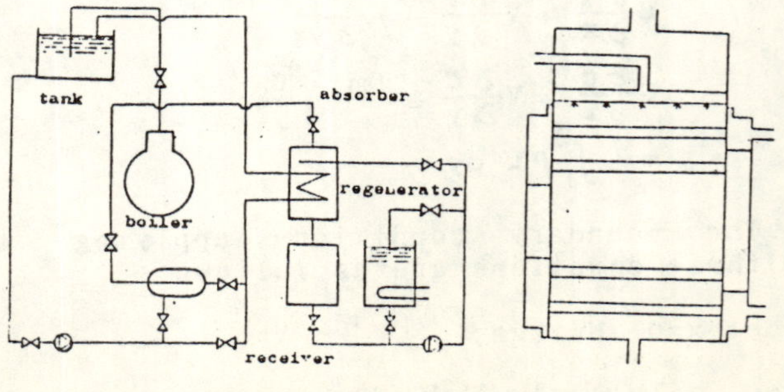

Fig.5 Structure of experimental setup

In view of the mixture's corrosiveness, all the components which contact with the solution have been made of stainless steel.

The tube bundles are composed of 10 stainless tubes with 18/14mm outer/inner diameter and o.4m length. The spraying tube is just beyond the bundles.

The dilute solution is concentrated to the definite concentration in the regenerator by electric heater, its temperature is controlled at the definite level by the temperature controller, then the solution is pumped into the absorber, sprayed through the spraying tube. Finally, the solution forms thin film on the tubes surface. The film flows downwards along the tube bundles, absorbs vapour into itself. In this process, the heat of absorption is generated, and the temperatures of the film and the cooling water inside tubes are elevated. The diluted solution flows into the receiver, and will be pumped back into the regenerator for the next one after each experiment. The absorbed vapour is delivered by the boiler.

The measured parameters during experiment include flow rate, concentration, pressure and temperature. Parameters of pressure and liquid depth are also measured to monitor the experimental process, but these two parameters have no effect on the result.

The flow rate is measured by individually calibrated LZB-15 rotameters; the concentration is calculated from the density and temperature of the sample solution, while the density and temperature is measured by the density meter ($\pm$0.001g/cm3) and mercurial thermometer ($\pm$0.1 ℃)respectively; the vacuum pressure is measured by U-tube filled with mercury, while the length of the U-tube is 2000mm. Other temperatures are measured by the thermocouple, as there are more than twenty temperatures that should be measured, HP3054 data-acqusition system is used to provide data logging, printing and plotting. Form the plotted curves of temperature, we can judge whether the performance is in steady state, and only the data measured in steady state is redarded as valid.

COMPARISON BETWEEN EXPERIMENTAL AND THEORETICAL RESULTS

Fig.6 shows the comparison between experimental results and calculation. As it is rather difficult to control every

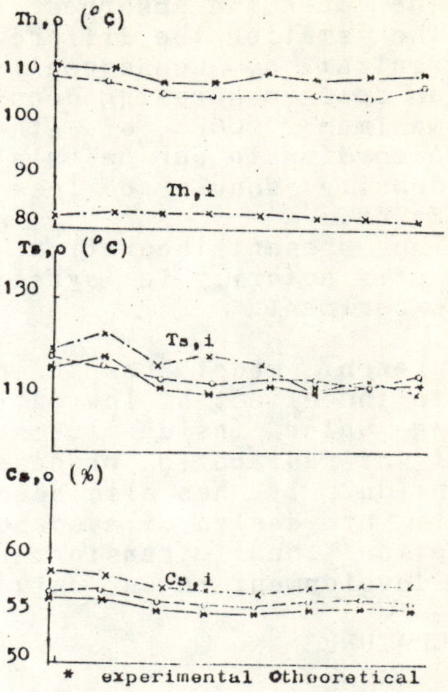

Fig.6 Comparison between experimental and theoretical results

inlet parameter accurately, some parameters are based on the real data measured in the experiment, so the inlet parameters are different from each other.

It can be observed from the Fig.6 that the predicted Tc,o is about 3 ℃ lower than the experimental results; and the predicted Cs,o is 0.2-0.4% larger. The relative error is 10% and 10-20% respectively. The deviation of Ts,o is less than 1 ℃.

So we can indicate that the present theoretical model is satisfactorily in agreement with the experiment.

CONCLUTION

According to the study of this paper, we can conclude that:

1) During film absorption, the concentration at interface is much lower than the average value. Using structured tube surface or adding surfactant into the solution to distuib or mix it is expected to enhance absorption effectively.
2) The liquid film is still subcooled at the end of absorption, this phenomenon should be carefully considered on the designing of absorber.
3) The variation of tube span has little effect on heat and mass transfer, so it could be as small as possible.

4) The larger the spraying density, the more the absorbed vapour and the smaller the difference of concentration. Consequently, there is an optimum spraying density for the maximum COP of the system. According to our calculation, this density should be less than 500 Kg/hr.m.
5) The present theoretical model is satisfactorily in agreement with experiment.

The present model has been used to analyze the effect of low pattern of the cooling water inside tubes; and the effect of adiabatic process on film absorption. It has also been used to assist the design of absorber for an absorption heat transformer that is under development in our university.

NOMENCLATURE

a	temperature diffusion coeffient m2/s
C	concentration of solution, %
Cp	specific heat of solution, J/Kg.K
D/d	outer/inner diameter of the tube, m
DG	amount of adsorbed vapour, Kg/hr.m
DH	Hv-Hs, heat of absorption, J/Kg
Dm	diffusion coeffient, m2/s
F	spraying density, Kg/hr.m
G	flow rate, Kg/hr
g	gravity acceleration, m/s2
H	enthalpy, J/Kg
L	length of tubes, m
m	mass transfer rate at the interface, Kg/s.m
N	number of the tubes
Pa	pressure at the absorber, mmHg
R	heat transfer resistance, m2.K/w
S	tube span, m
T	temperature, °C
Tw	T at outside surface of tube wall, °C
U/V	velocity in X/Y direction, m/s

GREEK SYMBOLS

λ	thermal conductivity of solution, W/m.K
ρ	density of solution, Kg/m3
ω	dimensionless streamfunction
δ	film thickness, m

SUBSCRIPTS

c	cooling water
i	interface
o	outlet
s	LiBr solution
x	average value at x
v	vapour
w	tube wall

REFERENCES

[1] Takao Kashiwagi et.al "Simulation of the performance characteristics of the absorption type heat pump for temperature boosting of waste heat", Trans. of JSME B, Vol.52 No.474 (1986), PP795-900 (In Japanese)

[2] P.D. Iedema, "Real process simulation of a LiBr/ZnBr2/CH3OH absorption heat pump". VDI-Ber 539(1984) PP931-944.

[3] G. Grossmam, Simulataneous heat and mass transfer in film absorption under laminar flow". Int. J. Heat Mass Transfer, Vol.26, No.2 (1983), PP357-371.

[4] Zhou Xinxi, Yu-chi Bin, Zheng Weixiang, "The analysis of heat and mass transfer of the LiBr-H2O solution over the horizontal tubes". Proc. of the 17th Int. Conf. on Refrigeration. Vol.E, July (1987), PP672-677.

[5] Lu Zhen, Gui Fuling, "Flow and heat transfer in the evaporation of falling film on horizontal tubes". Proc. of the 17th Int. Conf. on Refrigeration, Vol.E, July (1987), PP690-694.

[6] B.J. Cvan der Wekken, R.HWasennar and A.Segal, "Finite element method solution of simultaneous two-dimensional heat and mass transfer in laminar film flow". Warme-und stoffubertragung Vol.22 (1988), PP347-354.

[7] J. W. Andberg and A.C. Vliet, "Absorption of vapors into liquid film flowing over cooled horizontal tubes". Proc. of 2nd ASME-JSME Thermal Emgineering Joint Conf., Honolulu, Hawaii, Vol.2 (1987), PP533-553.

[8] Tang Jingsong, "Simultaneous heat and mass transfer in high temperature film absorption". M.Sc thesis of Shanghai Jiao Tong University, (1987), (In Chineses).

[9] Chang Xiuyuan, Chang Qang, "The numerical calculation of thermophysical properties of aqueous lithium bromide". Proc. of the 4th National Conf. of Refrigeneration Using Waste Heat Pump, Wuxi, China (1986), (In Chinese).

THE FEASIBILITY STUDY ON SOLID DESICCANT AIR CONDITIONER AND THE APPLICATION PROSPECT OF THIS SYSTEM IN CHINA

HAN HONGXING SHOU WEIDONG
College of Power Engineering
Shanghai Institute of Mechanical Engineering, China

ABSTRACT

This paper presents a simulation system of solid desiccant air conditioner using silica gel as the adsorbent and the experimental results on this system, the economic analyses under the condition of China are also carried out in the paper. Results of the simulation system and the economic analyses indicate this system can be successfully used in China, the questions as which climate is most appropriate for desiccant system operation, which system capacity is most optimum are also discussed.

At last, the paper points out that, the desiccant system utilizing solar energy has wide prospect in China.

INTRODUCTION

The solid desiccant cooling system utilizing solar energy or other low-quality energy has been developping very quickly over the past decade because of the energy crisis.

A national solar cooling program using desiccant processes has been made in the U.S., the COP of the system studied in the SERI was about 1.0. Similar cooling systems have also been finished in Japan, Austrilia and other countries.

In the recent years, the main research program of the solar desiccant air conditioner is system optimization to raise the COP.

In general, the solar desiccant cooling system can compete with vapor-compression one and the products of this system has been provided.

In China, the study on this system just began and the experience is very limited. According to the condition of China, silica gel was selected as the adsorbent in the simulation system, both the stationary adsorption and the desiccant wheel adsorption will be discussed in the economic analyses.

Desiccant cooling system cools air by a series of processes, the major step is adiabatic adsorption, this step is accomplished in the stationary columns or the desiccant wheel, then the dried air is cooled in the sensible heat exchangers and evaporative coolers to meet both sensible and latent air conditioning loads.

Desiccant cooling mainly has two operating modes. The ventilation mode introduces fresh air into the conditioned space and the recirculation mode recirculates air from the room, these two modes are shown in Figure 1. to Figure 4. In order to testify the feasibility of the desiccant cooling system with Chinese material on Chinese climate condition. We made a simulation system. Figure 5 illustrates the construction of this system.

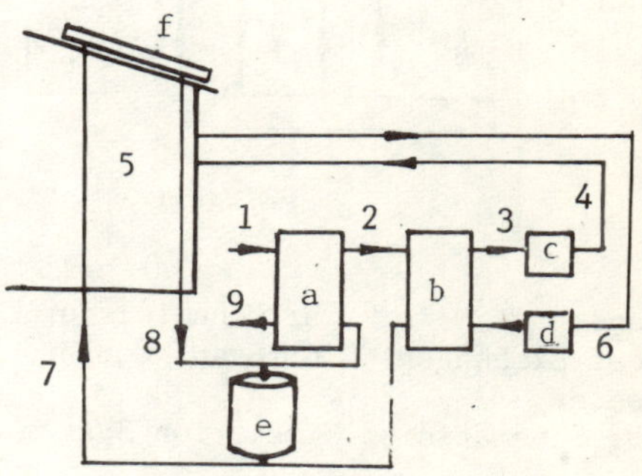

a: Dehumidifier b: Heat Exchanger
c: Evap. Cooler d: Evap. Cooler
e: Storage f: Collector

Figure 1: Schematic of Ventilation Mode

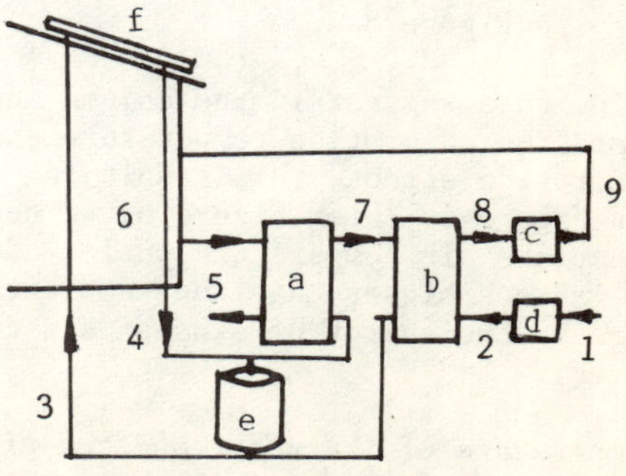

a: Dehumidifier b: Heat Exchanger
c: Evap. Cooler d: Evap. Cooler
e: Storage f: Collector

Figure 2: Schematic of Recirculation Mode

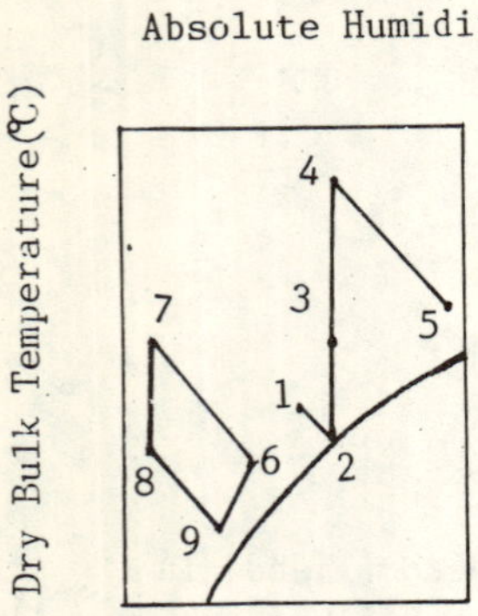

Figure 3: Psychrometric Diagram of Recirculation Mode

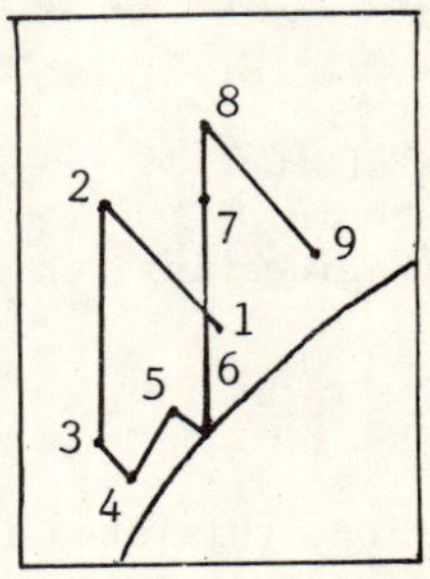

Figure 4: Psychrometric Diagram of Ventilation Mode

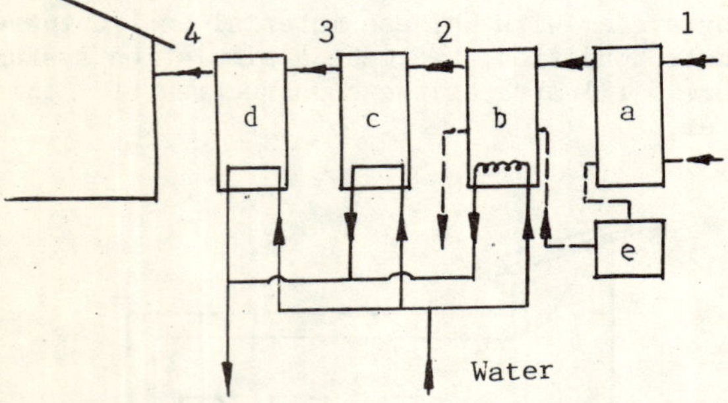

a: Fans b: Dehumidifier
c: Heat Exchanger d: Evap. Cooler
e: Heater

Figure 5: Schematic of Simulation System

EXPERIMENTAL RESULTS

The experiments include the system cooling operation test, system regeneration operation test, some results of the tests are presented in Figure 6 to Figure 8.

A cooling tube was set in the column and this decreased the adsorption temperature and raised the adsorption effect, this result can be seen in Figure 6. From Figure 6, we can also find that the air temperature outlet adsorption column is much higher than the ambient temperature so the energy of exhaust air can be recovered.

The temperature of the water and the efficiency of humidifier have great influence on the

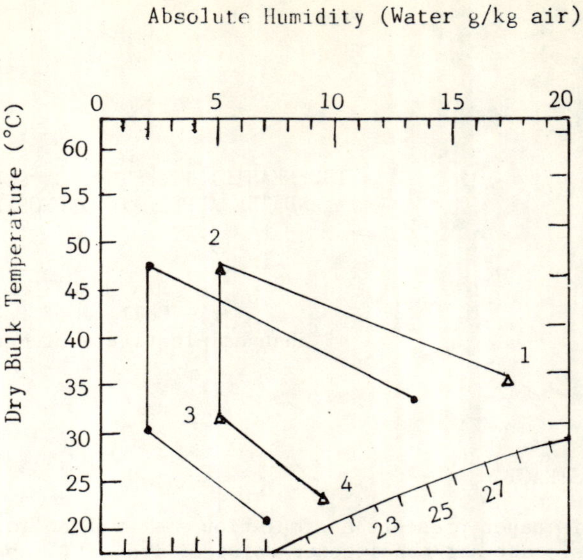

Figure 6: Psychrometric Chart of the System Cooling

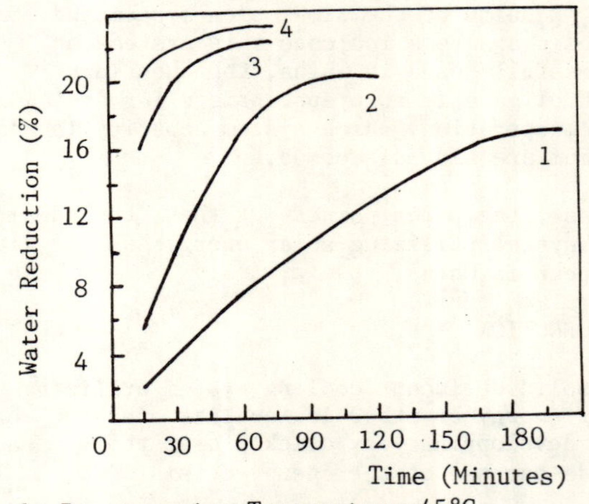

1: Regeneration Temperature=45°C
2: Regeneration Temperature=65°C
3: Regeneration Temperature=85°C
4: Regeneration Temperature=110°C

Figure 7: Water Reduction of Adsorbent VS. Regeneration Time

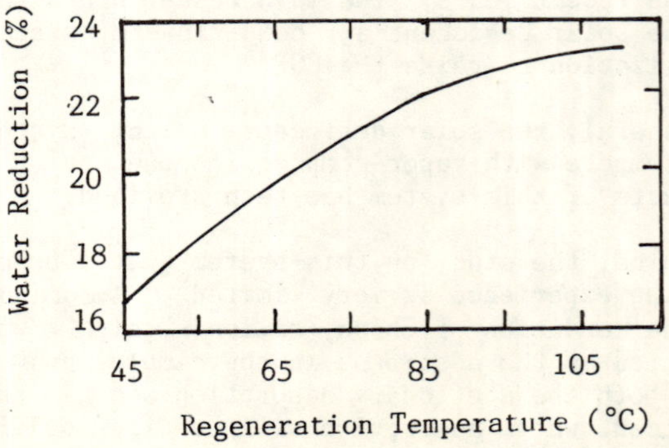

Figure 8: Water Reduction of Adsorbent VS. Regeneration Temperature

system. If the system cooling experinent had been done with lower temperature water, the results would have been better.

Figure 7 and Figure 8 show the regeneration properties of silica gel, the higher the regeneration temperature, the shorter the regeneration time, but the adsorbent can still be regenerated as long as the regeneration temperature is highter than 50° centigrade.

ECONOMIC ANALYSIS

The potentiality for using desiccant cooling in residential buildings depends on the economics of solar heating and cooling system as compared to conventional one. This system will be successful if the initial capital costs of the solar desiccant system over the conventional system can be recovered through energy cost savings within an acceptable time period.

Three cities whose climatic characteristics in summer are inllustrated in table 1 are chosen in the economic analysis program. For each city, both the ventilation mode and the recirculation mode with a series of cooling capacities are simulated.

Table 1: Cities used in the economic analyses

city \ parameter	Ta(°C)	Tw(°C)	Ø(%)
Shanghai	34	28.3	67
Bejing	33.8	26.5	60
Lanzhou	30.6	20.1	42

Several parameters required for the economic analyses are shown in the table 2.

Table 2: Parameters used in the economic analyses

System lifetime (years)	25
Interest rate of the bank(%)	10
General inflation rate (%)	6
Energy escalation rate (%)	10

In addition to the economic parameters of table 1 and 2, some assumptions are given here. There is no auxiliary cooling provided in the solar desiccant air-conditioned building, the regeneration heat is entirely solar energy.

By using these parameters and assumptions, the present worth of initial capital costs over the conventional one and the energy cost savings by use of the desiccant system can be calculated with the following equation (1) and (2)

$$Z_1 = NA\ (1+I)^{N-1}(1+K)^{-N} \quad (1)$$

$$Z_2 = P\left(\frac{1+I}{1+K}\right)^N \quad (2)$$

Once the Z_1 and Z_2 are calculated, the benefit/cost ratio can be determined. The benefit is defined as Z1, the cost is defined as Z2, thus, only does a case of benefit/cost ratio greater than 1.0 presents a successful system. The results of economic analyses are illustrated in Figure 9 and Figure 10.

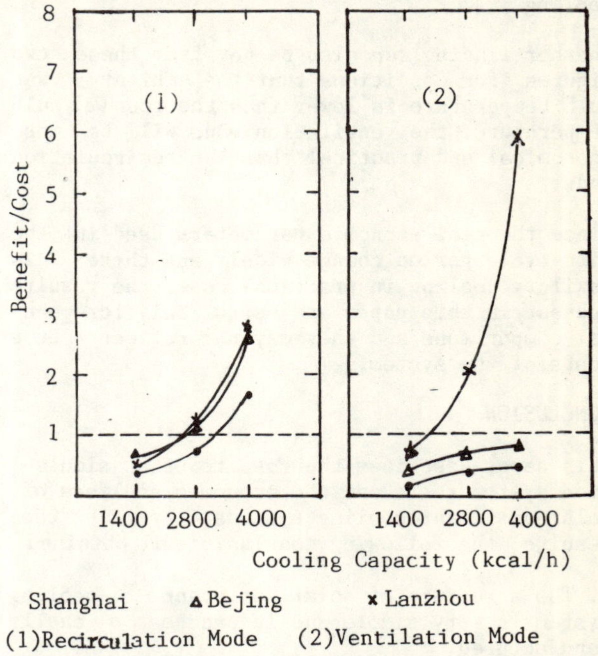

Figure 9: Benefit/Cost of Recirculation Mode and Ventilation Mode VS. Cooling Capacity (Adsorption Column)

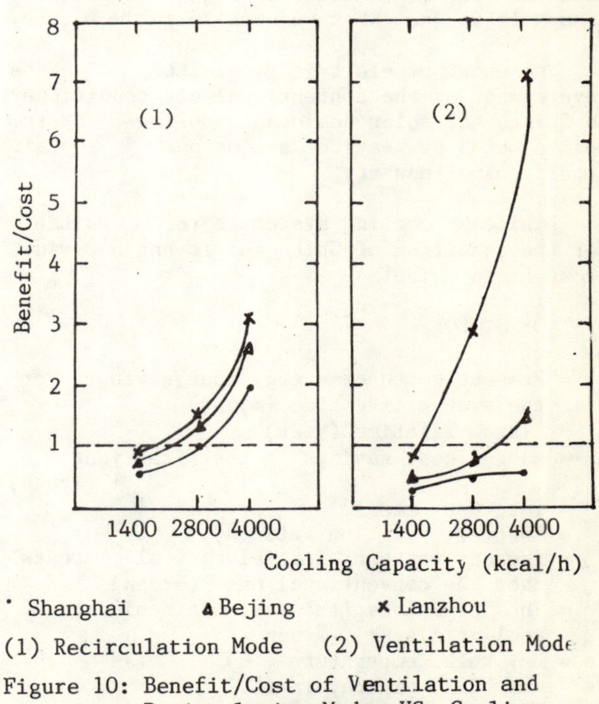

Figure 10: Benefit/Cost of Ventilation and Recirculation Modes VS. Cooling Capacity (Desiccant Wheel)

Solar heating and desiccant cooling system is

very suitable in the areas which experience both high loads in the winter or in the summer and dry climates. LANZHOU fits all these conditions. So it is a good place to use solar desiccant cooling.

Anather finding can also be got from these two figures, for conditions that the ambient wet bulb temperature is lower than the room wet bulb temperature, the ventilation mode will be more economical and practical than the recirculation mode.

Since the real economic parameters used in the life-cycle period change widely and there is auxilary cooling in practical case, the results present in this paper are useful only for general comparisons and they may not reflect true costs of the system.

CONCLUSION

This paper describes the results of a simulation system tests and the economic analyses of solar desiccant cooling systems. From the results, the following conclusions are obtained.

1. The structure of solar desiccant cooling system is very simple and it can be easily manufactured.

2. There are two methords to minimize the electrical power required to drive the fans: the using of low-drop adsorption column or desiccant wheel and the manufacture of high efficiency fans. It is just what we must do in the future.

3. The short of electricity limited the development of the conventional air conditioner in China, the solar desiccant cooling system points out a new way to develop our air conditioning industry.

4. Desiccant cooling system is very suitable for the condition of China and it has a wide prospect in China.

LIST OF SYMBOLS

Z_1 = Present worth of energy cost savings over the system life time (¥yuan)
N = System lifetime (year)
A = Energy cost savings in the first year (¥yuan)
I = Interest rate (%)
K = General inflation rate (%)
Z_2 = Present worth of initial capital costs over the conventional one (¥yuan)
P = The initial capital costs of desiccant cooling system (¥yuan)
Ta = Dry bulk temperature (°C)
Tw = Wet bulk temperature (°C)
Ø = Relative humidity (%)

REFERENCES

1. Benjamin C. Shelpuk 'Simulations and Economic Analyses of Desiccant Cooling Systems'. SERI/TR-34-090.

2. Lynn E. Bussey, 'The Economic Analyses of Industrial Projects'. 1978 by Prentice-Hall, Inc.

3. S.V. Szokolay, 'Solar Energy and Builing'. 1977 The Architecture Press Ltd. London.

4. Sheng Chengyu, 'Climate of China'. Science Press. 1986.

5. Chi, C.W. 'Fixed Bed Adsorption Drying'. AIchE Journal, Vol. 16(1970), pp.23-31.

6. Bullock, C.E. and Threlkeld, J.C. 'Dehumidification of Moist Air by Adiabatic Absorption' Transfaction, ASHRAE, Vol.72, (1966), p.301.

IDENTIFICATION OF THE SYSTEM STRUCTURAL PARAMETERS IN THE FROZEN ZONE OF NON-LINEAR BIOLOGICAL SYSTEM

Xiang Yu Li Haizeng Yie Dahai Guo Fangzhong
Dept. of Power Engineering
Huazhong University of Science and Technology
Wuhan, China

ABSTRACT

The heat conduction of lively biological tissues is strongly Non-linear. Obviously, the method of measuring thermal properties by the way of static state and small sample is not suitable for this problem. In this paper, the authors introduced the approach of Cybernetic Theory into the complex biological system in cryosurgery. It is found that the heat conduction characteristics can be reflected by system parameters, which replace the thermal properties of the system. In order to test feasibility of this method, experiments were conducted to simulate the cryosurgery process, and the off-line and on-line identifications were used to analyze the experiment results during the transient temperature recovery process. At last, the preliminary model of cryosurgery control is described.

INTRODUCTION

Cryosurgery is different from the normal surgery. It needs no anaesthesia, and the operation is bloodless and especially convenient [1]. Further more it can stimulate an antibody in the tissues, and people call this effect cryoimmunology. It is pointed out [2] that this effect is helpful in the treatment of malignant tumour. For the frozen zone in cryosurgery, early in 1948, Dr. H.H.Pennes and his colleagues [3], based on the research of forerunners on heat conduction characteristics in biological system, gave the following conduction for biological system:

$$D(\vec{x},T)\cdot C(\vec{x},T)\cdot\frac{\partial T}{\partial t}=\frac{\partial}{\partial\vec{x}}[K(\vec{x},T)\cdot\frac{\partial T}{\partial\vec{x}}]-\frac{\partial}{\partial\vec{x}}[H(\vec{x},T)\cdot F(\vec{x},T)]+Q(\vec{x},T) \quad (1)$$

and, in 1966, biologist P.Mazur [4] presented that there are two effects in damaging cells when they are cooled at different temperature dropping rate. One is by lacking of solution outside cells at low temperature dropping rate, and the other is by freezing inside cells at high temperature dropping rate. Although, much works has been done, the application of cryosurgery is very limited. The main reasons are that people can not accurately determined the size of the frozen zone and the optimum temperature dropping rate. These two points are the keys of destroying ill tissues fully while the good ones are prevented from being damaged. So the control of cryosurgery is very necessary. To realize this control, the heat conduction properties of biological tissues must be understood at first. But the physical and biological parameters $K(\vec{x},T)$, $D(\vec{x},T)$, $C(\vec{x},T)$, $H(\vec{x},T)$ and $Q(\vec{x},T)$ in tissues are the complex functions of position $\vec{x}$ and temperature T due to the complicate blood cycle, metabolism and different kinds of tissues. It is nearly impossible that all these parameters are accurately measured. Thus the conventional method of heat conduction has faced many difficulties in this problem.

SYSTEM STRUCTURAL ANALYSIS

The authors try to introduce the approach of Cybernetic Theory [5][6] into this problem. The predicted temperature distribution in the ill tissues and the optimum temperature dropping rate (we call these two factors output of the system) would be obtained if the quantity of cold (input of the system) produced by the cryoprobe is controlled reasonably. In order to obtain the relationship between the input and output of the system, we should analyze the heat conduction of the biological system. In the frozen zone in cryosurgery, the blood cycle and metabolism became slow obviously. So we ignored these two effects. Further more, we supposed that the frozen zone is homgemous, because such zone usually was composed of tissues without bones. Under this assumption, all of the thermal physical properties are only related to temperature T. Finally, based on the work of other people [8], we assume a spherical conduction model for the frozen region. Then the conduction equation (1) is simplified into the following form (2) by the above assumptions:

$$D(T)\cdot C(T)/K(T)\cdot\frac{\partial T}{\partial t}=\frac{\partial^2 T}{\partial r^2}+\frac{2}{r}\cdot\frac{\partial T}{\partial r}+\frac{1}{K(T)}\cdot\frac{\partial K(T)}{\partial T}\cdot(\frac{\partial T}{\partial r})^2 \quad (2)$$

Model (2) is still a Non-linear equation which is very difficult to tackle. The authors presented that the thermal property parameters $k(T)/(D(T) \times C(T))$ and $(1/k(T)) \times (\partial(k(T))/\partial T)$ can be replaced by the system structural parameters A and B, where A and B may be constant or variable. And, the system identification technique is applied to identify parameters A and B from the experimental results. Since not only the conduction characteristics of system but also the model error are considered in the system parameters, these parameters have the compensation effects in the cost of losing the conventional meaning of thermal property parameters. For the sake of convenience, we linearized the parameters A and B, and treated them as constants. This would not affect their compensation effects.

EXPERIMENT AND TRANSIENT OFF-LINE AND ON-LINE IDENTIFICATIONS IN FROZEN ZONE

From the above system structure analysis, the biological conduction equation (2) is converted into:

$$(1/A)\cdot\frac{\partial T}{\partial t}=\frac{\partial^2 T}{\partial r^2}+\frac{2}{r}\cdot\frac{\partial T}{\partial r}+B\cdot\left(\frac{\partial T}{\partial r}\right)^2 \qquad (3)$$

This equation (3) describes the heat conduction characteristics in the frozen part of bad tissues during cryosurgery. To verify the feasibility of the system method, we conducted an experiment to simulate the process of cryosurgery. Figure 1 shows the sketch of the experiment system. In figure 1, the cold probe of the cryosurgery instrument was soaked in quiescent water in a large pool. Five thermocouples embedded on a frame which was fixed on the probe, were distributed in the radius direction evenly. The distance between each two adjacent thermocouples is 3.0mm. When the liquid Nitrogen was supplied continuously to the inner part of probe, a ball like frozen zone began to form and expand till a quasi-steady state has been obtained. Then the liquid Nitrogen supply was cut off and the temperature recovery process began. From then on, transient temperature distribution was recorded by the single-board computer data acquisition system we developed. The experiment lasted 60 seconds. And table 1 gave the experiment results.

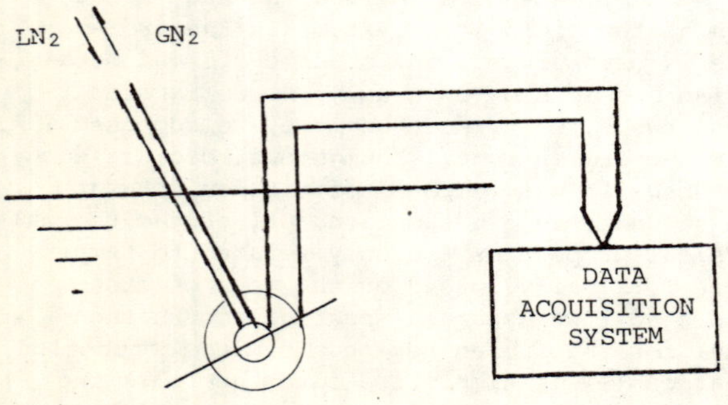

Figure 1

Two adjacent thermocouples No. 2,3

A(m^2/s)	1.28×10^{-6}	B(1./°C)	-0.02155
Two Points Exp. Tem.		Two Points Cal. Tem.	
52.00	42.00	50.80	40.98
52.00	42.00	50.48	40.76
52.00	42.00	50.25	40.63
51.00	42.00	49.91	40.51
51.00	41.00	49.51	40.38
51.00	41.00	49.22	40.22
50.00	41.00	48.84	40.06
50.00	41.00	48.40	39.88

Time: 24 seconds. Temperature: °C

Table 1

The experiment results were analyzed by the following two ways:

(A) Off-Line Identification

Identification is the process to establish the mathematical model or to find the unknown parameters in a mathematical model from the experimental results. For our problem, the mathematical model describing the heat conduction of biological system is established, but the thermal properties are not very clear. So the work is concentrated on identification of the unknown parameters.

Off-line identification is analyzing the data after experiment. The Least-Square method, commonly used by the people, was selected to identify the parameters A and B. Detail analysis are given below:

Firstly, the partial differential equation (3) was dispersed into the difference equation and summarized as (4):

$$(1/A)\cdot\frac{T_n^{m+1}-T_n^m}{\Delta t}-B\cdot\left(\frac{T_{n+1}^m-T_{n-1}^m}{2\cdot dr}\right)^2-\frac{T_{n+1}^m-2\cdot T_n^m+T_{n-1}^m}{(dr)^2}-\frac{T_{n+1}^m-T_{n-1}^m}{r\cdot dr}=0 \qquad (4)$$

This equation is linear and very simple in structure, as following:

$$(1/A)\times x1+B\cdot x2+x3=0 \qquad (5)$$

All the points in temperature field, recorded by the single-board computer system, can not accord with model (4) or (5). When a set of experiment data are inserted into the left side of equation (5), the result will most probably not be zero. This number is called node residue. The task of Least-Square method is to find the values of parameters A and B to make the square sum of all the node residues minimal. More detail procedures can be found in textbooks on numerical analysis. The parameter A and B id identified from the data of experiment (table 1) in this way are:

$$A = 1.28 \times 10^{-6} \quad (m^2/s); \quad B = -0.0215527 (1/°C)$$

If the parameters A and B are inserted into the equation (4), the computed temperature field which was listed in table 1 also, can be obtained. The maximum error between the computed and the recorded temperature field is less than +(-) 2.00°C.

It is very interesting that, A, 1.28×10^{-6} (m^2/s), is approximately equal to the average thermal diffusivity, 1.53×10^{-6} (m^2/s) [7] in the experimental range -60°C to 0°C. This coincidence shows that the meaning of system structural parameter is approximately the same as the thermal property parameter. The reason is that the system model error and other errors are very little in our experiment. When the errors are large, the system structural parameters still make the model agree with the experiment, but they lose the conventional meaning of thermal property parameter. Because they can not only reflect the thermal properties of system, but also compensate the model error and other errors. So the system parameters are superior to the thermal property parameters for such a complex system. Similarly, by definition, the system structural parameter B equal to $(1/K(T)) \times (\partial(K(T))/\partial T)$. From this we get:

$$K(T) = K0 \times EXP(B \times T)$$

here K0 = const; B = -0.0215527 (1/°C).

The heat conductivity of ice K(T) increases when the temperature drops. This phenomenon agree with the fact described in [7].

(B) On-line identification

Contrary to off-line identification, on-line identification treats the experimental data during the experiment. Because of the limit of experimental condition, on-line identification can not be carried out during the experimental process, so it is simulated by computer. Since the biological system is strongly non-linear with the temperature, and the temperature of system dropped when the experiment went on, the charactistics of biological system are variable with time. The recurring Least-Square Algorithm was adopted to change the mathematical model continuously so as to describe the varibility of the system. The concrete analysis of this method is ignored in the paper due to the limit length. But the identification results are listed in table 2, which shows the possibility of on-line controlling cryosurgery procedure. After making a comparison between the table 1 and table 2, we found that the on-line identification method is better than the off-line one in theoretical model's agreement with the experiment results. But the parameters A and B, identified by on-line method, are more in error in the conventional meaning.

THE PRELIMINARY MODEL OF CRYOSURGERY CONTROL

On-line cryosurgery control is that the moving rate of the phase-change boundary and temperature distribution in bad tissues can be determined by controlling quantity of cold of

One point of thermocouple No.2

Time	Exp. Tem.	Cal. Tem.	A(m^2/s)	B(1/°C)
3s	52.00	52.00	4.1×10^{-7}	-0.04649
6s	52.00	51.81	4.5×10^{-7}	-0.05732
9s	52.00	51.12	4.5×10^{-7}	-0.06029
12s	51.00	50.85	4.6×10^{-7}	-0.06238
15s	51.00	50.55	4.6×10^{-7}	-0.06502
18s	51.00	49.97	4.9×10^{-7}	-0.06346
21s	50.00	49.74	4.9×10^{-7}	-0.06835
24s	50.00	48.89	5.2×10^{-7}	-0.06710

Temperature: °C
Table 2

cryoprobe at any time during the process of cryosurger. Of course, there are many difficulties in realizing the control, because the knowledge of heat conduction characteristics of

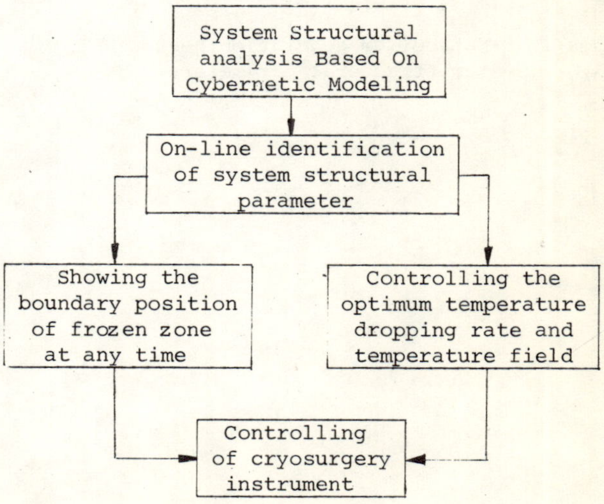

Figure 2

biological system is very little and the controlling method has defect. In spite of the serious situation, the authors, based on the above analysis, present a preliminary model in control as Figure 2.

The control has been initially realized in the YG-I model cryosurgery instrument system developed by the authors. The theoretical research is just beginning.

CONCLUSION

1) The system model (3) established in this paper agrees with the experimental results well.

2) The system structural parameters in the model are based on the thermal properties. When the model error is small, the system structural parameters are equivalent to the thermal properties. When the model error is large, not only can they compensate the error, but also show the conduction property of the system. For a complex system, the system structural parameters are superior to the thermal properties.

3) It is possible to realize the control of cryosurgery.

4) The system identification technique would be one of powerful weapons in researching complex non-linear conduction sytem.

LIST OF SYMBOLS

T···Temperature in the frozen zone

$\vec{x}$···Position in the frozen zone

$K(\vec{x},T)$···Conductivity of tissues

$D(\vec{x},T)$···Density of tissues

$C(\vec{x},T)$···Capacity of tissues

$H(\vec{x},T)$···Enthalpy of blood

$F(\vec{x},T)$···Flux of blood

$Q(\vec{x},T)$···Metabolism generation heat

A,B···System structural parameters

x1,x2,x3···Simplification of the factors in (4)

r···Radius in the frozen zone

dr···The distance between each two thermocouples

Subscript m···Number of Time point

Sub n···Number of thermocouples

ACKNOWLEDGMENT

The authors wish to thank Ms.J.Tian,Ms.L.Q.Zhang and Mr.J.Zhou, for many fruitful discussion on the subject, and thank for the help of Mr.G.Chen in our English writing.

REFERENCES

[1] Jame Fraser "Cryogenic Techniques In Surgery", Cryogenics, July 1979, P375.

[2] E.Zapdi "Cryo-Immunology: Surgical Approach And Thermal Regimen For Freezing The Elements Of The Male Rabbit Reproductive System", Heat Transfer, May. 1972. Page 129.

[3] H.H.Pennes "Analysis Of Tissue And Arterial Blood Temperature In Resting Human Fasten", I. app. Phy. Vol. 1, 1948, Page 93 - 122.

[4] P.Mazur "Physical And Chemical Basis Of Injury In Single Cell Micro-Organisms Subjected to Freezing And Thawing", Cryobiology, 1966.

[5] Arthur E.Bryson, Jr.Yu-Chi Ho, "Applied Optimal Control (OPTIMIZATION , ESTIMATION AND CONTROL)", JOHN WILEY & SONS 1975.

[6] Ernest O.Doebelin "SYSTEM MODELING AND RESPONSE — Theoretical and Experimental Approaches", JOHN WILEY & SONS 1980.

[7] Peter V.Hobbs "ICE PHYSICS" CLARENDON PRESS. OXFORD 1974, p357.

[8] H.Budman "Investigation of Temperature Fields Around Embedded Cryoprobe" I.Biomechanical Engineering Feb. 1986. Vol. 108/43.

A TECHNOLOGICAL MODULE FOR FOOD PRODUCTS REFRIGERATION STORAGE

St. Ditchev
Higher Institute for Food and Flavour Industries, Plovdiv, Bulgaria

T. Todorov
Institute of Refrigeration Technology - Sofia, Bulgaria

B. Bratanov
Higher Institute for Food and Flavour Industries, Plovdiv, Bulgaria

ABSTRACT

The design and construction of technological modules for fruits and vegetables refrigeration storage is an important and a present-day question as regards food and flavour industry. The proposed unit represents a sectional shell construction from profiled thin-sheet iron, heat-isolated, with an autonomous refrigeration unit, control and management of the technological process. The module can be utilized for completing units of technological systems for refrigeration storage of fresh agricultural production, having a capacity of 500 to 3000 tons.

INTRODUCTION

The problem for refrigeration processing and keeping the quality of food products is a complex one and has a great people's-economic significance.

According to FAO data, above 20% of agricultural production is being damaged worldwide annually, because of its untimely processing and unsuitable storage. The improvement of methods and units for refrigeration processing and storage of food products has a direct relation with their initial qualities preservation and the decrease of the mass losses.

The evaluation of the different methods and units for food products refrigeration processing (fruits and vegetables) should be complex, reporting the modern requirements for production organization, as well as the requirements for refrigeration technology and energy indices of the system.

The modern refrigerators, designed for cooling and storage of perishable food products, are characterized by the foolowing specific peculiarities:

1. They are used for storage of valuable, perishable food products, for which it is necessary to keep a definite temperature-humidity mode, and in a number of cases a definite air-circulation, too;

2. The refrigerators should be enclosed by special building constructions, which are heat and humidity-isolated;

3. The big stream of food products transported in the refrigerator and the big volume of loading/unloading activities require urgently special transport vehicles to be secured;

4. In the a refrigeratiors suitable sanitary-hygienic conditions are kept.

Resently a constant tendency towards the construction of refrigeration chambers and store-houses, can be seen. In order to satisfy in a short term the increased needs, we widly apply the module construction of sectional refrigeration store-houses (depositories) and chambers [1]. The advantage of the module principle is that by means of minimum nomenclature of main elements refrigeration chambers and store-houses of different volume can be constructed.

The sectional construction proposes a number of advantages:

1. The construction term is shortened many times;

2. Using of decreased number of working hands;

3. A high quality of production.

DEVELOPMENT PURPOSE

The present study gives description of a sectional technological module (depository) for cooling and storage of cooled down vegetables and fruits.

According to their purpose, the sectional refrigeration chambers (technological modules) can be designed for positive (from +2 to +8 °c), for intermediate (from -5 to +8 °c) and for negative temperatures (from -18 to -20 °c).

The modules are constructed from profiled thin-sheet elements (cylindrical shells - CS), on the basis of main modules, with dimensions CS 15 x 30 x 6 m, (Fig.1), CS 12 x 24 x5 m, and their combinations, which makes them suitable for mouting and combinations among the different sections. The technological modules are designed for cooling and storage of cooled down fruits and vegetables and can have a capacity of 500 to 3000 tons, at different combinations (Fig.2).

According to the way of production storage, two variants can be executed: for loose state storage and for pallet storage (Fig.3).

TECHNOLOGICAL MODES AND REQUIREMENTS FOR FRUITS AND VEGETABLES STORAGE

In order to preserve the freshness, nutritional value and biological value of fruits and vegetables, they should be cooled down immediately after their gathering. Depending on the species' peculiarities and their destination (a continuous storage in a refrigerator, transportation or direct supply in the market) different temperature-humidity modes for refrigeration storage of fruits and vegetables can be used.

The optimum temperature at which fruits and vegetables can be storaged in the modules is near the cryoscopic temperature, which depends on the chemical composition and varies in the limits from -1 to -4 °c.

The fruits and vegetables destined for a continuous storage can be cooled down in cooling tunnels, having an intensive air circulation, at a temperature from 0 °c to 1 °c and relative humidity in the range of 90-95% (Fig. 2a). The duration of cooling (from 25 °c to 20 °c) at these conditions, depending on the type of product and the way of ordering and arrangement, can be from 2 to 20 hours.

One of the main conditions for securing optimum mode of storage is to keep the air temperature and air relative humidity constant. The variations of these two parameters (especially of the temperature) can cause undesirable phenomena, like physiological disturbances, humidity condensation on the surface of the product, increased mass losses from evaporation, etc.

The automatic temperature and relative humidity regulation in the technological module is carried out in the range ± 1 °c as regards temperature and 2 % of relative humidity.

MODULE CONSTRUCTION DESCRIPTION

The proposed technological module for fruits and vegetables cooling down and storage has been developed on the basis of main modules and their combinationa at which there are: a refrigeration chamber, a manipulation room, an assembly room, transition tunnels, cooling tunnels.

The basic technico-economical indices of separate sectional module elements are:

1. A Refrigeration Chamber

Its construction is a cylindrical shell from profiled metal thin-sheet elements. Two main modules have been developed: CS 15 x 30 x 6 m and CS 12 x 24 x 5 m (Table 1). The refrigeration chamber consists of fundamentations from monolythic steel concrete, top enclosing construction from profiled thin-sheet iron with isolation, frontal walls and refrigeration doors.

Table 1

Main Module dimensions (m)	15x30x6	12x24x5
Built-up area (m2)	450	288
Effective area (m2)	431	272
Built-up volumetric capacity (m3)	2034	1310

The food products are cooled down and storaged in the refrigeration chamber at a temperature of 0 °c to +15 °c, depending on the species' peculiarities.

2. Manipulation Room

Its construction is a cylindrical shell from profiled metal thin-sheet elements. Its founda- tions are made of monolythic steel concrete, the top construction - from sheet iron, the frontal walls - from sheet iron, the woodwork - metal. In the manipulation room is done a preliminary production processing, at a temperature of +18 to +25 °c.

3. Assembly Room

It is constructed from metal supporting structure, with a roof and an enclosing wire-net. The refrigeration unit is mounted in it, because it needs free air circulation.

4. Transition Tunnels

They are cylindrical sheels with dimensions of 6 x L x 3 m. The temperature mode is from +18 °c to +25 °c.

5. Cooling Tunnels

They are cylindrical shell, with dimensions of 6 x L x 3 m. They serve for a preliminary cooling down of the production (if it is neces- sary). The temperature mode is from 0 °c to +2 °c.

The diagram of the technological module construction and the possibilities for combination of these separate elements are shown in Fig. 2.

REFRIGERATION AND VENTILLATION UNITS

The refrigeration unit of the technological module is freon, autonomous, completely automated, and it consists of the following elements:

1. Refrigeration units consisting of semipressurized compressors and air-cooled condensors;

2. Air-coolers;

3. Automatics and testing instruments.

In production storage in pallets, from 3 to 6 nb. are provided for each module (depending on the program) separate refrigeration circles, each of which consists of a freon semipressurized unit, having a cool-production of about 14 kW, and two air-coolers.

In production storage in loose state, 4-5 nb. of refrigeration circles are provided for, together with a multisectional air-cooler.

Depending on the concrete purpose, in the cooling tunnels can mounted 5-6 nb. of refrigeration circles, the same as those designed for the refrigeration chamber.

The ventillation in the refrigeration chamber, in production storage in pallets, should be realized by means of axial fans, mounted at both ends of the refrigeration chamber. In storage of loose state production, the refreshment should be done by means of the air-cooler fan and an additional axial fan.

Fig. 4 and Fig. 5 gives the refrigeration units diagram for both variants of production storage.

The designed technological modules so far can be equipped by loading/unloading operation devices, as well as by a specialized technological equipment,depending on the production programme.

CONCLUSION

The design and construction of a new type module for fruits and vegetables cooling down and storage is an important and live question for agriculture and food and flavour industry, whose successful settlement will lead to decrease in storage losses and to all-year population supply with fresh and high quality agricultural production. The technological module can be used for completing units of technological systems, having a different capacity and for its multiplication in preliminary fruits and vegetables cooling stations, as well as as in rich agricultural regions.

REFERENCES

1 Ditchev, St. and Kalev, St. Proektirane na hladilni instalatzii, DI "Technika", Sofia, 1986.

2 Maake - Eckert Pohlann. Taschenbuch der Kaltetechnik, Karlscruhe, Verlag C.F. Muller, 1978.

3 Danfoss. Automatization of Industrial and Commercial Refrigeration Plants'.

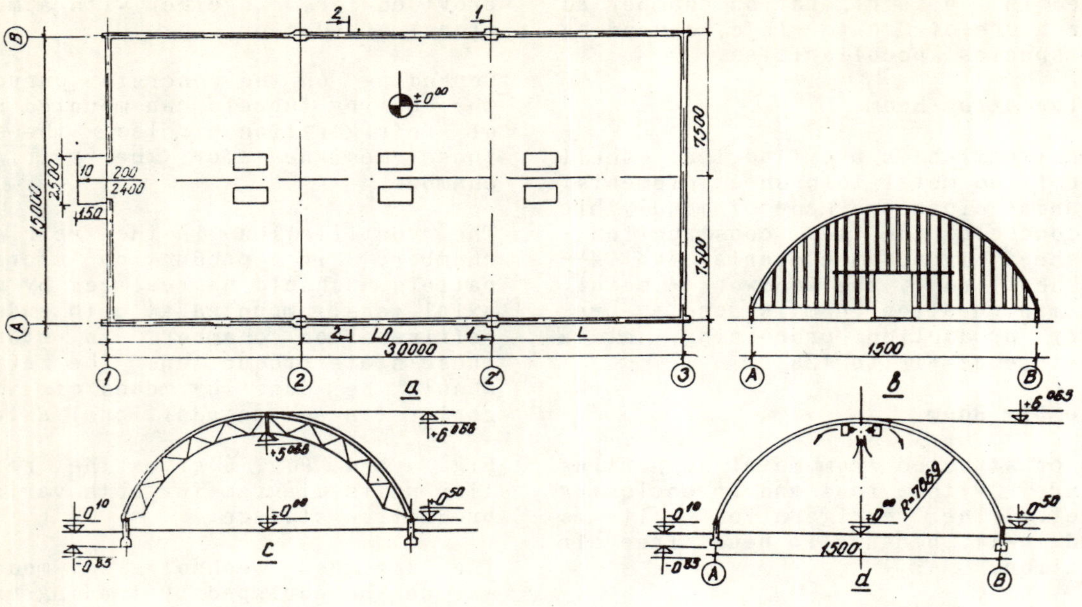

Fig.1 Technological module construction from profiled thin-sheet elements with dimensions 15/30/6 m
a - plan;
b - facade to "1-1";
c - cross-section to "1-1";
d - cross-section to "2-2".

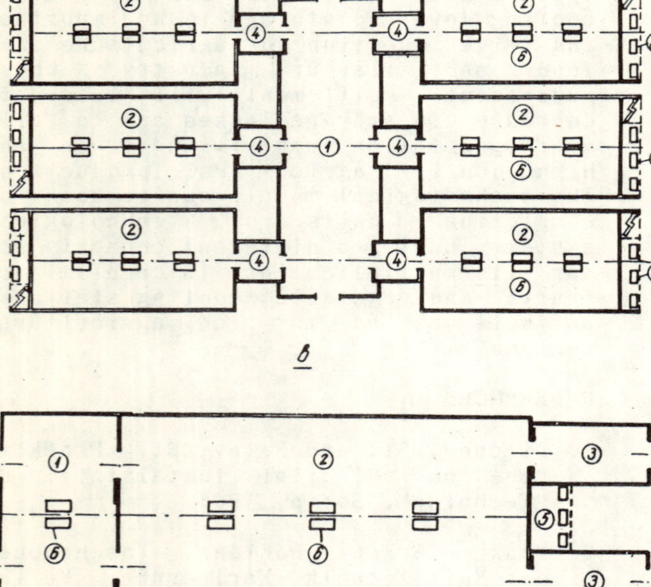

Fig.2 Different variants for module coordination
1 - manipulation room;
2 - refrigeration chamber;
3 - cooling tunnel;
4 - transition tunnel;
5 - unit; 6 - air-coolers.

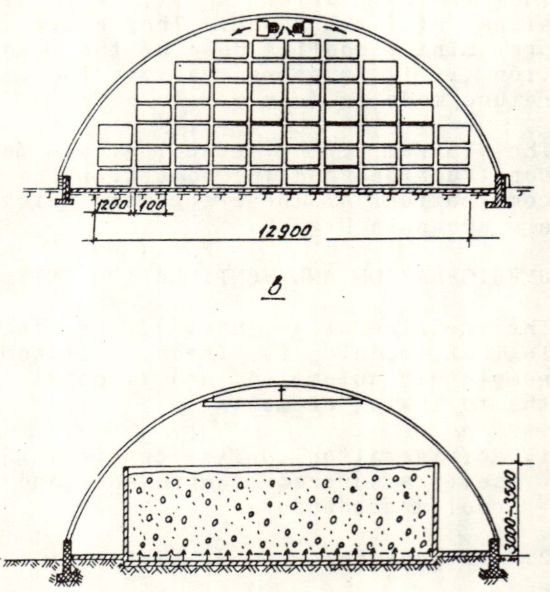

Fig.3 Variants for production storage
a - in a loose state;
b - in pallets.

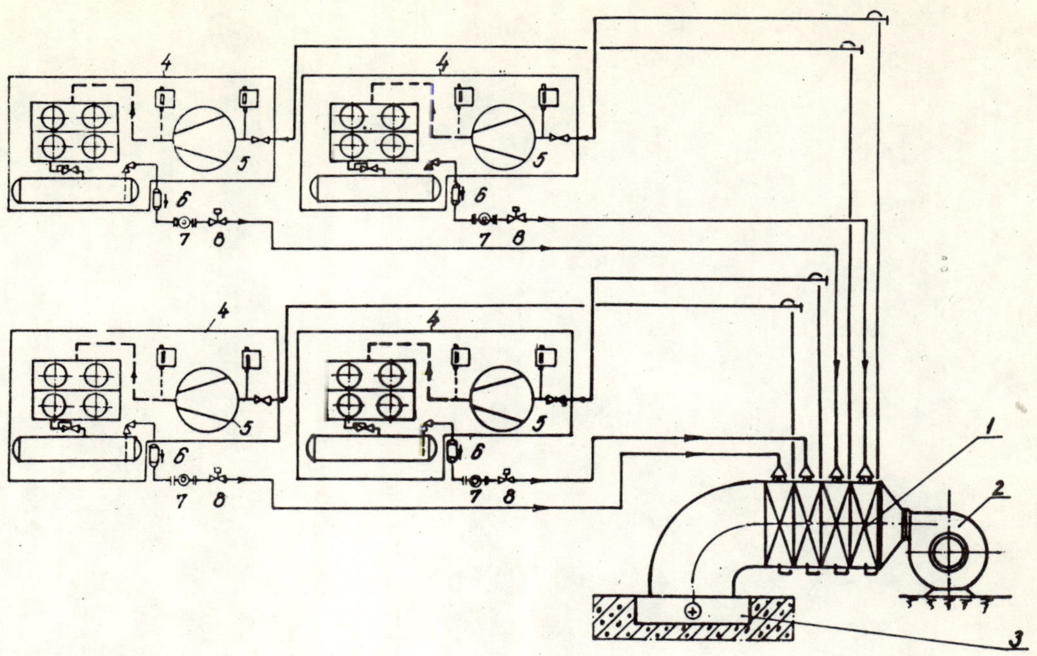

Fig.4 Diagram of the refrigeration unit for loose state storage
1 - air-cooler; 2 - fan;
3 - supercharging air-channel; 4 - refrigeration freon unit;
5 - semipressurized compressor; 6 - filter-dehydrator;
7 - inspection window; 8 - magnetic valve.

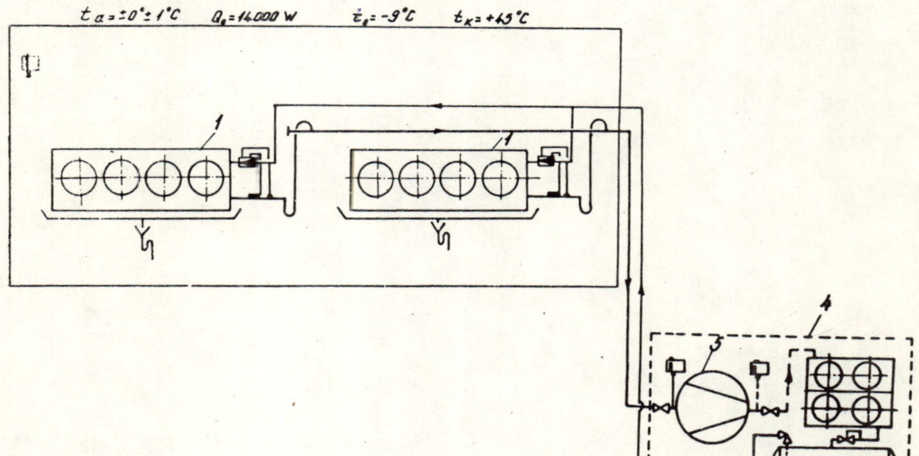

Fig.5 Diagram of the refrigeration unit for pallet-storage.
(see designations in Fig.4).

6. CRYOGENIC INSTRUMENTATION AND MEASUREMENT TECHNIQUES

THE REPRODUCTION OF LAMBDA TRANSITION TEMPERATURE OF LIQUID ^{4}He

Song Naihao* Mao Yuzhu Lin Peng Zhang Qinggeng Zhang Liang

Cryogenic Lab. Academia Sinica, P.O. Box 2711, Beijing 100080, China

ABSTRACT

This paper presents a practical method for the reproduction of lambda transition temperature T_λ of liquid ^{4}He as a thermometric fixed point. A device, called the λ-point cell, has been made and its qualities are very satisfactory. The principle of this design is based upon the dramatic change of the thermal conductivity and the extremely large specific heat of liquid helium while its temperature is passing through the λ point. Experiments show that with the λ-point cell T_λ can be reproduced with a precision of 0.1 mK or better.

INTRODUCTION

From the point of view of low temperature thermometry, the λ-transition of liquid ^{4}He provides a fixed point far superior to the currently used fixed points based upon the superconductive transitions. The width of the λ-transition and its reproducibility are typically 10 μK [1], and are not affected by ambient magnetic field, inhomogeneity and strain as in the superconductive transitions. The presence of ^{3}He as an impurity in commonly used ^{4}He has a negligible effect on the T_λ of liquid ^{4}He, the shift in T_λ is only a few μK [1]. The effect of gravitational field is also negligible for a column of liquid ^{4}He which is only a few cm in height [2]. In addition, T_λ=2.1768K [3] is almost in the middle of the temperatures of superconductive transitions of indium and aluminium, which are used in the SRM767a made by NBS [4]. The reproduction work of T_λ has practical meaning.

THE PRINCIPLES OF REPRODUCTION

The thermal conductivity of liquid ^{4}He has a sharp change when we heat or cool the liquid through its λ-point. In the HeI region, its thermal conductivity is typically 0.5 W/m.K, but in the HeII region it can reach as large as 7,700 W/m.K even in the mutual friction state, which is 16,000 times that of HeI (reference to the thesis of the first author). This dramatic change provides an excellent quality needed for a switch. Based upon this a sealed cell, called the λ-point cell, has been made as shown in Fig.1.

It consists of three parts:

I: An oxygen-free copper cell, V_1, its cavity is about 6.8 cm^3, filled with ^{4}He gas at ambient temperature and 70 atm.

II: A stainless steel tube, V_0, 3 mm o.d. and 2 mm i.d., 45 mm long, connected with V_1.

III: Another oxygen-free copper cell, V_2, its cavity is about 3.1 cm^3, filled with ^{4}He gas

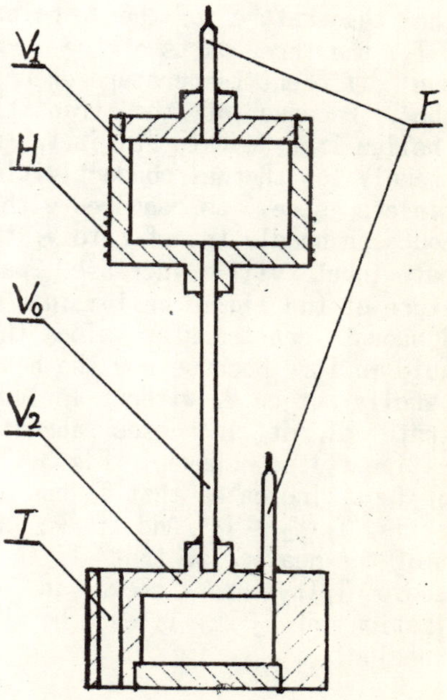

Fig. 1 Structure of the λ-point cell
V_1, V_2: copper cell; H: heater; V_0: stainless steel tube; T: hole for mounting thermometer; F: filling and sealing copper tubes.

*Present address: Institute of Electric Engineering, Academia Sinica, P.O. Box 2703, Beijing 100080, China at ambient temperature and 70 atm. The to-be calibrated thermometers are installed in this block of high-conductivity copper.

All joints were silver soldered. And a 114 ohm manganin heater was wound around V_1. There was enough liquid to fill up the stainless steel tube when the ^{4}He gas were condensed, and V_1 was partially occupied by liquid ^{4}He too.

The λ-point cell was suspended in a vacuum can from the 1.2 K pot in the cryostat [4] by means of three copper wires as shown in Fig.2. The suspension copper wires served as heat linkage between the 1.2 K pot and the λ-point cell. In the experiments there were two thermometers mounted on the λ-point cell, one on V_1, designated as T_1; the other one on V_2, designated as T_2. They were all germanium resistance thermometers.

The ideal process of reproduction is as follows: First, the λ-point cell is cooled down to below T_λ so that the liquids in V_1 and V_2 become superfluid(HeII). Then the heater H around V_1 is turned on so that the temperature of the λ-point cell becomes higher and higher. Because of the excellent thermal conductivity of copper and HeII the temperatures of the three parts of the λ-point cell are almost identical when their temperatures are still below T_λ. The Kapitza resistance can cause a small difference between the temperatures of V_1 and V_2, but it do not bring in extra errors as will be stated later.

When the temperature of the λ-point cell has reached T_λ, the three parts of the λ-point cell will have different temperature excursions from each other. Because of the transition of the liquid helium from HeII to HeI in V_1 and V_0, and the extremely low thermal conductivities of HeI and stainless steel as compared with HeII, the heat input can hardly transfer to V_2 through V_0. The heat input will increase mainly the temperature of the liquid helium in V_1, and only a small amount can transfer along the tube and the liquid in it. Because now the heat input is almost wholly put on V_1 without including V_2, the rising rate dT_1/dt increases abruptly at this moment (time t_2) as shown in Fig.3. The turning point of T_1~t indicates that T_λ has arrived for V_1, that is, $T_1(t_2)=T_\lambda$. And if we adjust the current of the heater, so that T_1 is controlled just above $T_1(t_2)$ as shown in Fig.4, the stabilization of T_2 at T_λ can be kept for a longer period.

Because the heat flux transfered along V_0 is very small and the heat capacity of the liquid helium in it is very large at $T \approx T_\lambda$, the V_0 tube becomes nearly an ideal delay-line for T_1 to transfer down to V_2. From the time when the temperature at the top of V_0 begins to rise above T_λ to the time when the temperature at the lower end reaches a temperature slightly higher than T_λ, the T_2 remains at T_λ —that is in the stabilized temperature stage. Supposing the temperatures of the wall of the tube and the liquid helium in it are identical at any cross-section along the tube, the delay time can be calculated (reference to the thesis of the first author).

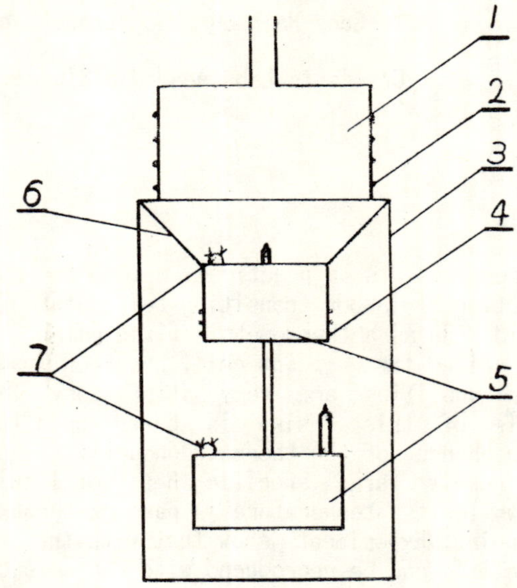

Fig.2 Schematic drawing of the inner portion of the apparatus
1: 1.2 K reduced pressure pot; 2: heater of the 1.2 K pot; 3: radiation shield; 4: heater of the λ-point cell; 5: λ-point cell; 6: suspension copper wires; 7: thermometers.

THE SPECIFIC HEAT OF LIQUID ^{4}He NEAR T_λ

As it is known, the specific heat of liquid ^{4}He becomes very large near the λ-point. Although this quality is not as good as the latent heat of other gases at triple points and boiling points for the reproduction of the transition temperatures, it has contributed a great deal to the realization of this design in two aspects. Firstly, the delay-line of the V_0 tube becomes possible because the heat capacity of the liquid ^{4}He in V_0 is very large, and the larger the heat capacity, the longer the delay time. Secondly, the large heat capacity of liquid ^{4}He in V_2 contributes a lot to the stability of the T_λ plateau (shown as a horizontal line from t_3 to t_4 in the T_2~t curve in Fig.3), as V_2 is not absolutely adiabatic during this period. For example, the heat input of the thermometer due to the measuring current is about 0.045 μW near T_λ, whereas the thermal capacity of the liquid ^{4}He in V_2 can approach 0.34 J/K near T_λ, the rising rate of T_2 is only

$$dT_2/dt = 0.0454/0.34$$
$$= 0.113\ \mu K/Sec = 0.008\ mK/Min.$$

So we can say the stability of the stabilized temperature is very good.

THE RESULTS OF EXPERIMENTS

The recorded temperatures T_1 and T_2 are shown in Fig.3 and Fig.4. At time t_1 the λ-point cell was in the HeII region and the heater was turned on, then T_1 and T_2 increased at the same rate. But T_1 and T_2 had a difference of 2.3 mK due to the Kapitza resistance, because the heat flux was relatively large. This temperature difference was figured out by considering the heat input and geometric size of the λ-point cell.

When t_2 was reached, T_1 would increase at a more rapid rate, the liquid ^4He at the lower end of V_0 was at T_λ, and the temperature difference caused by Kapitza resistance was $\triangle T_0 = T_\lambda - T_2 = 2.3$ mK, that is, the liquid ^4He in V_2 was 2.3 mK below T_λ. From t_2 to t_3 the temperature of liquid ^4He in V_2 was approaching T_λ while the temperature of liquid ^4He at the bottom of V_0 remained at T_λ. In this time the $\triangle T_0$ changed as follows

$$\triangle T = \triangle T_0 e^{-\frac{t}{\tau}}$$

which was an exponential attenuation process. The time-constant was $\tau = 3$ sec.. If $\triangle T$ changed from $\triangle T_0 = 2.3$ mK to $\triangle T = 0.01$ mK, the needed time was

$$t = -\tau \ln\frac{\Delta T}{\Delta T_0} = 16 \text{ sec.}$$

In experiments three successive data (about 30 second interval) showed clearly that the attenuation process had finished. So we could conclude that the Kapitza resistance had little effect on the reproduced T_λ, and we could eliminate its error by choosing the data after attenuation had finished, for example, from $t_3 = t_2 + 30$ Sec. From time t_3, the heat switch was turned off "completely" for some time and V_2 was in heat insulated state. During this time, a temperature gradient was gradually established along V_0 and T_2 increased very slowly (as shown in Fig.3) due to the heat input of the thermometer and later in addition due to the small heat transfered along the tube. T_2 had come into the stabilized temperature stage.

From time t_4 on, the T_2 rised more rapidly because the heat transfered to V_2 increased after the delay period and the heat capacity of V_2 decreased after the λ-point had passed. We stopped heating at t_5 and the λ-point cell was cooled down through the suspension copper wires. And after T_1 and T_2 had fallen down below T_λ (when T_1 and T_2 had almost the same cooling rate), we could repeat the procedure of the reproduction again.

Fig.3 showes a complete cycle for the operation and gave us a clear understanding of the process, but for the practical use of the λ-point cell the authors would like to recommend a controlling process for T_1 as shown in Fig.4.

THE ERROR

As stated above, the stabilized temperature is influenced mainly by the heat input of the thermometer. The rising rate is only 0.008 mK/Min under the heat input of 0.045 μW in Fig.4. The average resistances of thermometer T_2 during t_3 to t are listed in Table 1. We notice that the difference in average resistance is 0.016 ohm for $t - t_3 = 90$ sec. and $t - t_3 = 70$ sec. interval which corresponds to a temperature difference of 0.04mK. And as we have deduced in previous paragraphs that when the attenuation process finished at time t_3, T_2 had approached T_λ to within 0.01 mK, so the error of the stabilized temperature determined from the average resistance with the theoretical T_λ should be surely less than 0.1 mK. Moreover, we have calculated the reproducibility of T_λ reproduced with the λ-point cell. The measurement is carried out with a germanium resistance thermometer and the short-time reproducibility is about 0.09~0.11 mK.

Table 1. Mean Value, Standard Deviation of the resistance and temperature at T_λ

Time $t-t_3$ (S)	Mean $\bar{R}$ (Ω)	Standard Deviation $\sigma(\bar{R})$, Ω	$\sigma(T)$, mK
90	454.553	0.005164	0.013
180	454.553	0.006216	0.016
270	454.549	0.007584	0.020
360	454.547	0.007506	0.020
450	454.545	0.007761	0.020
540	454.543	0.009673	0.025
630	454.540	0.011677	0.030
720	454.537	0.013422	0.035

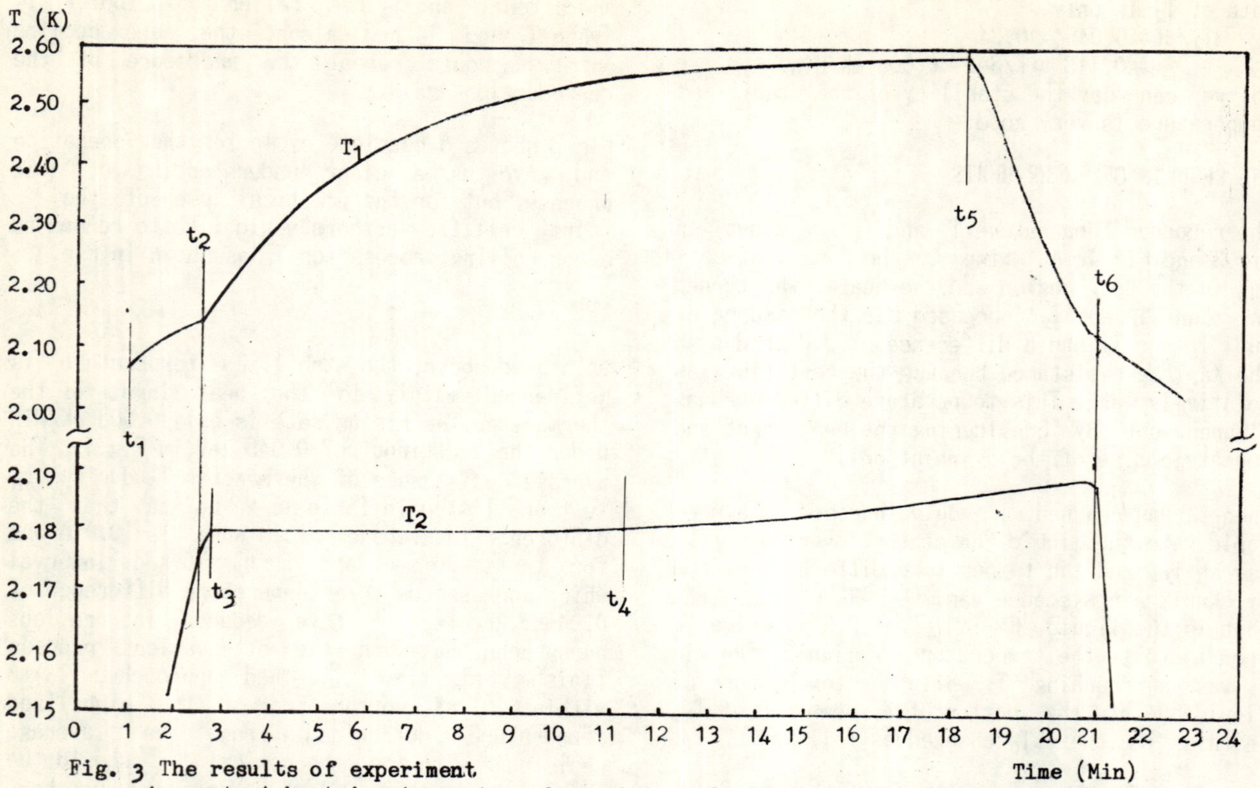

Fig. 3 The results of experiment
A constant heat input was turned on at t_1 and off at t_5. The data were collected at 15 seconds interval.

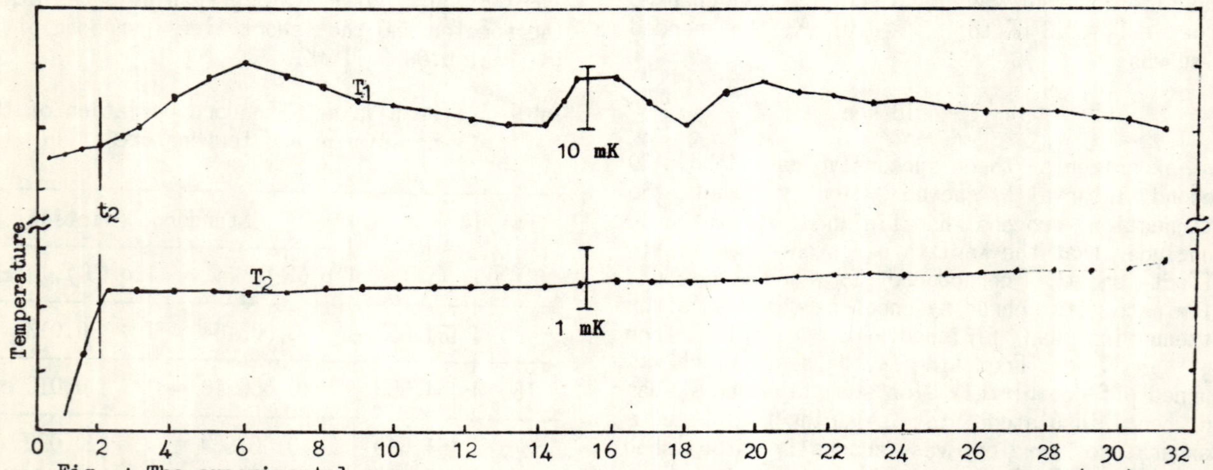

Fig. 4 The experimental curves
The heat input was controlled so that $5\ mK < T_1 - T_1(t_2) < 20\ mK$

CONCLUSION

The method of the reproduction of lambda transition temperature discussed is simple and the design is quite unique. The experiments have confirmed the practicality of the λ-point cell. It should be very useful in the precise calibration of low temperature thermometers.

LIST OF SYMBOLS

T_λ = lambda transition temperature (K)
T_1 = temperature of the V_1 cell (K)
T_2 = temperature of the V_2 cell (K)
τ = time-constant of the attenuation process (sec.)
ΔT = temperature difference (K)

ACKNOWLEDGEMENT

This work has been done in connection with the M.S. thesis requirement of the first author and he is grateful to Prof. C.S. Hong for the suggestion of constructing a λ-point cell making use of the heat conductivity change from He II to He I and the intimate guidance throughout this work. The λ-point cell experiments have been carried out at the Cryogenic Measurement Centre/Academia Sinica, operated jointly by the Cryogenic Laboratory and the Liao-ning Institute of Metrology.

REFERENCES

1 K.F.Hwang and B.M.Khorana, 'lambda transition of liquid helium as a thermometric fixed point'. Metrologia, 12(1976), pp 61-63.

2 G.Ahlers, 'effect of the gravitational field on the superfluid transition in He '. Physical Review, Vol.171(1968), pp 275-282.

3 Document of the Comité Consultatif de Thermométrie,CCT/87-2. 10 March, 1987.

4 Mao Yuzhu, et al, "checking equipment for thermometer in the range of 0.5 K to 30 K ", (in Chinese).

PRECISION SUPERFLUID DENSITY OF HE II IN POROUS MEDIUM

Zhu Yuqun, Guo Fanzhong, and Zheng Jingbao

Laboratory of Cryogenic Engineering
Huazhong University of Science and Technology
Wuhan, China

ABSTRACT

Acoustic resonators were used to measure the superfluid density of He II contained in Porous medium to a typical precision of 0.5%. Data were taken at temperature between 1.325K and the λ point. The temperature was determined to better than 0.1%.

INTRODUCTION

The possibility of phase transitions and in particular of superfluidity in systems of fewer than three dimensions has been widely investigated. An experimetal approximation to such systems in the case of helium is the absorbed film, which can vary in thickness from one atom (3Å) up to about 100 atoms. Another type of experiment which yield information on this system is the propagation, in the helium II region, of fourth sound, a density wave in the superfluid with the normal fluid locked in a stationary position by its viscosity.

In present paper, we described an experiment which uses acoustic resonantors to measure the density of superfluid helium in porous medium.

It can be shown[1] that the sound velocities (of first, second, and fourth) determine the most important thermodynamical properties of superfluid helium. In this experiment by using acoustic resonantors, the fourth sound velocity was measured better than 0.2% precision. So one can expect a accurate density of superfluid helium in porous medium. The paper is organized as follows: Section 2 contains a brief theory for the velocity of fourth sound. The experimental techniques are described in section 3. In section 4 we present the experimental results of density in porous medium.

THEORY

In order to calculate the fourth sound velocity for helium confined in porous medium the following linearized system of thermohydrodynamic equations[2] is considered:

$$\partial\rho/\partial t + \mathrm{div}\, j = 0 \tag{1}$$

$$\partial j/\partial t + \mathrm{grad}\, P = \eta(\Delta V_n + \frac{1}{3}\,\mathrm{grad}\,\mathrm{div}\, V_n) + \rho_{s0}\zeta_1 \mathrm{grad}\,\mathrm{div}(V_s - V_n) + \zeta_2\,\mathrm{grad}\,\mathrm{div}\, V_n \tag{2}$$

$$(1/n^2)\partial V_s/\partial t + (1/\rho_0)\mathrm{grad}\, P - \sigma_0 \mathrm{grad}\, T = \rho_{s0}\zeta_3 \mathrm{grad}\,\mathrm{div}(V_s - V_n) + \zeta_4 \mathrm{grad}\,\mathrm{div} V_n \tag{3}$$

$$\partial(\rho\sigma)/\partial t + \rho_0\sigma_0 \mathrm{div} V_n = (\kappa/T_0)\Delta T \tag{4}$$

$$j = \rho_{s0}V_s + \rho_{n0}V_n \tag{5}$$

$$\rho_0 = \rho_{s0} + \rho_n \tag{6}$$

The subscripts n and s refer to the normal fluid and superfluid components of He II , respectively. The subscript zero indicates that the quantity in question has to be taken at its equilibrium value, ρ and V are the density and the velocity, P,T and σ are the pressure, the temperature and the entropy per unit mass of helium, η and κ are the coefficients of first viscosity and thermal conductivity; ζ_1, ζ_2, ζ_3 and $\zeta_4 (\zeta_1 = \zeta_4)$ are the coefficients of second viscosity. The factor $1/n^2$ arises from the additional acceleration of the superfluid component around the powder grains in the sound field[3]. If the powder grains are assumed to be spheres and to be separated so that the velocity fields around them do not affect each other[4]. then the factor $1/n^2$ can be represented by

$$1/n^2 = (1+p)2p \tag{7}$$

where p is the porosity defined as the ratio of helium filled volume to the total volume of the porous sample. If the characteristic dimensions of the channel of porous medium are such that the normal fluid component is clamped (For this, it is necessary that either the length of the viscous wave $\lambda_V = \sqrt{2\eta/\omega\rho_n}$ or the length of the free path of the elementary excitations ℓ be much greater than the characteristic dimensions of the channel d). The velocity of fourth sound, obtained in this manner, is equal to

$$u_4^2 = \frac{\rho_s}{\rho} u_1^2 + \frac{\rho_n}{\rho} u_2^2 \quad (8)$$

where $u_1 = \sqrt{\partial p/\partial \rho}$ is the velocity of first sound, and $u_2 = [\sigma^2 \rho_s/\rho_n (\partial\sigma/\partial T)]^{\frac{1}{2}}$ is the velocity of second sound.

From Eq.(5) we have

$$\frac{\rho_n}{\rho} = 1 - \frac{\rho_s}{\rho} \quad (9)$$

so

$$u_4^2 = \frac{\rho_s}{\rho} u_1^2 + (1 - \frac{\rho_s}{\rho}) u_2^2 \quad (10)$$

and finally we get

$$\frac{\rho_s}{\rho} = \frac{u_4^2 - u_2^2}{u_1^2 - u_2^2} \quad (11)$$

EXPERIMENTAL TECHNIQUES

A. Cryostat and Thermometry

The experiment were performed in a standard metal cryostat. A mechanical pump located in the vapor space above the liquid helium level in the top part of the cryostat, was used to reduce the vapor pressure of liquid helium bath. With this device a temperature of 1.3K in the main helium bath could be obtained and maintained for more than 8 hours.

The temperature was determined to an absolute accuracy of ±1mk by measuring the vapor pressure of the helium bath and using The T_{58} vaper pressure scale of ^{4}He.

B. Resonator

The velocities of fourth sound were determined by measuring resonant modes of a cylindrical resonator. The resonantor is shown in Fig. 1. It is made of brass and consists of two transducers. i.e. electrostatic loudspeaker transducer and electrostatic microphone transducer, located at each end of the cavity. The fourth sound was excited at one end of the cavity by a 10V rms ac signal applied to the loudspeaker transducer that consisted of 2μm thick Makrofol SKG blue[5], covered with a thin layer of gold. It was detected by the electrostatic microphone transducer that was identical to the loudspeaker. The microphone was biased with a diffirent voltages ranged between 100V and 300V.

The resonator with the quality factor from 10 to 700[6] was packed with coarse 1μm aluminum oxide powders. The porosity is 0.678 determined by weighing the empty and filled cavity and from the formula

$$p = 1 - W/VD \quad (12)$$

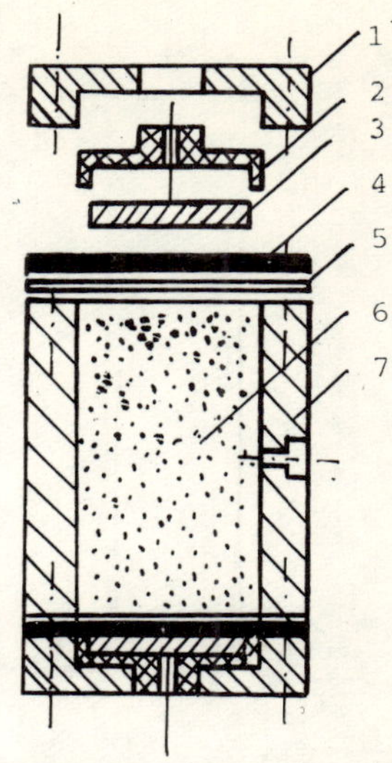

Fig.1 Sound cell for investigating fourth sound 1 - cover, 2 - insulating, 3 - massive electrode, 4 - membrane, 5 - gold film, 6 - fine grained power, 7 - cavity.

where W is the weight and D the density of the powder contained in the cavity of volume V.

RESULTS AND DISCUSSION

According to the Eq.(11) the value of ρ_s/ρ is determined by the sound velocities of first, second and fourth. The first sound velocity and the second sound velocity have been widely investigated, both theory and experiments on u_1 and u_2 have fairly success; So that one can use the precise results of fourth sound velocity measured experimentally to calculate the value of ρ_s/ρ.

In present paper, the velocity of fourth sound have been measured in liquid helium at temperature between 1.325K and the λ point along the vapor pressure curve. The measurements were made, using resonance technique and the helium was contained between small pores in porous medium (coarse 1μm)

Typical spectra of the resonators are shown in Fig.2. In which one can clearly find 6 harmonics were particularly strong and gave velocities which are shown in Fig. 3.[6] By using the velocities of fourth sound shown in Fig.3 and Eq.(11) we calculated the ρ_s/ρ and plotted the ρ_s/ρ against the temperature T in Fig. 4, some typical values of the ρ_s/ρ are presented is Table 1.

The accuracy of the data is limited by the error in the length of the resonators (0.1%). By using the method of standard error, we found, the accuracy of ρ_s/ρ is within 0.5% precision.

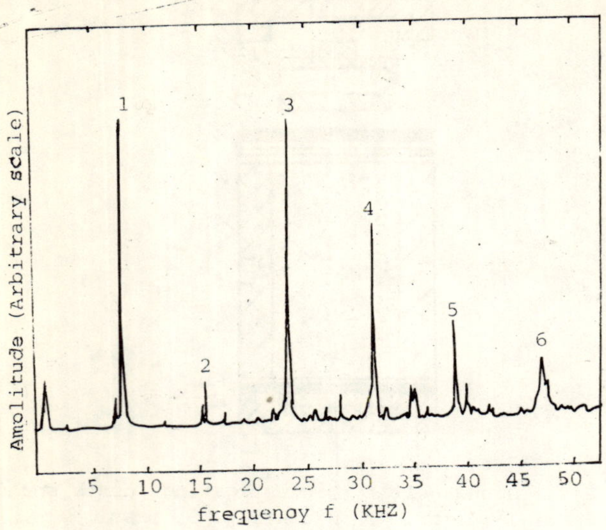

Fig.2 A typical spectrum of the resonance modes of fourth sound at 1.325K (SAmple 2). The labeled resonances correspond to the first six plane wave modes.

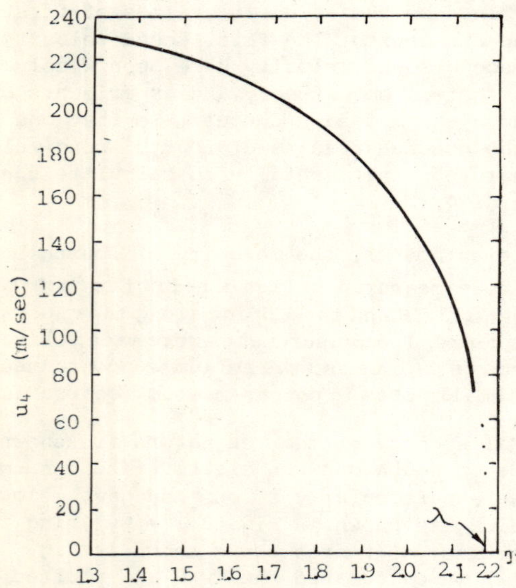

Fig.3 Velocity fourth sound vs temperature from the present work

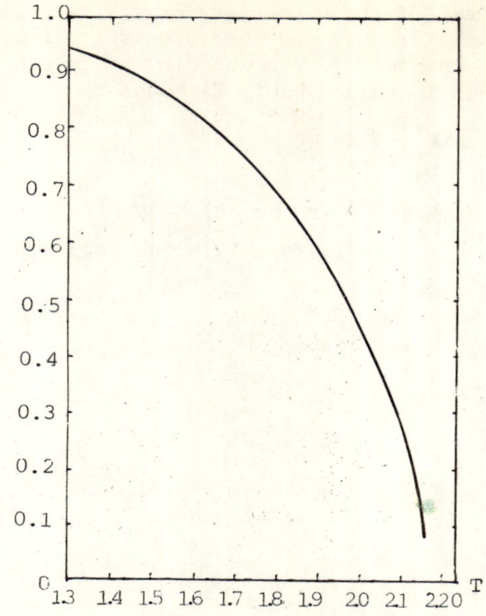

Fig.4 ρ_s/ρ in porous medium (P = 0.678) vs temperature.

Table I. ρ_s/ρ of He II in Porous medium (p=0.678)

T	ρ_s/ρ	T	ρ_s/ρ	T	ρ_s/ρ
1.300	0.9546	1.350	0.9389	1.400	0.9231
1.450	0.9045	1.500	0.8846	1.550	0.8600
1.600	0.8376	1.650	0.8055	1.700	0.7651
1.750	0.7274	1.800	0.6851	1.850	0.6354
1.900	0.5904	1.950	0.5062	2.000	0.4414
2.050	0.3604	2.100	0.2550		

REFERENCES

1 J.Maynard, phys. Rev. B14, 3868 (1976).

2 I.M.Khalatnikov, Introduction to the theory of superfluidity (Benjamin, New York, 1965).

3 M.Revzen, B.Shapiro, C.G.Kuper, and J.Rudnick, phys. Rev. Lett. 33, 143 (1974).

4 L.D.Landau and E.M.Lifsehitz, Lehrbuch der theoretischen physik, Vol. VI (Akademie-Verlag, Berlin, 1966). p34.

5 G.Schexnayder, Mobay chemical Co., Pittsburgh, PA 15205.

6 Y.Q.Zhu, to be published.

A SUGGESTION FOR A NEW LOW TEMPERATURE PRESSURE TRANSDUCER

Jin Xin, Chen Wuming, Ouyang Rongbai
Department of Physics, Nanjing University,
Nanjing, China

A.L.Thomson
Physics Department, University of Sussex,
Brighton, England, BN1 9QH

ABSTRACT

It is suggested that an open-ended microwave cavity can be used as a sensor for a pressure transducer to measure the pressure of a low temperature fluid. It has five advantages compared with the conventional low temperature pressure transducers.

INTRODUCTION

For the measurement of low temperature fluid pressures, many types of pressure transducer have been discussed[1,2]. They are mainly used to convert applied forces into displacement by means of a diaphragm, other means. They have to be calibrated in the working temperature range, and they all have a big problem of shifting calitration due to thermal cycling. An open-ended microwave cavity such as was used by Jerry Smetana et al for measurement of liquid hydrogen fluid has been developed by us as a transducer. Some of the authors of this paper have built this instrument and have discussed it in more detail[4,5].

THE PRINCIPLE OF THE TRANSDUCER

The principle of the open-ended microwave cavity is covered in detail in references 3 and 6. A diagram of it is shown in Fig.1. It consists of a circular waveguide of radius b, which terminates at each end with this coaxial cylindrical partitions of radius a that are separated by a distance L. Microwave energy is supplied to the cavity by small loops at the ends of a coaxial transmission line protruding through the side wall. If the cylindrical partitions are sufficiently long[6], they will act as perfect reflectors for the microwave energy over a frequency range that is dependent on the dimensions a and b. The cavity is resonant when the spacing of the partitions is adjusted so that the reflections are additive.

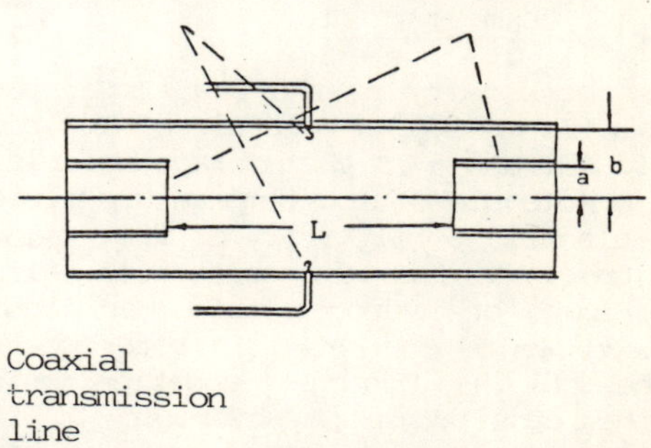

Fig.1 The principle diagram of microwave cavity

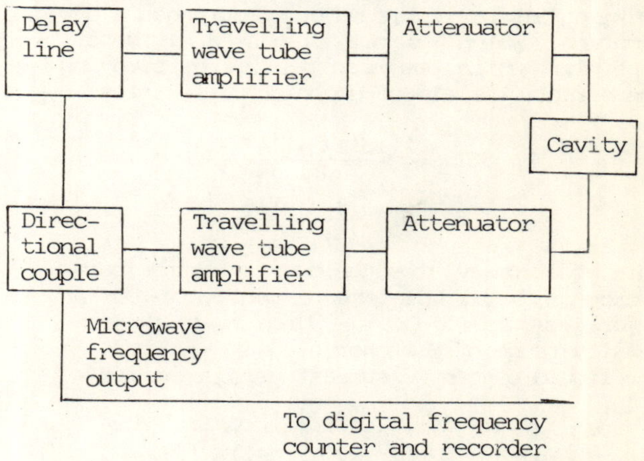

Fig. 2 The block diagram instrument of measuring and recording frequency of the microwave cavity

This condition occurs when the spacing L is approximately an integral mutiple of one-half a wavelength.
If the resonant frequency is measured for the case when the cavity is evacuated, it can be shown that

$$\varepsilon = \left(\frac{f_o}{f}\right)^2 \qquad (1)$$

where ε is the relative dielectric constant, f_o is the resonant frequency when the cavity is evacuated, and, f is the resonant frequency when it is filled with a dielectric material. In the cavity the TE_{011} mode of oscillation was selected.
For measuring and recording the resonant frequencies an instrument[5] similar to that of J.Smetana et al[3] was built. The block diagram is shown in Fir.2.
The well known Clausius-Mosotti equation is used to relate the density of nonpolar dielectrics to their relative dielectric constant. The form of this equation is

$$\rho = \frac{3\varepsilon_o m}{\alpha}\left(\frac{\varepsilon - 1}{\varepsilon + 2}\right) \qquad (2)$$

where ρ is the density of the low temperature fluid, m is the mass of a molecule of low temperature fluid, and α is the average polarizability of a molecule of low temperature fluid. ε_o is the electric permittivity of free space. The equation (2) is fairly accurate for helium fluid,[7] but for hydrogen and other liquid and gases needs to be corrected. In reference 3, they suggested the Böttcher equation which relates the density of parahydrogen to its dielectric constant with an error less than 0.001 over of the range of density from 2 to 80 kilograms per cubic meter. We propose a new equation, which is not discussed in detail here, but which has been derived by studying the effective field of nonpolar dielectrics.

$$\rho = 331.984\,\frac{15(2\varepsilon+1)(\varepsilon-1)}{4\varepsilon^2+37\varepsilon+4} - 32.567\,\frac{15(\varepsilon-1)^2}{4\varepsilon^2+37\varepsilon+4} \qquad (3)$$

This equation relates the density of the parahydrogen to its dielectric constant with an error less then 0.00004, which is in the same density range of Böttcher's equation.
For liquid oxygen, we suggest using equation(4)

$$\rho = A(T)\frac{(\varepsilon-1)(2\varepsilon+1)^2}{3\varepsilon\,[(n+1)-(n-1)\varepsilon]} - B(T)\frac{(\varepsilon-1)^2}{3\varepsilon\,[(n+1)-(n-1)\varepsilon]} \qquad (4)$$

where A(T), B(T) and n dependent on temperature showing in table 1.

Table 1 The values of A(T), B(T) and n dependent on temperature in equation (4)

T(K)	A(cm^3/mol)	B(cm^3/mol)	n
72	85.249	-201.627	2.10
90	85.311	-185.713	1.95
150	84.371	-404.477	4.70
200	84.308	-553.716	6.45
250	84.323	-591.719	6.95
300	84.268	-955.570	11.30

Equation (4) was derived by use of the 'Cage Model for Liquid' and the 'Cavity Spheres for Molecules' in a liquid dielectric, and through the study of molecular polarization and the influence of its neighbours. The density of liquid oxygen calculated by equation (4) conforms with the experimental results[8]. The maximum relative error is only 0.00004.
From equations of (1),(2),(3) and (4), we can use the open-ended microwave cavity instrument to measure accurately the density of low temperature helium, hydrogen and oxygen. If the temperature, T, of the low temperature fluid is measured at the same time as its dielectric constant, ε, we can obatin the pressure of the low temperature fluid at any given time from the equation of state of the low temperature fluid.
If we measure the pressure of the low temperature fluid, we can use a three parameter equation

$$p = \frac{RT}{\frac{1}{\rho} - b} - \frac{a}{T^n \frac{1}{\rho}\left(\frac{1}{\rho} + b\right)} \qquad (5)$$

where p is the pressure, R is the ideal gas constant, n,a, and b for helium and hydrogen as presented in reference[9]. For oxygen, n = 0.6, and

$$a = 287251.43\ (\times 10^{-5} Pa\ m^6 K^n/mol^2)$$

$$b = 2.1953\ (\times 10^{-5} m^3/mol)$$

It is difficult to find the equation of state for fluids, so that we suggest using the data of reference[10], [11], and[12] to obtain the pressure.

DISCUSSION

This new low temperature transducer has more than five advantages over conventional transducers. 1. Other transducers have to be calibrated in the working temperature range, but this is not needed for the new type, because we can calculate the pressure results from the theoretical anaylysis presented above. 2. The sensor and the microwave cavity of the new low temperature transducer are made from the same material, so that there is an absence of zero

shift due to thermal cycling. 3. It has an inherent high precision of measurement because the principle is based on equations(2),(3), (4) and the thermodynamic properties of the fluid. 4. The relaxation time (4) of this instrument is estimated to be about 1.63μs. It is better than with the other type of transducer which are based on the displacement of a diaphragm to measure the pressure of a low temperature fluid. 5. The new low temperature trnasducer is more reliable than others because it has no mechanical elements.
For measure the pressure of a two phase low temperature fluid, this new transducer may not be as good as the conventional ones.

REFERENCES

1. Bell, J.H., 'Cryogenic Engineering' Chapt.9 Prentic-hall, Inc.(1963)
2. Smelser, P. 'pressure measurement in cryogenic systems' Adv. in Cryo.Eng.8,Plenum Press Inc. New York (1963) pp 378-336
3. Smetana, J. and Wenger, N.C.'instrumentation for liquid hydrogen density measurements using an open-ended microwave cavity' NASA TN D-6415.
4. Jin X.,Chen W.M.,Ying Z.M. and Xu J.J., 'Development in measuring liquid hydrogen fluid density by using the microwave open-ended cavity tuned oscillator', Journal of Low Temperature Physics (Chinese), Vol.5 No. 1 (1984)
5. Jin X., Zhang H.H., Xu J.J., Chen W.M., et al 'a study in a cryogenic fluid of average density measurements using an open-ended microwave cavity', Journal of Refrigeration, (Chinese), No.4 (1984).
6. Wenger, N.C.,'theory of an open-ended circular cylindrical microwave cavity', NASA TN D-3514 (1966).
7. Atkins, K.R.,'Liquid Helium', Chapt.1, Cambridge University Press, London, (1959).
8. Chen W.M. and Jin X., 'a theory of the density of liquid oxygen -- accurate determination of the density of liquid oxygen through measurements of its dielectric constans', Journal of Low Temperature Physics (Chinese), Vol.8, No.1, (1986).
9. Jin X. and Xiaonong Xu and A.L. Thomson, 'The temperature effect of different working gas of Philips refrigerator' The Proceedings of ICCR, Hangzhou, China,(1986).
10. Roder, H.M. and Weber L.A. 'thermophysical properties',NASA SP-3071, C-2.
11. McCarty, R.D.,'thermophysical properties of helium-4 from 4 to 3000R with pressure to 15000PSIA, NBS Technical Note 622.
12. McCarty,R.D.,'hytrogen technological survey--thermophysical properties'. NASA, Report SP-3089.

STUDY OF THE CHARACTERISTICS OF A FLUON-COATED Au-Fe THERMOCOUPLE USED IN LOW TEMPERATURE MEASUREMENT

Wang Zhenhua

Chongqing Instrument Materials Research Institute, P.O.Box 1512, Chongqing, China

ABSTRACT

Since the Au-Fe thermocouple used for low temperature was investigated by R Bermen et al, the importance of which in low temperature measurement has well known. The study of this type of thermocouple in China and a Fouon-coated Au-Fe thermocouple used in cryogenic engineering are reported in this paper. The Analysis of composition in Au-Fe dilute alloy, the effect of Fe content on the themoelectric characteristic, the temperature-thermal electromotive force relation of the thermocouple wires in cryogenic range, and electric resisticivity and thermal conductivity of Au-Fe alloy in low temperature range are discussed. The stability of thermoelectric characterisic of the alloys at operating condition, the homogeneity of thermoelectric property along the length, effect of mechanical stress on thermoelectric property, and strenth of the alloys are determined. It was found that the possibility to probuct the Fluon-coated Au-Fe thermocouple. The design of construction, insulation resistance, voltage proof, sealing and wet-proof properties, noiseproof feature, mechanical strength and thermoelectric characteristic of this thermocouple are detailed. Finally, the applications of the thermometric probe, reference junction compensater and digital thermometer in cryogenic engineering are briefly introduced.

I. INTRODUCTION

Thermometric thermocouples have been used widely in the temperature measurement field, because of their characteristics of small-scale, rapid response, free self-heating, convenient fabrication and application, and sufficient accuracy. Since the higher sensitivity and stable indication of Au-Fe dilute alloy in cryogenic range was discovered by D.K.C. MacDonald in 1962[1], they have been used for thermoelement material for low temperature measurement. The Au-Fe thermocouple was investigated in more detail by R.Berman, Finnemore and the others[2,3] . For temperature measurement, reference tables of Au+0.03at.%Fe, Au+0.07at.%Fe, and Au+0.02at.%Fe thermocouples have been given respectively by UK, USSR and USA.

Owing to the requirements of cryogenic engineering, Au-Fe thermocouples have been widly investigated in China. The dilute alloys such as Au+0.03at.%Fe, Au+0.07at.%Fe, Au+0.02at.%Fe alloy and Chromel, Copper, Normal silver, PdCrRu thermoelement materials are made in China. The applications of the Au-Fe thermocouple in cryogenics promote standardisation, the experimental, Au+0.07at.%Fe alloy was determined as an impotant type of negative thermoelement and a Chinese Standard "Nickel-Chrome/Gold-Iron, Copper/Gold-Iron Low Temperature Thermocouple Wires and Their Reference Tables" was formulated[4].

For the requirements of low temperature measurement, the commercial Au-Fe thermocouple has many varieties such as silk-covered, paint-coated, stainless steel sheahed, and Fluon-coated. The characteristics of the alloy wire and the assembly of the Fluon-coated Au-Fe thermocouple are discussed in this paper.

II. THERMOELECTRICAL CHARACTERISTIC

A Au+0.07at.%Fe alloy was used in the test. After annealing, the composition of the alloy were analyzed by means of the Atomic Absorption Spectromery. The results indicate that Fe content in this alloy is 0.02 ± 0.0015wt.%, i.e. it is a Au+0,07at.%Fe alloy. The thermocouple wires were calibrated by menas of the apparatus given in Fig.1.A Type 5187U Rh-Fe resistance thermometer and a 5187L Pt resistance thermometer were used as Temperature Standards, and a type P306 d.c. potentiolmeter as an electric measuring instrument. The accuracy of the controled temperature in the adiabatic thermostat is 0.01K/15 min.

All thermocouple wires made of Au-Fe, Chromel, Copper and pure platinum were tied together followed by coating with insulator tubes. The measuring junction was welded with an argon arc welding machine, and it was then coated with copper foil followed by painting and inserted in the copper block of a calorimeter. A length of 20 cm of wire was wound on the copper block, which was inserted into an ice bath with a vacuum sealing. The Copper leads were connected to the potentiometer.

For the 2~4.2k range, the vacuum chamber was filled with gaseous helium at a pressure of 0.3 MPa. The measurements were made by decreasing the pressure of the liquidhelium. For the 4.2~70K range, liquid helium was used. For the 70-280K range, liquid nitrogen was used (decreased the pressure when below 77K). There were 25 calibrating points in the 2-280 K range, the total non-determinacy of the system is less than 0.05 K.

In order to be consistent with the reference table of practical standard themocouple, the Chromel (Copper) versus Au+0.07at.%Fe thermocouple reference table by NBS at 1972[3]was converted into a 0 ~ -273°C reference table, the reference junction is at 0°C, the form reference table, the reference junction is at 0° C, the form of which is similar to reference tables of some standard thermocouple in IEC65B.

The E-t values of Chromel versus Au-Fe were calculated from values of Chromel-Pt coupled with Au-Fe—Pt, the E-t curves of Chromel-Pt, Au-Fe—Pt and Chromel—Au-Fe are given in Fig.2, the EMF variation between our and NBS's values is within 1° C. The E—t values of Copper versus Au-Fe were computed by values of Copper-Pt coupled with Au-Fe—Pt and Copper—Au-Fe are shown in Fig.3, the EMF variation between our and NBS's values (1972) is within 1°C also.

Chromel and copper positive themoelements are commercial products of a type K and T standard thermocuple. Our purpose is to study the characteristics of the Au-Fe thermoelement.

III. RESISTIVITY OF A Au-Fe WIRE

The above adiabatic thermostat was used, a type 5187U Rh-Fe and type 5187L Pt resistance thermometer were used as Temperature Standards. The electric measuring instrument is the QJ18A bridge with an accuracy to $2X10^{-4}$.

The length and average diameter of the wire sample were measured. The sample was wound as a spiral and put into the adiabatic thermostat. The measurement was made by a four-wire system. The resistance of the sample was determined at various temperatures over 2.242 to 273.15K. The formula for calculating resistivity given by

$$\rho_1 = \frac{R_1 S}{L} \ X\ 10^{-6} (\Omega\ cm)$$

where ρ_1 is the temperature coefficient of resistance, R_1 is resistance value in the Au-Fe wire, S and L are the cross-sectional area and length of the wire, respectively.

The analysis of error indicates that the total non-determinacy in the resistivity measurement is about $3X10^{-4}$.

With least-squares fit, the relation curve of resistivity of Au+0.07at.%Fe thermoelement wire and temperature was computed and given as Fig.4, from which a characteristic plate in Kondo alloy is shown in ρ-T curve below 20K.

IV. THERMAL CONDUCTIVITY OF AuFe WIRE

The steady-state thermal conductivity of a rod type sample was determined. By the steady heat flow equation, the thermal conductivity λ is given as:

$$\lambda = \frac{Ql}{\Delta TA} = \frac{1}{A}\ \frac{P}{\Delta T}$$

where λ is a function of temperature T, Q=heat flow, l=lenth, A=cross-section ΔT shall be a small value.

To ensure the precise measurement of the various parameters in the heat flow equation, the following technical measures were taken. The sample was fabricated to be straight and homogeneous, so that the area A can be determined precisely. For the measurement of value a sheet of elastic alloy was pressed onto the sample and thermometers were put into two cutting points to determine the temperature and difference between the two points.

The sample was put in the adiabatic thermostat, and the heat flow Q was provided from a heater under the sample. The heating power P=IV. When the sample was in the adiabatic condition only, the heating power P can be used to provide a completely steady-state heat flow Q, i.e. P=Q. To prevent heat loss in the solid conductor, the lead for the current should be as thin as possible (Cu, wire, diameter 0.07mm), the lead for the voltage is a thin wire of Manganin with a low thermal conductivity. To prevent heat loss from the residual gas, the thermostat was be pumped to high-vacuum and filled with a degasser (activated Carbon), so that the pressure in the vacuum chamber was held at $133.32X10^{-5}$ Pa. To prevent loss of heat by radiation, a radiation screen was mounted. A subsidiary heater was used to hold the temperature difference between the thermostat and the sample to less than 1K.

The experimental temperature range was from 4.2K to room temperature. A least-squares multinomial was computed to fit experiment results and the λ—T curve is shown in Fig. 5.

The error analysis indicates that the measuring error of P and ΔT is less than 0.1%, the error due to heat loss is less than 0.5%, and the error in measurement of l and A, the total non-determinacy is less than 1%.

We have discussed the E—T, ρ—T and λ—T characteristics of the Au-Fe thermoelement wire already. As a practical thermoelement material, an Au-Fe alloy should possess other properties, for example, an excellent stability of thermoelectric characteristics of the alloy at operating conditions, the homogeneity of thermoelectric characteristics along the length of wire, sufficient mechanical strength and negligible effects of mechanical stress on the thermoelectric properties. A study of these properties was made in our work.

V. STABILITY OF Au-Fe THERMOCOUPLE WIRE

Stability is one of the most important properties of a practical thermometer. A stability test was made at 77K and 4.2K. In the long term stability test, a Au-Fe thermocouple wire was aged for 3 months, and the EMF of the thermocouple before and after aging were determined to obtain the ΔE value. In the thermal shock test, the thermocouple was raised from and lowed into a liquid nitrogen or liquid helium container. The temperatures at the measuring and reference junctions were determined by a respectively. After 20 cycles of thermal shock, the ΔE_{max} values are given in table 1. It was shown that the Au-Fe thermocouple wire is stable after long term aging and multiple thermal shock and similar results can be obtained in the liquid nitrogen and liquid helium test.

VI. HOMOGENEITY OF THE Au-Fe THERMOCOUPLE WIRE

A measuring error will be introduced when a thermocouple material has thermoelectric inhomogeneity along the length. It is therefore necessary to examine thermoelectric inhomogeneity in a thermocouple wire.

The begining and end of the wire, 2m long each, are taken out of a coil (weight over 200g) and the EMF of each of them versus Pt were measured at 77K and 4.2K, respectively. The purity of the Pt thermoelement, R_{100} R_0 =1.3925. The inhomogeneity, i.e. the difference of EMF in the begining and end of the wire, is given in table 2.

It was shown that the inhomogeneity of the thermocouple wire is similar at 77K or at 4.2K. Therefore, the inhomogneity test was made only at 77K.

The experimental items include:
1. Inhomogeneity in the begining and end of the wire.
2. EMF dispersion degree for 5 pieces of wire, 1.5m long each, which were taken arbitrarily from a wire 50m long.
3. The wire 3m long was double-wound on a glass bar, then it was inserted into a liquid nitrogen bath slowly and the EMF variaton was measured. The experimental results are shown in Table 3.

With the three methods, the EMF variation in the whole wire can be shown. The results indicate that the degree of dispersion thermoelecteric characteristics is lower.

VII. MECHANICAL STRESS TEST

The thermocouple wire will suffer from mechanical stress when it is coated or assembled. It is necessary to examine the effect of stress on the thermoelectric characteristics.

The examination procdure was as follows. Sample wire was cut into two parts equal, one was as-cut, another was under a mechanical stress condition. Then the EMF difference of the two parts of the wire versus Pt at 4.2K was determined.

In the test, three sample wires each 4m long were cut from a coil. The hardening treatment method on the wire 500mm long includes:

1. The sample wire was wound on a hard rod 5 times repeatedly to induce significant work hardening.
2. It was twisted once 10mm long.
3. It was wound on a hard wire 2mm in diameter and wound off from it lightly, so that general work hardening was incuce:

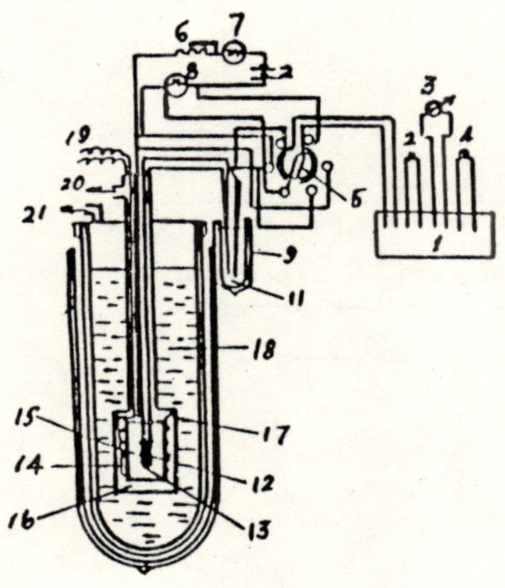

Fig. 1-Temperature measurement apparatus

1. d.c.potentimeter 2. accumulator 3. galvanometer 4. standard cell 5. muitipoint switch 6. varistor 7. milliammeter 8. standard resistance 9. ice bath 10. Dewar flask 11. thermocouple reference junction 12. thermocouple measuring junction 13. standard Pt resistance thermometer 14. heating wire 15. calorimeter 16. vacuum chamber 17. insulating support 18. liquid N_2 or He 19. circuit 20. vacuum system 21. decreasing pressure system.

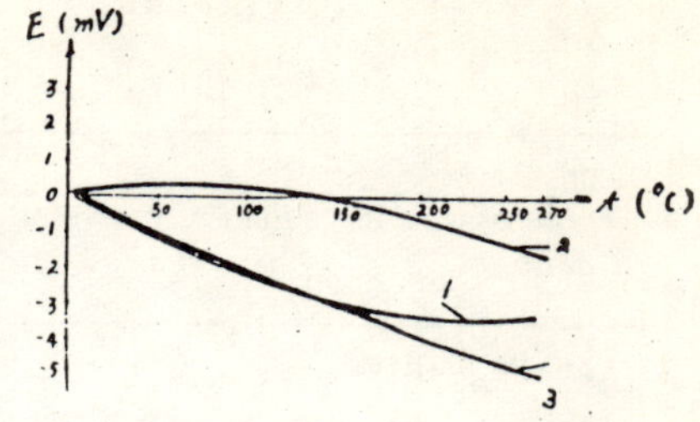

Fig.2 E-t curves
1. Kρ-Pt
2. Au+0.07at.%Fe-Pt
3. Kρ-Au+0.07at.%Fe

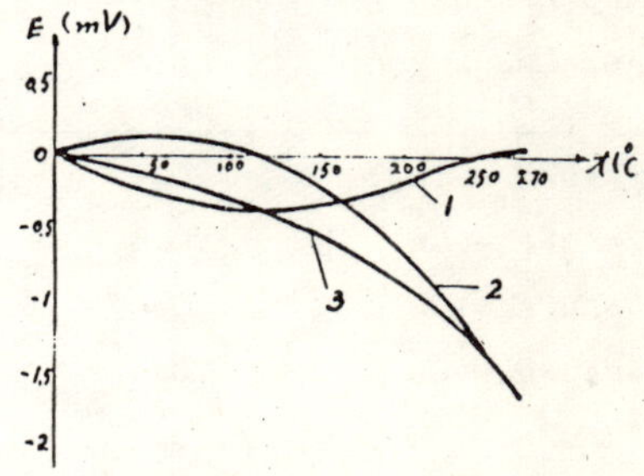

Fig.3 E-t curves
1. Cu-Pt
2. Au+0.07at.%Fe-Pt
3. Cu-Au+0.07at.%Fe

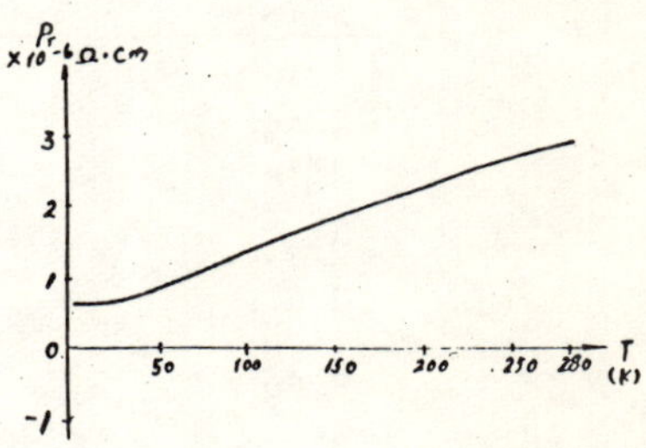

Fig.4 ρ -T curve of Au-Fe wire

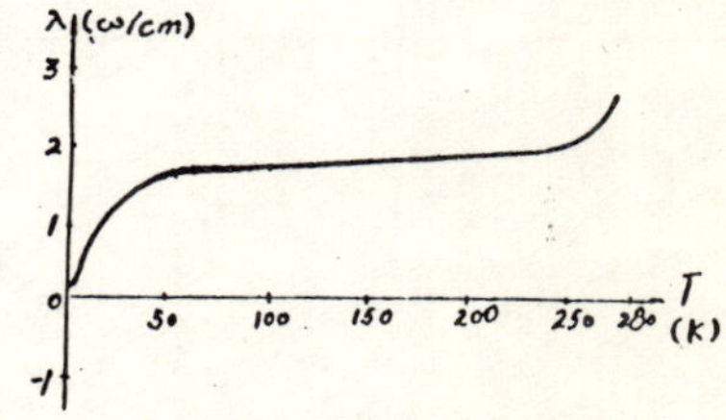

Fig.5 λ-T curve of Au-Fe wire

Table 4. The effect of work hardening on thermoelectric Characteristic of Au+0.07at.%Fe wire

	ΔE (μV)	ΔE_b (μV)
Significant work hardening	15.2	14.6
Twist	4.4	5.5
General work hardening	1.0	1.0

Table 1. Emax in stability test (μV)

Temperature Sample	77K After ymonths aging	77K After thermal Shock*	4.2K After 3 months aging	4.2K After thermal Shock*
1#Au-Fe_7--NiCr	0.1	0.2	0.3	0.0
2#Au-Fe_7--Nicr	0.1	0.2	0.3	0.7
3#Au-Fe_7--NiCr	0.5	0.2	0.3	0.4
4#Au-Fe_7--NiCr	0.4	0.2	0.4	0.1
5#Au-Fe_7--NiCr	0.1	0.2	0.6	0.6
6#Au-Fe_7--NiCr	0.0	0.1	0.3	0.9
1#$AuFe_7$--Cu	0.1	0.0	0.0	0.0
2#$AuFe_7$--Cu	0.1	0.1	0.2	0.3
3#$AuFe_7$--Cu	0.0	0.0	0.1	0.6
4#$AuFe_7$--Cu	0.0	0.0	0.1	0.3
5#$AuFe_7$--Cu	0.1	0.3	0.2	0.3
6#$AuFe_7$--Cu	0.2	0.2	0.5	0.

*20times from room temperature to 77K. **5times from room temperature to 4.2K

Table 2 Inhomogeneity in the begining and end of the wire -- difference in EMF (μV)

Sample	77K*	4.2K*
1#$AuFe_7$	0.5	0.4
2#$AuFe_7$	0.5	0.4
3#$AuFe_7$	1.7	2.0
1#NiCr	0.6	0.8
2#NiCr	10.4	12.2
3#Cr	7.0	7.0
1#Cu	0.0	0.0
2#Cu	0.3	1.4
3#Cu	0.2	0.7

*The thermocouple wire was measured at this temperature

Table 3. Inhomogeneity in the thermocouple wire

Sample No.	Inhomogeneity* (μV)	Dispersion Degree** (μV)	Inhomogeneity*** (μV)
1	0.5	0.5	1.0
2	0.6	0.8	1.0
3	0.4	0.5	0.5
4	0.8	1.1	1.0
5	2.3	1.1	2.0
6	1.7	1.5	1.5
7	1.6	1.0	1.9
8	1.4	1.4	2.9
9	1.0	2.0	2.0
10	1.1	2.3	2.4
11	0.9	2.4	1.6
12	0.1	0.1	0.5
13	0.8	0.1	0.4

*in the begining and end of the wire.
**in 5 pieces taken arbitrarily from wire 50m long.
***measured succesively in wire 3m long.

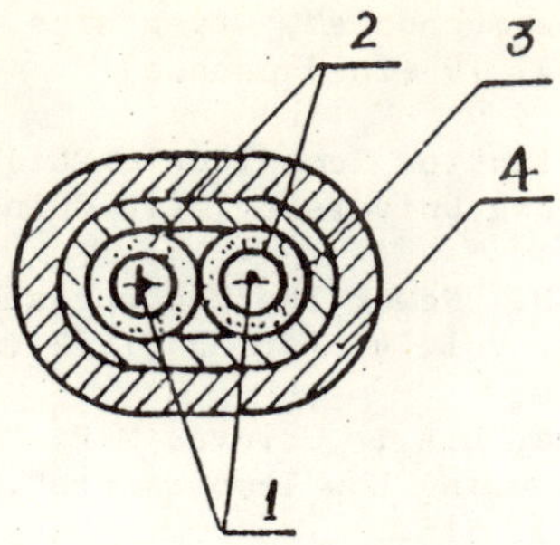

Fig.6 Cross-section of Fluon-coated Au-Fe thermocouple 1. thermocoaple wires 2. insulator 3. screening layer 4. outside coat

Two sample coil were taken from two factories. The measured results at 4.2K are given in Table 4. From Table 4, the effect of general work hardening on the thermoelectric characteristics of a Au-0.07at.%Fe wire is less than 0.1^{o}C. Therefore, when the thermocouple wire was coated, the mechanical stress did not affect the EMF values of the Au-Fe wire to a different extent from the values in the standard reference table. Moreover, the error produced by the assembly of the wire will not be notable.

VIII. DESIGN AND MANUFACTURE OF A FLUON-COATED Au-Fe THERMOCOUPLE

A Au-Fe thermocouple possesses many advantages such as a wide operating temperature range, good repeatability, stable indicated values and good homogeneity. It is an ideal thermocouple for cryogenics. Because Fluon (F.) has excellent strength, insulation, corrosion and aging-resistant properties, the Fluon-coated Au-Fe thermocouple is better than a conventional Au-Fe thermocouple.

The cross-sectional diagram of a typical Fluon-coated Au-Fe thermocouple is shown in Fig.6. The insulator and the outside coating material are Fluon. The screening layer is woven out of silver plated coupper wire, and the weaving density is more than 90%. The diameter of the thermocouple wire is 0.3mm, the insulator thickness is 0.05mm, and the size of coat is 0.4X1.2mm.

For the requirement of good thermocontact, the thermocouple can be used without a screening layer and an outside coat. In some cases, the thermocouple may be made in a strip without a screening layer.

The strip of Fluon film was wound on the thermocouple, then it was sintered to get an insulating coat layer. Because the thermocouple wire resisted stabilizing treatment and the effect of mechanical stress was small, the EMF variation of the thermocouple is less than 0.2^{o}K after processing. Therefore, the EMF-temperature relation of the Fluon-coated thermocouple is in accord with the standard reference table.

IX. CHARACTERISTICS OF A FLUON-COATED THERMOCOUPLE

The method of examination for the Fluon-coated thermocouple is the same as that for the wire. The results indicate that the EMF—tempreature relationship has not changed, but the stability and homogeneity are rather better.

After immersion for 8 h in water, the insulation resistance between the wires and wire and screening layer was measured. The insulation resistance is more than 100 MΩ at 20°C. In the volatge-proof test, it did not break down after 1 min under 500V.

The Fluon-coated thermocouple has a high tensile strength, higher than that of a thermocouple wire, and the joint does not twist in use, so that it has a high reliability in engineering applications.

It did not disply leakage in the gas-tightness test.

For a practical Fluon-coated Au-Fe thermocouple, the EMF—Temperature relation must be in accord with the GB2904-82 reference table, the tolerance should be ±1°C and ±0.5^{o}C, the homogeneity should be 1.5 ~ 0.3μV and the stability should be less than 2.5μV. This thermocouple is a reliable and cheap low temperature thermometer.

X. CONCLUSIONS

The Fluon-coated Au-Fe thermocouple has the advantage of both a Au-Fe thermocouple and a Fluon-coated cable, and its application will increase in cryogenics.

For practical requirements, we have developed a temperature measuring probe with a temperature conmpensator. The probe has a food compatibility and can save thermocouple material for distant measurements. We have produced a series of digital thermometers also. Their measuring ranges are from 27°C to -270°C, resolution 0.1^{o}C, accuracy is (0.5+indicated value X0.03%)^{o}C, and the thermometer can print automatically and use easily.

XI. ACKNOWLEDGEMENT

The other investigators in the present work are, Huang Chengxin, Ma Yangqiu, Yu Caizhen, Cen Sanmei, Qian Yitang, Gao liming, Lu Gao, Wang Lixia, Zheng Guozhen, Wang Yuzhi and Yang Furong.

REFERENCES

1. Mac Donald D.K.C., Pearson W.B.and Tempeton

I.M.,Proc. Roy. Soc. (London), Vol. A266(1962), 161.

2. Berman, R. "Gold-Iron Alloys for Low Temperature hermocouples", Temperature, Vol. 4, Part 3 (1972), 1537.

3. Powell, R. "Low Temperature Thermocouples KP, Normal Silver and Copper versus Au+0.02at.%Fe and Au+0.07at.%Fe", J. of Research of NBS, Vol. 76A, No. 3(1972), 263-283.

4. GB 2904-82, "Nickel-Chromium/Gold-Iron,copper/Gold-Iron low temperature thermocouple wires and their reference tables", 1982, Tech. Standards Publishing House. (in Chinese)

5. Study on Low Temperature Thermocouple, 1971, Chongqing Instr. Mat. Res. Inst.(in Chinese)

6. Takano Masayasu, "Au (Fe) Alloys for Temperature Thermocouple", Cryogenics, Vol.9, No. 2 (1974), 79 (in Japanese)

7. The Application Technique of Liquid Helium, 1973, Zhejiang University. (in Chinese)

8. Zysk, E.D. "Newer Thermocouples Materials", Temperature, Vol. 4, Part 3 (1972), 1697.

9. Medvedeva, L.A., Orlova, M.P. "Thermocouple for Measuring Low Temperature",Cryogenics, No. 8 (1971), 316.

10. The Cretirication Specification on Chromel-AuFe Thermocouple, JJG 334-84,1984, China Metrology Publishing House.(in Chinese)

11. The Plastics Application Handbook, 1980, Shanghai sci. and Techn. Publishing House (in Chinese)

AN INSTRUMENT WITH AUTOMATICAL TEMPERATURE CONTROLLING AND DATA COLLECTIHG OF LOW VALUE RESISTANCE

He Zhenhui, Liu Jiquan*, Cai Mingfu*, and Zhang Qirui
Department of Physics, *Department of Advance Physics,
University of Science and Technology of China,
Hefei, Anhui, P.R. China

ABSTRACT

This instrument consists of BCM-81 Single Board Computer, IEEE 488 controller, DAC temperature controller, SOLARTRON 7150 digital multimeter and KEITHLEY 181 NANOVOLTMETER. It can control temperature automatically in the range of temperature from 55 to 250K and collect data automatically. It can also print the measured data in digital form and simultaneously show the standard errcrs of each measured point, which can be regarded as the criterion of the reliability of the measured results. The highest precision of temperature controlling can reach 5mK. This instrument is successful in the application of cryogenic measurement of low resistance about 10^{-5} Ω with sample current 1mA, such as superconducting samples. It can be used instead of manual operation to a great extent.

For resistance-temperature dependence measurement, automatical data collection is in commen use, with temperature changing contiueously. But it is not widely used in the experiment with stable temperature for data collection, i.e., involved automatical temperature control, especially in our country. Here, we demostrate an instrument with automatical temperature controlling and data collecting of low value resistance.

This instrument consists of a BCM-81 Single Board Computer, a IEEE 488 controller, a DAC temperature controller, a relay switch circuit for current reversal, and tunnel selection, a SOLARTRON 7150 DIGITAL MULTIMETER and a KEITHLEY 181 NANOVOLTMETER. When it is connected with a cryostat with a sensor and a heater in it, it can measure 8 samples (at most) at a stable temperature (simutaneously). (see Fig.1) The values of accuracy of temperature control, temperature step, proportional-integral-differential (PID) parameters, temperature region, sample current and tunnel order are keyboarded into the single board computer. The measured values are calculated, then the resistances and their standard errors are printed in unit of ohm. Data are finally transmitted to Apple II computer for curve printing and disk recording.

The SOLARTRON 7150 DIGITAL MULTIMETER with GPIB interface is employed for temperature control and measurement. Data of temperature is read from a sensor of resistance. The data are then calculated with PID method and back feed to the heater controller, the heating power is then controlled automatically according to agiven value. The moving window method is used to determind the temperature stablility. When temperature is within a given scale, the average of thirty points heating power is taken as stable heating power when sample voltages are read. This can avoid the errors due to the variation of thermo-electric power with temperature and heating current in the feedback process. About 10 to 30 seconds after such heating power is supplied, the 181 NANOVOLTMETER begins working. Fig.2 shows the circuit for temperature control and heater control.

Two cryostats, with sensor of Ge resistance and Pt resistance respectively, are used to test the temperature controller from 55K to 250K. Only one set of PID parameters is needed for the one with Ge resistance, but two for another with Pt resistance, in the temperature region from 55K to 120K and from 90K to 250K respectively.

The precision of temperature control depends mainly on the selection of PID parameters and the precision of the digital meter which reads the temperature.

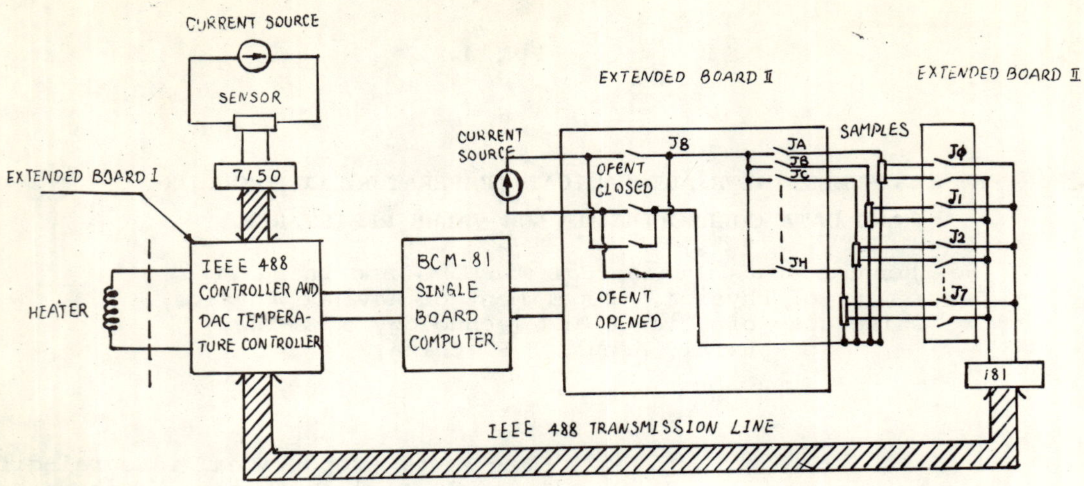

Fig.1 Shematic drawing for the instrument

Our system can reach a precision of 5mK.

The KEITHLEY 181 NANOVOLTMETER with GPIB interface is used for sample voltage reading. Relay switch circuit is used for current reversal and tunnel selection as relay switch allows relatively larger current to flow through. In this way, suitable relay switch should be selected to avoid the contact power. The data are collected n times, then the average of which, thus the resistance and its standard error are calculated and printed.

Because of the limit of current for some samples and the measurement circuit as well, the sample current is usually 1mA.

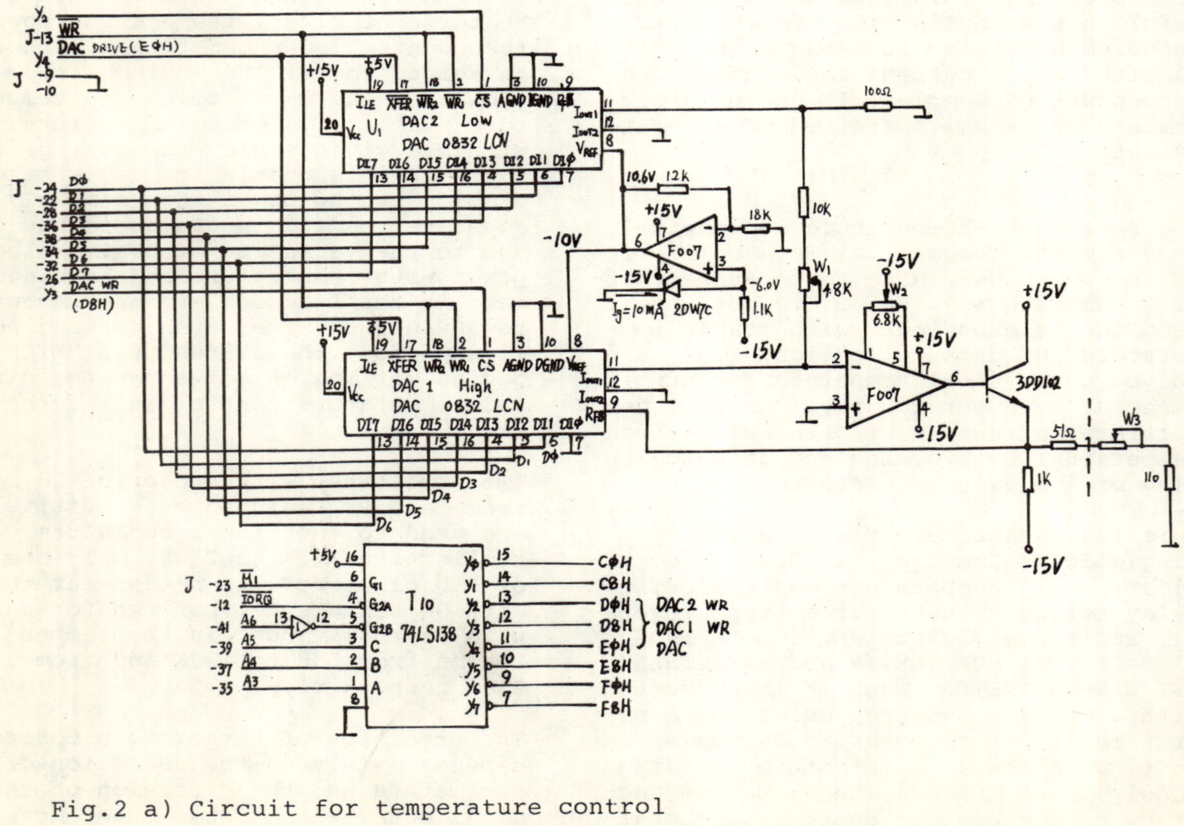

Fig.2 a) Circuit for temperature control

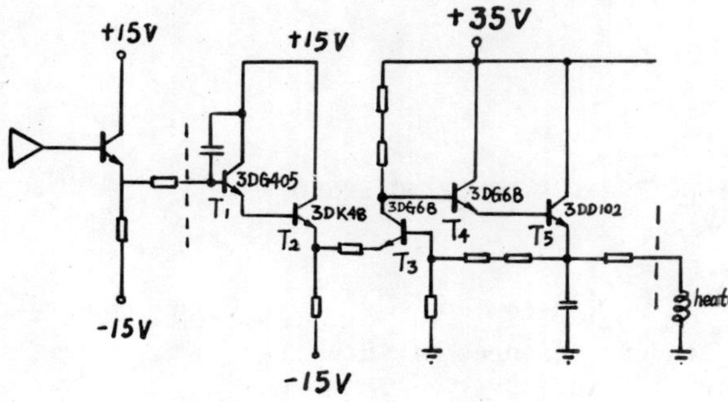

Fig.2 b) Circuit for heater control

Thus, the precision of resistance measurement in this system reaches 10^{-5} ohm. Tests indicate that most of the data can reach or be prior to this precision. The precision also depends on the stability of temperature and lead contact. So, the standard error of resistance is used to determind the reliability of measurement, If standard error/resistance 10^{-6}, it is reliable, otherwise, not.

Usually, it takes about three hours to measure one set of sample. (Two measurements with different temperature step have to be carried out if any sample exhibits superconducting,) It is comparable to that of the manual measurement. Therefore, it remains to be improved that temperature step can be modulated automatically, according to the resistance-temperature coefficient (1/R)(dR/dT).

A CRYOGENIC INSTRUMENTATION WITH MULTI-FUNCTION FOR FREEZING BREED

Kang Xibin
Xinjiang Institute of Physics Academy Sciences of China
Wurumuqi Xinjiang of China

ABSTRACT

This paper will discuss the design principles of a cryogenic instrumentation with multi function for freezing breed. The device consists of a sensor temperature, an A/D translator and digital display, a network of measurement translation--a bridge circuit with stable electric voltage and current, alarm system and printing data system. A range of the temperature from -199.9°C to +55°C is measured and printed with a sensitivity of 0.1°C within maxium error of ±0.5% in full scale region. A sound, light or electric signal can be sent out in the same temperature region when it is needed.

INTRODUCTION

The artificial semen breeding is a new technology in the process of semen breeding. Its advantage is: more fine off spring can be gotten with little semen of good breed. This new technology for freezing semen breeding may keep succession of animal breeding. It is very important for breeding of domestic animals and develop the animal husbandry. In recent years experts have pay more attention to it and have done much research on the technological process. The key for popularization of the new technology is to control optimum temperature for freezing semen by which we can get high quality of freezing semens and increasing of ratio of impregnancy. It is very important for productive development of domestic animals.

THE PRINCIPLE OF THE WHOLE CIRCUIT DESIGN

From use'r using the instrument effectively, from the instrument's practicality, economisation and high dependence, we use of COMS circuit in hand ware. The frame structure is shown in figure 1.

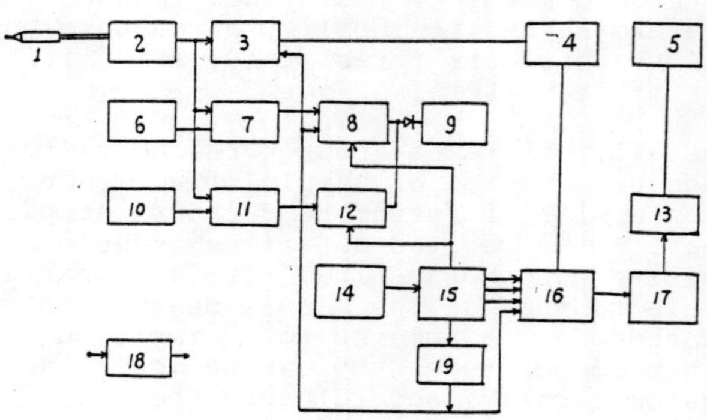

Fig. 1 WDW-J-1 low temperature measurement instrument

1. Sensor
2. Transfrom network
3. A/D translation
4. LED display
5. Print out
6. Toplimit temperature basic standard
7. Compare out
8. Intermission vibration circuit
9. sound and light out
10. Low limit temperature basic standard
11. Compare out
12. Intermission vibration circuit
13. Print out port
14. Time clock circuit
15. Count/divide
16. Multiple translation switch
17. 5/4 line translation
18. Power supply
19. Print control

The detector of temperature measurement is made of temperature sensitive component and the pretect tube cover made of heat non--

conduct materials. The measurement signals passed through the detector measurement transform circuit, a bridge circuit constituted by steady voltage and permanent flow sources, enter a analog digital converter(A/D). The A/D converter is the large integrated circuit ICL7117. The outs of the A/D converter ZCL7117 drive LED display directively to show the digital temperature numbers and are set in multi-channel converted switch to cutover it orderly in control of the time order controller made of the time clock pulse generator and counter/frequently-divider at the same time. The multi-channel transform switch sets the different bit LED digital display's data in 5/4 line transform circuit. The controller sends off read operation time order pulses that print codes. The code device read in the stable data on data lines. To display data in stable state the controle system send out keeping signals to A/D ICL7117 in the meantime of printing data. To avoid mistake operations the printing code device transtales 5/4 line transfer codes into printor operation codes. Which control printer's key array. The large scale interated circuit computer core boand (D1848C) manages and drives printer to record the temperature data. As shown in the figure, the print fixed-time pulses take from frequency divider. The length of dividing frequencies pulse cycles determine the printing delay cycles. So as the user puts forward demands. The three detemined delay time are send off. The user oneself sets the high and low levels of alarm by screwing the potentiometer on the instrument panel. The setting data transfer and exchange the switch by pressing and stopping butten, by which the setting temperature readings are displayed on the LED display direct visibly as well as can be printed and recorded, New We discuss a few of questions in the whole circuit design principle as follows:

1. The temperature sensor

The detection of freezing semens temperature demands the temperature detector to be provided with higher sensitivity, very small delay property, detect temperature region over 200°C, resolution equals 0.1°C as well as the measurement errow is less then 0.5% under the full scale. So the low temperature sensor is an important element in the instrument. We examined a few of thermistors in using initially. Example as Pt-thermistor-couple, there is nonlinear errow that cann't be received in it's ideal R--T property besides the time delay and sensitivity etc. can not be received. For the platinum thermistor couple, it's R--T property is shown as follows:

0--630.74°C Rt=RO(1+At+Bt^2)
-190--0°C Rt=RO(1+At+Bt^2+(ct+100)t^3)

here RO is equal to the volue of the couple resistance on the zero degree C.

A=3.96847 x 10^(-3)/°C
B=-5.896 x 10^(-7)/°C
C=-4.22 x 10^(-12)/°C

If we use of a single thermistor, because of the exponential function relation in its R--T property curve We hardly use of it in so wide temperature region. For the negative temperature coefficient the relation of resistance volue R and it's temperature T (expressed in °K) can be shown in following expression:

$$Rt=Rt_2 exp(Bn/T1-Bn/T2)$$

here B is the material constant

Therefore we know that a single thermistor can not be used in such wide temperature region and high sensitivity because of it's serious nolinear property.

The thermocouple alse is usually used as temperature measurement sensor and adopted widely. But it's using is confined by it's measurement connecting wire and zero-point compensation as well as weak signal produced by the sensor. these questions above discussed must be overcome and adopted new temperature measurement sensor.

We choose the temperature sensor of Xin Jiang Institute of physics of Chinese Scientific Academy. The temperature sensor is suitable for the temperature measurement region of -199.9-- +50°C. It's maximum nolinear errow is less than +0.5%-- -0.5% of the full scale. This sensor and other resistant elements make up a network which is suited to our demands for the detector.

2. A/D Converter and Digital Display

The A/D converter's main element is CMOS ICL7117 which has read keeping function. The IC piece is manufactured in our country factories now. The element's feature is it's high performance and low power consumption. It is consisted of seven section code device, display driver, 3 1/2 bit data locking circuit, integrator buffer and a few of analog switches which are all integrated on a single si substrate. By it we can design a instrument with locking out circuit, basic voltage level invariably is referenced to the COM pin's low level. In this way the circuit performance can be accurate to within the top accuracy. That is, it's itself-calibrated accuracy accurate to within 15 uv, zero shift is less than 1 uv/°C, input resistance is 10 M, transform errow is less than one bit. The multipurpose of a accurate differentiate input expresses as the feature in which it, it can be connected directly with measurement loading unit, multi-measurement circuits and other standard sensor's circuits. Therefore it has a special advanced performances. In the economy of the unit, it can work well by using a single battery and easily be constituted up a digital panel meter with low production cost and light make up a temperature measurement system.

3. Measurement Transform Network---a Steady Voltage and Permanent Flow Source Electric Bridge

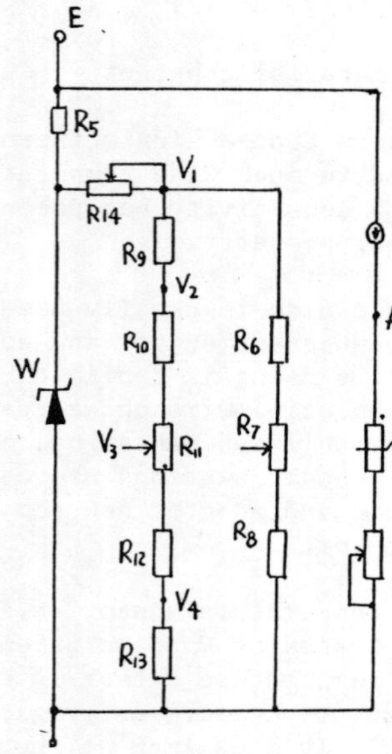

Fig. 2

The measurement transform network with a steady voltage and permanent flow source electric bridge is illustrated in picture 2. Here E is power supply, W is stable diode by which the basic standard level and sensor's transform pins can be attained to a common stable volt source. Rr is a resistance that corrects the stablizer's working current. R9. R10, R11, R12, R13 are dividing volt resistance. The temperature over limitation alarm signals are out from the circuit V2 is the top limitted standard signal and V4 the low one. R7 is the full scale ajustor. The ajusted precision can be raised by making use of R6 and R8. Rd is the detector's compensation resistor. R11 is the zero setting potentiometer. Rt is the sensor to heat, is a steady current source. Illustrated as above, the network make up a deformed electric bridge circuit which is consisted of sensor to heat Rt and the branch circuit where there is a stead flow source, as well as other elements. A terminator and the setting pointer of R11 from the signal out end. The bridge is compared with other usual bridge the difference between them is the bridge's arm is in the stablizer circuit, in which there is also a voltage compensator's input end of the alarm circuit.

G point on the figure 2 is connected to the ICL7117's COM pin. Known as above, steady voltage source bridge circuit and ZCL7117 have a common ground which float and isolated from the source's ground, i n such a way the measurement accuracy can be araised. So the intrument's accuracy is near by the detector's accuracy.

4. The Top and Low Limit Alarm Besic Network

As mentioned above, the alarm network's power supply is provided with the two points of V2 and V4. In other words, the alarm region is determined between V2 and V4. Neither of over-wide or over-narrow alarm regions are our hope. We need to attain a properly alarm voltage. Hence we need to set the resistors in the figure 2 properly according to the alarm voltage. In order to set the alarm temperature straight in the meter, the shifting switch is adopted. So the setting values of top and low temperature limit can be displayed by the LED display. It's principle as follows:The stadard volt level respond to given top and low limit temperature compare with the lever respond to practical temperature on the voltage comparator. The out state hange of the voltage comparator will drive or stop the chopping oscillator. Then the different frequency pulses take out by using of the frequency divider are set into the top and low limit temperature chopping oscillator respectively. Therefore the top and low limit temperature chopping oscillator oscillate in the different frequencies. This faction distinquish the top limit temperature alarm from the low one. The chopping oscillator's alarm signals are sent into voice and light alarm circuit. Which send off light and voice of alarm. The voltage comparator's output, the alarm signals, are sent into the terminators on the instrument's binde panel. They can be used for temperature control signals.

5. Data Record System---Seven Input, Five Line Four Output Circuit

The low temperature instrument is a data record system of convenient, cheep and complete function. We seleted read-made HR-8 (or HR-7) exterior system to be printing record equipment. The data record system is consisted of clock circuit, sequence instruction generator, printing instruction corde and printer interface. The clock circuit is formed by one operated part and four resistors. The clock puls e cycle is approximate colculated by following formula.

$$T=2Rc\ln(2R1/R2+1)$$

The sequence instruction generator is one binary counter. The bit numbers are determined by the needed steps in a cycle of the sequences. The needed sequences controls of low temperature instrument require eight steps. Therefore, it is used a three bit binary counter. The circuit is shown as figure 3. performances. We make use of these advantages to Low temperature instrument adops multichannel analog switch array to form printing instruction code. The data lines and control instructions to print and record set on the input pin on the multichand analog switch. To implement the sequence output of the printing instruction, the analog switch array is output connect printing interface circuit, the array's address lines connect sequence instruction generator.

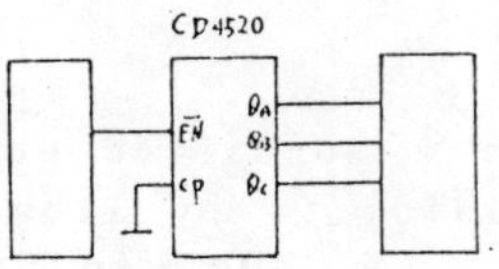

Time clock happen   printing Instruct plait code

Fig. 3 Count sequencer control

Printing interface circuit is a middle circuit in which the data from A/D are transformed into the printer and the printer print and record these data. Major characteristic of the circuit is to reduce seven display code lines to four data lines, and hardware numbers of the interface circuit. At the same time, it reduces cost, increases running reliability. Its principle is shown as figure 4. In figure 5, five/four lines translation circuit is consisted of two OR gates It should transform a,b,g,e,f signals to a+b, g+b,e,f signals. The signals are imported to printing code circuit, as figure 4.

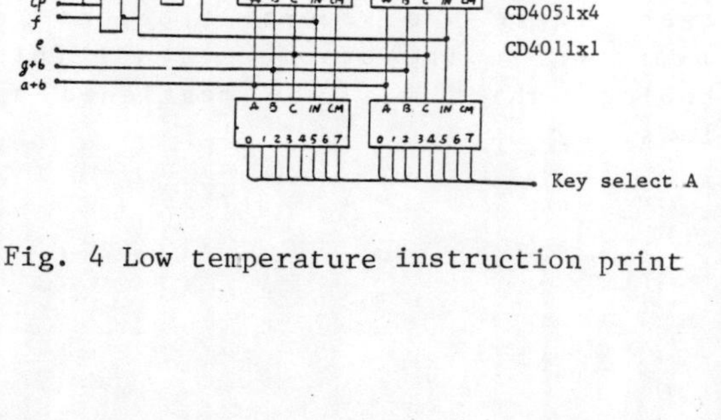

Fig. 4 Low temperature instruction print

The clock pules CP which is consisted of 4 CD4051 analog switch and two NAND gate is connected on the two NAND gate. The f line, one of four data lines is connected two NAND gate. Pre-half-circle (low voltage) of clock pules makes two NAND gate's output to emerge high voltage. The analog switchs are high impedance, so that printer does not recieve instable data signals and operate wrong. Back-half-circle of the clock pules (high voltage) makes two NAND gate's output is only related to the f line's state. So that the operation instructions implement correctly. Figure 6 shown printing time's sequences state.

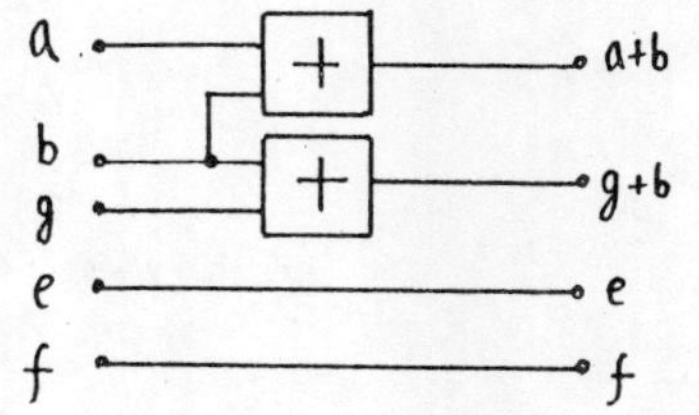

Fig. 5 5/4 line translation

The analog switches divides two groups. every two CD4051 switches composed one analog switch of double blade and eight touch points. The data lines f,e a+b, g+b are analog switch addresses and select sone one setting of the switch to analog printer's key array and implement data sequences printing control.

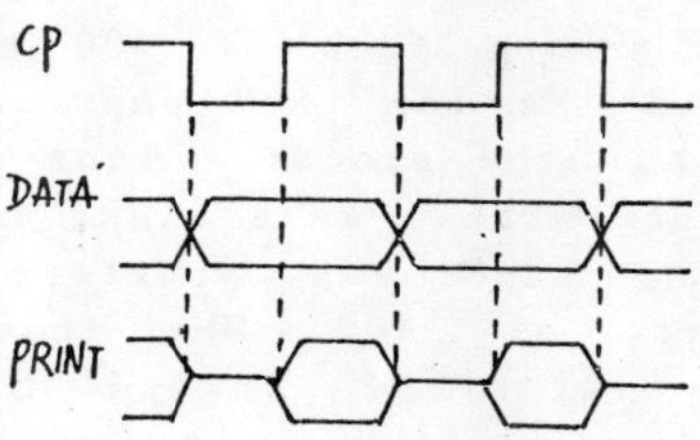

Fig. 6 printing time order state

End

At present more than ten sets low temperature instruments have been operated in practice work. The test result showed this device characte of determination accuracy and high sensitivity. It is easy to carry. It really is a practical instrument of high-quality but cheape. It has win acclaim of user.

Cryogenic Acoustic Microscope

Y. Okuda
Department of Appl. Phys.,
Tokyo Institute of Technology,
O-okayama, Meguro-ku, Tokyo 152, Japan

and

K. Karaki
Olympus Optical Co. Ltd.,
Kuboyama-cho, Hachioji, Tokyo 192, Japan

ABSTRACT

We have started to develop the superfluid ^{4}He acoustic microscope. This technology is to get an ultimate spatial resolution in acoustic microscope using superfluid ^{4}He as an acoustic coupler. In a preliminary test, we obtained a big echo at 400 MHz from superfluid ^{4}He for the temperature below 0.3 K. The acoustic lens was made of sapphire with c-axis along the sound direction. ZnO was sputtered as a transducer of 400 MHz sound. A matching layer of W and SiO_2 was deposited on the lens surface to increase sound transmission. Lens scanning and focussing were performed by using superconducting and permanent magnets.

INTRODUCTION

Superfluid ^{4}He can be used as a coupler of an acoustic microscope with a very high spatial resolution, which was first developed in Stanford University (1,2). The spatial resolution of the water coupler acoustic microscope is limited by the attenuation of the high frequency sound. Whereas in the superfluid the sound attenuation can be very small.

Liquid ^{4}He is a so-called quantum fluid, which remains to be liquid even at T=0 K. It is due to a violent zero-point motion coming from its small mass and weak van der Waals force be tween atoms. Below 2.17 K it goes into the superfluid phase (λ transition) and loses the viscosity, being an ideal fluid. It is well known that sound attenuation in this superfluid phase is proportional to ωT^4, where ω denotes the sound frequency and T temperature. (3) So for sufficiently low temperature the attenuation is negligible even for very high frequency. Another important feature is its low sound velocity which is 240 m/s. The spatial resolution of the microscope of limited by the sound wavelength in the coupler. Suppose we can inject 24 GHz sound into helium, which is not impossible in the recent technology, the wavelength is about 100 A. As far as the contrast is concerned, 70 A can be expected. even at 400 MHz the wavelength is 0.6 μm, which is already the wavelength of the visible light.
However, there are several problems to be overcome in the development of this technology which will be mentioned as follows.

(a) Anomalous dispersion

As the phonon dispersion in the superfluid ^4He is known to be anomalous,(3) the ω - k curve is downward bent, a single acoustic phonon easily decays into two thermal phonons. This problem, however, can be removed by lowering temperature below 0.1 K using a modern dilution refrigerator, since attenuation is decreased as ωT^4. This anomalous dispersion turns out to be normal when pressure above 15 bar is applied. Therefore pressurized ^4He is also a tempting coupler for the acoustic microscope.

(b)Mismatch of the acoustic impedance between liquid ^4He and sapphire

As the density of liquid ^4He is 0.148 g/cm^3 and sound velocity is 240 m/s, the acoustic impedance of ^4He is very small compared to most materials. Most acoustic power is reflected at the interface. To increase the transmission efficiency, it is necessary to deposit an anti-reflection matching layer. It is difficult to find a good material as a matching layer because the impedance of liquid helium is so large. Here we deposited a double-layer of W/SiO_2 of the quarter wavelength.

(c)Nonlinear effect

The nonlinear effect in the propagation of high frequency sound occurs very easily in liquid ^4He, because of its low density and low sound velocity. This gives rise to the saturation of received signal, or even depletion due to parametric damping,(4) when input power exceeds some critical value. So it is necessary to reduce the input power as low as possible and receive the small signal with low noise preamplifier. This phenomenon is enhanced as the frequency goes up above several GHz. So a kind of cryogenic preamplifier will have to be installed.

EXPERIMENTALS

Low temperature below 1 K is obtained by using a dilution refrigerator (model 200 of Oxford Instruments), which has a

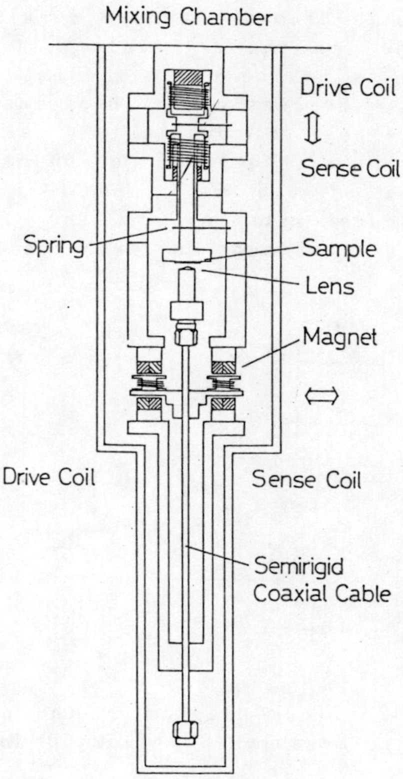

Fig. 1. Schematic drawing of the cell.

cooling power of 2000 erg/s at 100 mK. This is enough for the present purpose. The cryostat is sit on the heavy anti-vibration stage (1 ton). The stage is very important for the performance of the microscope, which was realized in the preliminary test at LN_2 temperature.

Figure 1 gives a schematic drawing of the experimental cell. Acoustic lens is scanned in xy-directions (parallel to the lens surface) by superconducting magnets between SmCo permanent magnets. Focussing is performed by displacing the sample holder in z-direction with the technique of the voice coil (superconducting coils). The position is controlled within 0.01 micron by this technique.

Figure 2 is a block diagram of the pulsed acoustic measurements from 50 MHz to 415 MHz. Amplifier total gain is 90 dB above the noise. Both amplitude and phase signals can be obtained so that amplitude and phase images should be available. Conventional 3.2 mm outer-diameter copper coaxial cable is used in the cryogenic part to make the loss in the cable as low as possible at the cost of the heat leak to the sample cell. Acoustic lens was made of sapphire with aperture of 60 degree and radius of 0.5 mm. As for acoustic impedance matching at the interface between lens and ^{4}He, we deposited a

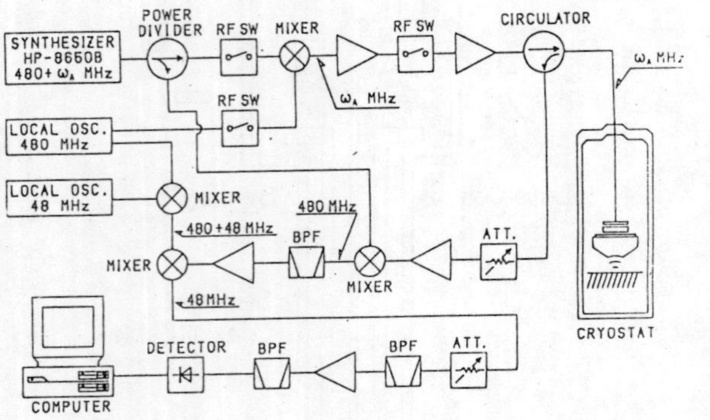

Fig. 2. Block diagram of the pulsed acoustic measurements from 50 MHz to 415 MHz.

quarter wavelength matching layer of tungsten and SiO_2 for the center frequency of 400 MHz. It is almost impossible to make an effective matching layer with a single material because of very low acoustic impedance of liquid helium.

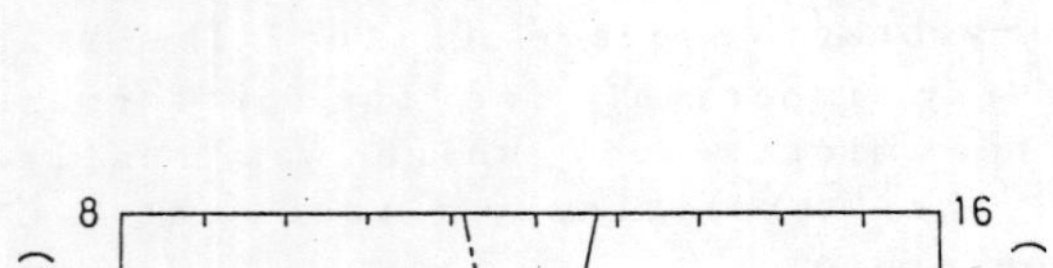

Fig. 3. Calculated reflection and transmission loss in the case of sapphire /W/SiO_2/^{4}He.

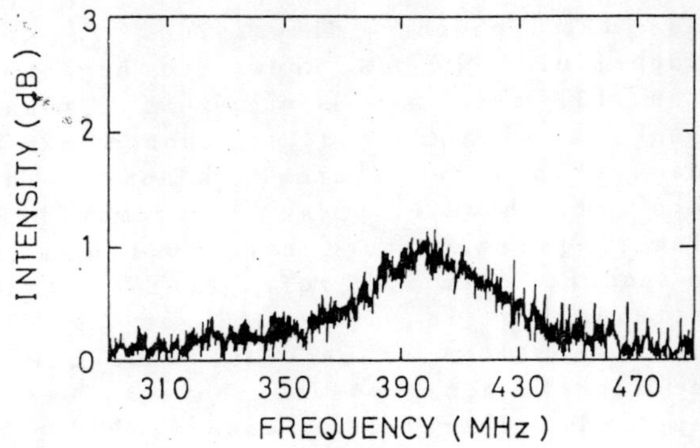

Fig. 4. Transmission of acoustic energy to Ar gas of 60 Bar at room temperature.

Figure 3 gives calculated reflection in the case of tungsten and SiO_2 against liquid ^{4}He.

Figure 4 is the test result of this matching layer against Ar gas of 60 bar. The acoustic impedance of 60 bar of Ar gas at room temperature is almost the same as that of liquid ^{4}He. As is seen in the figure, some power of about 1 dB is transmitted to Ar gas near 400 MHz.

RESULTS AND CONCLUSIONS

Figure 5-a shows a typical echo from liquid ^{4}He. The temperature was 0.3 K and the pressure of the ^{4}He liquid was 11.8 bar. The input power was -10 dBm. As temperature was raised up above 0.5 K, signal decreased rapidly, which indicates that loss in the superfluid ^{4}He is most serious for 400 MHz. When we increased the power above 10 dBm, we did not see the nonlinear effect, that is, signal was increased as the power increased, whereas the noise level increased. That means we did not gain any but lost some with higher input power. We have not observed any signal depletion up to 25 dBm, as was reported for higher frequencies.(4) When the pressure of liquid ^{4}He was decreased down to 4.2 bar, signal was depleted down to almost one-third as was expected. This is shown in Fig. 5-b. So pressurized ^{4}He can be a better coupling fluid, since its sound velocity does not get so high.

When the operating frequency was decreased below 370 MHz, the signal disappeared. As the band width of the ZnO transducer is rather wide, the matching layer is found to be very important.

Unfortunately, due to some trivial problems in the lens scanner, we could not get an image at present. As it should not be a serious problem, it is straight forward to get an image.

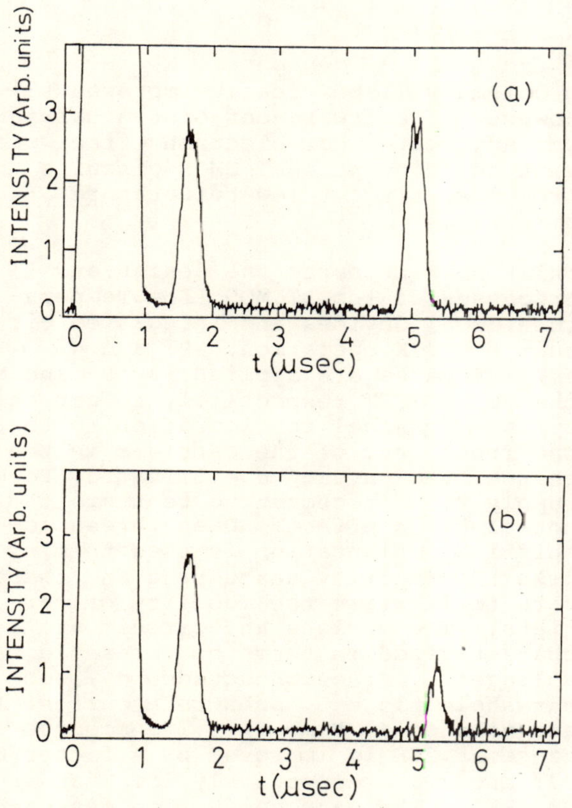

Fig. 5. Typical echo from ^4He. The first big signal is the one reflected at the transducer and the second is the one from the interface between lens and ^4He. The third one is the signal from ^4He. Frequency is 408 MHz and T=0.3 K. The pressure is 11.8 bar (a) and 4.2 bar (b), respectively.

In conclusion, we have seen a big acoustic echo from superfluid ^4He with 400 MHz lens below 0.3 K. It has been revealed that the double-layer of quarter wavelength matching layer, which is W/SiO_2, is very effective and essential. It is also very important to decrease the temperature below 0.5 K. We have seen a saturation of the received signal for the input power above 25 dBm, with increasing the noise level. For higher frequencies we will have to decrease the input power much more and to install a low noise cryogenic preamplifier. The pressurized liquid ^4He is found to be a better coupler.

At the conference, the image and the test results of the contrast with varying various parameters will be presented.

ACKNOWLEDGEMENTS

The author would like to thank Prof. N. Chubachi of Tohoku University for promoting the present work.

REFERENCES

1 Hadimioglu B. and Foster J.S. 'Advances in superfluid Helium acoustic microscopy'. J. Appl. Phys., vol. 56 (1984), pp 1979-1980.

2 Foster J. S. 'High resolution acoustic microscopy in superfluid helium' Physica, vol. 126B (1984), pp 199-200.

3 Maris H. 'Phonon-phonon interaction in liquid helium'. Rev. Mod. Phys., vol. 49 (1977), pp 341-359.

4 Foster J. S. and Putterman S. 'Parametric self-enhancement of the spontaneous decay of sound in superfluid helium' Phys. Rev. Lett., vol. 54 (1985), pp 1810-1813.

1μm CRYOGENIC CMOS INTEGRATED CIRCUIT WITH HALF NANOSECOND GATE DELAY TIME*

Li Yuanxiong and Zhang Min
Microelectronics Division, Shanghai Institute of Metallurgy, Academia Sinica,
800 Yishan Road, Shanghai, China

ABSTRACT

The design and processes of 1μm CMOS devices and integrated circuit for liquid nitrogen temperature (77K) operation are described. P^+ and N^+ poly-Si gates are applied for P and NMOSFET respectively to get the surface channel configuration so that the freeze-out effect of carriers in the channel region at 77K can be avoided. The power supply voltage is chosen to be 2.5V to restrain the hot-electron effect which is aggravated at 77K. Deep threshold-shifting implantation is used to minimize the dominant impurity scattering at 77K. Excellent device characteristics at 77K are obtained. The gate delay time is less than 500ps at 77K and is two times shorter than that of the room temperature circuit at 300K.

INTRODUCTION

CMOS is the leading technology of VLSI (Very Large Scale Integration) at present as well as in the future because of its high noise margin, very low static power dissipation, high radiation immunity, and compatibility with other technology [1]. As the feature size of the devices is shrunk to 1μm or smaller, the improvement of operating speed of the conventional CMOS circuit by scaling-down [2] is limited due to the short channel effect, series resistance and latch-up effect. Lowering the operating temperature (e.g. in liquid nitrogen T=77K) is a new way of achieving high speed operation. At low temperature, the mobility of the carriers, the sub-threshold leakage current, the interconnecting resistance, the latch-up effect, heat dissipation and the related reliability problems will be greatly improved [3]. However, the freeze-out of the carriers and aggravated hot-electron effect prevent the conventional CMOS circuits from operating at low temperature directly.

* Project supported by the National Natural Science Fundation of China.

This paper presents the design and processes of 1μm LTCMOS (Low Temperature CMOS) devices and integrated circuit for 77K operation. P^+ and N^+ poly-Si gates are applied for P- and N-channel MOSFET respectively to get the surface channel configuration so that the freeze-out of the carriers in the channel region can be eliminated. Power supply 2.5V is chosen to restrain the hot-electron effect. Deep threshold-shifting implantation is used to minimize the impurity scattering and therefore to increase the mobility of carriers. The devices and circuit are characterized in terms of threshold voltage V_t, transconductance g_m, subthreshold slope S, substrate current I_b and gate delay time T_d. Transconductance at 77K is improved by a facter of 3.1 and 3.3 respectively for N and PMOSFET. The gate delay time measured from 1μm 21-stage ring oscillator is less than 500ps at 77K and is two times shorter than that of the conventional CMOS circuit at 300K.

DEVICE DESIGN

1. Elimination of freeze-out of the carriers

The ionized impurity concentration in Si can be expressed as

$$\text{N-Si} \quad N_d^+=N_d/\{1+2\exp[(E_f-E_d)/kT]\} \quad (1)$$
$$\text{P-Si} \quad N_a^-=N_a/\{1+4\exp[(E_a-E_f)/kT]\} \quad (2)$$

where N_d^+ and N_a^- is the ionized doner and accepter concentration respectively, E_d and E_a is the doner and accepter level respectively, E_d=0.044eV for phosphorus and E_a=0.045eV for boron, E_f is the Fermi level determined by both temperature and doping level, k is the Boltzmann's constant, T is the temperature in K. For Si substrate with doping level common in CMOS circuit, at room temperature (300K), E_d-E_f and E_f-E_a >>kT and almost all the impurities in Si are ionized. However, at 77K, the Fermi level moves toward the bottom of conduction band in N-Si (or the top of the valence band in P-Si) and the value of E_d-E_f or E_f-E_a approaches that of kT. Thus part of the impurities will not ionized (freeze-out) from Eq.(1) and (2).

In the channel region of MOSFET, not only the temperature and doping level, but also the band bending caused by the gate voltage V_g affects the ionization situation of impurities. For PMOSFET in the conventional CMOS circuit with N^+ poly-Si gate, a layer of boron is implanted in the channel region to shift the threshold voltage V_{tp} so that $|V_{tp}|$ is near the same as the threshold voltage of NMOSFET V_{tn} and maximum noise margin is obtained. Thus the PMOSFET is a burried channel device while NMOSFET is a surface channel one. As the result of one dimensional numerical simulation [4], the energy band diagrams in the channel region of both P and NMOSFET at 77K are shown in Fig.1. For NMOSFET (surface channel), $E_f-E_a \gg kT$ in the channel region, and all the impurities are ionized even at 77K. However, for PMOSFET (burried channel), when the channel is in strong inversion with large V_g, E_f-E_a is small compared with kT and a large part of the implanted boron will not be ionized. Therefore V_{tp} can not be shifted. Moreover, when the channel is in sub-threshold region with small V_g, E_f-E_a>>kT and the impurities which were frozen-out before become ionized, resulting in the large sub-threshold leakage current. Therefore, to avoid the freeze-out of carriers in the channel region, PMOSFET should have the surface channel configuration as NMOSFET. This means the P^+ poly-Si should be used as gate material for PMOSFET. As can be seen from Fig.1, for surface channel PMOSFET, E_d-E_f>>kT in the channel region and all the doners are ionized.

2. Selection of the power supply voltage

In the short channel MOSFET, the carriers in the channel will gain energy large enough to break the Si-Si bond and generate electron-hole pairs.

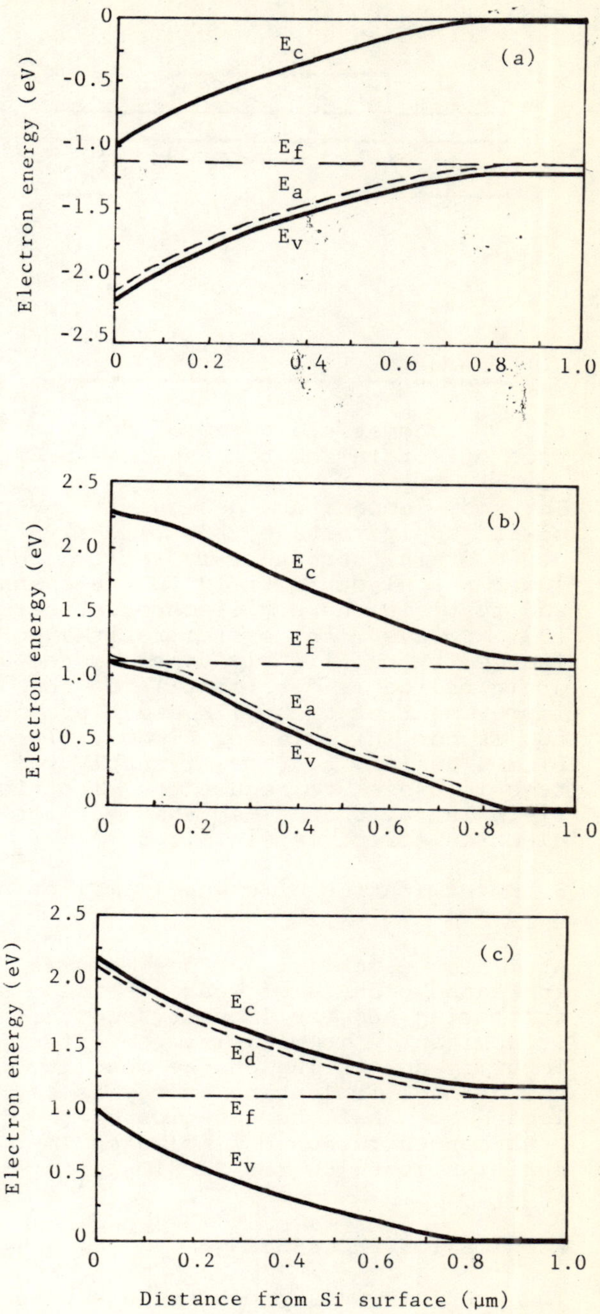

Fig. 1 The energy band diagrams of surface channel NMOS (a), burried channel PMOS (b) and surface channel PMOS (c) in strong inversion at 77K.

For NMOSFET, the generated electrons will surmount the Si-SiO_2 barrier and be captured by the trap centers in SiO_2; the holes form substrate current, as shown in Fig.2. The threshold voltage will increase and the transconductance will decrease due to the electron trapping in SiO_2. At 77K, the mean free path of electrons is increased and hot-electron effect aggravated.

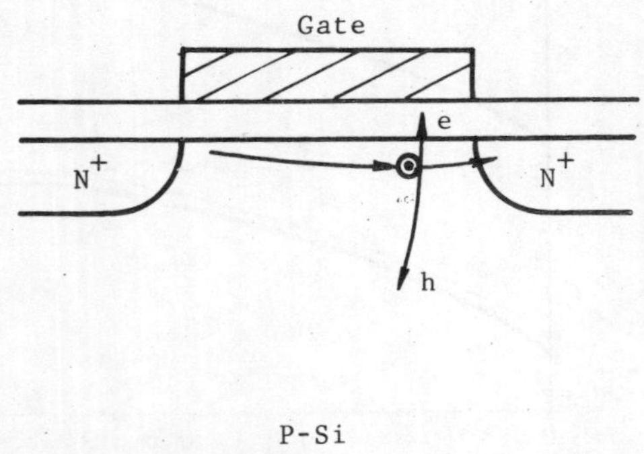

Fig. 2 Schemetic diagram of hot-electron effect in short channel NMOSFET.

For room temperature design with 5V power supply voltage, LDD (lightly-doped drain) structure is applied to lower the electric field in the channel and restrain the hot electron effect [5]. However, the series resistance induced by the lightly doped source and drain region is detrimental to the improvement of g_m at 77K and therefore LDD is not suitable for LTCMOS. In our LTCMOS design, 2.5V power supply voltage is chosen to reduce the electric field in the channel so that the hot-electron effect is minimized.

3. Determination of channel implanting depth

At 77K, the mobility of carriers is increased because of weak lattice scattering and the impurity scattering is dominant. To minimize the effect of impurity scattering on the mobility, the channel implanting energy is selected so that the location of the peak concentration of implanted impurities is below the Si-SiO_2 interface.

FABRICATION PROCESSES

The simple P-well processes are used in the fabrication of the LTCMOS devices and circuit. The starting material is N-Si with 5×10^{15} doping concentration. P-well with 2×10^{16} surface concentration is formed by boron implantation and high temperature driving in. A LOCOS field oxide of 8000Å thickness is thermally grown in wet O_2 atmosphere. Then the active areas of devices are opened and threshold shifting implantation is performed. A 300Å gate oxide is thermally grown in dry O_2 atmosphere. Next a 5000Å thick poly-Si is deposited by CVD as gate material and the gates are delineated by AZ1350 positive photoresist and RIE (Reactive Ion Etching). The poly-Si gate is doped at the same time when source and drain self-align implantation is performed with As and BF_2 for N and PMOSFET respectively. This is the simplest way of obtaining P^+ and N^+ doped poly-Si gate for P and NMOSFET respectively [6]. After the activation of implanted impurities by thermal annealing, a layer of BPSG of 8000Å thickness is deposited as dielectric layer. Then the contact holes are opened and the Al-Si metallization pattern is delineated. A layer of PSG is deposited on the top of the wafer as passivation film. Fig.3 is the cross section view of LTCMOS circuit.

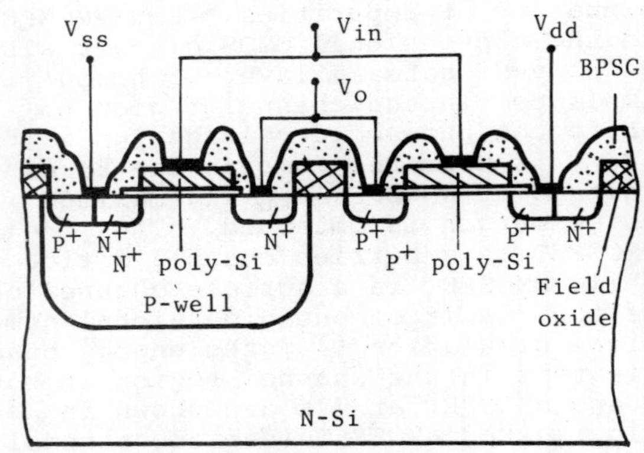

Fig. 3 Cross-section diagram of LTCMOS circuit.

After fabrication, the chips are diced and mounted in DIP (dual in-line package) without sealing, allowing the chip to contact directly with liquid nitrogen when measured at 77K.

DEVICE CHARACTERISTICS AND CIRCUIT PERFORMANCE

The device parameters of 1μm LTCMOS are shown in Tab.1. The effective channel length is derived from the dependence of drain conductance on V_g with different channel length at 50mV drain voltage V_{ds}. V_t is obtained from the intercept of the extrapolated drain conductance versus V_g at 50mV drain voltage.

Tab.1 The device parameters of LTCMOS

	NMOS	PMOS
Substrate doping (cm^{-3})	2×10^{16}	5×10^{15}
Gate oxide (nm)	30	30
Junction depth (μm)	0.25	0.25
Leff (μm)	1	0.9
V_t(77K/300K) (V)	.44/.15	.4/.1

1. Threshold voltage

Over the range of temperature studied the long channel threshold voltage for NMOSFET with homogeneous substrate doping is

$$V_{tn}=V_{fb}+2\psi_f+T_{ox}(4qE_oE_sN_a\psi_f)^{\frac{1}{2}}/E_oE_{ox} \quad (3)$$

where V_{fb} is the flat-band voltage, T_{ox} is the thickness of gate oxide, E_o is the permittivity of free space, E_s, E_{ox} is the relative dielectric constant of Si and SiO_2 respectively,

$$\psi_f=kT\ln(N_a/n_i)/q \quad (4)$$

is the Fermi potential, n_i is the intrinsic concentration of carriers. At low temperature, V_t will increases due to the increase of ψ_f resulted from the decrease of n_i. As shown in Tab.1, $|V_t|$ at 77K is about 0.3V larger than that at 300K, which is in good agreement with values calculated from Eq.(3) and (4). However, for burried channel PMOSFET, $|V_t|$ is about 1V greater than that of surface channel one at 77K because the implanted accepters are not effective due to the carrier freeze-out.

2. g_m improvement at 77K

For small drain voltage, the FET operates in linear regime and g_m is

$$g_m=\partial I_{ds}/\partial V_g=\mu_sE_oE_{ox}WV_{ds}/T_{ox}/L \quad (5)$$

where μ_s is the surface mobility of carriers, I_{ds} is drain current, W, L is the channel width and length respectively. In Eq.(5) only μ_s are related to temperature.

Fig.4 is the g_m dependence on effective V_g of 1µm LTCMOS devices at 77K and 300K with 50mV drain voltage. The 77K improvement factor of maximum g_m is 3.1 and 3.3 respectively for N and PMOSFET. The improvement of g_m means the increase of mobility of carriers, which resulted from the weakening of lattice scattering at low temperature. The decrease of g_m with large V_g is the result of mobility degradation at high electric field.

When MOSFETs operate in the saturation regime with large V_{ds}, the saturation of drifting velocity of carriers at high electric field is the main factor which determines the g_m of devices.

$$g_m=V_{max}E_oE_{ox}W/T_{ox} \quad (6)$$

where V_{max} is the maximum velocity of carriers and is equal to 1.5×10^7 cm/s

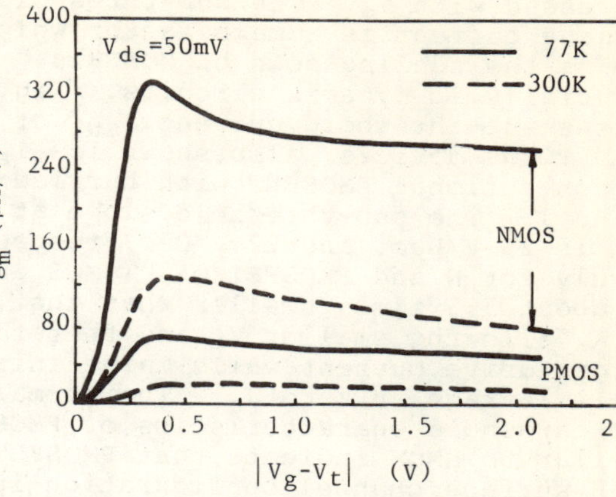

Fig. 4 Dependence of g_m on effective V_g of MOSFETs in linear regime at 77K and 300K.

at 77K and 1×10^7 cm/s at 300K respectively for electrons. Fig.5 is the g_m of N and PMOSFET versus V_g at 77K and 300K with 2.5V V_{ds}. The 77K improvement factor of g_m of NMOS is 1.5 and is in accordance with that of drifting velocity. The velocity saturation of PMOSFET is not pronounced because of smaller mobility of holes. For burried channel PMOSFET, the maximum g_m is only 2/3 of that of surface channel one because of the greater $|V_{tp}|$.

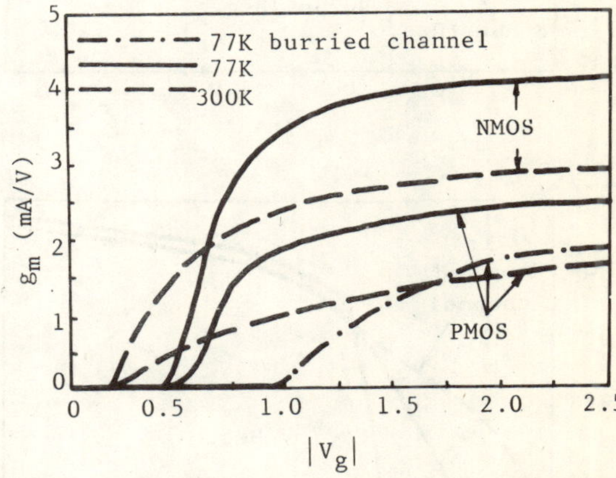

Fig. 5 g_m versus V_g of P and NMOSFET with 2.5V V_{ds}.

Because MOSFETs are in either linear or saturation regime when CMOS circuits operate in switching state, the improvement factor of operating speed resulted from mobility increase at low temperature is between 1.5-3.5.

3. Sub-threshold current

When the channel of MOSFET is in weak inversion with $V_g<V_t$, the drain current results from the carrier diffusion from

source to drain and is exponencially decreased with V_g. The sub-threshold leakage current is a main factor which limits the scaling-down of MOSFETs, especially in dynamic circuits. Fig.6 is the sub-threshold current I_{sub} of 1μm LTCMOS devices, also shown is I_{sub} of conventional PMOSFET with burried channel. The sub-threshold slope at 77K is 25mV/Dec. and 22mV/Dec. respectively for N and PMOSFET of LTCMOS and is about 3.5 times smaller than that at 300K, allowing smaller V_t and therefore larger drive current while maintaining small leakage at zero V_g. The normal sub-threshold characteristics of PMOS similar to NMOS indicate that PMOSFET with surface channel configuration is free of the freeze-out of carriers indeed. The slow variation of I_{sub} with V_g of burried channel PMOSFET is the result of re-ionization of carriers being frozen-out previously.

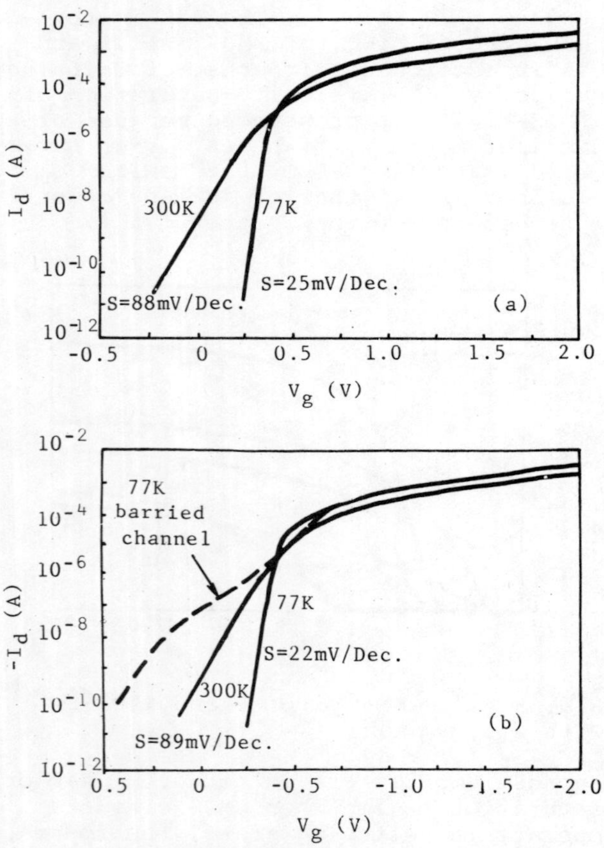

Fig. 6 The sub-threshold current of 1μm NMOS (a) and PMOS (b) at 77K and 300K.

4. Substrate current

Substrate current I_b results from the impact-ionized holes swept into back contact electrode. I_b is a good monitor of hot-electron effect. Generally speaking, the situation is worse in NMOSFET than in PMOSFET because of larger impact ionization effeciency of electrons. Fig.7 is the measured I_b of 1μm NMOSFET at 77K with 2.5V V_{ds}. The peak I_b is 8nA/μm and is two order of magnitude smaller than that at 300K with 5V V_{ds}. Stress experiment shows that the life time of device (defined as the time needed with 10mV V_t variation at the stress condition of $V_{ds}=V_g=2.5V$) at 77K is about 10 years. The result indicates the hot-electron effect at 77K is restrained effectively by lowering the power supply voltage.

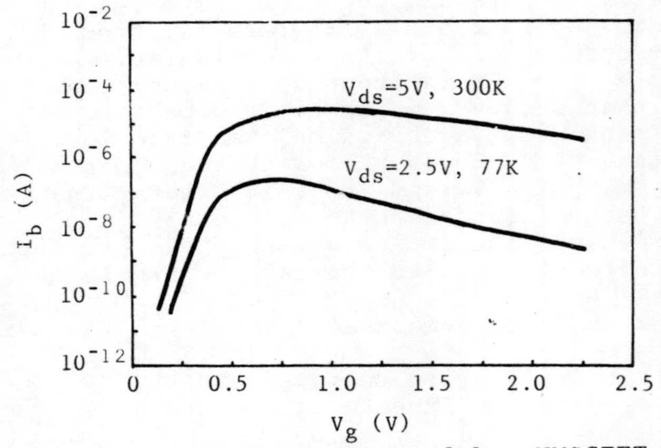

Fig. 7 Substrate current of 1μm NMOSFET at 77K.

5. Reducing of interconnecting resistance at 77K

The sheet resistances of poly-Si gate, source and drain diffusion region and Al-Si metallization at 77K and 300K are listed in Tab.2. As cab be seen, the series resistances of interconnecting lines are lowered at 77K. The smaller improvement of doped poly-Si, source and drain region than Al-Si metallization may be the result of impurity scattering at low temperature.

Tab.2 The sheet resistance $R_{\square}$ (Ω) of interconnecting line

	300K	77K	300K/77K
N^+ poly-Si	25	20	1.25
P^+ poly-Si	38	35	1.08
N^+ diffusion	58	35	1.65
P^+ diffusion	80	45	1.83
Al-Si line	0.06	0.008	7.5

6. Suppressing of latch-up

Due to the positive feedback in the

parasitic P-N-P-N structure formed by the P source, substrate, well, and N source, latch-up occurs in CMOS circuit if the product of the gain of parasitic pnp and that of parasitic npn transistor is greater than 1 [7]. Latch-up usually disrupts the function of the circuit. The large excess current will cause permanant damage of the circuit. In the room temperature design, the circuit density will be greatly limited due to the required spacing between P and NMOSFETs to prevent latch-up effect. At 77K, because the gain of the parasitic bipolar transistors is well below 1, there is no latch-up observed in our LTCMOS circuit.

7. Circuit performance

It is expected that the operating speed of CMOS circuit will be improved because of increasing of mobility and reducing of interconnecting resistance. Fig.8 shows the gate delay time of LTCMOS measured from the 21-stage ring oscillator at 77K and 300K. The gate delay time is less than 500ps and is 2 times shorter than that of the conventional circuit at 300K. However, for circuit of room temperature design, the 77K improvement facter is about 1.2-1.5 because the greater $|V_{tp}|$ resulted from the carrier freeze-out leads to small drive current of PMOSFET.

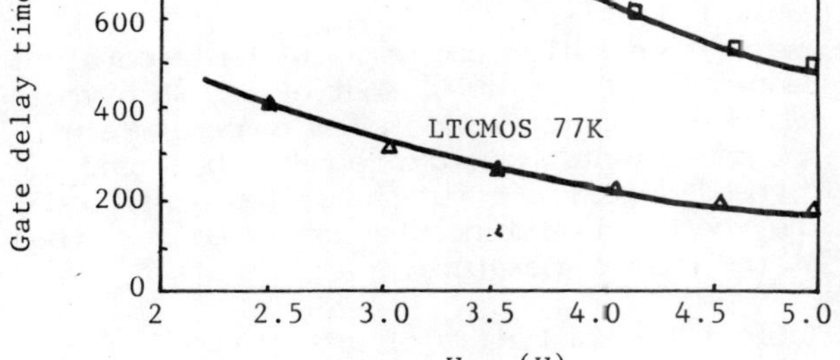

Fig. 8 The delay time of 1μm LTCMOS

The improvement of operating speed can be made further by properly optimization of design and processes of LTCMOS. This improvement of speed is very attractive as the scaling-down of MOS devices become more and more difficult. Moreover, the rapid development and progress of liquid nitrogen temperature superconductive materials and devices provide great promise for CMOS circuit which operate at 77K.

CONCLUSION

1μm CMOS devices and circuit for 77K operation have been designed and fabricated. By using P^+ and N^+ poly-Si gate for P and NMOSFET respectively, surface channel configurations are obtained and the freeze-out of the carriers at 77K is avoided. The aggravated hot-electron effect at 77K is restrained by lowering the power supply voltage to 2.5V. Excellent device characteristics at 77K are obtained. The gate delay time of 1μm LTCMOS is less than 500ps and is two times shorter than that of conventional CMOS circuit at 300K.

REFERENCES

1 K. Yu, et al., "HMOS-CMOS—a low power high performance technology", IEEE J. Solid-State Circuit, Vol.SC-16 (1981), pp.454-460.
2 R.H. Dennard, et al., "Design of ion-implanted MOSFETs with very small physical dimensions", IEEE J. Solid-State Circuit, Vol.SC-9 (1974), pp.256-268.
3 F.H. Gaesslen, et al. "Very small MOSFET's for low temperature operation", IEEE Trans. E.D., Vol.ED-24 (1977), pp.218-229.
4 Li Yuanxiong, et al., "Investigation of the carriers in CMOS devices at low temperature", to be published.
5 S. Ogura, et al., "Design and characteristics of the lightly doped drain-source (LDD) insulated gate field effect transister", IEEE Trans. Electron Devices, Vol.ED-33 (1980), pp.1359-1367.
6 D.W. Wenocur, et al. "Fabrication and characterization of sub-micron thin gate oxide p-channel transistors with P^+ polysilicon gate", in IEDM Tech. Dig. (1985), pp.212-214.
7 F.S.J. Lai, "A highly latch-up-immune 1-μm CMOS technology fabricated with 1-MeV ion implantation and self-aligned $TiSi_2$", IEEE Trans. Electron Devices, Vol.ED-33 (1986), pp.1308-1319.

A NEW TYPE LIQUID NITROGEN LEVEL MONITOR

Meng Linghua
Hefei institute of cryogenics and Electronics
Hefei China

ABSTRACT

A new type monitor of liquid nitrogen level has been developed. The sensor using magnetic buoy and dryreed switch is very reliable, and almost without energy consumption when sensor works. It can be used to monitor liquid nitrogen level continuously.

INTRODUCTION

Up to now, many sorts of useful low-temperature liquid level indicators have been developed. Their principles are that the cryogenic liquid level is sensed by the difference of physical property of gas and liquid state of of low-temperature liquid. Resistance liquid level indicators such as carbon resistor[1,2] thermal resistor, semiconductor[3], copper[4] and superconductive wire[5] were used as a sensor sensitive to the difference of thermal conductivity of the gas and liquid state, can indicate liquid level continually or at some set points. But these indicators must be provided by a necessary heat-power increasing liquid extra consumption. Capacitance liquid level indicators[6,7] on principle of the difference of dielectric constant of liquid and vapour, can indicate continually, accurately and without energy consumption, but it is only used for measurements of liquid helium level because of its expensive price. The characteristic of the manometric liquid level indicator[8,9] on principle of the static pressure of low temperature liquid is visualized, but it has to be put in before a cryogenic equipment is assembled. Ultrasonic liquid level indicator[10], Buoy liquid level indicator and heat-oscillation liquid level indicator [11] are also only used for liquid Helium. Now a new type liquid nitrogen level monitor making use of magnetic buoy and dryreed switch has been developed, It can mainly used in small and middle type liquid nitrogen Dewar. The effect is satisfactory.

THE PRINCIPLE AND STRUCTURE OF THE NEW TYPE MONITOR

The sensor of this liquid lever monitor is shown schematically in Figure 1. The magnetic buoy 6 can float up and down along the guide tube 4 with the change of liquid nitrogen level. When the magnetic buoy reaches the alarm height, the dryreed switch inside the guide tube will be turned on by the magnetic field of the buoy. At that time, the monitor begins to work and alarm sound and light signals.

The dryreed switch is a sensitive magnetic-switch device. It takes a short time to react. Reed switch is made of iron-nickel alloy. Spring contact with protective film is put into a air tight glass tube full of inert gas to prevent oxygenation or corrosion. It possesses less contact resistance and a long working life. We chose JAG-4H small dryreed switch with an outside diameter of 3mm. It is easier to put it into the guide tube. A JAG-4H type dryreed switch is shown schematically in Figure 2, and its performance is shown in table 1

The dryreed switch can work in the temperature range of -10°C to 40°C. Most of JAG-4H dryreed switches can also work in low-temperature with apt management. When the magnetic buoy and the dryreed switch are put into liquid nitrogen, the contact resistance becomes even less than at the room temperature.

Table 1 The parameters of JAG-4H type dryreen switch

turning on A	25 - 50
turning off A	8
holding period ms	0.9
contact resistance	0.15
capacity of contact VA	$12V_{DC}$ x0.05A
working life, No. of time	10^7
working temperature range	-10°C - + 40°C
insulation resistance Ω	10^8

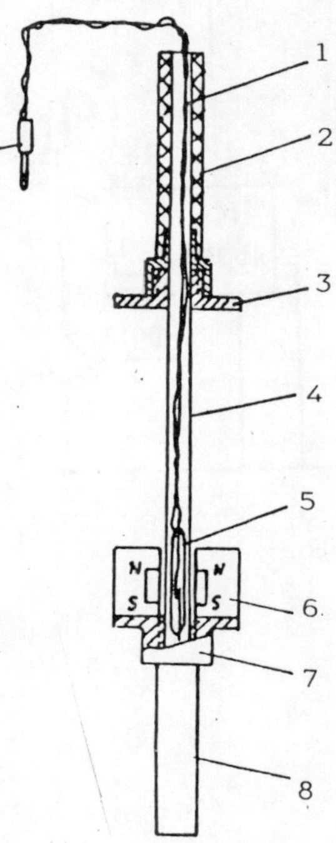

Figure 1. Diagrammatic sketch of the sensor

1. lead wires of constantan
2. insulation tube
3. high limit collar
4. guide tube
5. dryreed switch
6. magnetic buoy
7. low limit collar
8. support rod
9. plug adapter

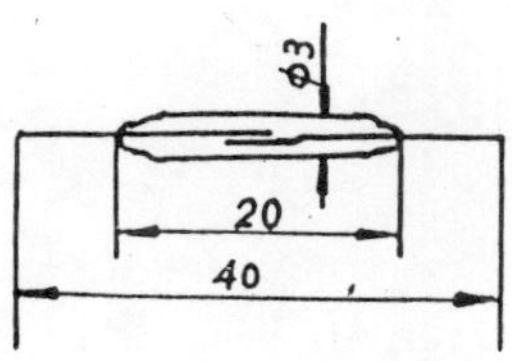

Figure 2 JAG-4H type dryreed switch

Magnet buoy consists of a piece of foamed plastics and a permanent magnet.The permanent magnet is hid in the buoy after foaming.It can be demagnetised at low temperature,so we chose it made from cementation strontian ferrite.Its magnetic flux density of surface is about 500 gauss at room temperature.The effect is very satisfactory.

There is a low limit collar 7 near the alarm height. When liquid nitrogen level falls beneath the alarm height, the magnetic buoy will stay on the limit collar and the dryreed switch will turn on all the time until the level rises higher than the alarm height again.

In order to reduce the heat leakage caused by the sensor, the lead wires of the dryreed switch is made of enamelled constantan wires 1. Copper wires 10 connect constantan wires 1 passing through the insulation tube 2 with the plug adapter 9. Because the dryreed switch is normally turned off, there is no current in the circuit until the buoy contacts the alarm position. Even the buoy comes into operation, the Joule heat caused by the dryreed and the constantan wires is still less than 1mw. So it is possible for the device to monitor liquid nitrogen level continuously day and night.

MONITOR CIRCUIT

Fig 3 shows the circuit diagram of the monitor A multivibrator consists of transistors BG1 BG2, capacitors C_1 , C_2 , Light-emitting diode D_1 and resistors R_1 , R_2 , R_3. As the contact of dryreed switch is closed the base electorde and the collector of the transistor BG3 is made up, the transistor BG3 and audio circuit KD9561 begin to work at once. Alarm sound signals coming from KD9561 are magnified by transistor BG4, and can be heard from the speaker Y. The dryreed switch also connects with the multivibrator that makes the light-emitting diode D_1 to flash. Because BG3 is a silicon transistor, the current of penetration I_{ceo} is less than 1-2μA, the power of the alarm which is not at work is little.

EXPERIMENT RESULTS

The monitor has been experimentally studied. The maximum echo power of the alarm was 60-80 mw. When the sensor was put into liquid nitrogen, the extra conduction heat leakage caused by the sensor was less than 0.03W, extra consumption of liquid nitrogen was less than 0.016 liter daily. When the sensor alarms, the Joules heat was less than 1mw. It is possible to monitor a longer distance by adding lead wires. The length of wires can not lengther over 100 meter unless the entire resistance of the lead wires is less than 100 Ω.

CONCLUSION

Based on the results of these experiments, several conclusions can be made. The monitor of liquid nitrogen level is success. It is apparently sensitive to the response, and very reliable. There is only a little influence on the vaporation loss of the liquid nitrogen.
According to the difference usage, the system can be made to alarm at high-limit, at low limit, or both. When liquid transfer is proceeding, the high-limit can alarm a signal before the Dewar is full of liquid nitrogen to avoid waste of liquid.

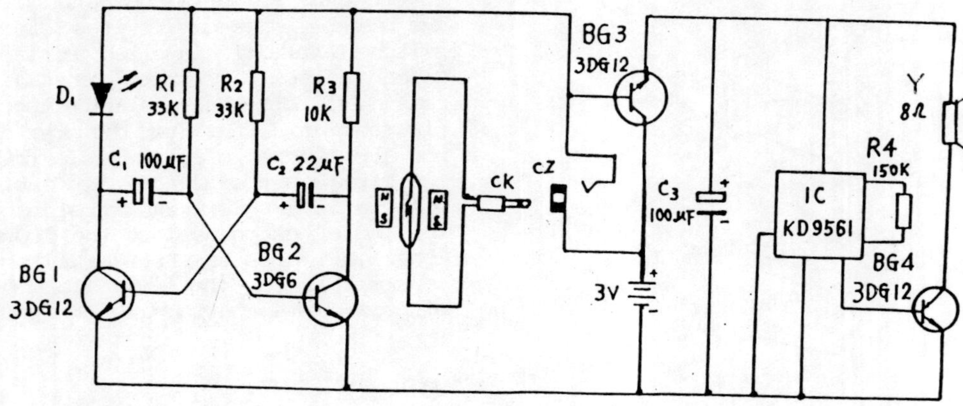

Figure 3 the circuit diagram of the monitor

Because the sensor is at normal off contact, it is easy to fit with an automatic feed liquid system to control liquid level.

Though this monitor has been especially developed to monitor liquid nitrogen level, it is also possible to monitor other cryogenic liquid. If specific gravity of the buoy were less than 0.1, It could also monitor liquid helium level.

ACKNOWLEDGEMENT

The auther gratefully acknowledges the help and advice of my companions; Lin Xiou Min, Chen Mei Lan, Jia Jun, Wu Tong, Li Wei. Thanks Hefei Institute of Cryogenics and Electronics for offering all financial support.

REFERENCES

1. Bapna S.C.Prarinkumar M.Thiumaleshwar, Cryogenics Vol.23 (1983) P275
2. Vander, kooy J.Kany C.S, Rev.Sci.Instr. Vol.49 (1978) P1746
3. Cosier,J. Cryogenics,Vol.26 (1986),P45
4. Akintud,J.A., Cryogenics, Vol.27 (1987)
5. Efferson, K.R. Advances in Cryogenic Engineering, Vol.15 (1970) P124
6. Yuan Yaozhong, Journal of the Chinese Association of Refrigeration No.1 (1979) P22
7. Ma Liyao, Tang Weiwei, Journal of Chinese Association of Refrigeration No.1 (1979) P7
8. D.N.Lyon and J.J Gillch: Rev Sci Instr 34 (1963) P 811
9 W.L.Rope and F.F. Melanghlin J.Sci Instr: 43 (1956) P260
10. Selig K.P., Wener H.G., Cryogenics Vol.21 12 (1981) P680

AN EXPERIMENTAL INVESTIGATION OF THERMAL CONDUCTIVITY OF A HIGH T_c SUPERCONDUCTING MATERIAL AT LOW TEMPERATURE

Zheng Zhengquan, Zhang Daiguang, Wu Chaomei
Huazhong University of Science and Technology
Wuhan, China

ABSTRACT

This paper reports an experimental investigation of the thermal conductivity of a high Tc superconducting material at low temperatures. The thermal conductivity apparatus adopted a new type of floating heat sink to ensure the accuracy of the measurements. The thermal conductance of Y-Ba-Cu-O was calculated from the experimental data. Finally, these results are analyzed and explained.

INTRODUCTION

Recent research on high Tc superconducting materials has entered a stage of thorough and careful study. As a part of this the thermal conductivity of Y-Ba-Cu-O has been experimentally studied and was found to be very small at low temperatures. This characteristic is very important in the applications.

MEASUREMENT PRINCIPLES AND EXPERIMENTAL APPARATUS

From Fourier's law, the thermal conductivity of a material λ can be given by

$$\lambda = \frac{L}{A} \cdot \frac{\phi}{\Delta T} \quad (1)$$

where L is the length of the specimen, A is the cross section of the specimen, ΔT is the temperature difference between two points on the specimen and ϕ is the heat flow through the specimen[1],[3] At the steady state the λ of the specimen can be given by

$$\lambda = \frac{L}{A} \cdot \frac{P}{\Delta T} \quad (2)$$

where P is the electric power of the heater fitted on the hot end of the specimen[1]
Using equation (2) we measured the thermal conductivity of a Y-Ba-Cu-O ----- type of high Tc superconducting material by using an axial heat-flow method in its steady-state. Because λ reaches a limit when ΔT approach zero, the key technique of the method is to establish a minimum temperature difference.
Because Y-Ba-Cu-O is a type of ceramic, its thermal conductivity is rather poor. Therefore the specimen chosen is short and thick. The schematic drawing of the experimental apparatus shows a new type of cryostat with a floating heat sink to ensure the accuracy of the measurement and convenience of the temperature adjustment between 77-200K[2]

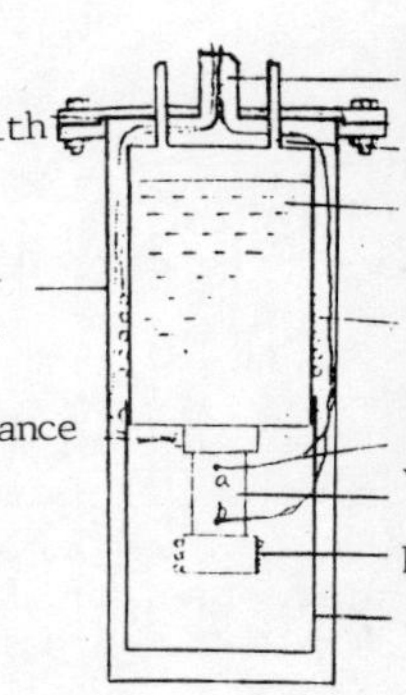

Fig. 1 Schematic Drawing of The Experimental Apparatus

The floating heat sink was filled with liquid nitrogen. To adjust temperatures of the heat sink , a heater H1 was fitted on to its outer casing. One pole of the specimen was fitted on to a floating heat sink, the other end was fitted on to a heater H2. The differential thermo-couples were fitted on to the surface of the specimen in order to measure temperature differences between the points a and b. The temperature of the floating heat sink is measured by a platinum resistance thermometer. A steady DC power supply was used for heaters H1 and H2. A microprocessor voltmeter with $0.1\mu V$ sensitivity is used for measuring the voltage of the platinum resistor, the differential thermocouples and the heaters.

EXPERIMENTAL RESULTS AND ANALYSES

First, the specimen was cooled down to To(77k) then a small amount of power supplied to heater H2. After the system had become completely steady, The temperature difference T1 between a and b on the specimen the electric power P1 of heater H2 and the temperature To of heat sink were recorded. Repeating the step mentioned above, ΔT2,P2 and T1 can be obtained. It took about 120 minutes to reach the steady state in each run.The temperature difference is about 0.3k.The results are shown in Table 1 and Fig.2.

Table 1 Thermal Conductivity of Y-Ba-Cu-O at Different Temperatures

temperature(k)	86	90	106	174
λ(w/m k)	0.5	1.8	6.2	11.2

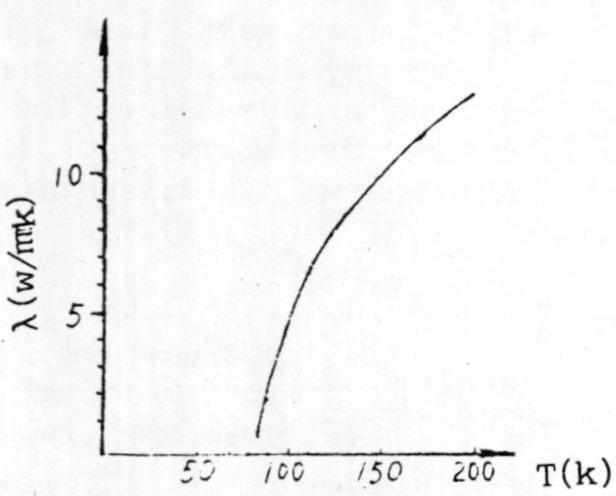

Fig. 2 Thermal Conductivity of Y-Ba-Cu-O vs. Temperatures

It is obvious that the thermal conductivity of Y-Ba-Cu-O is very small in its superconducting state (Tc=92k). The λ of Y-Ba-Cu-O sharply increases with use of temperature. This is similar to the R-T characteristics of superconducting materials near to their critical temperature.

Because the Y-Ba-Cu-O is a type of ceramic, heat is mainly conducted by lattice waves. The heat conduction process is limited by scattering. The relationship of thermal conductivity λ and heat resistance Wp can be given by

$$Wp = \frac{1}{\lambda} = Wpp + Wpi + Wpb + Wpd \qquad (3)$$

where Wpp is the scattering heat resistance between lattices, Wpi is the heat resistance due to impurity atom scattering and isotope defect scattering, Wpb is the heat resistance caused by the boundary of the specimen and the boundary of crystalline grains, Wpd is the heat resistance caused by the scattering owing to place disorder. Because the heat resistance caused by the scattering between lattices, and the scattering caused by impurities at low temperatures are very small, the heat resistance is mainly dependent on the boundaries of the specimen the boundaries of the crystalline grains and the scattering caused by place disorder. Therefore, when the temperature is rising, the heat resistances will sharply decrease, i.e the thermal conductivity will rapidly increase. Compared with other materials, the thermal conductivity of Y-Ba-Cu-O is similar to that of nylon and PTFE when $T \leq Tc$. In its normal state, the λ of Y-Ba-Cu-O is similar to that of stainless steel, which is 32 times greater than that of nylon.

CONCLUSION

The Y-Ba-Cu-O superconducting material is a good insulator when in its superconducting state. This characteristic is harmful to the stability of a high Tc superconductor, so it would be necessary to increase the heat exchange surface for the stability of a high Tc superconductor.

LIST OF SYMBOLS

λ = thermal conductivity
ϕ = heat flow through the specimen
ΔT= temperature difference
L = length of the specimen
A = cross section of the specimen
P = electric power of heater H2
Wp= heat resistance
Wpp= scattering heat resistance between lattices
Wpi= heat resistance caused by impurity atom scattering and isotope defect scattering
Wpb= heat resistance caused by boundary of the specimen and the boundary of crystalline grain
Wpd= heat resistance caused by the scattering owing to place disorder

REFERENCES

1. Yan Shousheng, Luo Guo 'The Principles and Methods of The Physical Experiment At Low Temperature' Science and Press (1986) Cryogenics, PP170-175 (in Chinese)
2. Zhu Xian 'Vacuum and Cryogenics' No.4(1987) pp.28-36 (in Chinese)
3. Reed Richard P., and Clark Alan F.'Materials at Low Temperatures' American Society for Metals (1983) PP.133-144

THE DESIGN AND SOME APPLICATIONS OF A SENSITIVE CAPACITANCE MICROMETER

Ruan Yaozhong Li Liping

Dept. of Phys. Uni.of Sci.& Tech. of China
Hefei, P.R. China

ABSTRACT

A general account is given of both the electrical and the mechanical aspects of the design of capacitance micrometer. Its applications are illustrated by the measurements of thermal expansion coefficient (TEC), dielectric constant in the polymer and high Tc superconductor systems. Some further developments are proposed.

INTRODUCTION

An important aim in solid state physics is the understanding of lattice dynamics, which can be revealed by the thermal expansion coefficient, elastic constants, and specific heat. Of these properties, the first two can be measured by the determination of the displacement. Recently, there are some methods for measuring the displacement, these include the interferometer (1), X-ray diffractometer (2) etc., but these methods meet some very complicated troubles at low temperature condition. In the article, We propose a method for measuring the small displacement at low temperature. We have designed and constructed a capacitance dilatometer and applied it tc some researches. This provides a very useful tool for the physical properties at low temperature.

DESIGN:

The capacitance cell is shown schematically in Fig.1. It is similar to White's (3). Where S is a sample, one side of which is fixed to base B. Another side of the sample is free, a copper piece E (diameter 12mm) is stuck to it for high plate. The low plate P (made of printedcircuit plate) is fixed to the base B by three copper columns D. The gap between E and P is 0.2mm. The heater H is wound around copper cell G. The temperature is controled by a apparatus adapted from DWT-702 controller (4).

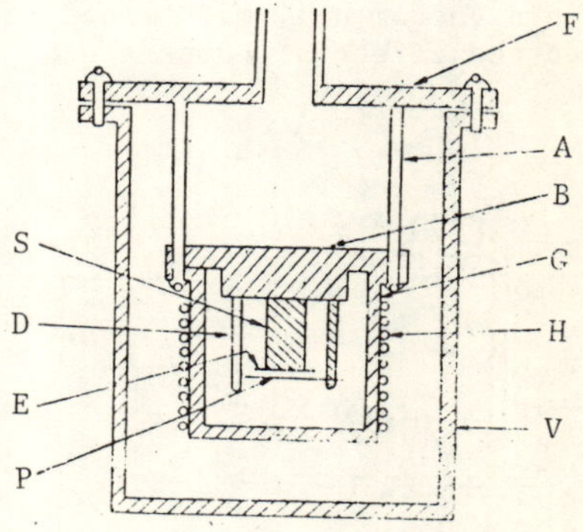

Fig. 1 Three-terminal capacitance cell

The elementary capacitance vs plate separation L_g is:

$$C = \frac{\varepsilon_0 A}{L_g} \qquad (1)$$

where A is the area of plate, ε_0 is the vacuum dielectric constant. From the (1), $L_g = L_c - L$ can be obtained by measuring the capacity C. Where L_c is the length of the copper column D, L is the length of sample. The TEC of the sample is got as an average from the length change over a finite temperature change:

$$\alpha = \frac{1}{L}\left(\frac{\alpha L}{\alpha T}\right)_p$$

$$= -\frac{L_c - 2L_g}{L_c - L_g}\,\alpha_{cu} + \frac{L_g}{L_c - L_g}\cdot\frac{\Delta C}{C\Delta T} \qquad (2)$$

Where the α_{Cu} is the TEC of copper, We adopt the data of Kroeger (5). In fact, because of $L_c >> L_g$:

$$\alpha \simeq \alpha_{Cu} + \frac{L_g}{L_c - L_g} \cdot \frac{\Delta C}{C\ \Delta T} \qquad (3)$$

In our experiment, the resolution of capacitance is higher than 10 E-6PF, so the resolusion of TEC is higher than 10E-8K. Because the capacitance of detector at low temperature is small ($\sim$10PF), the variation with temperature is very small (10^{-3}-10^{-6}PF), less than the capacitance of the circuit and its variation. So it is very important to avoid the capacitance of the circuit. The ratio transformer capacitance bridge and the three terminal connections are used in order to avoid the influence of other capacitance (6). The diagram is shown in Fig. 2. The balancing condition is detected with 186A lock in amplifier, the resolution of ratio transformer is 10E-6. The frequency of genarater is various and continuous. The frequency applied in this experiment is 9KHz. The standard capacitance is a common air variable capacitance keeped in a vaccum cell maintained in 77K to eliminate the effect of temperature.

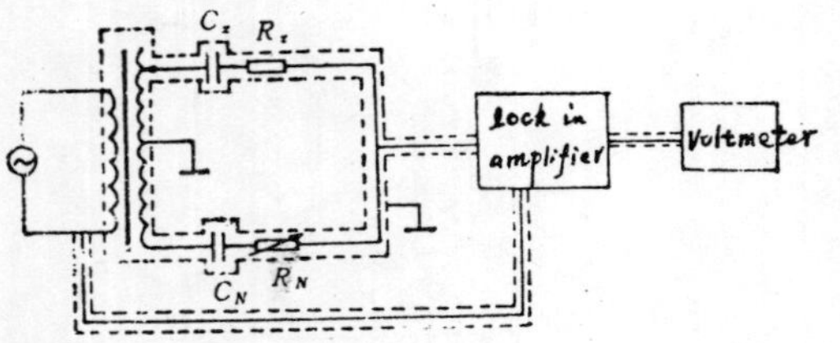

Fig. 2 Schematic diagram of the capacitance bridge

APPLICATIONS

By the means of the equipment, we have measured the thermal expansion coefficient and dielectric properties of the organic polymer. The sample is a kind of polydiacetylenes which can be prepared as the large single crystal. It has an ideal quesi-one dimension structure (7), and special charge-transfer and optical character. At first, we measured the TEC along the chain direction (b direction). The interval of temperature is about 4K, the time is 15min. Fig. 3. is our results, it is observed from Fig. 3 that there is no anomaly from 77K to 190K, the TEC increase with temperature. But the TEC have a sudden change at 195K, then decrease rapidly, near 200K, the TEC become a negative value. Above 210K, the TEC become positive again.

To ascertain the cause of the negative TEC, we further measured the TEC of PTS normal to the chain direction (a, c direction). From the result of c direction (Fig. 4.), We find that there exists a bump between 170K and 220K. It indicates that the coupling between

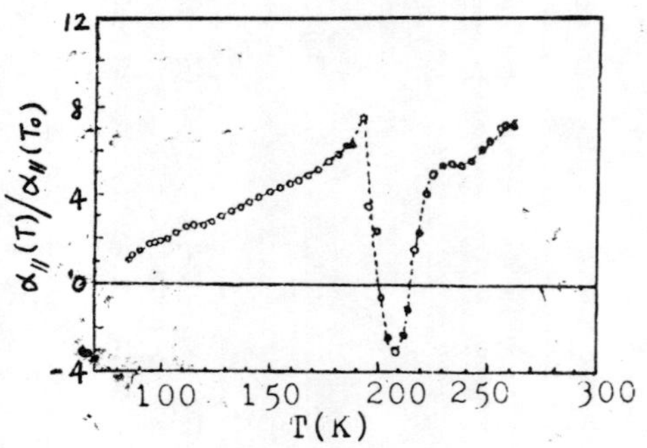

Fig. 3 The temperature dependence of TEC for PTS along Chain direction, To=85K

chains along c direction becomes weakened suddenly near 200K and there is a change of coupling during the phase transition. (7). Since the advent of high Tc oxide superconductor, the structure anomalies and instability

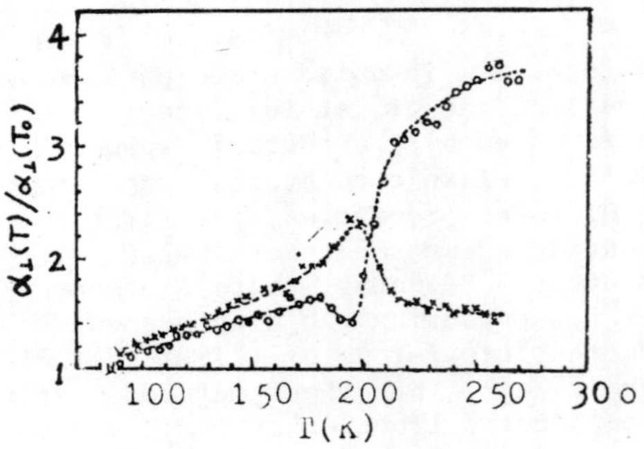

Fig. 4 The temperature dependence of TEC for PTS normal to the chain direction. To=86K.

have been studied with the apparatus. The thermal expansion coefficients of $YBa_2Cu_3O_{7-x}$ with orthorhombic and tetragonal phases (8), and Tl-Ba-Ca-Cu-O system (9) have been measured and analysed. Fig. 5 shows the temperature dependence of $\Delta L/L$ of orthorhombic phase $YBa_2Cu_3O_{7-x}$. It can be seen from Fig. 5 that there is an obvious decrease of volume near 205K.

Fig. 6 is the enlarged section of the Fig. 5 near Tc. it indicates that the thermal expansion coefficient has a small jump at the superconducting transition temperature 90K.

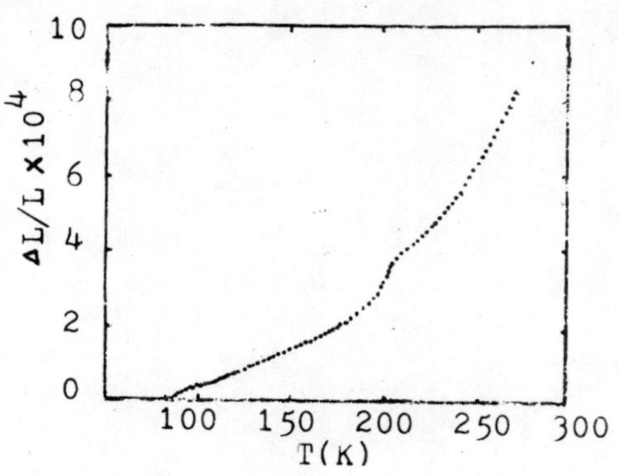

Fig. 5 The temperature dependence of length of single phase $YBa_2Cu_3O_{7-x}$

It means that there is a second order phase transition at Tc. From the jump, the pressure effect on Tc can be estimated (10).

Furthermore, the equipment can be used in studying the piezoelectric effect, internal friction and Young's modulus. The measurement of the high pressure at low temperature, and the temperature under high magenatism can be made with the equipment.

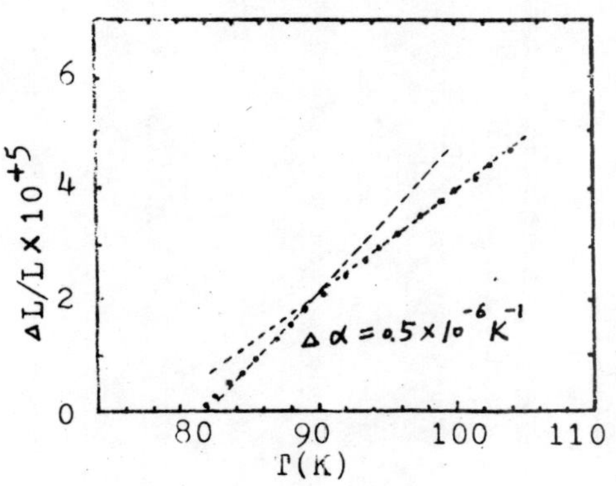

Fig. 6 The amplification of the Fig. 5 near Tc

REFERENCES

1. Rubin, T., Altman, H.W., and Johnston, H. L. J.Amer. Chem. Soc.76(1954) 5289

2. Simmons, R.O., and Balluffi, R.W. Phys. Rev. 108(1957) 278

3. White, G.K., Collins, J.Low. Temp. Phys. 7 (1972) 43

4. Ruan Yaozhong, Lin Peng, Wang Jiang, yao Lin. ACTA Phys. Temp. Hum. Sinica 8(1986) 68

5. Kroeger, F.R., C.A. Swensson, J.Appl. Phys. 48 (1977) 853

6. Ruan Yaozhong, Li Guihe, Wang Guishi, Wang Wei, ACTA Phys. Temp. Hum. Sinica 1(1979) 44

7. Ruan Yaozhong, Li Liping, Wu Zhiqiang, and He Pingsheng. ACTA Phys, Sinica 36(1987) 1219

8. Ruan Yaozhong, Li Liping, Hu Xuelong, Peng Dingkun, Hu Junbao, and Zhang Yuheng. Mod. Phys. Lett. B (to be published)

9. Ruan Yaozhong, Li Liping, Cao Liezhao, Zhu Changfei, Zhang Yuling, Huang Jiuqi, and Xie Sisheng. Mod. Phys. Lett. (to be published)

10. Ruao Yaozhong, Li Liping, Hu Xuelong, Mao Xianglei, Meng Guangyao, Peng Dingkung, Hu Junbao, and Zhang Yuheng Chin. Phys. Lett. 5(1988) 153.

7. APPLICATION OF CRYOGENIC TECHNIQUES

APPLICATION OF THE FUNCTIONAL MAGNETIC MATERIAL FOR CRYOGENIC ENGINEERING IN JAPAN

T. Hashimoto
Department of Applied Physics, Tokyo Institute of Technology,
Meguro-ku, Tokyo 152, Japan

ABSTRACT

Recently, two kinds of works to improve the performance of the small refrigerators have been developed by several groups in Japan:

One is the Carnot magnetic refrigerator in the temperature range from ∿20 K to 4.2 K. It is thought that this refrigerator possesses in principle many kinds of superior characteristics and high efficiency in comparison with the present gaseous refrigerator commonly used. Therefore, this refrigerator will be used for the cascade refrigerator with the gaseous refrigerator above ∿20 K.

The other is development of new regenerator which works effectively in the temperature range below 20 K. The present regenerator is effective only above 10 K and this limitation is originated from the steep decrease of the specific heat of the present regenerator materials, that is, lead and its alloy, below ∿10 K.

Presently, several kinds of new magnetic materials applicable for the regenerator below 10 K have been developed and also the regenerator systems using these material have been investigated.

In the present paper, several select topics about those two kinds of investigations will be introduced.

I INTRODUCTION

Recently, several series of high Tc oxide super couductors have been discovered and, however, considering their physical properties and the present status of the investigation on those materials, it has been thought that the application of the high Tc superconductor especially to the large scale superconducting magnet is very difficult.

This situation strongly impresses upon us the importance of the present practical metal superconductors.

While, we have to use those metals in the low temperature about ∿4 K because those transition temperatures from the super phase to the normal one are in the range of ∿25 K. Therefore, the convenient refrigerators for He liquefaction with high efficiency are necessary to make use of those metals.

In the current small He liquefiers, however, practical performance of those are considerably inferior to the ideal value. One of the important reasons why those have low efficiency is to use the Joule-Thomson (J-T) expansion process instead of the thermodynamically reversible process, such as the isothermally depressuring process, in the range below ∿20 K.

However, there is no choice at present, because of no existence of the effective regenerator in this range, which is necessary for composition of an ideal cycle satisfying the Carnot condition.

Therefore, it has been thought that one of the most important objects in the present cryogenic engineering field is to develop the simple He liquefier with high efficiency.

In order to improve the performance of the small refrigerator below 20 K, two kinds of investigations have been advanced by several groups in Japan.

One is the Carnot magnetic refrigerator in the range from ∿20 K to 4.2 K which is used for the cascade refrigerator with the gaseous refrigerator above ∿20 K.

The other is development of new regenerator material which works effectively in the temperature range from 20 K to 4.2 K.

In the present paper I will show several topics about those two kinds of investigations during past a few years in Japan.

2 THE CARNOT MAGNETIC REFRIGERATOR

The Carnot magnetic refrigerator is thoght to be one of the simplest applications of the magneto-caloric effect for the refrigeration.

The Carnot magnetic refrigeration cycle is composed of the following four kinds of processes; the isothermal magnetization (releasing heat to heat bath at the high temperature ∿20 K), the adiabatic demagnetization, the isothermal demagnetization (absorbing heat from the low temperature heat bath) and the adiabatic magnetization processes, as shown in Fig.1.

In principle, those four kinds of processes are thermodynamically reversible processes and, as the results, it is expected that this refrigerator has high efficiency.

The most important subjects to actualize this system are selection of the magnetic refrigerant and the development of heat exchange mechanism between the magnetic refrigerant and the heat baths (high and low temperature).

As for the magnetic refrigerant for the Carnot, we have to select the materials in consideration of the following condition at least (1,2):

(A) Its phase transition temperature is in the range below 1 K for the Carnot cycle liquefing He.
(B) In order to obtain the large refrigeration power per cycle, the magnetic material must include the large numbers of the large localized spins.
(C) Its thermal conductivity has to be very high so as to make the high speed cycle operation.

From these standing point $Gd_3Ga_5O_{12}$(GGG) has been selected as the refrigerant in the low temperature(3).

Appling the magnetic field to the spin system, the Zeeman energy is given by $g\mu_B B$, whcre g and μ_B are g-factor and Bohr magneton number. While, the thermal excitation energy is given by ∿kT, where k is the boltzman constnat and T is temperature.

In the temperature range above ∿15 K, where $kT > g\mu_B B$ on condition that $B \leqq 6$ T and $g = 2$, GGG is ineffective, although it's effective below ∿15 K. Therefore, in order to make the Carnot cycle with the wide temperature span from 20 K to 4.2 K, we have to use the spin system with large g factor or strong magnetic field.

In the anisotropic materials the g-factor is strongly anisotropic and the g-value parallel to crystalline principal axis becomes frequently of $g_{\parallel} \cong 19$ in contrast to $g = 2$ in the isotropic magnetic material(4).

As a typical example, the temperature variation of the magnetic entropy S_J in the anisotropic material $Dy_3Al_5O_{12}$(DAG) is shown as a function of the applied magnetic field in Fig.2 (5). This figure clearly shows that the refrigeration from 20 K to 4.2 K is possible by using DAG.

As for the heat transfer mechanism, the heat pipe operating at 4.2 K has been developed (6). The conceptual figure of the heat pipe is shown in Fig.3.

From the experiments of the heat pipe similar to Fig.3, it is made clear that the thermal conductivity K of the heat pipe is superior to that of the commercial pure copper and the

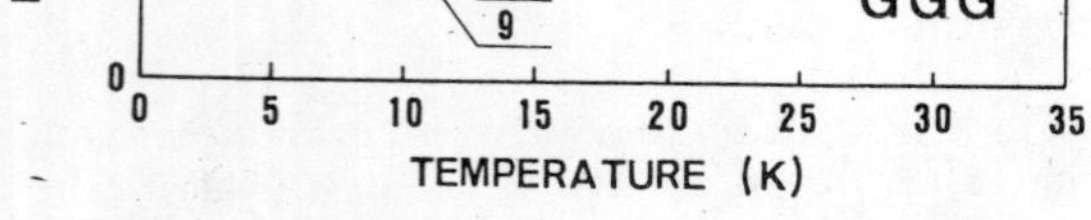

Fig.1 The Carnot magnetic refrigeration cycle (ABCD) shown on the entropy v.s. temperature plane in GGG. (A→B:Isothermal magnetization, B→C:Adiabatic demagnetization, C→D:Isothermal demagnetization, D→A:Adiabatic magnetization process.)

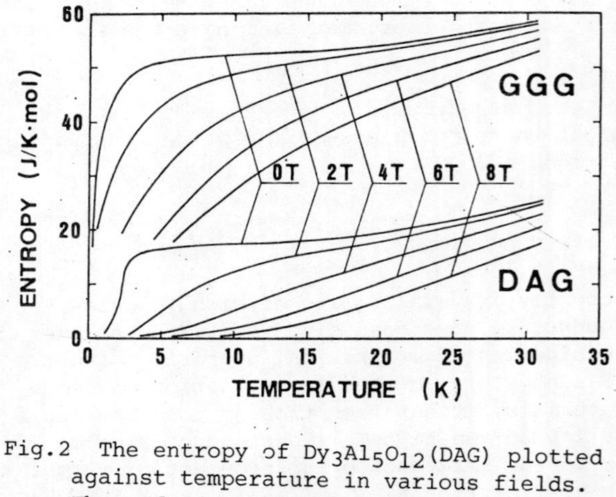

Fig.2 The entropy of $Dy_3Al_5O_{12}$(DAG) plotted against temperature in various fields. That of $Gd_3Ga_5O_{12}$(GGG) is also shown for comparison.

switching ratio (the ratio of K_{on} to K_{off}) is more than 5×10^2. The performance of this heat pipe is summarized in Table I.

On the basis of the fundamental experimental results as mentioned above, several types of the Carnot magnetic refrigerators have been developed. The most typical prototype test machine is shown in Fig.4 (7).

In this machine, by the reciprocating motion of the pistons, the magnetic refrigerants are moved alternatively between the zero magnetic field and the strong field area, produced by the persistent current mode superconducting magnet.

Therefore, during the course from the strong field to the zero the adiabatic and the isothermal demagnetization processes are made and in the course inverse to the above the adiabatic and the isothermal magnetization processes are produced.

In transfering the strongly magnetized magnetic materials from the strong magnetic field to zero field, strong power is needed. While, in the spacially gradient magnetic field the magnetized material is attracted to the strongest magnetic field spot.

Therefore, in this machine, two pistons with the magnetic refrigerants (GGG) make the antiphasic reciprocating motion so as to cancel out those two kinds of forces.

The refrigeration power and efficiency of this machine in operation of the range from 12 K 4.2 K are ∿1.2 watt and ∿30 % carnot, respectively, though the performance of this machine strongly depends on the refrigeration cycle span.
The performance of the magnetic refrigerator also depends on the magnetocaloric characteristics of the refrigerants.

In Fig.5 are shown comparison between the refrigeration powers of two kinds of refrigerant, GGG and DAG, in the reciprocating test machine (8). In this figure refrigeration powers [J] in those two cases are shown as a function of the heat reservoir temperature at the high temperature side.

This result clearly reflects difference of the magnetic characteristics between GGG and DAG.

In GGG, G_d^{3+} ions have the largest spin value $J = 7/2$ in the rare earth ions, though the orbital spin L is equal to zero. Therefore, in the low temperature range, $g\mu_B B \geq kT$, Fig.5 shows that GGG is a powerful refrigerant.

Table I Characteristics of the heat pipe

	Thermal conductivity(W/cm·K)		K_{on}/K_{off}
	K_{on}(on-state)	K_{off}(off-state)	
heat pipe no fin	14.6	0.022	664
heat pipe with fin	34.0	0.022	1545
pure copper metal	~ 15		

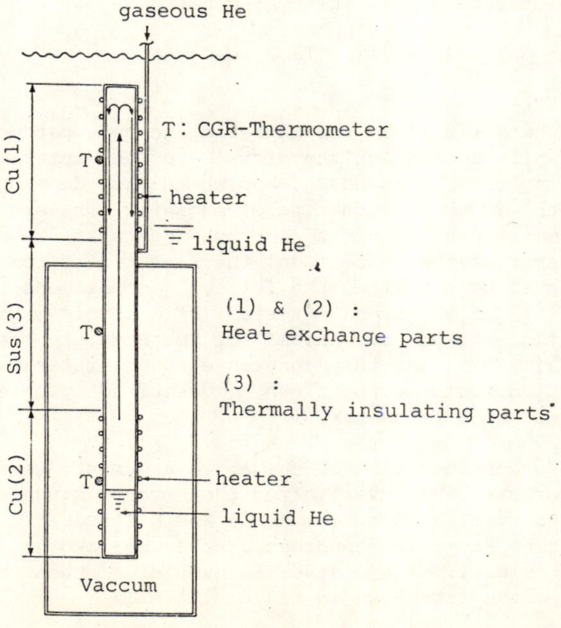

Fig.3 The conceptual figure of the heat pipe.

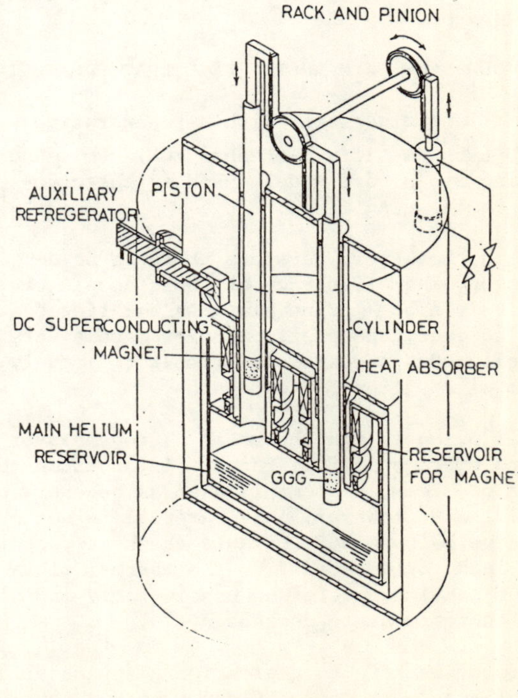

Fig.4 A typical prototype Carnot magnetic refrigerator.

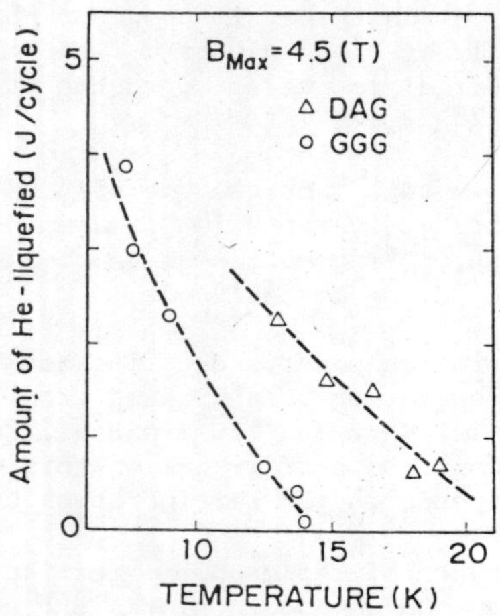

Fig.5 Comparison between the refrigeration powers of GGG and DAG refrigerants in the reciprocating test machine.

On the otherhand, in the range above ∿12 K, anisotropic magnetic refrigerant DAG, whose effective g-factor paralle to <111> is 10.8, becomes to be effective.

Until now, many kinds of the Carnot magnetic refrigerator have been developed (9). From view point of those operation, those are roughly divided into three types, that is, static, reciprocating and rotating refrigerator. Each of those three types of refrigerators possesses peculiar characteristic and, however, as far as considering those efficiency, the most promissing machine is thought to be the reciprocating refrigerator.

3 DEVELOPMENT OF MAGNETIC REGENERATOR MATERIAL

The small refrigerators with a regenerative cycle, such as Gifford-McMahon (GM) cycle and Stirling cycle, are widely used in the cryogenic engineering field.

Since in those refrigerators there is no J-T expansion, the efficiency of those should be relatively high in principle. As a matter of fact, however, these efficiencies become very low below 20 K and it is difficult to liquefy Helium without the J-T expansion.

This is owing to that there is no regenerator able to operate in this range. In an usual regenerator, lead is commonly used as regenerator material and, however, this material is not effective below ∿15 K, because the heat capacity of lead, originated from the thermal vibration of lattice, decreases considerably in this range as temperature decreases.

Therefore, we have to search for a new material having large heat capacity due to other physical origins.

It is well-known that a phase transition of the system, such as order-disorder transition, is accompanied with abnormal increase of the specific heat due to its entropy change.

One of the most promising phase change is the magnetic phase transition, because in this temperature range the entropy change with the order-disorder transition in the spin system is the largest in whole phase change.

The entropy change S_J with the magnetic phase change is given by the following formula,

$$S_J = Nk\ln(2J + 1) \tag{1}$$

, where N and J are the numbers of spins and the locarized spin value, respectively.

While, the magnetic specific heat C_J is given by

$$C_J = T(\partial S_J/\partial T) \tag{2}$$

where T is temperature.

These two equations teach us the followings:

(i) In order to obtain the high specific heat C_J per unit volume near Tc, we have to select the magnetic system which includes the large numbers of the large spins per unit volume from Eq.(1).

(ii) From Eq.(2) it is clear that the peak value of C_J near Tc is determined by the temperature dependence of S_J, $(\partial S_J/\partial T)$, which closely correlates to the sharpness of the phase transition. The magnetic order-disorder transition is governed by the magnetic exchange interactions in the system expressed as

$$H = -2 \sum_{i>j} \lambda(r_{ij}) J_i J_j, \tag{3}$$

where $\lambda(r_{ij})$ is the magnetic exchange parameter which depends on the distance of the interacting spins, J_i and J_j, and which also depends on the binding mechanism of magnetic ions and the configuration of the surrounding ions. To sharpen the transition, the magnetic system must be satisfied the following conditions; (a) a homogeneous material (b) a simple magnetic exchange mechanism, for instance, in the simplest case the magnetic exchange interactions between spins can be described by unique exchange parameter.

In consideration of the above argument, several groups have investigated the several kinds of heavy rare earth compounds and discovered some promising candidates as summarized in table II. Also, several examples of those compounds are shown in Fig.6 (10).

On the otherhand, the investigations on the new regenerator system using the magnetic regene-

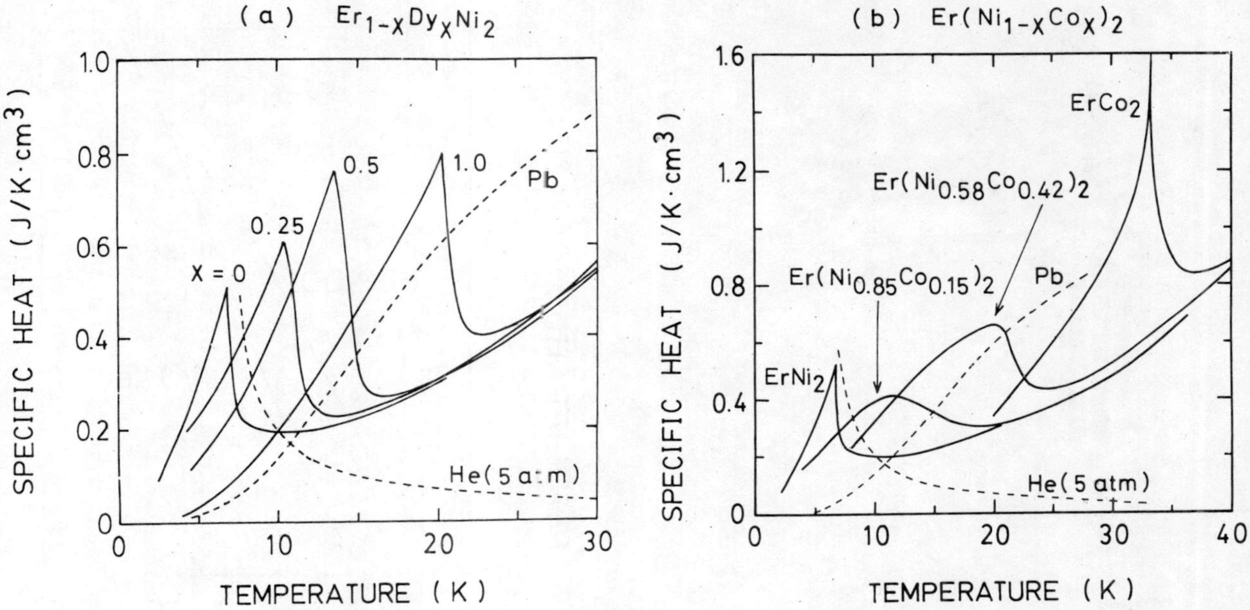

Fig.6 The temperature variations of the specific heats in several magnetic regenerator materials. (a) $Er_{1-x}Dy_xNi_2$, (b) $Er(Ni_{1-x}Co_x)_2$

rator materials have been begun from last year by two groups.

Their aims are slightly different each other: One is the development of high power refrigerator in ~5 K range, which is used for the cooler of the thermal shield in the large scale cryogenics apparatus, such as NMR-CT. The other is the development of the small power He liquefier.

Therefore, the former group (Tokyo Inst. Tech. -Toshiba) has investigated two stage regenerator, usable for the GM refrigerator, whose materials are lead and Er_3Ni as shown in the schematic figure (Fig.7). The picture of this regenerator are shown in Fig.8.

The typical experimental result are shown in Fig.9. In the Figure are shown the refrigeration powers of the new regenerator and lead regenerator plotted against the 2nd stage

Table II Several promising regenerator materials below ~15 K

Compounds	Saturation Moment $\mu_{eff}(\mu_B/ion)$	Transition Temperature (K)
$Er_{1-x}Dy_xNi_2$	Er:6.8, Dy:9.2	6.7 - 20.7 *
$Er(Ni_{1-x}Co_x)_2$	Er:8.9	6.7 - 32.9 *
$Gd_{0.5}Er_{0.5}Rh$	Er:9.4 Gd:7.9	~ 5
Er_3Ni	Er:7.1	7.1
Er_3Co	Er:6.9	12

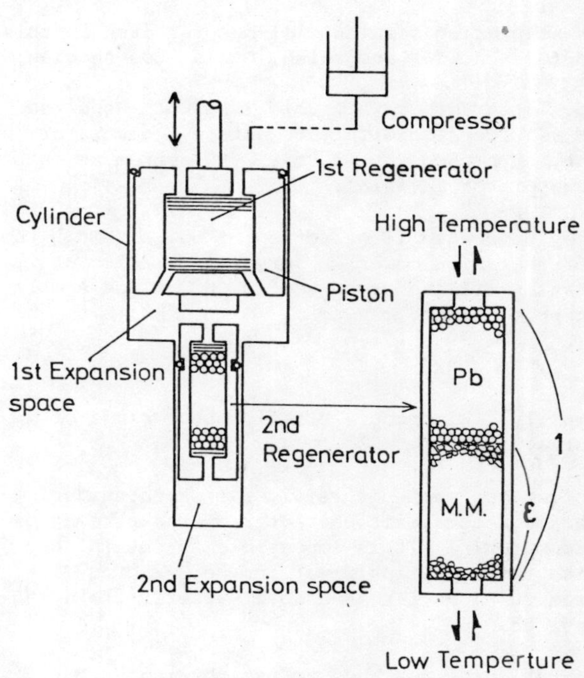

Fig.7 Schematic figure of two stage regenerator. (M.M. in the expanded figure of the 2nd regenerator stands for the magnetic material).

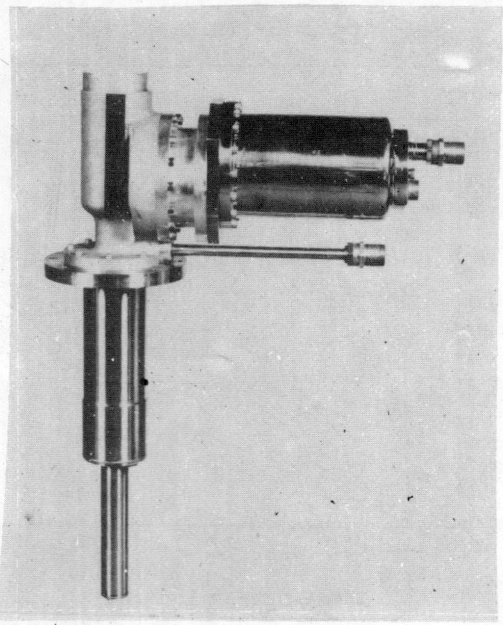

Fig.8 The picture of the two stage regenerator usable for the GM refrigerator.

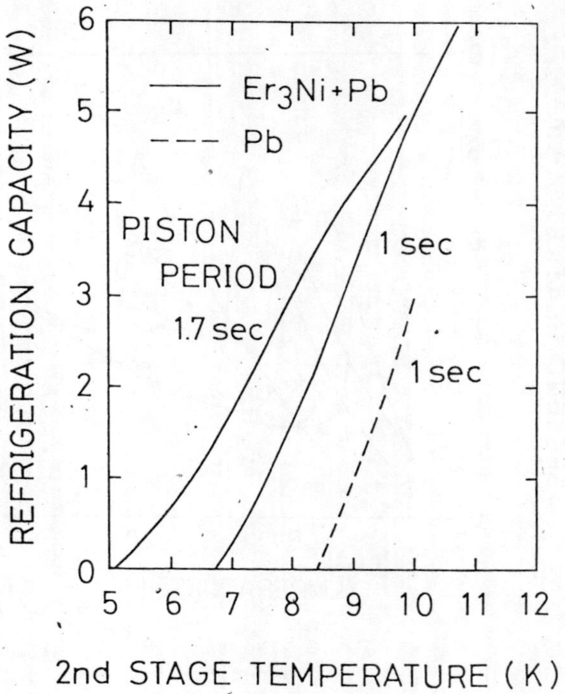

Fig.9 The refrigeration power of the GM refrigerator using the new magnetic regenerator. For comparison, the power of the common GM refrigerator (dotted line) are also shown.

(cooled-off) temperature. This result shows that this new regenerator has higher ability in the ∿5 K range compared with the common lead regenerator.

These precise results will be presented in this conference ICCR and, also, CEC at Los Angeles.

The latter group (Mitsubishi El.Co.) used the GM refrigerator with three stage regenerator. They have used the $Gd_xEr_{(1-x)}Rh$ system as the regenerator material.

They succeeded liquefaction of He, although the refrigeration power is very small. Also, this results will be presented in CEC at Los Angels in this summer.

4 OTHER WORKS

Finally, I introduce briefly other kinds of works.

As for the magnetic refrigeration, considering the magnetocaloric character of the paramagnetic material, it is impossible to obtain the high refrigeration power in the Carnot cycle from 20 K, as far as usable magnetic field is less than 6 T.

In this case, we have to use the magnetic Ericsson cycle instead of the Carnot cycle. The fundamental investigation of those has been performed by the group in Tokyo Inst. Tech (11).

Another hopeful application of the magnetic refrigeration is the cooler for infrared senser in space. Presently, it has been investigated by the group (Toshiba-Osaka Univ.) and it seems that the most important problem for this application is to find out the magnetic refrigerant usable in the temperature range from 4.2 K to 0.1 K.

As for a new application of the magnetic material, the active regenerator has been investigated, which may be thought the mixture of the magnetic refrigerator and the regenerator. A theoretical investigation about this problem will be also presented in the present conference ICCR.

As mentioned above, nowadays many kinds of application of the magnetic substances in the cryogenic engineering field have been actively developed by some groups in Japan.

REFERENCES

1 Hashimoto, T. et al. Cryogenics, vol.21 (1981), pp 647
2 Hashimoto, T. Adv. Cryog. Engnr., vol.32 (1986), pp 261
3 For instance, Hashimoto, T. et al. Proc. ICEC 9 Kobe, (1982), pp 26

Numazawa, T. et al. Proc. ICEC 9 Kobe, (1982), pp 30
4 Barclay, J.A. Cryogenics, vol.20 (1980), pp 467
5 Li, R. et al. Adv. Cryog. Engnr., vol.32 (1986), pp 287
6 Numazawa, T. et al. Adv. Cryog. Engnr., vol.31 (1986), pp 771
7 Nakagome, H. Teion-Kohgaku, vol.20 (1985), pp 281
8 Yazawa, T. et al. Proc. ICEC 11 Berlin, (1986), pp 275
9 Here, I will list the recent typical report in Japan.
Nakagome, H. et al. Adv. Cryog. Engnr., vol.29 (1984), pp 581
Numazawa, T. et al. Adv. Cryog. Engnr., vol.29 (1984), pp 589
Fujita, T. et al. Adv. Cryog. Engnr., vol.31 (1986), pp 763
Nakagome, H. et al. Proc. ICEC 11 Berlin, (1986), pp 246
Hakuraku Y. et al. Jpn. J. Appl. Phys., vol.25 (1986), pp 140
10 For instance,
Tomokiyo, A. et al. Adv. Cryog. Engnr., vol.32 (19860, 279
Li, R. Proc. ICEC 12 (1988)
11 Matsumoto, K. et al. Adv. Cryog. Engnr., vol.33 (1988), pp 743

THERMAL CONTROL OF CRYOPRESERVATION OF BIOMATERIALS *

T.C. HUA
Cryobiological Engineering Laboratory
Shanghai Institute of Mechanical Engineering
Shanghai, 200093, China

ABSTRACT

Thermal control of the freezing-thawing procedures is the key of cryopreservation. This paper presents the significance, the analyses and the methods of thermal control of cryopreservation, including, especially, the research in author's laboratory in resent years.

INTRODUCTION

Since Polge discovered accidentally in 1949 that glycerol protected the spermatozoa when it was frozen to -79°C (1), cryopreservation has had 40 years history. Thus far, the important cells and tissues which can be preserved at low temperature are erythrocytes, leucocytes, lymphocytes, platelets, marrow stem cells, hepatocytes, spermatozoa, embryos, islets of langerhans, cornea, skin, pancreas and some plant cells. The preservation of these biological materials has brought great practical benefit to medicine, agriculture, animal husbandry and to biological studies.

Nevertheless, the biomaterials which can be cryopreserved are limited. Even for those cells mentioned, the survival probability is not high in many cases. The main reason may be the lack of thermal control study of the freezing-thawing procedures.

As we know, the basic effects which influence the survival of cryopreservation are:

1, the cooling procedures and their cooling rates;
2, the warming procedures and their warming rates;
3, the cryoprotective agent (CPA), its concentration,and the addition and removal procedures.

Some common CPA are glycerol, dimethylsulfoxide (DMSO), polyvinylpyrollidone (PVP), hydroxyethyl starch (HES), and others. Although the mechanism of CPA in cryopreservation is still not very clear, the following conclusions are evident:

1, Different kinds of cells favor distinctive kind and concentration of CPA. For example, bull spermatozoa favor 4-9 % glycerol, and cow embryos favor 1.5 M DMSO (2).
2, There is an optimal cooling rate for each kind of cell in some solution. The optimal cooling rate for one type of cell may differ by several order of magnitude when compared with other type of cell. For example, the optimal cooling rates for red cells with 10 20 % glycerol and mouse embryos with 1.5 M DMSO are 100°C/min and 0.2-0.8°C/min, respectively (3,4).

In 1972, Mazur proposed the "two factor" hypothesis of freezing damage, assuming that there are two independent major damage modes, i.e. intracellular ice formation and "solute damage" (5).

INJURY AT SUPRA-OPTIMAL COOLING

Supra-optimal cooling is the process heat transfer dominated. Rapid freezing will create quickly high extracellular osmotic pressures. Due to the low permeability of the membrane and the short time elapsed during which the temperature drops, the intracellular water has no enough time to transport through the cell membrane, then becomes supercooled and forms ice.

The intracellular ice formation during freezing and its recrystallization during thawing have been thought as the mechanism of rapid cooling by most scientists. Our cryomicroscopic observations of human leucocytes (6), cow embryos (7), and fish embryos (8) also confirmed this point of view. It is found from our experimental study on gold fish embryos that the survival percentage is exactly equal to the unfrozen cell percentage. This shows that the intracellular ice formation is actually playing a direct role in producing freezing damage to embryos (8).

INJURY AT SUBOPTIMAL COOLING

Suboptimal cooling is the process mass

* Project Supported by National Science Foundation of China

transfer dominated. In slow freezing, intracellular water has enough time to transport through the cell membrane so that the intracellular and extracellular solutions are essentially in a condition of mutual equilibrium. Several results are induced simultaneously from the freezing of extracellular solution: the rise in solute concentration, the decrease in fraction of unfrozen water and unfrozen solution, the severe dehydration of cells.

The mechanism of suboptimal injury is still unclear. Meryman proposed a "minimum volume" hypothesis for slow freezing injury in1968, which suggested that when a critical minimum cell volume be reached by dehydration, an irreversible and lethal increase in membrane permeability occur for normally impermeable solutes (9). Mazur proposed a "solute damage" hypothesis in 1972, in which he thought that suboptimal cooling would imply the extended exposure time of cells to concentrated solutions at higher temperatures, and would result in the injury of cells (5).

Recently, however, Mazur and his colleagues have reported that the survival of human red blood cells in glycerol solution during slow freezing depend more on the decrease of unfrozen fraction of water than on the increase of solute concentration (10). To determine whether Mazur's hypothesis is applicable to other cells, we have studied the cases of leucocytes suspended in DMSO-NaCl-H_2O solution and Glycerol-NaCl-H_2O solution at cooling rate of 1°C/min on the cold stage of our cryomicroscope. It is found that the survival of leucocytes is very weakly affected by the high concentrations of salt produced by freezing, at least up to 3.0 Molal of NaCl, and is also not strongly affected by the temperature or by the CPA concentrations. On the other hand, the survival of leucocytes is almost totally dependent on the unfrozen fractions of water and solution when more than 80% of the water in solution is frozen (11), as shown in Fig. 1.

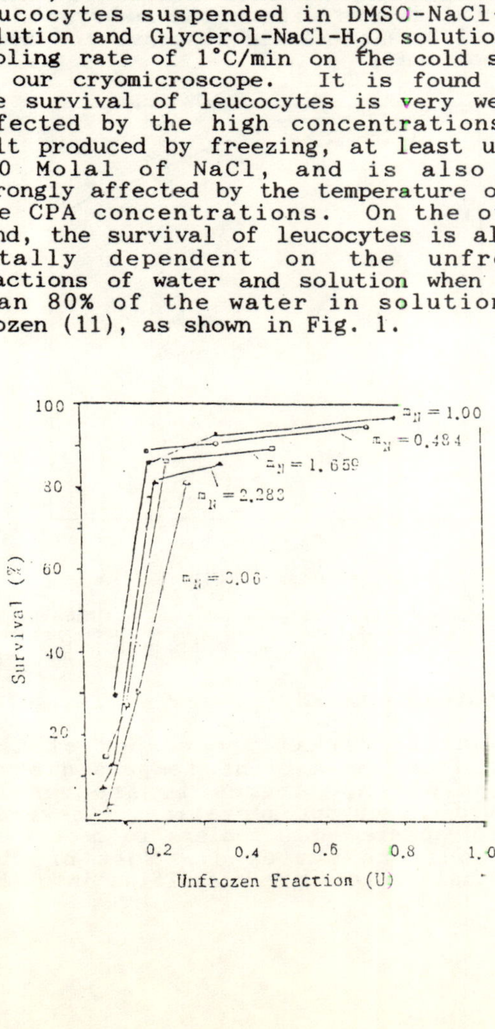

Fig. 1. Survival of leucocytes vs. unfrozen fraction of water (12) (Solution: DMSO-NaCl-H_2O, R=5. M_N: molality of NaCl)

PHASE DIAGRAM AND CALCULATION OF TERNARY SYSTEM

Cocks and others established the phase diagrams of the ternary systems of Glycerol-Sodium Chloride-water and DMSO-Sodium Chloride-Water (13), by using differential thermal analyzer (DTA) and differential scanning calorimetry (DSC).

During freezing, pure water is removed from the solution as ice, and this causes the residual unfrozen solution to concentration in the manner, in which R (the weight ratio of CPA to NaCl) remains unchanged, as shown in Fig. 2 by the curve T'T".

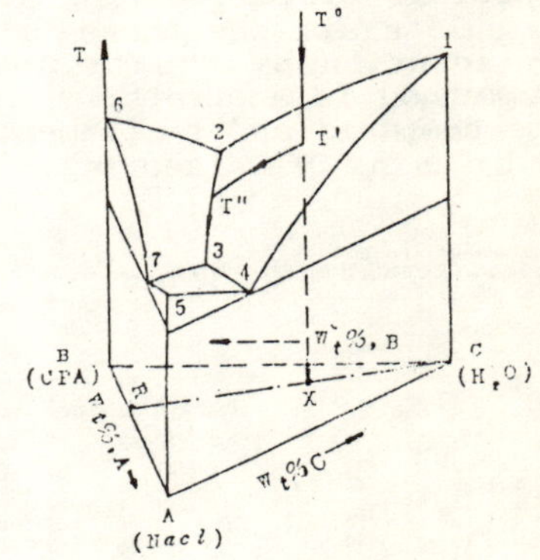

Fig. 2. Phase diagram of ternary system

Based on the experimental data from Cocks, some simple equations have been derived by Pegg recently for the calculation of the melting points of the ternary solutions.

For Glycerol-NaCl-H_2O system (14):

$$T=A(100W_T)+B(100W_T)^2 \quad \{1\}$$

$$A=1/(-1.6-1.27R-0.25R^2)$$

$$B=-0.010$$

For DMSO-NaCl-H_2O system (15):

$$T=A(100W_T)+B(100W_T)^2+C(100W_T)^3 \quad \{2\}$$

$$A=-0.6+0.17arctg(R)$$

$$B=(arctg(R/2))/132-0.001$$

$$C=-4.5*10^{-4}$$

Here, T - the melting point, °C;

W_T- total solute concentration, Wt %;

$$W_T=W_N+W_C=1-W_H; \quad \{3\}$$

$$R=W_C/W_N. \quad \{4\}$$

All the phase properties of ternary system can be calculated by phase diagram or by the equations {1} - {4}, if the temperature T and the initial composition $W^{\circ}_H, W^{\circ}_N, W^{\circ}_C$ are known. The composition of unfrozen fluid W^L_H, W^L_N, W^L_C at temperature T can be calculated by the following equations:

$$f_L=(1-W^{\circ}_H)/(1-W^L_H)=W^{\circ}_N/W^L_N$$
$$=W^{\circ}_C/W^L_C \qquad \{5\}$$

$$U=f_L*W^{\circ}_H/W^{\circ}_H=(1-W^{\circ}_H)/(1-W^L_H)$$
$$*W^L_H/W^{\circ}_H \qquad \{6\}$$

A new experimental method to get the phase properties of ternary system has been being developed in our laboratory (16). Using cryomicroscopy and image processing technique, this new method can get the liquid fraction f_L directly from the cryomicroscopical picture during freezing, and can calculate all the phase properties of ternary solution by equation {5} and {6}. This method is accurate, visual, and can also be compared and complemented with the method by using DTA. (see Tab. 1)

Tab 1. The values of fraction residual liquid f_L of the ternary systems $CPA/NaCl/H_2O$ by two different methods (16)

Initial Composition of solution			Temperature	f_L	
CPA	W°_{NaCl}	R	°C	DSC & DTA	Cryomicroscope
DMSO	0.45 %	20	-35	0.21	0.27
glycerol	0.9 %	20	-20	0.53	0.49
glycerol	0.9 %	10	-5	0.47	0.49
glycerol	0.9 %	10	-10	0.33	0.35
glycerol	0.9 %	10	-30	0.185	0.225
DMSO	1.8 %	10	-10	0.79	0.82
DMSO	1.8 %	10	-40	0.43	0.47

CRYOMICROSCOPIC OBSERVATION AND IMAGE ANALYSIS

The desire to view and record the significant events during freezing and thawing has led directly to attempts to create a variable low temperature environment for cell on a microscope, i.e., cryomicroscope system. The first satisfactory cryomicroscope was built by Diller and Cravalho in M.I.T. in 1976 (17). The first cryomicroscope in China, set up in our laboratory in 1985, is an inverted one and has new structure (18,19). Fig. 3. shows the total view of this cryomicroscope.

A primary design consideration for the low temperature stage in this system is satisfying the good thermal performance and optical performance. The temperature of the sample in the stage, controlled by a microcomputer system, could be varied from +40°C to -180°C. The cooling rate can be up to -50°C/min, and thawing rate to +100°C/min.

With this cryomicroscope system, researchers could observe and record the events of solution solidification and of cell's changes during freezing and thawing. This system could provide the

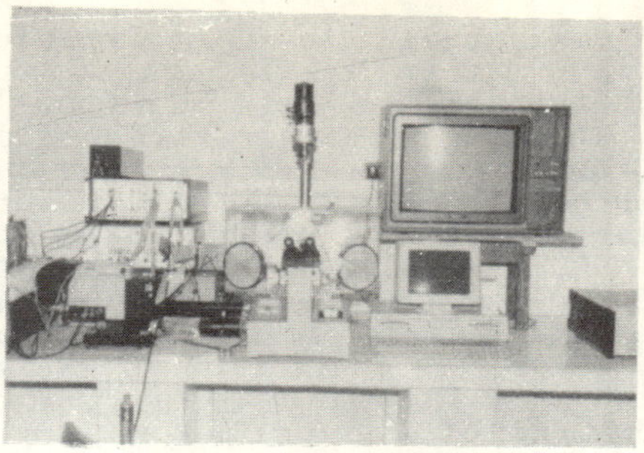

Fig. 3. Total view of the cryomicroscope in Shanghai Institute of Mechanical Engineering

phenomenological basics for interpreting experimental trials. It has been used by us in the cryopreservation studies of human semen (20), leucocytes (6,11,12), skin (21), fish embryos (8) and antibiotic producing strains (22).

The sequential photomicrographs obtained from the cryomicroscope system can also be transferred to the computer image processing system. Then, we could calculate quantitatively the volume change of the cell and get the gray-level histograms. In our study of fish embryos, for example, we have found from the behavior of their volumes varying that the membrane system of fish embryo has a very low permeability of water and CPA, and the different layers within the embryo have quite different osmotic response to them, as shown in Fig. 4. (8)

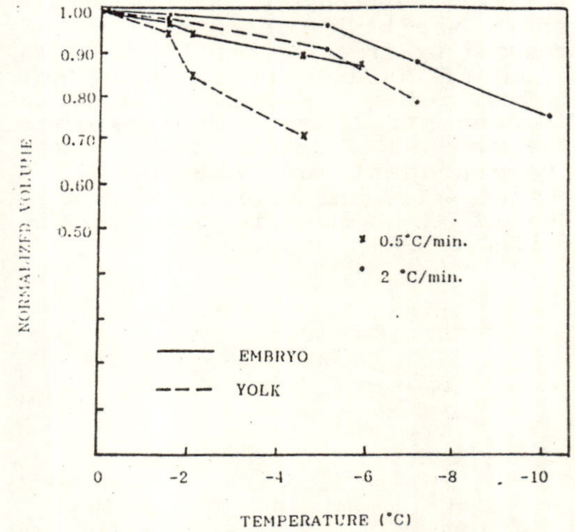

Fig.4 Normalized volume vs. temperature for gold fish embryo and yolk at heart stage during freezing at cooling rates of 0.5 °C/min. and 2 °C/min.,with 7.5% DMSO.

PROGRAMMABLE COOLERS

To control thermal procedures of the sample from the ambient temperature to cryogenic temperatures is frequently required in cryopreservation. Several kinds of programmable coolers to meet this need could be surveyed. Most of the commercial coolers are classified into the spurt type, in which liquid nitrogen is delivered into the inner chamber through a solenoid valve, and the temperature within the chamber is controlled by adjusting the spurt rate of the liquid nitrogen and the heat released from the electric heater (23). There are some shortcomings of the spurt type cooler. when it is used for very slow cooling procedure, large amount of liquid nitrogen would be consumed, and refilling the liquid nitrogen into the Dewer vessel would interrupt the cooling procedure.

A new cooler developed recently by our laboratory, as shown in Fig. 5, consists of the following components: 50 l liquid nitrogen Dewer vessel with 21 cm inner diameter neck tube and cylindrical attachment; a step-motor with 1.5 stepping angle and 1.2 Kg-cm static torque; a horizontal circular platform suspended within the neck tube; a control system with single component 8-bit microcomputer. Using this new cooler, the cooling (or warming) procedure can be realized in several (up to eight) subprograms with rates from 0.01°C/min to 20°C/min, and the liquid nitrogen consumption is very small, only the natural vaporization, so that this cooler can continuously operate for ten days without refilling.

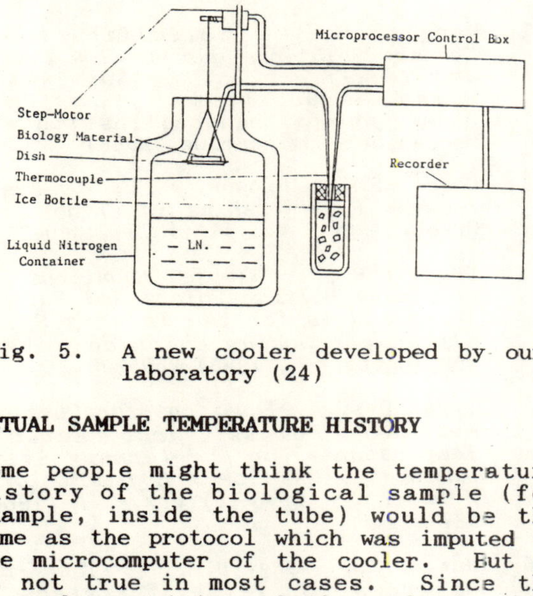

Fig. 5. A new cooler developed by our laboratory (24)

ACTUAL SAMPLE TEMPERATURE HISTORY

Some people might think the temperature history of the biological sample (for example, inside the tube) would be the same as the protocol which was imputed to the microcomputer of the cooler. But it is not true in most cases. Since the thermal capacities of the solution and tube are not small enough, the heat transfer coefficient outside the tube is not large enough, and especially since the solution would release latent heat during freezing, the sample temperature usually could not follow the protocol which is put in the programable cooler. In many cases, there is large deviation between them. For instance, from our experiment, the sample temperature is 90°C higher than that in the chamber of cooler, and the cooling rate of sample after solidification is much higher than that desired. (26)

A technical difficulty encountered in the freezing of biomaterials having large volumes is how to keep the optimal cooling rate through the sample during solidification. Usually, a predetermined constant cooling rate is imposed on the outer surface of the sample by using a programable cooler. But, in the frozen medium, particularly on the phase change interface of solidification, the cooling rate deviates far away from the optimal one. A new cooling protocol is suggested that the optimal cooling rate be given at the phase change interface rather than at the outer surface. The above problem is concerned with the so-called inverse Stefan problem, wherein the moving of the phase change interface is given and the thermal conditions of the outer surface are to be predicted.

The heat-balance integral method has been developed to solve the inverse Stefan problem just mentioned (25).

Assuming that the phase change interface movement with time follows a power law, namely

$$S(t')=1-(\beta t')^{1/n} \qquad \{7\}$$

where $t'=a*t/r_o^2$, is dimensionless time;

$S(t')=s(t)/r_o$, is the nondimensional position of the phase change interface;

The dimensionless temperature at the outer surface can be controlled as below

$$Ur'_o=[T(r'_o,t')-T_{ph}]/[T_o-T_{ph}]$$
$$=-4H/(1-S)-(1-S)S^i(dS/dt')/Ste$$
$$+(1-S)2S^i(dS/dt')^2/(6Ste) \qquad \{8\}$$

where, T_o -reference temperature external;

T_{ph}-temperature of solidification;

$Ste=C_p(T_o-T_{ph})/L$;

$U = (T-T_{ph})/(T_o-T_{ph})$;

$r'= r/r_o$; $h=r'^i*U$;

$i = 0$, or 1, for slab or sphere, respectively.

$H = \int_1^S hdr'$, is the solution of the following ordinary differential equation

$$dH/dS=12H/[(1-S)^2(dS/dt')]+6S^i/Ste$$
$$-(1-S)S^i(dS/dt') \qquad \{9\}$$

A NEW COOLING TECHNIQUE

For many cases of cryopreservation, a uniform cooling rate is needed for large temperature range. If a plastic bag containing biomaterial and solution is put into a freezer with a constant temperature (e.g. -60°C), the temperature of biomaterial would decrease with time on negative exponential manner, not on linear manner. To solve this problem, we have developed a new technique (26), putting the plastic bag into a gauze envelope which is soaked with some intermediate freezing medium having much lower freezing point (T_{f2}) than that (T_{f1}) within the plastic bag. When the temperature of intermediate medium, T_2, is higher than T_{f2}, the medium is still in liquid state and has large heat capacity, so that the cooling rate of the biomaterial in the bag at higher temperatures would be slower, and the freezing plateau would be shortened than usual. But when the intermediate solution is frozen and has smaller heat capacity than at liquid state, the cooling rate of biomaterial at lower temperatures would be faster than usual.

When the plastic bag with a gauze envelope is put into a freezer of -60°C, the nearly uniform cooling rate of 1.5°C/min in biomaterial is achieved from 0°C to -40°C if 23.1 % brine is used as the intermediate medium; whereas the constant cooling rate of 1°C/min is achieved from 0°C to -50°C if 30% alcohol-H_2O is used as the intermediate medium. This new cooling technique can also decrease the post freezing plateau which is beneficial to preventing cell injury. This new technique has been used successfully in cryopreservation of human semen and skin in our laboratory (20,21).

REFERENCES

1 Polge, C., Smith, A.S. *Revival of Spermatozoa after Vitrification and Dehydration at Low Temperature* Nature, 164, (1949), 666.

2 Polge, C., *Freezing of Spermatozoa* in Low Temperature Preservation in Medicine and Biology ed. by M.J. Ashwood-Smith, J. Farrant, University Park Press, Baltimore, 45-64, 1980

3 Rowe, A.W., Lenny, L.L., and Mannoni, P. *Cryopreservation of Red Cells and Platelets* in Low Temperature Preservation in Medicine and Biology ed. by M.J. Ashwood-Smith and J. Farrant , University Park Press, Baltimore, 285-310, 1980

4 Wittingham, D.C., Leibo, S.P., and Mazur, P. *Survival of Mouse Embryos Frozen to -196°C* Science, 178, (1972), 411-414.

5 Mazur, P., Leibo, S.P., and Chu, F.H. Y. *A Two-Factor Hypothesis of Freezing Injury-Evidence from Chinese Hamster Tissue Culture Cells* Exp. Cell Res. 71, (1972), 345-355.

6 Chen, X.H., Chen, E.T., Hua, T.C., and others *Cryomicroscopic Observation and Basic Investigation on the Effect of Cryopreservation of Human leucocytes* Chinese Journal of Cryogenic Engineering (1987), 4, 1-8

7 Sun, Z.Q., Hua, T.C., Cravalho, E.G. *Cryomicrosopic Study on Cryopreservation of Bovine Embryos* Chinese Journal of Refrigeration (1987), No.4. 18-25.

8 Chen, X.H., Hua, T.C., and others *Experimental Study of Freezing Damage Factors and Feature Analysis of Microscopical Images of Fish Embryos* 25th Annual Meeting, Society for Cryobiology, July 11-15, 1987 Aachen, West Germany.

9 Meryman, H.T. *Modified Model for the Mechanism of Freezing Injury in Erythrocytes*, Nature, 218, (1968), 333-336.

10 Mazur, P., and others, *Relative Contribution of the Freezing of Unfrozen Water and of Salt Concentration to the Survival of Slow Frozen Human Erythrocytes* Biophys. J. 36, (1981), 653-675.

11 Wang, D.R., Hua, T.C., Wu, S.Y., *Influence of Unfrozen Fraction on the Survival of Leucocytes during Slow Freezing* 25th Annual Meeting, Society for Cryobiology, July 11-15, 1988, Aachen, West Germany.

12 Wang, D.R., *Solidification of Ternary Solutions during Slow Cooling and Its Effect on the Survival of Human Leucocytes* (Ph.D. Dissertation, Shanghai Institute of Mechanical Engineering, 1988)

13 Cocks, F.H., Brower, W.E., *Phase Diagram Relationships in Cryobiology*, Cryobiology, vol.11 (1974), 340-358

14 Pegg, D.E., *Simple Equations for Obtaining Melting Points and Eutectic Temperatures for the Ternary System Glycerol / Sodium Chloride / Water* Cryo-Letters, (1983) 259-268

15 Pegg, D.E., *Equations for Obtaining Melting Points and Eutectic Temperatures for the Ternary System Dimethyl Sulphoxide / Sodium Chloride / Water* Cryo-Letters, 7 (1986), 387-394.

16 Hua, T.C., Wang, D.R., *The Phase Diagrams of Ternary Systems in Cryobiology* Submitted to Chinese Journal of Refrigeration.

17 Diller, K.R., *Quantitative Low Temperature Optical Microscopy of Biological Systems* J. Microsc., 126 (1982), 9-28.

18 Wu, S.Y., Yao, K.M., Hua, T.C., *The New Structure of Cryomicroscope* Cryobiology, 6, (1987), 553-554.

19 Wu, S.Y., Yao, K.M., Hua, T.C., *An Inverted Cryomicroscope System* Chinese Journal of Scientific Instrument, vol.1, (1988), 90-93.

20 Chen, E.T., Hua, T.C., and others, *Cryopreservation for Human Semen* Chinese Journal of Biomedical Engineering vol.7, (1988), No.1, 20-26.

21 Hua, T.C., Chang, Z.H., and others, *Clinical Skin Banking methodology by Cryopreservation* Submitted to Chinese Journal of Biomedical Engineering

22 Zhao, L., Zhu, H.Y., Hua, T.C., *Low Temperature Preservation of Producing Strains of Streptomycin and Gentamycin* Cryo-Letters, 9, (1988) 144-151

23 Hua, T.C. *Heat Transfer Problems in the Freezing Process of Biomaterials Cryopreservation* Heat Transfer Science and Technology, (1987), 57-64, ed. B.X. Wang, Hemisphere Publishing Corp.

24 Chen, E.T., Zhao, L., Hua, T.C. *A New Cooler with Very Low Consumption of Liquid Nitrogen* 25th Annual Meeting, Society for Cryobiology, July 11-15, 1988, Aachen, West Germany

25 Ren, H.S., Hua, T.C., Sun, W.D. *New Solution of Inverse Problem for Freezing and Thawing* 25th Annual Meeting, Society for Cryobiology, July 11-15, 1988, Aachen, West Germany

26 Wu, S.Y., Hua, T.C., Ren, H.S., Gen, Y. *A New Cooling Technique in Cryopreservation of biomaterials* Chinese J. of Refrigeration (1987), No.4, 11-17

THE PROSPECTS FOR DEVELOPMENT AND UTILIZATION OF LIQUID NITROGEN IN CHINA

Hu Deying
Engineering institute of the
Ministry of Textile Industry
Fuwai Ganjiakou, Beijing China

ABSTRACT

This paper provides statistical and economic data, in support of the importance of the development of multipurpose utilization for products in the air separation industry it presents the current situation of development and utilization, and the prospects for further development of liquid nitrogen. Taking an air separation plant as an example, this paper workout in detail the method of cost accounting for liquid nitrogen.

Up to 1984, 4726 sets of different kinds of air separation plants had been set up in our country. Their total capacity was 13.2 billion Nm^3 per year. There are 282 sets with the capacity over 1000 Nm^3/hr each, occupying 84.6 percent of the total capacity.(1) In the process of cryogenic air separation, the production ratio of oxygen and nitrogen is normally 1:1.1. In 1984, 12 billion Nm^3 of nitrogen was produced in China. It is very important to the economy that such a huge amount of nitrogen should be recovered. The use of nitrogen is so wide spread that it is said to be a "panacea". It is widely used in food preservation, the materials, industry machinery, electronic and petrochemical and engineering construction. The nitrogen output in U.S.A in 1983 was 16.46 billion Nm^3, 60 percent more than the oxygen output. The nitrogen consumption will also increase by 11 percent every year in the coming 10 years.(2) Most of the nitrogen consumption is used for farm production preservation.

At present, most of the huge amount of nitrogen produced in China is not made use of, so that the loss to the economy is astonishing. According to the technical condition of air separation plants in China, the average energy consumption is 0.5 Kwh/Nm(O_2). If the oxygen and nitrogen shared the energy consumption, that is 0.25 Kwh/Nm^3 (O_2+N_2). As the nitrogen from the air separation plant is not used, 3 billion Kwh of electrical energy is lost, which is equivallent to standard coal of 1.2 million tonnes. If all the nitrogen is recovered and liquified, the loss is 7.4 billion yuan RMB at the price of 0.50 yuan/kg. For example, nitrogen is an ideal medium for food preservation. At present, there is not an effective means for keeping farm produce fresh in storage and transportation in China, leading to widespread spoiling of farm produce. According to statistics, the loss of farm produce in storage and transportation in 1984 was 2.35 billion yuan RMB. Our current situation is that the nitrogen is not used in air separation plants. On the other hand, it is very difficult for many departments to get liquid nitrogen, causing enormous losses to the national economy. The reasons are as follows. First, the price of liquid nitrogen is too high, which limits the application of liquid nitrogen. Second is the unreasonable system of enterprise management. Many air separation plants do not have their own power to make business decisions and "the communal rice bowl" blunts the activity of the enterprise management.

Now, a real example to explain cost accounting of liquid nitrogen in detail and so show why the current price of liquid nitrogen is unreasonably high compared with its real cost. In 1982, two nitrogen production plants with the capacity of 2000 Nm^3/hr each and a nitrogen liquification plant with the capacity of 1000 Nm^3/hr were set up in a certain company. After two year's construction, the three plants were operating. Their main technical design indices were reached and the index of the final accounts of the capital construction of the three plants met the budget of the project, so the data needed for the cost accounting of nitrogen gas and liquid nitrogen are available

According to "Fixed Assets of State-owned Enterprise Depreciation try out Regulations" issued by the statecouncil (1985) No.63 "About the Depreciation specific Regulation of the Units and Buildings" and referring to the method of cost accounting of nitrogen in the power plant of Liao Yang Oil-Chemical Company, based on the data above, the cost of nitrogen gas and liquid nitrogen of the company has been calculated, and the results of the cost accounting in detail are as follows:

I THE ORIGINAL DATA

1. The original value of the fixed assets is 85 percent of the total investment, including equipment and erection costs. The scrap value is 5 percent of the original value of the fixed assets.

2. The period of depreciation of the air separation plant and the nitrogen liquification plant is 15 years. While one of the two air separation plants is in operation, the other one stands by. The production cost is calculated as half of the depreciation. The heavy repair fee is 50 percent of the depreciation. The depreciation period of the building is 50 years according to the regulation mentioned above.

3. According to calculation of total investment of each plant, the apportion of the company is 1.5 million yuan RMB.

4. The price of electricity and water

price of electricity ....
0.08 yuan/kwh
price of production water ....
0.08 yuan/t
price of circulation colling water.......
0.025 yuan/t

5. The apportion ment of the system lines of the overall plan of the power factory is 0.7 million yuan RMB.

6. The operation time of the plant is 8000 hours per year.

7. The number of the management personnel
air separation plant 43
nitrogen liquification plant 12

8. The shop management fee is 7 percent of the total fee per year.

9. The wage per person per year is 1000 yuan.

10. The cost accounting of liquid nitrogen is calculated according to the two situations as following:

(1) The cost of nitrogen gas is not calculated.

(2) 50 percent of the cost of nitrogen gas is taken into the calculation.

11. The consumption of water and electricity in the air separation plant is as follows:

the compressor for raw material air
DH-35 N=700kw

the compressor for nitrogen.......
ZD-7/4.5-15 N=140kw

water pump and oil pump ....
N=30kw

the consumption of water ....
W=25t/hr

circulation cooling water ....
W=180t/hr

12. The consumption of electricity and water in the nitrogen liquification plant.

the nitrogen circulation compressor-
2MCL406+2MCL356 ..... N=1270kw

main oil pump .... N=7.5kw

refrigerator .... N=40kw

circulation cooling water
.... W=200t/hr

The original value and the scrap value is calculated according to the following equations:

X = 0.85A ....(1)

Z = X.(1-y) ....(2)

A ...... the total investment

X ...... the original value of fixed assets

Y ...... the scrap value of original value fixed assets.

Z ...... the depreciation value of fixed assets

Every item is calculated from the original data above. (3)

II. THE RESULTS OF THE COST ACCOUNTING (according to data in 1982) ARE AS FOLLOWS:

fee classification	air separator	nitrogen liquified plant	Unit
unit+erection fee	7063960	4218000	¥
building fee	1501010	578653	"
apporition of co.	1200000	300000	"
total investment	9764970	5096653	"

original value	8300225	4332155	¥
depreciation value	7885213	4115547	"
depreciation {unit	222438	24319	"
depreciation {building	24241	9345	"
repair fee	123339	126281	"
wage	43000	12000	"
unit operation fee	608800	878720	"
production fee per year	1021817	1269564	"
shop management fee	71527	88869	"
total production fee of nitrogen gas per year	1093344		"
total output of nitogen per year	16000000		M^3
the cost of nitrogen gas	0.068		¥/m3
total produc. fee of LN_2		1358434	¥
total output of LN_2		10000000	kg
the cost of LN_2 per kg		0.135	¥/kg

note: when the cost of nitrogen is not counted, the cost of liquid nitrogen is 0.135 ¥/kg. when the cost of nitrogen gas is counted as 50%, the cost of liquid nitrogen is 0.16 yuan/kg.

The result shows that the cost of liquid nitrogen is 0.14 yuan/kg. At first sight, the cost of liquid nitrogen from the accounting seems too low, but through further analysis, it can be found that the cost of liquid nitrogen is still too high.

The main reason is as follows:

(1) The power factory was built in several stages. In the first stage the project and the auxiliary facilities were built in an open field and the company's apportion ment occupied too high a proportion of the total investment of the power plant.

(2) The capacity of the nitrogen liquification plant is on the small side, and its efficiency is low. According to the data from Linde Company,(4) if the capacity of liquid nitrogen plant increases to above 4000 Nm^3/hr, the cost will be reduced by 25 percent at least compared with the cost of liquid nitrogen in this company.

Wuhan steel works is an old enterprise with a good utility condition. Since the investment on nitrogen liquification plant is low, the cost of liquid nitrogen could be much lower.

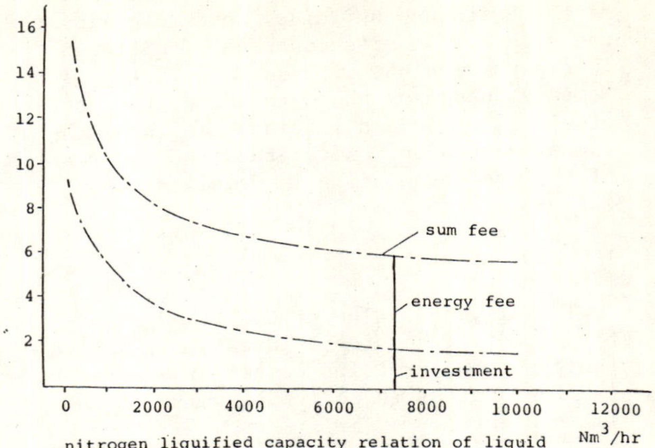

nitrogen liquified capacity relation of liquid nitrogen fee to the capacity of unit

A lot of liquid nitrogen with low cost produced in the nitrogen liquification plant of this company has paved the way for the development of the air separation industry of China. It has particular significance for capital construction and has interested specialists in liquid nitrogen technology.

The price of liquid nitrogen in the present market is 1.5--4.0 yuan/kg. It is really unreasonable. The facts show that the current price of liquid nitrogen should be greatly reduced. To meet the social needs, with liquid nitrogen, every department in the national economy could make products which were difficult to be made in the past. An entrepreneur with foresight and sagacity knows well the close relation between the price of the products and the development of them. The main reason that the application techniques of liquid nitrogen develop slowly is that the price of liquid nitrogen has been too high for many years. To reduce the unreasonable price of liquid nitrogen is an urgent priority.

The cost of liquid nitrogen is very low, but because of the management system of enterprises, as mentioned above, "the communal rice bowl", the low cost of liquid nitrogen cannot be sustained . Many of the enterprises know that the cost of liquid nitrogen is very low, but when the liquid nitrogen is in excess they would rather let the nitrogen gas escape to the atmosphere than sell it at a lower price, because working more does not mean receiving more. Therefore, to realize both low cost and low price, the problem of the management system in air separation plants must be solved first.

The large air separation plants concentrate on the metallurgical and petro-chemical industries, which are state-owned enterprises with the common malady , the so-called "communal rice bowl". The activity of enterprises is restricted. Quantities of energy and natural resources have been thrown into the eastward following stream, and it has been a common sight for many years.

The air separation plants in the metallurgical industry need to make a break-through in the reform of management, to obtain more rights of management decision, including independent economical accounting, and to be strict and fair in meeting out rewards and punishments. If most nitrogen were recovered and reused in the air separation plants in the metallurgical industry, the price of liquid nitrogen would be different from the present price, changing from some yuans per kilogram to some jiaos per kilogram. This will bring about a profit of as much as some billions of yuan per year.

In order to realize both low cost and low price of liquid nitrogen, it seems that the development of liquid nitrogen in tax revenue and interest rates at the bank need reforming besides the reform of the management system. In the present market of liquid nitrogen, tax is not stipulated in detail. It is considered that for the cost of liquid nitrogen to be low, the endurable limit for the food industry is also low. The suggestion is that if the price of liquid nitrogen is lower than a certain price It should be tax free, but if the price exceeds this value, then the regulation tax should be levied.

Financed by Hua-Neny Company as a shareholder, managed by the Nanjing authority with capital as a shareholder, and joined by Yanzi Company with the production of liquid nitrogen, Our corporation has been set up to make, use and develop liquid nitrogen in recent years. Now the corporation is preparing for wholesale food transport and processing by liquid nitrogen freezing.

Geographically, the Nanjing zone is favourable to this, with its variety of products, convenient communication and transportation and good industrial conditions. There are especially some people enthusiastic about the application and development of liquid nitrogen there. Also, people of the same trades or occupation of air separation industry in other cities such as the Tianjing zone, Beijing zone, Shanghai zone, are working hard for the development and the wider application of liquid nitrogen. Because of the development of the commodity economy, the situation of the price competition of liquid nitrogen is improving now in the Nanjing and Shanghai zones, and at present, the price of liquid nitrogen has a downward trend

We can foresee that in the following years level of skill of management will rise greatly, and the price of liquid nitrogen will go down sharply.

It becomes possible for liquid nitrogen to be used in the further processing of products in many trades and professions, which were hard to be processed in the past.

With liquid nitrogen used for the preservation of farm produce, the loss by decomposition of farm produce can be greatly reduced, and the supply to the market will be much improved.

To sum up, the price problem and the problem of enterprise management in the development of using liquid nitrogen will be solved in coming years. The demand of enterprise sectors for liquid introgen will be met before long. The prospect of application and development of liquid nitrogen in China is bright. In the air separation plants, nitrogen, that has been discarded for many years will play a more and more important role in the construction of the material civilization in China.

REFERENCE

1. "The energy saving and reform in oxygen and nitrogen production, central supply of oxygen and nitrogen". Tang Changping. Cryogenic technology 1988, No.3.
2. Cryogenic technology 1984, No.1
3. Cryogenic technology 1987, No.3
4. "The liquification of oxygen, nitrogen and Argon" Linde reports on science and technology Oiplhys, Helmut Sprigman Gerethsried 28/1978.

THE CRYO-PUMPING SYSTEM OF THE HEAVY ION RESEARCH FACULITY (LANZHOU) CYCLOTRON

Zhang Shuxiu
Institute of Modern Physics, Academia Sinica
Lanzhou, China

ABSTRACT

The Heavy Ion Research Faculity(Lanzhou)(HIRFL) cyclotron is a double cyclotron for accelerating heavy ions. The cryo-pumping has been chosen to offer a powerful tool for the HIRFL cyclotron. This paper mainly describes the vacuum system of the new cyclotron.It requires a ultra-high vacuum system to be provided with a high pumping speed and large pumping capacity. A total effective pumping speed of 140 ^{3}m /s produced by eight 800 mm diameter cryo-pumps is needed to obtain a pressure in the 10^{-6} Pa region in a 100m^3 vacuum chember.Liquid nitrogen required for the shielding of the cryopumps is supplied by a semi-open circuit. Under the steady state,liquid nitrogen from a 5m^3 store tank is transferred to each point through 90m transfer lines.A KLN 20Y Philips gas refrigerator is installed to re-process liquid nitrogen from the exhausted cold gas.Some smaller cryo-pumps are used for the two-beam lines. A new plan for making the HIRFL 500 and 800cryo-pumps at the laboratory is discussed.

1 INTRODUCTION

Heavy Ion Research Facility Lanzhou is a double cyclotron for accelérating heavy ions. It consists of two cyclotrons, a spiral sector focusing cyclotron (SFC) converted from an existing 1.5 m classical cyclotron as an injector, and a new separated sector cyclotron (SSC) with an energy constant of 450. With this combination, lighter ions such as carbon can be accelerated to an energy of 100 Mev/A and heavier ion such as tantalum to 5 Mev/A with the beam currents in the range of 10^9 to 10^{12} particles per second.

The vacuum system of the HIRFL cyclotron is divided into four sub-systems, namely the SSC, the SFC and two beam lines. For the SSC, because of a large volume, quite a number of accelerating elements involved almost impossible baking conditions and the intricate shape of the chamber to be evacuated, it requires an ultra-high vacuum system with a high pumping speed and a large pumping capacity. After assessing each pump, it was evident that only cryo-pumping would be able to meet the needs of the design.

Originally, a Claude cycle refrigeration system with a cooling capacity of 200W at 20K was being considered for building, but it was rather complicated for us to construct and operate.Later we had available some commercial standard cryo-pumps made up as miniature refrigerators so as to ensure a simple design, economy of operation and reliability in service. However, a standard product does not completely accommodate itself to the cyclotron. So many modification have have had to made improve the cryo-pumps.

2. VACUUM SYSTEM OF THE HIRFL CYCLOTRON

The four sub-vacuum systems of the HIRFL cyclotron connected to each other demand slightly different continuous working pressure but quite different pumping speed. Figure 1 shows the schematic diagram of the cryo-pumping system of the cyclotron.

The vacuum chamber of the SSC is a monolithic stainless steel vessel with a volume of 1 00m^3 and a weight of 65t. more than 300 t accelerating elements are contained in it, and the total surface area exposed to the vacuum amounts to about 1100m^2 . The estimated gas load of $1.8\cdot10^{-3}$ Pam^3/s after pumping 10 hours needs a total effective pumping speed of 140m^3/s to produce an ultimate vacuum in the 10^{-6} Pa region, and to maintain a working pressure of $1.3\cdot10^{-5}$ Pa in the vacuum chamber with the beam. Eight 800 mm diameter cryo-pumps modified from a Balzers RKP 800 commercial standard product have been chosen as the main pumping equipment for the SSC. All cryo-pumps which can withstand an air inrush accident are directly assembled on the vacuum chamber so that the expensive large valves are omitted.

The auxiliary pumping system for the regeneration of the cryo-pumps and leak testing is composed of four Pfeiffer TPH 5000 turbomolecular pumps with a pumping speed of

4.3m ^{3}/s each for nitrogen. The rough pumping system includes two ZJZ 600 Roots pumping units arranged in parallel. During one hour's pumping, a pressure in the vacuum chamber can be reduced to 10 Pa from atmosphere, after which the turbo-molecular pump can be brought into operation.

The vacuum system with a volume of 15m^3 of the SFC consist at present of the original oil diffusion pumps which are wide to maintain a pressure in the 10^{-4} Pa region. Two HIRFL 500 cryo-pumps are to be built for this use.

The smaller cryo-pumps without liquid nitrogen are needed on the two beam lines for dealing with the somewhat higher gas loads. The first beam line, 10cm diameter and 60m long, uses four RKP 161, two RKP 251 cryo-pumps and ten FB 450 turbo-molecular pumps. The second beam line, 7cm diameter and 100m long, uses six RKP 161 cryo-pumps and twenty L300 sputter ion pumps.

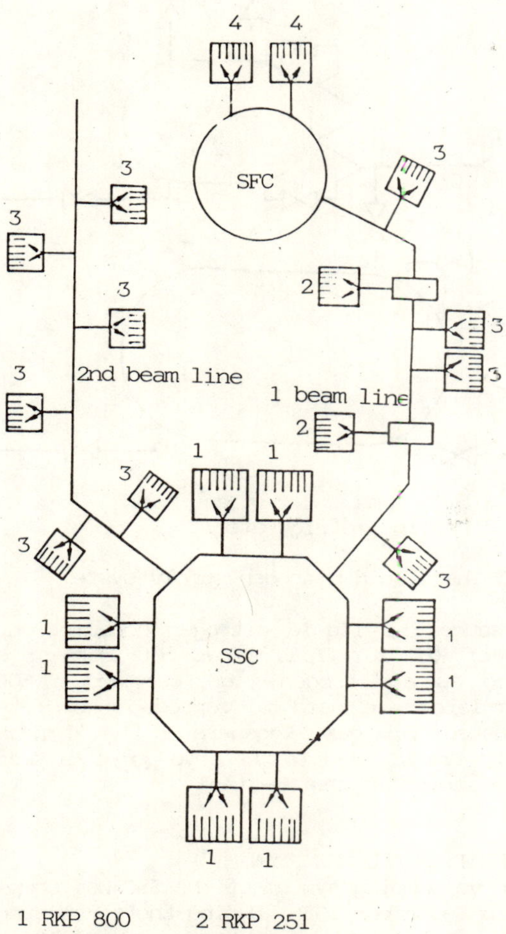

Fig.1 Cryo-pumping system of the HIRFL cyclotron

3 MODIFIED RKP 800 CRYO-PUMP

The type of RKP 800 cryo-pump manufactured by Balzers Ltd. combines an Air Products & Chemical DE 208L expander, a Solvay cycle, with a Balzer UC 010 380-T helium compressor that gives a cooling capacity of 10w at 20k. A pumping speed of 14m^3/s in the molecular flow range for nitrogen for a standard RKP 800 is rather lower than we expected. Fortunately, the cold panel and the baffle could be easily changed and improved in order to get better properties of the cryo-pump. Experiments and calculations have been made with the help of a Monte Carlo simulation computer program. The outer part of the chevron baffle of the standard RKP 800 cryo-pump has been removed to increase the transmission probability. The 600mm diameter cold panel is composed of a plate and an inverted cone on which is glued about 400 grams of activated charcoal grains which are provide enough pumping capacity for an extended time of running of the cyclotron. Because of the limitation in the cooling capacity of the refrigerator, the first stage of the RKP 800 cryo-pump is cooled by liquid nitrogen, which is helpful and stable for the operation of the second stage.

An arrangement of performance test for the RKP 800 cryo-pump is shown in Figure 2. The test dome is made according to the molecular pump standard DIN 28428. As a result, the performance characteristics of the modified RKP 800 cryo-pump are summarized as follows:

Ultimate vacuum	Pa	$5\cdot10^{-8}$
Pumping speed		
nitrogen with the baffle	m^3/s	21
without the baffle	"	38
hydrogen	"	20
water vapour	"	65
Liquid nitrogen consumption		
continuous operation	1/h	4

The modified baffle directly covers the 20k cold panel and has little effect on the pumping speed of the hydrogen. On the test dome, the baffle can be removed while there is a temperature of 16k on the cold panel. But all baffles have remain on the vacuum chamber of the SSC, because of the large radiating surface emmissivity for the unpolished inner surface of the vacuum chamber, in which the cold panel without the baffle can not reach a temperature lower than 20 k.

All compressors are installed outside the cyclotron hall for the sake of radioactive safety. A compressor is connected to a refrigerator in the cyclotron hall through a pair of 60m long 1" stainless steel tubes. It is filled with 99.995% pure helium, The pressure drop was measured between two ends of a tube was found to be negligible.

Two platinum resistance sensors and a electronic unit are used for maintenance of a constant liquid nitrogen level in the container of the cryo-pump with the help of a solenoid

valve on the inlet.
Liquid nitrogen consumption was measured to be 4 1/h in the early continuous operation by means of a weight change.

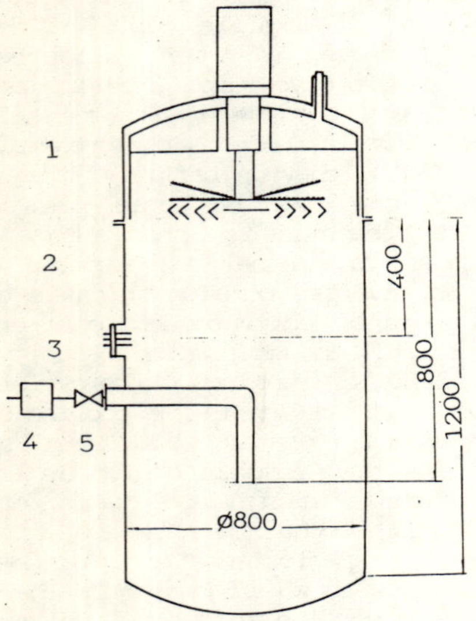

1 cryo-pump	2 test dome
3 vacuum gauge	3 gas flow meter
5 gas inlet valve

Fig.2 Arrangement of performance of the cryo-pump

4 LIQUID NITROGEN SUPPLY

A simple semi-open circuit of liquid nitrogen was built in the cyclotron building for all cryo-pumps. As Figure 3 shows it consists of a $5m^3$ store tank and 90m transfer lines. The main transfer line, made up 20·1.5mm inner copper tube and 100·5mm outer stainless steel tube, is a vacuum jacket with multilayer insulation. The Teflon spacers provide enough flexibility to accommodate thermal expansion and permit moderate bends for the inner tubes. The loose flanges on the outer tube enable the line to assemble and disassemble easily. A semi-static vacuum pumping system including two smaller oil diffusion pumps can be operated in case vacuum is lost in the jacket. The heat load of the line was measured to be 3.3 w/m in a model test in the laboratory. To simplify the construction, the return tubes from the cryo-pumps are made of polyurethane foam insulation. The ratio of the diameters of the inner and the outer tubes is 1.5.
Liquid nitrogen is produced by an industrial air separation equipment 11-800 with an output of about 100 1/h, and transported to the $5m^3$ tank by a $1m^3$ trolley tank. Liquid nitrogen makes use of static pressure vapourized by an ambient temperature exchanger to be transferred to each point under steady state. The capability of supplying liquid nitrogen designed for the vacuum system is about 50 1/h.

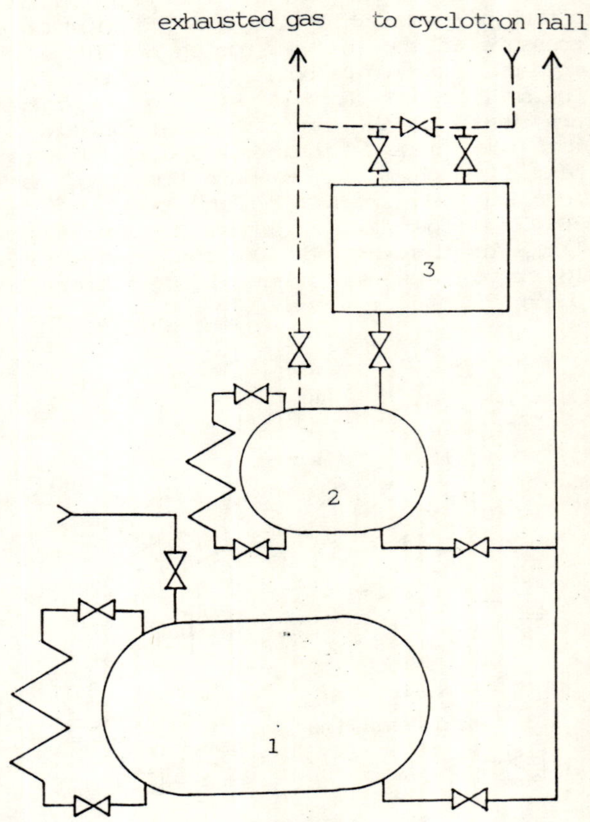

1 $5m^3$ tank	2 $0.5m^3$ tank
3 Philips refrigerator

Fig.3 The Liquid nitrogen supply system

To reduce the liquid nitrogen consumption, a type of KLN 20Y Philips gas refrigerator employed as a nitrogen liquifier and a 500 1 accumulator are used to reproduce completely exhausted cold gas. Regenerated liquid nitrogen may be transferred to the cryo-pumps directly or be stored in the $5m^3$ tank.

5 OPERATION AND NEW PLAN

The cryo-pumping system of the SSC was completed in September 1986. During the first evacuation, a pressure of $8 \cdot 10^{-6}$ Pa was obtained only using four modified RKP 800 cryo-pumps. Now only two or three cryo-pumps are operating to support a vacuum for the beam adjustment of the HIRFL cyclotron that is better 10^{-5} Pa. Two beam lines have already been brought into operation. In our case, liquid nitrogen is rather costly. As it is known, there is no a commercially available 800mm cryo-pump without liquid nitrogen at present, so we are planning to make a new HIRFL 800 cryo-pump at the laboratory. It will be composed of two refrigerators, in which a APC DE 208L expander and a Balzers compressor will be used but a CVI modified CGR 012M single expander and a CBST 6.0 scroll compressor will be chosen as a single stage refrigerator with a cooling capacity of 150 to 250w depending on the frequency interval from 72 to 140cpm of the expander. The model test has proved that the cooling capacity for the first stage of this refrigerator is enough to replace the liquid nitrogen supply system. In the meantime the HIRFL 500 cryo-pump used for the SFC is being built.

FEATURE ANALYSIS OF MICROSCOPICAL IMAGES OF FISH EMBRYOS DURING FREEZING*

X.H.Chen T.C.Hua
Cryobiological Eng. Lab.,Shanghai Inst.of Mech.Eng.
Shanghai 200093, P.R. China

ABSTRACT

In this paper,the image processing techniques developed for automated feature analysis are used for microscopical images of gold fish embryos during freezing, and the freezing damage factors affecting the survival of embryos are studied quantitatively. The conclusions drawn from the feature analysis and experimental results are that the ice formation within embryos actually plays a direct role in producing freezing damage and the toxicity of cryoprotectant(CPA)to embryos is found under pretreating and freezing conditions.

INTRODUCTION

The cryopreservation of fish eggs and embryos would open a new prospects in fish culture, and frozen eggs and embryos could provide a good means of basic studies in genetics. Due to the large size of fish eggs, the complicated membrane system, a low water permeability (1,2),cryopreservation of fish eggs and embryos is still an unsolved problem in the world(3,4).

Our cryobiological engineering laboratory has made some progresses in cryopreservation of fish embryos in past few years(5). A few of carp fish embryos could be survival after freezing-thawing,in which the cooling rate is about 0.07°C/min, and some of them had developed to normal fry after being stored at liquid nitrogen (6). However, the experiments could not be reproduced well (7). This shows that the major factors affecting the survival of fish embryos are to be carefully studied and clearly defined. Therefore, the inverted cryomicroscope and the image processing techniques have been used to analyze quantitatively the osmotic response of embryonic membrane, the crystallization phenomenon occurred within embryos during freezing, and the changes of embryonic structure caused by the toxic effect of CPA solution during pretreating. Besides, several cryopreservation experiments are conducted in a new type cooler.

MATERIALS AND METHODS

(a) Embryos for experiments

All the experiments were performed on gold fish embryos, which were breeded in the pools in Shanghai Heping Park. The embryos spawned naturally were obtained with palm fibre. In each experiment, the embryos were incubated in distilled water at room temperature (20±4°C) until the required embryonic stage for the experiment.

(b) Survival experiments in the cooler

The freezing tests were performed with a new type cooler (8). The cooler was developed in our laboratory for the cooling rates in the range 0.01--20°C/min. In the tests, the gold fish embryos, developed to some stage between muscular-contracting stage and heart-beating stage, were initially exposed to 10% DMSO or 10% glycerol at room temperature for required pretreatment time. About ten embryos were placed in a 25-mm-diameter tube containing 2 ml of 10% DMSO(or 10% glycerol) solution. The solution was first cooled to the seeding temperature (-4°C),then frozen at constant cooling rate from 0.2 to 0.7°C/min. Upon reaching -10°C, -20°C,-25°C and -30°C,we counted the number of frozen embryos,in which crystallization occurred,and calculated the frozen percentage of tested embryos. Thawing procedure was carried out in water bath at room temperature. Then, the embryos were washed and incubated for survival test.

(c) Assessment of embryo survival

* Project supported by National Natural Science Foundation of China

Frozen-thawed embryos survival assessments were used to test the viability, the developing ability and the hatching ability after freezing-thawing. The ratio of the number of surviving embryos after thawing to the number of total embryos tested yields the survival percentage, and the ratio of the number of alive embryos after incubating for more than two hours to the number of total tested embryos yields the developmental percentage. The ratio of the number of hatched fry to the number of total tested embryos also yields the hatched percentage of post-thawed embryos.

(d) Freezing tests on the cryomicroscope

The freezing tests were performed on the inverted cryomicroscope in order to precisely observe the entire freeze-thaw process and further analyze by the digital image processing system.The inverted cryomicroscope, built in our laboratory in 1985, was controlled by a micro-computer coupled to a video recorder(9). The sample's temperature was monitored between 50°C and -130°C by the controlling system using a microcomputer. The entire process of sample structural variation and solution phase change during freezing and thawing was recorded by a video recorder and a camera.

The tests were conducted with gold fish embryos at heart stage in the solution of 7.5% DMSO. A volume of 0.3ml of the solution containing several embryos was introduced into the cold stage of the inverted cryomicroscope. After certain equilibration period, the embryos were precooled to -4°C and frozen to -10°C with constant cooling rate 0.5°C/min(or 2°C/min), then warmed in room temperature air.

(e) The cryoprotectant toxicity tests

The embryos with 12.5% DMSO(or 7.5% glycerol) were held on sample stage of the cryomicroscope to be observed and recorded during pretreating. The tests were performed at room temperature to study the variation of embryos caused by the toxicity of CPA.

(f) Microscopical image processing methods

The sequential photographs,taken by the cryomicroscope, are converted to digital format on a video digitizer which is coupled to a PC-1000 digital image processing system (Numelec Company, France). The system is used to process the digital images and display the information visually in a color monitor with a 256*256 pixel(image element) and 256 gray-levels(8bits). Digital image analysis algorithms have been developed and applied in the Microscopical image processing system. The algorithms are detailed as follows:

<u>Object extraction</u>: the first step of the image analysis is to recognize the specific embryo from the overall image. Then, the image is segmented into two regions:"embryo" and"background". This process takes place in four phases: edge detection, noisy removal, shape recognition and boundary tracing(10). First, the edge detection is performed using a Laplacian filter, which is a two-dimensional discrete convolution filter in the mask of 3*3 pixel. After removing noisy edge and recognizing shape, the rough outlines of the embryo and yolk are completed using a boundary tracing operation.

<u>Feature analysis</u>: the final step of the computational process is the analysis of the morphological features of the embryo. From the boundary of the embryo or yolk, the two-dimensional features such as cross-sectional area and gray-level histogram for the embryo or yolk, are determined. Further, the normalized volume of embryo or yolk is approximately calculated from the measured area by a simple geometric relationship based on the assumption of spherical configuration.

RESULTS AND DISCUSSION

(a) Microscopical image analysis during freezing

The freezing tests were conducted in the cryomicroscope for the heart stage embryos of gold fish in 7.5% DMSO.The freezing process, cooled to -10°C at the rate of 0.5°C/min or 2°C/min, was recorded into video tape and transferred to the digital image processing system. Table 1 shows the results of cryomicroscopical observation of embryo cooled to -10°C. Further, some results of image processing are illustrated in Table 2 and Fig.1,2.

As shown in Table 1, the survival is markedly higher at cooling rate of 0.5°C/min than at 2°C/min. There was evidence of embryo damage during freezing: the embryonic layer darkened and the yolk syncytium disrupted although the whole embryos still had a normal shape, intact chorion and clear yolk. Also, the gray-level histograms within the embryo show obviously the difference between survival one and dead one, as shown in Fig.1. Especially, the dead embryo is marked as the double-peak gray-level histogram. Moreover, the ice crystal formed initially within the yolk part of embryo and developed to the whole embryo during slower cooling, and most embryos hardly underwent to the surrounding temperature below -10°C at which the ice would formed within the embryos (see Fig.2). Therefore, a conclusion is drawn here that the ice crystal formation within embryo plays a

Table 1. Cryomicroscopical observation of gold fish embryos at heart stage cooled to -10°C at cooling rates of 0.5°C/min and 2°C/min in 7.5% DMSO respectively.

Cooling rate(°C/min.)	Crystalization phenomenon	No. embryos	No. survival(%)	Morphology assessment after warming
0.5	At -1.45°C,ice cryst formed in the solution and grew slowly in the form of dendrites.At -5.8°C, darking occured within the yolk part of embryo and slowly developed to the whole embryo. When warmed to 0°C or so, the darking embryo turned clear again.	7	3(43%)	Exect 3 survived embryos which was unfrozen during cooling, the others having the normal shape, the intact chorion,the clear yolk,but the darking embryonic layer and the disruption of yolk syncytium.
2.0	At -5.0°C, ice cystal formed in solution and grew fast in form of dendrites. At -8.0°C, darking occured immediately within almost whole body of embryo. When warmed to 0°C or so,the darking embryo turned clear again.	4	0(0%)	All embryos having the normal shape, the intact chorion, the clear yolk, but the darking embryonic layer and the disruption of yolk syncytium.

Table 2. Image processing results of gold fish embryos at heart stage during freezing at cooling rates of 0.5°C/min and 2°C/min in 7.5% DMSO respectively.

Cooling rate(°C/min.)	Temperature (°C)	Measured area			Normalized volume		
		Embryo	Yolk	Frozen-zone	Embryo	Yolk	Frozen-zone*
0.5	0.0	7069	2893	0	1.000	1.000	0
	-1.45	6951	2789	0	0.975	0.951	0
	-2.0	6835	2601	0	0.950	0.852	0
	-4.5	6605	2299	0	0.903	0.708	0
	-5.8	6491	---	2629	0.880	---	0.258
	-9.0	---	---	---	---	---	---
2.0	0.0	6571	2614	0	1.000	1.000	0
	-1.45	6541	2589	0	0.993	0.986	0
	-5.0	6436	2466	0	0.969	0.916	0
	-7.0	6057	2225	0	0.885	0.785	0
	-8.0	5857	---	4199	0.842	---	0.607
	-10.0	5447	---	4216	0.755	---	0.681

*: Normalized volume of frozen-zone is the ratio of frozen-zone volume to embryo volume.

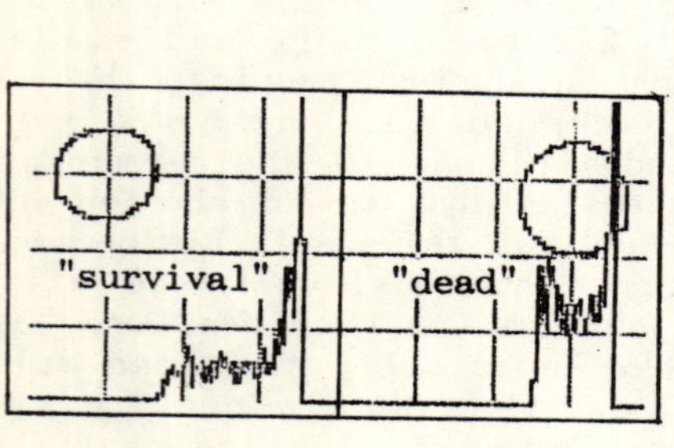

Fig.1 Gray-level histograms within the frozen-thawed gold fish embryos, cooled to -10°C at a cooling rate of 0.5°C/min in 7.5% DMSO.

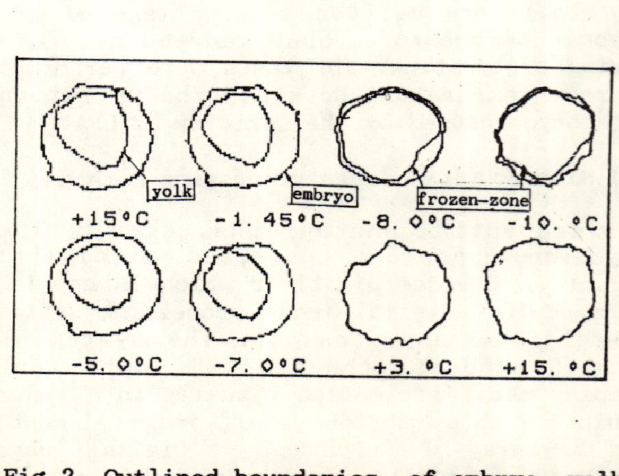

Fig.2 Outlined boundaries of embryo, yolk and frozen zone of gold fish embryos during freezing at a cooling rate of 2°C/min in 7.5% DMSO.

direct role in producing injury during freezing, which will be further confirmed by the survival experiments.

But why did so? One reason may be that the embryonic membrane has such a low permeability of water and CPA that the embryo has insufficient time to dehydrate water and permeate CPA during cooling at the applied rate. Another reason, as shown in Table 2, may be that the different layers of embryo have much different osmotic response and the inner layers form a barrier to water and CPA movement so that a lag of dehydration is created between the inner yolk part and the outer previtelline space. When cooled at the slower cooling rate, this lag will have several important consequences: the CPA concentration in the outer part may be so excessive that CPA has some toxic effect on the embryo when it is barely adequate toward the inner part of embryo, and the inner part may not have enough time to dehydrate sufficiently to prevent crystallization within the yolk. This is why the ice crystal occurred at the higher temperature and formed initially within the yolk part during freezing. Then for fast cooling, both the inner yolk part and the outer previtelline space have insufficient time to response osmotically to the surrounding medium, so that the ice crystallization occurred within almost whole body of embryo, indicating insufficient water loss at the applied cooling rate. The frozen zone within the embryos at 2°C/min is larger than that at 0.5°C/min, which may further confirm above point of view.

(b) Microscopical image analysis during pretreating

The toxicity of CPA, such as DMSO and glycerol, to the gold fish embryos was tested. Fig.3 shows the computer print out of some image processing results

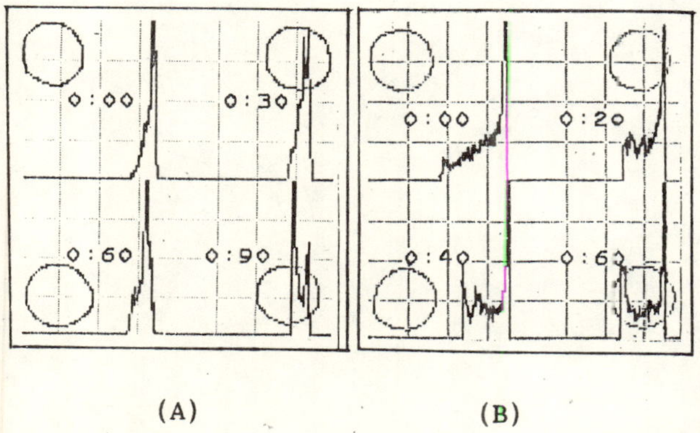

(A) (B)

Fig.3 Gray-level histograms within the gold fish embryos during pretreating with (A)12.5% DMSO for 0, 30, 60, 90 min and (B)7.5% glycerol for 0, 20, 40, 60 min at room temperature.

for gold fish embryos during pretreating with 12.5% DMSO and 7.5% glycerol, respectively. First of all, it is observed from the source and enhanced images that some changes took place within embryo obviously, such as the blackening of yolk and the disruption of yolk syncytium, which are reflected on the gray-level histograms. After the embryo was pretreated with 12.5% DMSO for 90 min or 7.5% glycerol for 20 min, the histograms were changed from one-peak gray-level to double-peak gray-level, which is similar to that of the dead embryo after freezing-thawing. Therefore, the CPA solutions are much toxic to the gold fish embryo during pretreating at room temperature. The longer the embryo was pretreated with CPA, the more toxic effect on them. Furthermore, for 12.5% DMSO, the changing of gray-level histogram was less visible until 90 min, but for 7.5% glycerol, the changing was much distinct after 20 min. That shows that 7.5% glycerol is more toxic to the fish embryo than 12.5% DMSO.

(c) Survival experiments in the cooler

As shown in Table 3, the survival percentage of post-thawed embryos decreased as the temperature decreased, and the declining trend was much marked when cooled below -20°C. In contrast, the frozen percentage of embryos increased with the temperature declining, and the survival percentage was exactly equal to the unfrozen percentage. This shows that the ice formation within embryo was actually playing a direct role in producing freezing damage to the gold fish embryos.

CONCLUSIONS

The microscopical image analysis of gold fish embryos during freezing and pretreating show that the membrane system of gold fish embryos has a very low permeability of water and CPA, and the different layers of embryo have much different osmotic response and the lag of dehydration is created between the inner yolk part and the outer previtelline space. Some conclusion are drawn as below within the inner yolk part of embryo during slow freezing, and it plays a direct role in producing freezing injury. And for survival assessment, the gray-level histograms within embryo show obviously the difference between survival one and dead one. Especially, the dead embryo is marked as the double-peak gray-level histogram. Also, when the embryo is affected by the toxicity of CPA solution, the histogram is similar to that of the dead embryo after freezing-thawing. The CPA solution, such as DMSO and glycerol, are toxic to gold fish embryos during pretreating, and glycerol produces more toxic influence on embryos than DMSO. Furthermore, the survival experimental

Table 3. Cooling of gold fish embryos, developed to the stage between muscular contraction stage and heart beating stage, to -10°C, -20°C, -25°C and -30°C at different cooling rates in 10% DMSO and 10% glycerol.

No.	CPA	Cooling rate(°C/min.)	Temperature (°C)	No. embryos	Percentage frozen(%)	Percentage survival(%)	Percentage development(%)	Percentage hatched(%)
1	10%DMSO	0.2	-10	11	45.5	54.5	54.5	54.5
			-20	11	54.5	45.5	9.1	9.1
			-25	13	92.3	7.7	7.7	0
			-30	13	100	0	0	0
2	10%DMSO	0.3	-10	16	25	75	75	75
			-20	15	40	60	60	60
			-25	14	71.4	28.6	28.6	21.4
			-30	17	100	0	0	0
4	10%DMSO	0.5	-10	12	50	50	50	50
			-20	11	54.5	45.5	45.5	45.5
			-25	16	87.5	12.5	12.5	6.25
			-30	12	100	0	0	0
5	10%DMSO	0.7	-10	12	50	50	50	50
			-20	13	53.8	46.2	46.2	30.8
			-25	14	100	0	0	0
			-30	14	100	0	0	0
6	10%glycerol	0.3	-10	20	0	30	0	0
7	control	---	---	17	---	---	---	100

results show that the survival of embryos depends on the unfrozen percentage of embryos. The conclusions drawn from the above experimental results are in accordance with ones obtained from the image feature analysis.

REFERENCES

1 Harvey, B. and Ashwood-Smith, M.J., Cryoprotectant penetration and supercooling in the eggs of salmonid fishes, Cryobiology: 19, 29-40, 1982.

2 Harvey, B., Kelly, R.N., Ashwood-Smith, M.J., Permeability of intact and dechorconated zebra fish embryos to glycerol and dimethyl sulfoxide, Cryobiology: 20, 432-439, 1983.

3 Stoss, J., and Edward, M.D., Studies on cryopreservation of eggs from rainbow trout (*Salmo gairdneri*) and coho salmo (Oncorhynchus kisutch), Aquaculture: 31, 51-65, 1983.

4 Harvey, B., Cooling of embryonic cells, isolated blastoderms and intact embryos of the zebra fish (*Brachydanio rerio*) to -196°C, Cryobiology: 20, 440-447, 1983.

5 Hua, T.C., Cryopreservation of Biomaterials, Science (Chinese) Vol.39, No.1, 37-42, 1987.

6 Zhang, X.S., Hua, T.C., and others, Cooling of embryos of common carp *(Cyprinue carpio)* to -196°C, Journal of Shanghai Institute of Mechanical Engineering, Vol.9, No.2, 83-90, 1987.

7 Chen, X.H., Experimental study of freezing damage factors and feature analysis of microscopical images of fish embryos, Thesis of M.Sc., Shanghai Institute of Mechanical Engineering, 1988.

8 Chen, E.T., Zhao, L., and Hua, T.C., A new cooler with very low consumption of liquid nitrogen, 25th Annual Meeting of the Society for Cryobiology, Aachen, West Germany, 1988.

9 Wu, S.Y., Yao.K.M., and Hua, T.C., The new structure of cryomicroscope, Cryobiology: 6, 553-554, 1987.

10 Dietz, T.E., and others, Computer recognition and analysis of freezing cells in noisy, cluttered images, Cryobiology: 19, 539-549, 1982.

THE FREEZING OF OIL AND WATER IN PIPLINES

Wang Xinwen, Zhang Rujie, Yao Fengming, Qu Jianhua
Guangming Research Institute of Chemical Industry
P.O.Box 481, Dalian, China

ABSTRACT

This paper describes the experimental study of the freezing period, sealing pressure and liquid nitrogen consumption when freezing crude oil and running water in pipes of dimension ∅114x6.5mm, ∅219x8mm, ∅ 273x6.35mm and ∅ 720x9mm. It also calculates theoretically the freezing period. The experiment shows that when liquid nitrogen is used as a coolant long oil transporting pipes and pipes containing other fluids can be rapidly frozen. This greatly helps maintenance repair, hydrostatic test and other construction work on long transporting pipelines.

PREFACE

Since there are more than ten thousands kilometers of oil pipe lines in our country, it is important to ensure the safety of oil transportation. The planned maintenance repair is one of the major means of guaranteeing the transporting of crude oil without any leaking. Among the several sealing methods cryogenic pipe freezing is a new technique. It possesses the advantage of a shorter freezing period and can be carried out without the problems of sealing pressure, shape changing of pipes and of the terrain. Furthermore it requires simple instruments, less personnel, a shorter time and has an economical cost. The principle of pipe freezing is to take liquid nitrogen as a coolant and freeze the fluids socalled in pipes so that they will not flow out during repair work.

According to what has been reported in the available literature,up till now pipe freezing has only been applied to pipes containing fresh-water or sea water. Attempts to freeze crude oil has not been made. We chose four specifications of carbon steel pipes to carry out pipe freezing experiments on water and crude oil and to determine the freezing period, the strength of the freezing plug and the consumption of liquid nitrogen. This paper will focus on the relationship between the freezing period, the seal pressure and the liquid nitrogen consumption and the different liquids, liquid initial temperatures and pipe diameters.

Experimental Instruments and Methods

Fig. 1 shows the experimental instruments. Liquid nitrogen is used as a coolant. The frozen fluids are crude oil and sea water.When freezing crude oil the pipe specifications were ∅ 114x6.5mm, ∅ 219x8mm, ∅ 273x6.35mm.When freezing running water the pipes were ∅ 114x 6.5mm, ∅ 219x8mm, ∅720x9mm. The pipe material is carbon steel. A thermocouple was placed in the pipe to measure the plug temperatures at different points in the pipe and observe the length of the plug and to inspect the freezing periods. A pressure gauge was fixed on each end of the experimental pipe to test the strength of the plug.

Essential points of the experiments: heat the oil in the oil tank to the expected temperature, open the inlet valve of the experimental pipe, start the pump to establish circulation until the fluid in the pipe reaches the experimental temperature, then, open the by-pass valve, chose the inlet valve in order that the fluid in the pipe keeps a certain pressure. After all these are done, the experiment begins.

The Result of the Experiment

This part focuses on the formation time of the crude oil plug and the running water plug, the sealing pressure and the consumption of liquid nitrogen.

a) The plug forming time

First we consider what happen when freezing crude oil and running water in the four kinds of carbon steel pipes by means of liquid nitrogen as coolant.

For the same diameter, there is a considerable time difference between the freezing of crude oil and running water. The major reason is that the heat conductivity of ice is about thirteen times that of crude oil. The freezing time not only relates to the property of the frozen fluids but alsc to the fluid initial temperatures and final center temperatures of the plug.

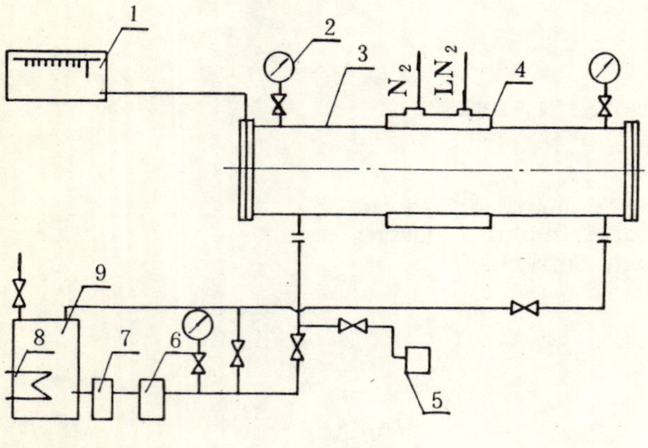

Fig.1 Experimental instruments

1. XWC-300 A refitted auto temperature recorder
2. Y100 pressure gauge (1.5 grade)
3. experimental pipes
4. freezing chamber
5. SY-350B forcing motor pump
6. RCY-2/14.5-1 gear oil pump
7. oil filter
8. steam heater
9. oil tank

Take the freezing of crude oil as an example: in pipes of the same diameter the initial temperature of the oil and the center temperature of the plug have an influence on the freezing time. As the diameter increases the freezing time increases with it. If a two-components linear regression is made using the experimental freezing data of the crude oil a regression curve is observed as fig.2. when the pipe diameter is ø114x6.5mm the influence of the initial temperature of the oil and the center temperature of the plug on the freezing time is not significant. But when the pipe diameter is ø219x8mm the influence of the initial temperature of the oil and the center temperature of the plug on the freezing time appears to be significant. Thus, freezing time is proportional to the initial temperature and is in inverse ratio to the plug center temperature. When using a mathmatical formula to calculate freezing time a curve relating freezing time and pipe diameter at the time of water and crude oil freezing can be drawn.
The mathmatical formulae are as below:

Liquid phase $$C_l\frac{\partial T}{\partial t} = \nabla q_l \quad (1)$$

Solid phase $$C_s\frac{\partial T}{\partial t} = \nabla q_s \quad (2)$$

On the interface $$T_s = T_l = T_f \quad (3)$$

$$\frac{d\xi}{dt} = \frac{1}{\rho L}\left(\lambda_s\frac{\partial T}{\partial r} - \lambda_l\frac{\partial T}{\partial r}\right) \quad (4)$$

where: L: liquil phase s: solid phase
T: temperature t: time
λ: thermal conductivity
ρ: concentration
L: solidifying heat
ξ: location of the interface
f: freezing point r: radius

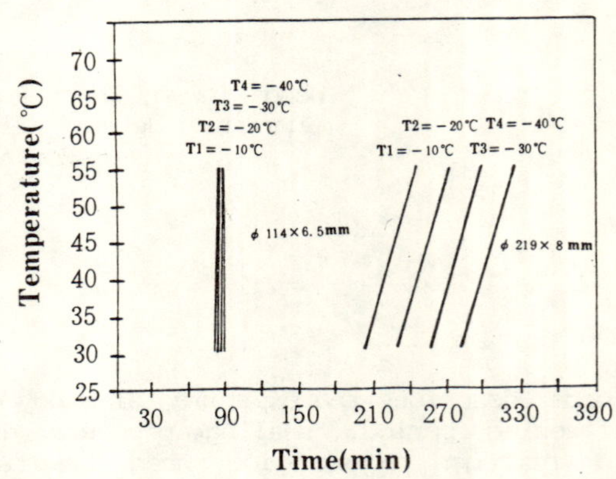

Fig.2 The curve relating initial temperature and freezing time

Fig.2 is the curve relating initial temperature and freezing time when freezing crude oil in the pipe of ø114x6.5mm and ø219x8mm.
Fig. 3 and Fig. 4 are respectively the curves observed from the calculation of freezing time of crude oil and water. From the curves we see the results of calculation are very close to the results of experiment.

b) Plug strength

The results of the experiments indicate that when pipes of the same diameter bear the same pressure, the oil plug is longer than the ice plug, or in another word, when the crude oil is frozen the freezing chamber is longer and when water is frozen the freezing chamber is shorter. For example, when freezing water in a pipe of ø219x8mm, the freezing chamber length is 400mm and the seal pressure is above 9.8MPa. When freezing oil the freezing chamber length is 800mm and the seal pressure is 8.1 MPa. Thus, the plug strength is related to the frozen fluids. It is directly proportional to the chamber length and inversely proportional to the pipe diameter.

c) Temperature distribution and forming of the plug

Fig.5 shows the installation position of the thermocouple in the pipe of ¢273x6.35mm. Fig.6 indicates the temperature change in the plug in the freezing process. The freezing curve shows that ice was formed in the internal wall of the pipe and as it began to expand to the center the plug was gradually formed. At the same time the two ends of the plug slowly grow in to the directions. In the process of crude oil freezing, since the volume contracts the curve oil should be replenished continuously. But in the process of water freezing when freezing takes place the water volume expands, so close observation is necessary in case the pressure inside becomes more great

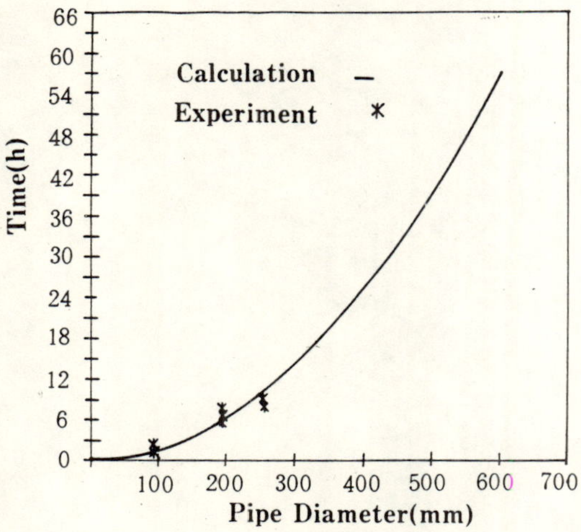

Fig.3 The time curve of freezing of crude oil

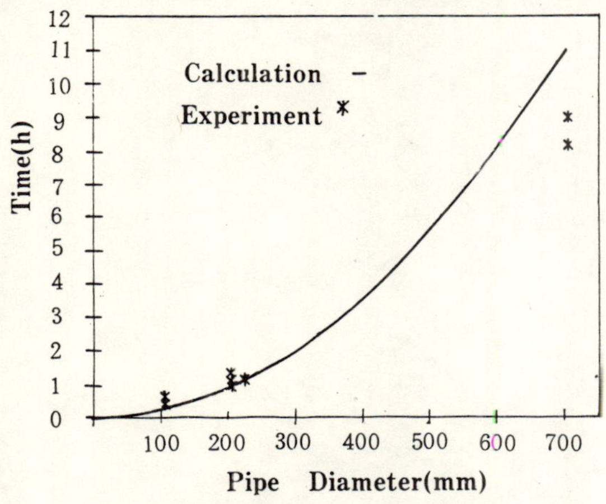

Fig.4 The time curve of freezing of running water

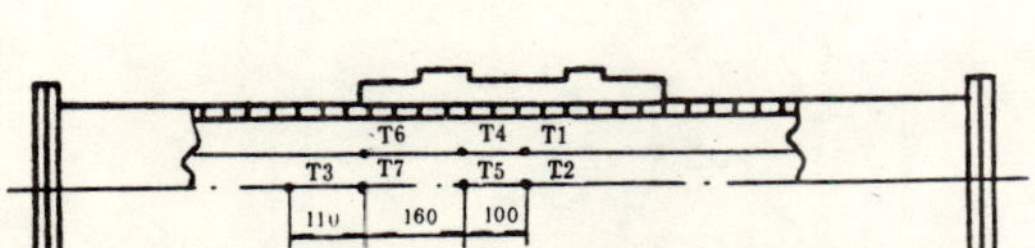

Fig. 5 Temperature testing points

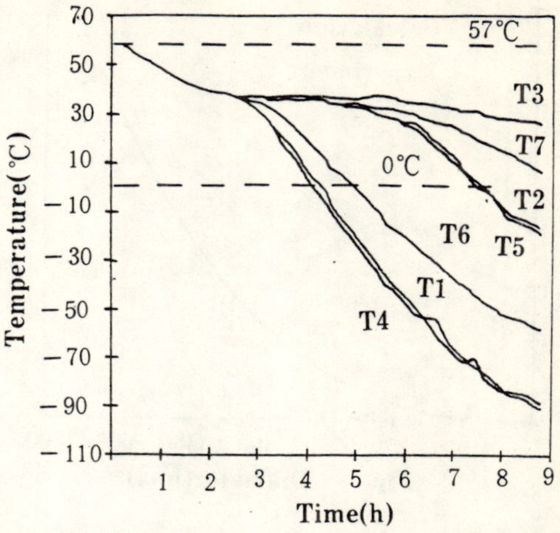

Fig.6 The temperature curve of the temperature points as a function of time

d) Liquid nitrogen consumption

From the result of the experiments the conclusion can be drawn that liquid nitrogen consumption is related to several factors. They are: heat capacity of the liquid, heat of solidification, the initial temperature of the liquid, the center temperature of the plug, the length of the freezing chamber and the effect of the freezing insulation. In another word, the liquid nitrogen consumption is proportional to heat capacity of the liquid, the heat of solidification and the initial temperature of liquid, the length of freezing chamber and is universely proportional to the center temperature of the plug and the insulating effect of the freezing chamber. Thus,as the conditions permit the initial temperature of the liquids should be reduced as low as possible and the center temperature of plug and the effect of freezing insulation should be raised as much as possible.

In the calculation the freezing chamber lengths of-various pipes are taken as twice the pipe internal diameter, Fig. 7 shows the relationship between liquid nitrogen consumption and pipe diameter. This can be used in planned maintenance work.

APPLICATION

There is an oil transporting line which is 218km long. The diameter is ø273x6.35mm. Because of the imperfect construction work, the line of a certain length appeared cambered. In order to enlarge the oil transporting quantity it was decided to replace the faulty line. In the repair work, the pipe freezing techniques was utilized to isolate the crude oil in the pipe. This work was accomplished satisfactory according to plan and the client was satisfied.

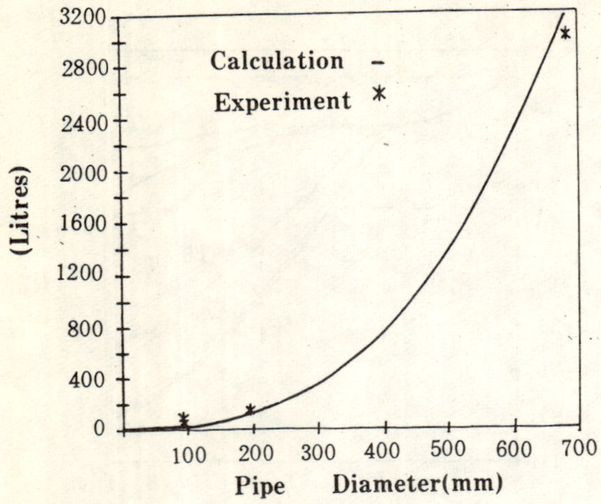

Fig. 7 The relation curve of the liquid nitrogen consumption in the pipes of different diameters

CONCLUSION

a) The freezing time is inversely proportional to the fluid thermal conductivity, and proportional to the heat of solidification and the specific heat and it increases as the square of pipe diameter.

b) When large diameter pipes are to be frozen the initial temperature of the fluids should be reduced as low as the work permits and the center temperature of the plug should be raised as high as possible in order that the freezing time can be shortened.

c) The seal pressure of the oil plug is lower than that of ice plug.

d) The liquid nitrogen consumption of crude oil freezing is higher than that of water.

e) The application exercise showed clearly that it is possible to freeze crude oil using a freezing method outside the pipe on the oil transporting lines.

REFERENCS

Tantam, D.H. and Bishop, C. 'Proc.7th Int. cryogenie Eng. Conf. (ICE7)' Guildford (1978), PP 365-372.

Buckman, D. 'Pipe Line Industry', No.2(1980), PP52-54.

STUDY AND APPLICATION OF CRYOGENIC TREATMENT ON INCREASING THE USE LIFE OF TOOL AND DIE

Huang Luguan, Yin Yanguo, Hu Xianguo
Tribology Research Inst., Hefei Univ.of Technology,
Hefei, Anhui, China

ABSTRACT

After cryogenic treatment, the wear-resistance of all kinds of tool and die steel are increased, their size stability improved, the early damage of tool and die avoided and their use life are obviously increased.[1] Tool and die steel D2 (AISI) is the main object to study in this paper. It's texture property and wear-resistance mechanism are investigated after different temperature cryogenic treatment by means of SEM and other micro-analysis meter.

INTRODUCTION

In the process of being used, tool and die are usually in the state of complex stress. Wear is their main form of failure and also exists some ratio of early stripping and tipping. In order to increase the wear-resistance and the use life of tool and die, research workers have made unremitting efforts in the world. Especially in the early 1980's, the development of cryogenic treatment technology has made the excellent effects. The use life of tool and die are increased by a big margin.[2] Cryogenic treatment is the follow-up process after usual quenching. After the quenched workpiece is tempered at 100°C, the workpiece is put into the cryogenic treatment equippment refrigerated by liquid nitrogen. It shows that the best cryogenic treatment temperature of D2 is at -160 — -140°C. After D2 is cryogenically treated at -160-- -140°C, the residual austenite is about 1.5 -- 3% by means of X-ray diffractometer. From the SEM diagram, the texture fines obviously, and lots of micro-fine carbides dispersion -educe along the grain boundary of the original austenite and martensite and inside crystal grain of martensite. After the M12 nut cold-forging die made of D2 is cryogenically treated at -160 -- -140°C, its use life increases by 6 - 8 times at high stress. The wear data from MM200 wear tester show the good property of D2 after cryogenic treatment. It has an important economic value and theory significance to get clear about the texture transforming property and wear-resistance mechanism of cryogenic treatment, and put to use and popularize this new technology in tool and die industry.

EXPERIMENTAL RESULTS AND DISCUSSION

1. The process curves of heat and cryogenic treatments of D2 wear sample are shown in Fig.1.

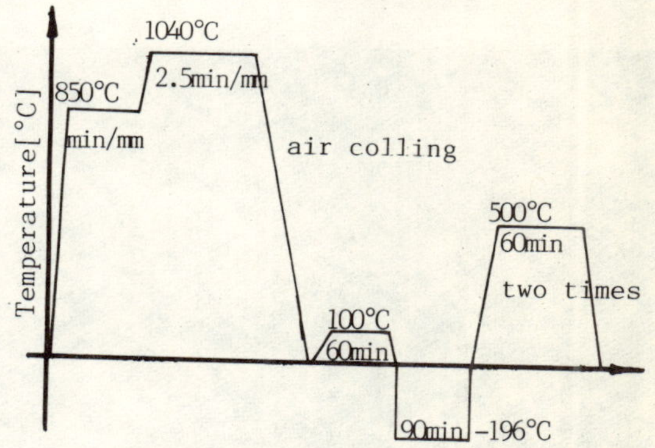

Fig.1 Heat and Cryogenic Treatments Process Curves of D2.

The various property targets after cryogenic treatment are shown in Table.1.

These show that cryogenic treatment increases the wear-resistance, especially at -160°C. When M12 nut cold-forging die made of D2 and with 300 tons of punch force isn't cryogenically treated, it disables after functioning 20000--30000 times. The cryogenically treated one disables after 200000---250000 nuts are made. The actual use life of die increases by 6--8 times.

Table 1. Property Targets of D2 Wear Samples after Different Temperature Cryogenic Treatment

Property / Treatment Temperature	Hardness (HRC)	Impact Toughness (kg.m/cm²)	Wear Volume (mm³)	Residual Austenite (%)
Uncryogenic Treatment	60	3.60	0.313	6 - 8
-80°C	60.5	3.60	0.202	4 - 5
-120°C	60.8	3.55	0.142	3 - 3.5
-140°C	60.8	3.58	0.103	2.0 - 2.8
-160°C	61	3.70	0.095	1.5 - 2.0
-196°C	61.5	3.60	0.107	≤ 1.0

2. Table 1 shows that the hardness and impact toughness have little change before and after cryogenic treatment, only hardness increasing slightly. But wear-resistance increases sharply. It is found out that the result isn't better when the temperature of cryogenic treatment becomes lower. Excessive low temperature causes power waste and increases wear. The best temperature of increasing D2 wear-resistance is at -160-- -140°C. Fig.2 (a) and (b) are SEM diagrams of the actual disabled die under two different treatment processes.

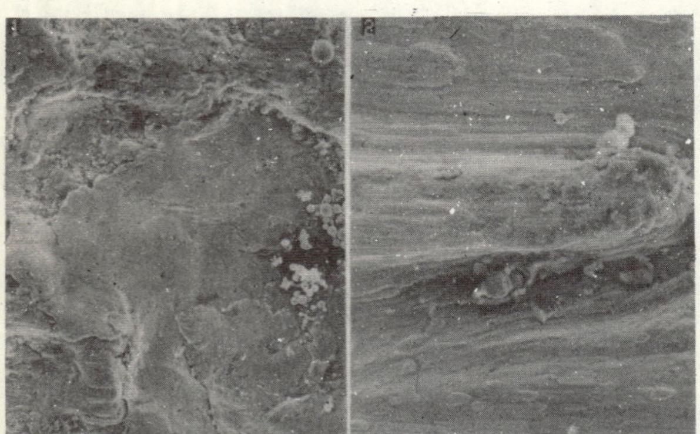

(a) (b)

(a) Uncryogenic Treatment
(b) Cryogenic Treatment at -160°C

Fig.2 SEM Diagrams of The Actual Disabled Die

It is shown that worn surface of uncryogenic treatment die is rougher and there are lots of stripping pits and block of wear particles to be delaminated. After cryogenic treatment, the failure surface of die is more smooth and stripping pits decrease obviously. It is indicated that M12 cold-forging die made of D2 and cryogenically treated at -160°C is difficult to produce early stripping, and has a high wear-resistance and long use life under high frictional force and impact force.

3 To sum up, wear-resistance of wear experimental sample or actual used die after cryogenic treatment is increased. This depends on the texture property caused by cryogenic treatment. Fig.3 is the micro-structure of D2 after cryogenic treatment at -160°C.

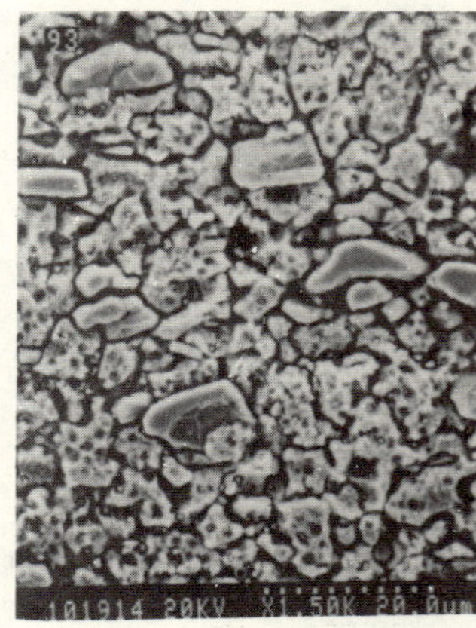

Fig.3 Microstructure of D2 after Cryogenic Treatment

As shown in Fig.3 the structure is very fine with dispersion-educing lots of even fine carbides from martensite grain and original austenite boundary after cryogenic treatment. The texture structure of the sample of uncryogenic treatment is coarse and the educed carbide is little. As shown in Table 1, by using X-ray diffractometer the residual austenite (r_R) in texture reduces with the decrease of temperature of cryogenic treatment. r_R is around 2% after D2 is cryogenically treated at -160°C and will be reduced to below 1% at -196°C. But in Table 1, the material toughness doesn't decrease due to the decrease of r_R after cryogenic treatment. Fig.4 is SEM fracture topography of impact sample fracture at -160°C. On the fracture surface there are some ductility fosses. It shows that it belongs to quasi-cleavage fracture of ductile rupture. Cryogenic treatment transforms most residual austenite into martensite. On the other hand, the fundamental reason why the material toughness doesn't decrease is that even fine carbides are dispersion-educed and the texture micro-fined. So the D2 tool and die don't lead to early tipping and fracture with the toughness decrease after cryogenic treatment. After D2 is quenched, there are lots of residual austenite (r_R =25%) in texture. When these residual austenites are transformed into martensites at the cryogenic treatment temperature, very fine martensite are gained because of the formation of many crystal nuclei. After quenching at high temperature, there remain lots of texture defects (dislocation and vacancy. etc.). These texture defects stabilize further and form the region where carbides are educed. The follow-up subtexture can

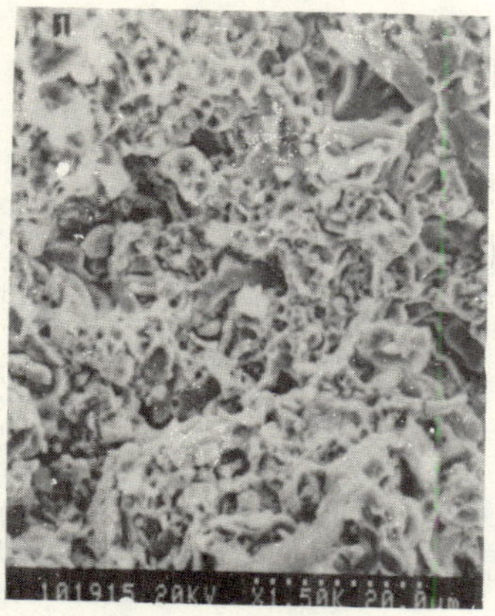

Fig.4 SEM Fracture Topography of D2 after Cryogenic Treatment

accelerate the ageing reaction under a temperature below zero, that is, at low temperature the carbon atoms gathered have a rapid forming tendency near the follow-up dislocation. It is the result of the combination between the carbon atoms and the stress around dislocation.[3] The more carbon atoms gather, the bigger and finer the dispersion of carbide. So the resistance force of dislocation movement and migration of grain boundary is increased and the abliaty of anti-tempering strengthen. The grains are difficult to grow up, so the block strength and toughness of material are increased.

The normal failure of tool and die is wear. The hardness and micro-structure of material have a crucial effect on wear-resistance. The hardness of D2 doesn't increase obviously after cryogenic treatment, but the residual austenite is little, the texture is improved and the carbides are dispersion-educed. Thus the strength and toughness and wear-resistance of D2 are increased. The residual austenite is a fatal weakness to tool and die. Because it is a kind of soft phase, it is easy to cause fatigue delamination or be cut off under the high periodic stress, thus making the carbides relieve and seperate from matrix to produce stripping pits. These carbides intensify the wear of tool and die. At high stress residual austenite may transform into martensite, and bring out the new texture stress. The stress can't be eliminated and tipping and fracture are produced in the process of working. The use life of tool and die decreases.[4] Under work-condition, the more and finer the carbide, the smaller is the force acting on every carbide and the better the wear-resistance might be. It has a higher combinaing force between the carbide wound by high dense dislocation and matrix. The carbide isn't easy to seperate from matrix. So the wear-resistance increases. The less the mean free path, the better the wear-resistance under equal bulk hardness.[5] Thus we can conclude from these that adequately decreasing residual austenite, increasing the dispersion degree of carbide and the combining strength with matrix, and preventing the carbide from early stripping are all effective ways of increasing the wear-resistance and use life of tool and die.

While the D2 is studied through cryogenic treatment, the $W18C_r4V$(AISI T1), $W6M_o5C_r4V2$(AISI M2), GC_r15(AISI E52100) and other alloy steel tool and die are investigated. It is found out that their use life increase by more than two times. So cryogenic treatment as a new technology has a wider application prospect in tool and die industry.

CONCLUSION

1. The main reason of increasing use life of D2 tool and die at -160°C cryogenic treatment is that most residual austenites in steel are transformed into martensite, texture microfined, carbide dispersion educed and the combining strength between carbide and matrix increased.
2. After the toc and die steel are cryogenically treated, the residual austenites are almost eliminated, the combining strength between dispersion-educed carbide and matrix is improved. The carbide is difficult to seperate from matrix so wear-resistance is increased. The carbides in tool and die of uncryogenic treatment are easy to seperate from matrix, form stripping pits and intensify wear.
3. Cryogenic treatment can fully exploit the material potentialities of most tool and die and therefore can increase the use life of tool and die by a big margin.

REFERENCES

1. Shigeo Owaku 'cryogenic treatment'. Metal, no.6 (1981) pp 98-101.
2. Donald R.Dreger 'forcast of metal heat treatment'. Machine Design, vol.53. no.2 (1981), pp73-78.
3. T.Z.Len 'contributions of steel tempering'. Machine Industry Press, (1986), pp 74-85.
4. Shigeo Owaku.'die materials and their heat treatments'. Heat treatment. vol.27. no.4 (1987) pp 65-69.
5. Zum Gahr.K.H. 'wear of materials'.ASME, (1979), pp 266.

ACCURATE CALCULATION FOR EXPANSION FIT

Bu-Yu Yan
Sichuan Air Separation Plant Manufactory
Jianyang, Sichuan, People's Republic of China

ABSTRACT

Expansion fit has been used widely in mechanical assembling, but there is still lack of an accurate method to perform the necessary calculations. Formulas found in standard mechinery's handbooks can be off by as much as 40% [1,2,3]. Others are only good when liquid nitrogen or dry ice is used as the coolant [4,5]. In this article, the author has developed a new algorithm for the calculation of expansion fit, by taking into account the behavior of the thermal expansion coefficients of materials under low temperatures. Based on an integral expression, a far superior but still simple formula for the expansion fit calculation is obtained. The errors introduced by this new method is analyzed in detail. It is also shown that the traditional shrink fit calculation is only a special case under this new formulation. An illustration of the calculation procedure is also given.

INTRODUCTION

Among the interfered fit techniques available, press fit is the simplest to perform, although it requires that the surface of the work piece be carefully machined, and it cannot be used when the sample is to deliver a large torque. On the other hand, shrink fit does allow the delivering of a large torque, but the fit procedure is considerably more complicated. The work piece may suffer loss of hardness, tensile, toughness and even destruction during shrink fit operation if it is over heated. The expansion fit, to a large extend, combines the advantages of both press and shrink fits and hence has been increasingly substituting for the above two and becoming a major interfered joint method in mechanical assembling.

It should be noted, however, that there exists no easy and reliable way to perform the necessary calculations for an expansion fit. As can be seen later in this article, some of the methods can be off by as much as 40%. Others have made some improvements but are still limited to the cases where liquid nitrogen and dry ice are used as coolants. Therefore, it has become a key point to look for a more accurate procedure for the expansion fit calculations.

CURRENT EXPANSION FIT CALCULATIONS

The Russian "Handbook on Industry Technology" [1] recommends this following equation for both shrink and expansion fits

$$T > \frac{\delta}{\alpha D} + T_0 \quad , \qquad (1)$$

where T is the heating temperature (in K) of the outer work piece for shrink fit or the cooling temperature of the inner work piece for expansion fit, α is the linear expansion coefficient (in 1/K) of the work piece to be heated or cooled, T_0 is the temperature of the surrounding, δ is the interference of the fit (in mm), and D is the nominal dimension of the fit (in mm). It is also recommended by the Handbook that a correction of 20%, i.e., raising the heating temperature of the outer work piece or lowering the cooling temperature of the inner work piece, can be imposed to allocate an assembly clearance and to compensate for the temperature change of the work piece during operation.

All of the available Chinese engineering handbooks [2,3] are unexceptionally based upon Ref. 1. The only improvement, if any, is to allocate the assembly clearance and temperature compensation by increasing the interference (about 1% of the nominal dimension), instead of changing the temperature.

Studies have shown that Eq. (1) is accurate only in the high temperature region, i.e., for the shrink fit processes. For the expansion fit the result becomes rather poor that operation failures are sometimes inevitable. In order to improve the reliability of calculations, some of the handbooks published in the United States [4,5] use Table I as a guide. For example, to see if a 2-inch SAE steel shaft can be successfully fitted into a 1.997-inch diameter hole under liquid nitrogen temperature, one first finds from Table I that the contraction of SAE steel is 0.0022 inch per inch, so that the diameter after cooling will be 2 - 2 × 0.0022 = 1.9956, which is less than the inner diameter of the hole, indicating that the fit is possible. When the coolant is dry ice instead of liquid nitrogen, 1/3 of the values in Table I are to be used.

Table I. Metal shrinkage with nitrogen cooling

Metal	Shrinkage, inch, per inch of shaft diameter
Mg Alloy	0.0046
Al Alloy	0.0042
Cu Alloy	0.0033
Cr-Ni Alloy (18-8 to 18-12)	0.0029
Monel Metal	0.0023
SAE Steel	0.0022
Cr Steel (5 to 27% Cr)	0.0019
Cast Iron (not alloyed)	0.0017

With the rapid advance of cryogenic technologies, more and more low temperature devices are available for various applications. Some of them allow the operation temperature to be precisely controlled anywhere between the room and liquid nitrogen temperatures [6]. The constraint that an expansion fit be performed at dry ice or liquid nitrogen temperature no longer exists. In addition, from strength and wear resistance considerations, steels with high content of carbon and alloy elements are widely used. These materials may result in cracks if emerged into liquid nitrogen directly. Therefore, modifications of the existing methods are necessary in order to meet the growing demand of expansion fit applications.

THERMAL EXPANSION COEFFICIENTS AT LOW TEMPERATURES

By looking into the behavior of thermal expansion coefficients of materials as a function of temperature T, it is straight forward to find out why the above formulas give good results at high temperatures but not at low temperatures. Using length L instead of diameter D, Eq. (1) can be rewritten as

$$\Delta L = L \alpha \Delta T \quad (2)$$

Here $L\alpha\Delta T$ represents the amount of change in length, ΔL, corresponding to a temperature change of ΔT.

Fig. 1 shows the relation between linear expansion coefficients α with temperature T above the room temperature [7]. As can be seen from this figure, α is almost a constant for the wide temperature range of 293K to 793K. Fig. 2 is the same α-T relation but for a temperature range of 0K to 250K. Unlike the high temperature region, the linear expansion coefficients vary rapidly with temperature, approaching zero as T goes to 0K. Therefore Eq. (1), which does not incorporate the temperature dependence of α into account, is inadequate in describing the thermal expansion properties of materials at cryogenic temperatures.

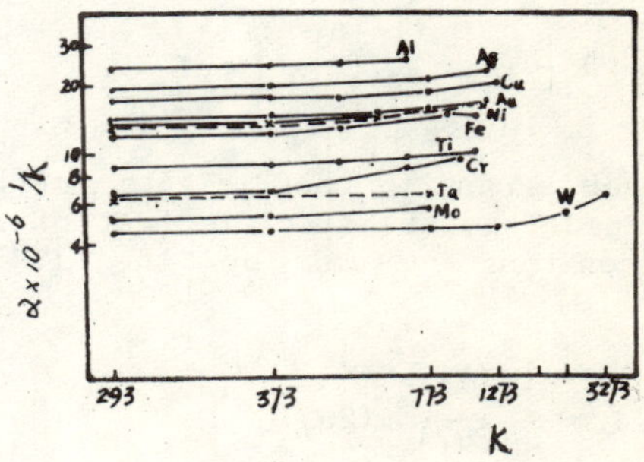

Fig. 1. The linear expansion coefficient of materials as a function of temperature for temperatures above 293K.

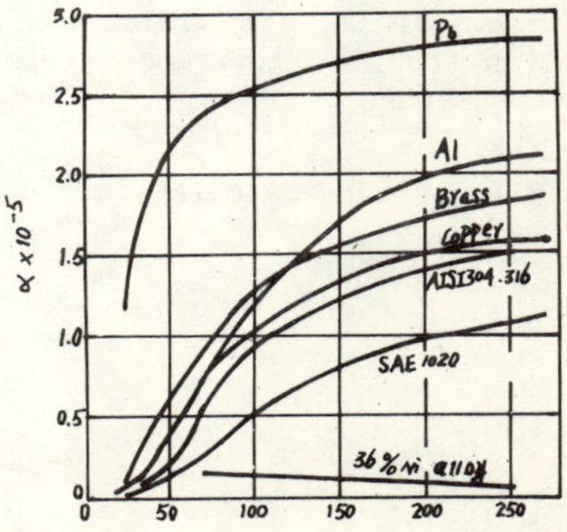

Fig. 2. The linear expansion coefficient of materials as a function of temperature for temperatures below 293K.

NEW METHOD OF EXPANSION FIT CALCULATION

From the definition of linear expansion coefficient, the following differential equation holds:

$$dL = L\,\alpha(T)\,dT \quad . \tag{3}$$

because

$$L\,\alpha(T)\,dT = [L_0 + L_0\,\alpha(T)]\,\alpha(T)\,dT$$

$$= L_0\,\alpha(T)\,dT + L_0\,\alpha^2(T)\,dT \quad , \tag{4}$$

the second term in the right hand side can be neglected, so

$$dL = L_0\,\alpha(T)\,dT \tag{5}$$

Where L_0 is the length of the work piece at a reference temperature, i.e., room temperature (293K). An integration of Eq. (5) gives

$$\Delta L = L_0 - L_T = L_0 \int_T^{T_0} \alpha(T)\,dT \quad . \tag{6}$$

The linear expansion coefficients, $\alpha(T)$, can be obtained from the specific heat of metals under constant volume by the following relation [8]

$$\alpha(T) \approx \frac{C_v(T)}{C_v(293)}\,\alpha(293) \quad .$$

Take copper as an example,

$$\alpha(T) \approx \frac{C_v(T)}{C_v(293)}\,\alpha(293) = 2.916\times10^{-7}\,C_v(T).$$

Substituting α into Eq. (6) yields the amount of contraction from the reference temperature T_0 to any temperature T.

To further simplify the calculation, let

$$R = \int_T^{T_0} \alpha(T)\,dT \quad , \tag{7}$$

then

$$R = \frac{\Delta L}{L_0} \quad . \tag{8}$$

Where R is the ratio of the amount of contraction to its original length, and is called the contraction ratio. Eq. (6) is then

$$\Delta L = L_0 - L = L_0 \int_T^{T_0} \alpha(T)\,dT = L_0\,R \quad . \tag{9}$$

The contraction ratios can be and have been measured experimentally. Table II shows the contraction ratio, R, for some materials [9].

Table II. Contraction ratio $\frac{L_{293} - L_T}{L_{293}} \times10^{-6}$

Name	Temperature (K)							
	20	40	60	80	100	150	200	250
Al	41.5	41.3	40.5	39.1	37.0	29.6	20.1	9.7
Ag	41.2	40.5	38.9	36.6	33.9	25.9	17.3	8.2
Cu	32.6	32.4	31.6	30.2	28.3	22.2	14.9	7.1
Fe	19.8	19.7	19.5	18.9	18.1	14.8	10.2	4.9
Mg	49.9	49.6	48.9	47.2	44.6	33.7	24.8	9.95
Ni	23.8	23.7	23.4	22.6	21.5	16.5	12.2	5.09
Pb	70.0	66.7	62.4	57.7	52.8	39.9	26.3	12.4
Ti	15.5	15.4	15.3	14.9	14.1	10.8	8.09	3.39
Zn	6.8	56.0	54.1	51.2	47.6	34.9	25.7	10.6
1	21.0	20.9	20.6	20.0	19.0	14.7	10.9	4.59
2	23.7	23.6	23.2	22.5	21.3	16.4	12.1	5.09
3	-	26.4	25.8	24.7	23.2	18.3	12.1	5.9
4	38.3	38.0	36.8	35.0	32.6	25.3	16.9	8.0
5	29.7	29.6	29.0	27.8	26.0	20.3	13.8	6.6
6	29.8	29.6	29.4	28.2	26.7	21.3	14.3	4.3
7	26.0	26.2	25.4	24.5	23.1	17.6	13.1	5.39
8	105	102	98	94	88	71	50.5	25.0
9	7.4	7.3	6.9	6.5	6.0	4.7	3.3	1.5
10	3.0	2.9	2.7	2.5	2.3	1.7	1.1	0.55
11	211	206	201	193	185	160	124	75
12	-	39	-	-	35	28	19	10
13	138	135	131	125	117	95	67	34

Note on Table II:
- 1 = SAE1020
- 2 = Inconel [Ni(80),Cr(14),Fe(6)]
- 3 = Constantan [Cu(60),Ni(40)]
- 4 = Brass [Cu(65),Zn(35)]
- 5 = Stainless Steel [304,316]
- 6 = Stainless Steel [347]
- 7 = Monel [Ni(67),Cu(15),Fe(1),Mn()]
- 8 = Epoxy
- 9 = Graphite [Parallel to Stretch]
- 10 = Graphite [Normal to Stretch]
- 11 = Teflon
- 12 = Bakelite
- 13 = Nylon [Nylon 66]

Under this new formulation, an expansion fit satisfies the following criteria,

$$\Delta L \geq \delta + \Delta \quad , \tag{10}$$

where δ is the interference and Δ is for assembly clearance and temperature compensation. A recommended value of Δ is between 0.05 to 0.1% of L (or D). For a given L and δ, Eq. (10) gives ΔL, which is then substituted into Eq. (9) for R, and the required cooling temperature can be obtained from Table II.

ESTIMATION OF ERRORS

It is clear from the above discussion that the critical part of an expansion fit calculation is the determination of ΔL. The precision of ΔL from Eq. (9) depends on the uncertainties of the contraction ratio, R. Since the errors of Table II are less than 10^{-6} mm per mm, they can be ignored. Therefore Eq. (9) is essentially error free.

Fig. 3 is the α-T relation for iron. When the material is cooled down from T_A to T_B, area ABCDA represents the integral of ΔL as given by Eq. (9). Whereas the prediction of Eq. (2) also depends on where the linear expansion coefficient, α, is evaluated. If $\alpha = \alpha_A$ is chosen, Eq. (2) gives the area of rectangle ABED. If $\alpha = (\alpha_A + \alpha_B)/2$ is chosen, the result is the area of rectangle ABFG. The relative error of Eq. (3) can then be expressed in terms of α and R as

$$\varepsilon_1 = \frac{\Delta L_1}{\Delta L} - 1 = \frac{\alpha_A}{R} \Delta T - 1 \qquad (11)$$

for $\alpha = \alpha_A$ (shown on Fig. 3 as the shielded area) and

$$\varepsilon_2 = \frac{\Delta L_2}{\Delta L} - 1 = \frac{\alpha_A + \alpha_B}{2R} \Delta T - 1 \qquad (12)$$

for $\alpha = (\alpha_A + \alpha_B)/2$ (shown on Fig. 3 as the double-shielded area).

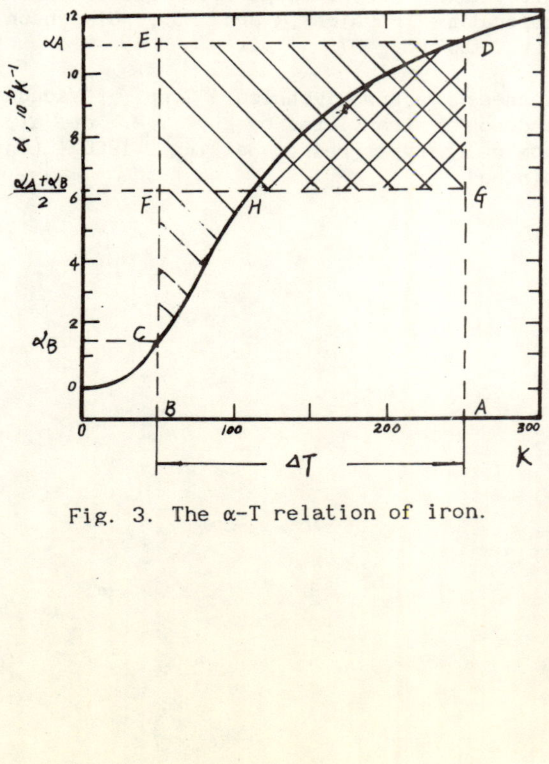

Fig. 3. The α-T relation of iron.

Table III contains the values of ε_1 and ε_2 for 3 cooling temperatures: dry ice (195K), liquid nitrogen (77K) and 50K. The largest deviation is found to be 42.2%, for SAE steels at 50K. These large errors strongly suggest that Eq. (2) can not be safely used for cryogenic applications.

Table III. Percentage errors in ΔL when $\Delta L = \alpha$ T is used in an expansion fit calculation.

Name	Cooling Temperature					
	195K		77K		50K	
	ε_1	ε_2	ε_1	ε_2	ε_1	ε_2
1	10.6	-0.2	31.0	-9.3	38.2	20.4
2	10.5	≈ 0	17.6	-11.4	22.9	-19.7
3	3.7	1.2	21.5	-10	29.3	13.7
4	7.0	-0.4	28.5	-13.7	37.0	-21.3
5	6.7	-0.8	26.8	-13.0	34.9	-21.3
6	8.7	-0.5	3 3.1	-12.9	42.2	-20.4
7	6.5	-0.2	23.8	-13.7	30.8	-25.1

Note on Table III:
1 = Beraloy [Be(2),Co(0.3),Cu(97.7)]
2 = Brass [Cu(65),Zn(35)]
3 = Constantan [Cu(50),Ni(50)]
4 = Inconel [Ni(80),Cr(4),Fe(6)]
5 = Monel [Ni(67),Cu(30),Fe(1.5)]
6 = SAE1020 [C(0.18),Mn(0.33),Si(0.01), rest Fe]
7 = AISI304 [C(0.08),Mn(2),Si(1), Cr(18-20),Ni(8-11)]

THE RELATION BETWEEN α and R

The contraction ratio is defined in terms of the linear expansion coefficient by

$$R = \int_{T}^{T_0} \alpha(T)\, dT \quad ,$$

which reduces to $R = \alpha\ \Delta T$ when α is a constant. In this case Eq. (6) also reduces to Eq.(2). This implies that Eq. (2) is accurate when α does not change with T very much, i.e., at high temperatures. However, if α is not a constant, Eq. (6) will be different from Eq. (2), and the difference becomes larger as ΔT increases (i.e., at lower temperatures).

As shown in Fig. 4, the contraction ratio versus temperature curves have steep slopes between 300K to 400K, but are more or less flat in the region of 0K to 100K. It indicates that, for a given temperature difference ΔT, ΔR is larger in the higher temperature region. For example, SAE1020 steel has $\Delta R = 2.5\times10^{-4}$ when T changes from 250K to 230K and 1.0×10^{-4} from 100K to 80K. Because the new method has taken this characteristics of R-T relation into account, it tends to provide "just enough" cooling, whereas conventional expansion fit usually over cools the sample. Therefore this new method can prevent the work piece from building up the unnecessary high strain as introduced by the large temperature difference. The coolant consumption is also reduced accordingly. These features are not available from any of the existing methods [4,5].

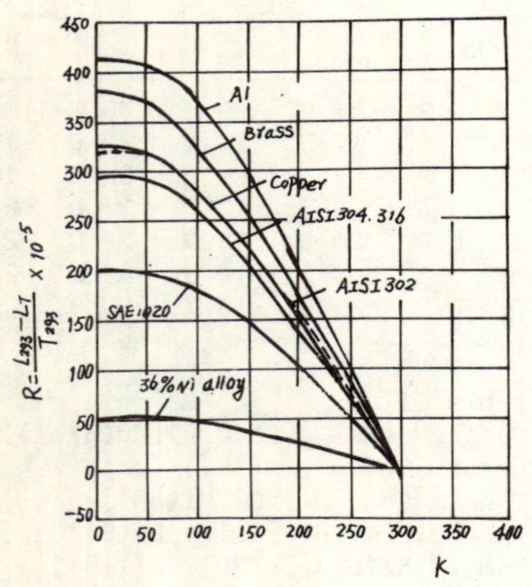

Fig. 4. The contraction ratio, R, as a function of temperature.

EXAMPLE

Consider the fit for ϕ60H/r. Inner work piece is SAE1020 steel. The maximum interference is δ = 0.054 mm, assembly clearance is chosen to be 0.075%.

$$\Delta = 60 \times 0.0075\% = 0.0405 \quad ,$$

$$\Delta L \geq \delta + \Delta = 0.0945 \quad ,$$

$$R = \frac{\Delta L}{L_0} = \frac{0.0945}{60} = 15.75\times10^{-4} \quad .$$

An interpolation of Table II gives the required cooling temperature to be 138K.

If one of the traditional methods were used instead, only liquid nitrogen temperature would be available. There would be an additional 61K of "over-cooling". The coolant consumption would also be considerably higher.

REFERENCES

1 Kovan, B.M. et al, "Handbook on Industry Technology", in two volumes. Mashgis, 1959. (in Russian)

2 Shen, H. et al, "Handbook of Mechanical Engineering", vol. 9, Industrial Engineering Publishings, Beijing, 1978. (in Chinese)

3 The Editors, "Metal Machining Handbook", Shanghai Science and Technology Publishings, Shanghai, 1982. (in Chinese)

4 Hicks, T.G. "Standard Handbook of Engineering Calculations", McGraw-Hill, New York, 1972.

5 Oberg, E., Jones, F.D., and Horton, H.L. "Machinery's Handbook", 21st Ed., Industrial Press, New York, 1980.

6 Yan, B.-Y. "Cryogenic Treatment Unit with Adjustable Temperatures Using Liquid Nitrogen", Sichuan Cryogenics, vol.1 (1987) pp.40-46. (in Chinese)

7 Smithells, C.J. "Metal Reference Book I", Butterworths Scientific Publishings, London, 1962.

8 White, G.K. "Experimental Technique in Low Temperature Physics", 3rd Ed., Clarendon Press, Oxford, 1979.

9 Japanese Vacuum Technical Corp., "Vacuum Handbook", translated by Lu, Y.M. et al, Atomic Energy Press, Beijing, 1986. (in Chinese)

8. SUPERCONDUCTING TECHNIQUE AND CRYOGENIC SYSTEM

CRITICAL CURRENT DENSITY OF THE HIGH T_c SUPERCONDUCTING OXIDES: $LnBa_2Cu_3O_{7-y}$

Zhang Qirui, Zhang Tao
Department of Physics
University of Science and Technology of China
Hefei, Anhui, P.R. China

ABSTRACT

Based on the observation of the microstructure and the measurements of the effects of superconductive glass state in single crystal and single phase polycrystalline $LnBa_2Cu_3O_{7-y}$ samples and combined with the theoretical analysis of type-II superconductors, the major factors related to the critical current density are discussed. Some suggestions about how to raise J_c are proposed in Physical view and the prospects for using the new materials in high field applications are given.

INTRODUCTION

It is well known that Tc, Hc_2 and Jc are the three most important basic parameters of the practical superconducting materials. As for the current widely-used material of Nb_3Sn, Tc=18.5K, Hc_2(T=4.2K)=24Tesla and Jc(T=4.2K,H=5 Tesla)=$1.6x10^6$A/cm^2. Since the significant breakthrough in high Tc superconducting oxide, it has been discovered that Tc=90-100K and $Hc_2(0)$=150-200Tesla in $YBa_2Cu_3O_{7-y}$. Hence, if its Jc can reach 10^5-10^6A/cm^2, the large scale application of superconduting electrical technology will be much easier to realize, in view of technology, and also it will be much cheaper to use liquid nitrogen as coolant instead of liquid helium, for the rich resource, low price and high-efficient refrigeration of nitrogen. So, there is no doubt that the development of superconducting electrical technology at liquid nitrogen temperature will bring about the revolutionary changes in many industrial departments.

In the past two years, many research laboratories in the world have focused on improving the superconducting properties of $YBa_2Cu_3O_{7-y}$ bulk materials, making wire or thin film for practical purpose, and improving the techniques to raise Jc. However the situation now is that the maximum Jc of the bulk or wire materials is only of the order of 10^3A/cm^2 at 77K and in zero magnetic field. Although some researchers have reported the Jc of 10^4A/cm^2, the result can hardly be repeated stably. Things seem much better for the single-crystalline thin film, grown on the substrates of single-crystalline $SrTiO_3$ or MgO by the methods of epitaxy, magnetron-sputtering and plasmon-sputtering, which Jc can reach 10^6A/cm^2 at 77K and remain at 10^5A/cm^2 under the magnetic field of 1 Tesla, but the quality and characteristics for practice are still needed to be improved. Therefore, how to raise Jc of the new superconducting oxides, especially for the bulk materials, is the key problem in the application of both the superconducting powerful electricity and electronics at liquid nitrogen temperature.

Based on the amount of experimental results got on the single-phase polycrystalline $LnBa_2Cu_3O_{7-y}$ (Ln=Y and rare earth elements) bulk samples, and from their structure, basic properties and effects of superconductive glass state (SGS), we have made a detailed discussion on the factors that limit the raise of Jc and given some proposals to raise Jc in physical view.

STRUCTURAL CHARACTERISTICS

It has been known that $LnBa_2Cu_3O_{7-y}$ belongs to the distorted orthorhombic perovskite structure with typical lattice parameters as: a=3.820Å, b=3.893Å, c=11.688Å. It is composed of three overlapped perovskite unit cells with the compacted layers repeated in the order of Cu(1)-O, Ba-O, Cu(2)-O, Ln, Cu(2)-O, Ba-O, and Cu(1)-O. The two central Cu(2)-O planescompose octohedrons with Cu(2) at the center, and the Cu(2)-O bands

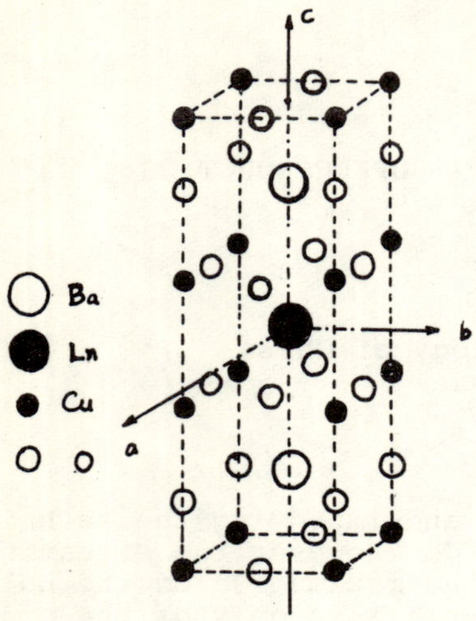

Fig.1 Crystal structure of orthorhombic $LnBa_2Cu_3O_7$ (ordered vacancy model)

along the direction of c-axis are longer than those in the a-b planes. This is because that the absence of all the oxygen atoms in Ln layer makes it a positive charge surplus layer, thus drawing the surrounding oxygen atoms more nearly to it which cause a distortion of the Cu(2)-O planes. The Ba-O layers are insulative for their ionic bands. In the Cu(1)-O planes (between two Ba-O layers), since the oxygen atoms in the direction of a-axis are easy to lose, an ordering structure with one-dimension chains in the direction of b-axis is established. It is widely believed that the Cu(2)-O planes and Cu(1)-O chains play the predominant role for the high Tc superconductivity and also lead to its different characteristics from the conventional superconducting materials.

High resolution electron microscope (HREM) observations of the microstructure of the single phase polycrystalline $YBa_2Cu_3O_{7-y}$ indicate that there exist three regions of distinctly different structure (Fig.2). Region A is the grain area, where the atoms are orderly arranged to form the lattice. It is of orthorhombic structure and is commonly considered to be responsible for 90-100K superconductivity. Region B is the lattice distorted area just in the neighbourhood of the grains, where the atoms are arranged only to form short range order. It is probably low Tc or nonsuperconducting metal phase. Region C is the intergrain area of amorphous structure. There is lower Y (Y:Ba:Cu=1:3:4) and exist $BaCuO_2$ and CuO phases. As for its electrical property, it is probably semiconductive or even insulative. The picture of high Tc phase surrounded by low Tc or nonsuperconducting metal phase and semiconductive or insulative phase is just the physical foundation of the SGS in polycrystalline $YBa_2Cu_3O_{7-y}$.

FOUNDAMENTAL PHYSICAL CHARACTERISTICS

Quite a lot experimental results have indicated that $LnBa_2Cu_3O_{7-y}$ are superconductors with Tc=90-100K. In normal state, it shows typical metallic behavior with linear temperature dependence of resistance. Its main charge carriers are holes with density of the order of $10^{21}cm^{-3}$. While in superconducting state, it is discovered the electrons exist in pairs by the measurement of magnetic flux quantum and a.c. Josephson effect. It is also believed that there are $2\Delta/kT_C \sim 5$ and incomplete isotope effect in this material. According to the experiments on single crystal, $LnBa_2Cu_3O_{7-y}$ shows strong anisotropy in superconductivity(such as Tc , Hc_2 and Jc) and the superconducting coherent length in a-b plane and in the direction of c are $\xi_{\parallel}$=30-70Å and $\xi_{\perp}$=7-10Å, respectively. So it is the dirty-limit type-II superconductor with superconducting parameters as follow: $\kappa \simeq 100$, $Hc_1(0) \doteq 800G$ $Hc_2(0) \simeq 150\text{-}200$ Tesla and $dHc_2/dT \simeq -2.5$ Tesla/K. In magnetic field ($H > Hc_1(T)$), the magnetic flux distribute in triangle figure, together with the pinning of magnetic flux and creep and flow of those driven by thermal activation or Lorentz force.

The physical properties of $LnBa_2Cu_3O_{7-y}$ both in normal and superconducting state are very sensitive to the number and the ordering level of distribution of the oxygen vacancies in the lattice. The high Tc superconductor with orthogonal structure transfers into tetragonal one when y increases from 0.1 to0.6, with the vanish of its superconductivity. In the meantime, R(T) change into a semiconducting one and magnetism abides by the Curie-Weiss law instead of the weak paramagnetism.

In polycrystalline $LnBa_2Cu_3O_{7-y}$ samples, the superconductive glass state (SGS) behaviors which introduced first by Bednorz and Müller[1] and vertified by the experiments of inner Josephson effect and magnetic hysteresis in bulk and thin film samples are often observed We have proposed a series of experiments, using the method of electrical measurement, and observed many new phenomena originated from the SGS in the single phase polycrystalline $LnBa_2Cu_3O_{7-y}$ sam-

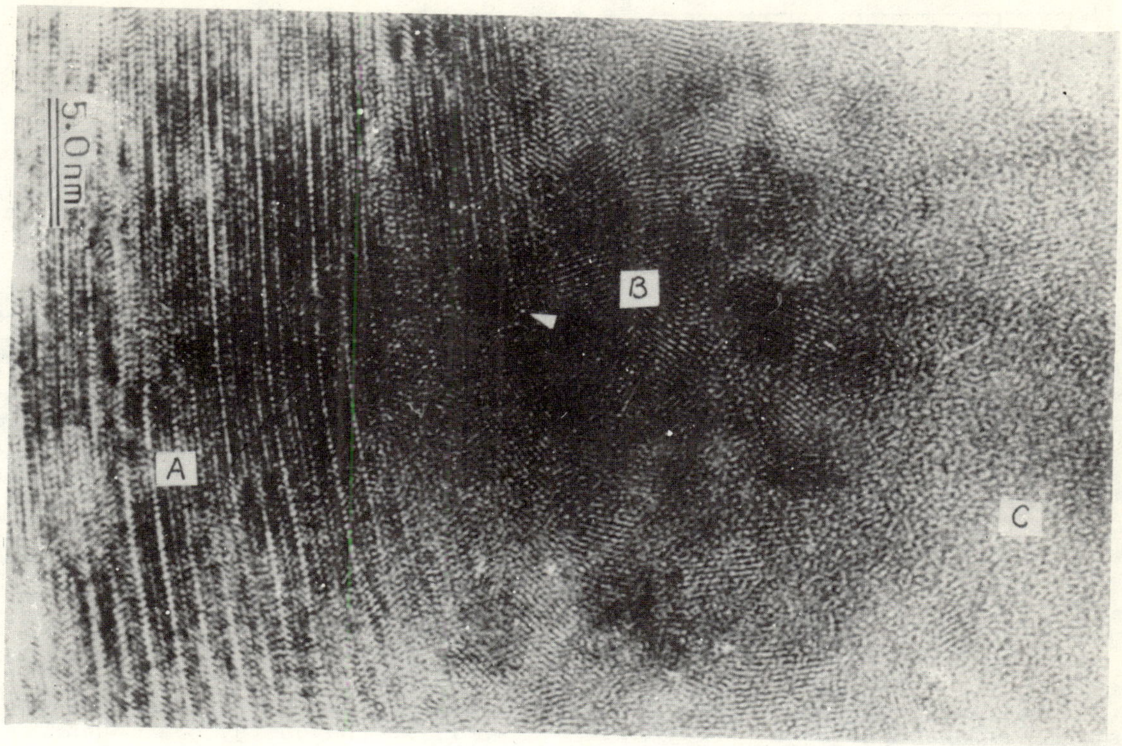

Fig.2 HREM micrograph of the single phase polycrystalline $YBa_2Cu_3O_{7-y}$

ples, such as the influence of weak magnetic field on the transition from normal state to superconducting state, negative magnetoresistance, two types of hysteresis, behavior of thermal activation, anomaly of Hall effect and characteristics of the critical current, etc. And based on these results, we introduced a 3-D random weak-link Josephson junction network to give them a satisfactory explanation. It can be seen afterwards that the large amount of weak-link junctions in the bulk material is one of the most important key factors that limit their Jc.

CRITICAL CURRENT DENSITY (Jc)

a. Bulk polycrystalline materials

The Ic(H) curve of the single phase polycrystalline sample of $GdBa_2Cu_3O_{7-y}$ at T=80K is shown in Fig.3. Its major feature is that the critical current is very sensitive to the magnetic field. Whem the magnetic field increases from 0 to 60G, the critical current drops by about 80%. Further increasing field only leads to slow decrease of the critical current. The regularity of the behavior of Ic(H) at different temperature is almost the same and similar results are also observed in the single phase polycrystalline $YBa_2Cu_3O_{7-y}$.

The above-mentioned phenomena can be explained by the model of 3-D weak-link junction network. We all know that for a single Josephson junction, the response of the critical current to the magnetic field satisfies the superconducting diffraction condition (quantum interference effect):

$$Ic(H)=I_o|\sin(\pi H/H_o)/(\pi H/H_o)| \qquad (1)$$

$$H_o=\Phi_o/(dL) \qquad (2)$$

where H_o is the superconducting diffraction period, Φ_o is the magnetic flux quantum, $d=\lambda_1+\lambda_2+d'$, λ_1 and λ_2 are the penetrating depths, d' is the thickness of the junction, L is the width of the junction. The corresponding curve of Eq.(1) is shown in Fig.3 (curve 1). It has a definite diffraction period and peaks. However, it is obvious that the experimental results do not satisfy (1). And we should make at least two of the following amendments. First, in the real network system consuituted by a large amount of junctions, i.e. the period H_o and the coupling intensity I_o follow some distribution function $f(H_o,I_o)$. This makes the diffraction period of the network system obscure and difficult to be distinguished. Second, the random distribution and orientation of the grains of approximate size 10μm make such junc-

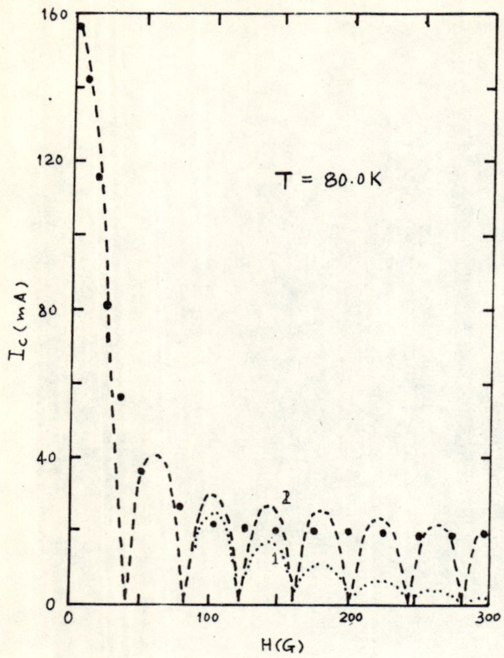

Fig.3 Ic(H) curve of single phase polycrystalline $GdBa_2Cu_3O_{7-y}$ at 80K. Solid circle represents the experimental result.

tion be non-ideal Josephson junction with inhomogeneous current density distribution (Fig.3, curve 2). In the simplest case, Ic(H) of the real weak-link junction network may be describe as follow:

$$Ic(H)=\int_0^\infty dIo\int_0^\infty dHoIc(H,Ho)f(Ho,Io) \quad (3)$$

where $I_c(H,H_o)$ denotes the critical current of a nonideal junction with diffraction period Ho, f(Ho,Io) is the distribution function for the network system.

The Ic(T) curve of the single phase polycrystalline sample of $GdBa_2Cu_3O_{7-y}$ in zero field is shown in Fig.4. It is found that within the range of our measuring temperature, Ic(T) always satisfies $d^2I_c/dT^2>0$. The behavior of Ic(T) in the applied magnetic field is similar to that in zero field with changes only in the magnitude of the critical current. The regularity of Ic(T) we measured is similar to the behavior of S-N-I-N-S junction(Josephson).

For a S-N-I-N-S junction, if the thickness of the metal layer(N) is smaller than or equal to the mean free path l of the electron, its Ic(T) characteristic is similar to that of the Josephson junction, which can be described by Ambegaokar-Baratoff theory[2]:

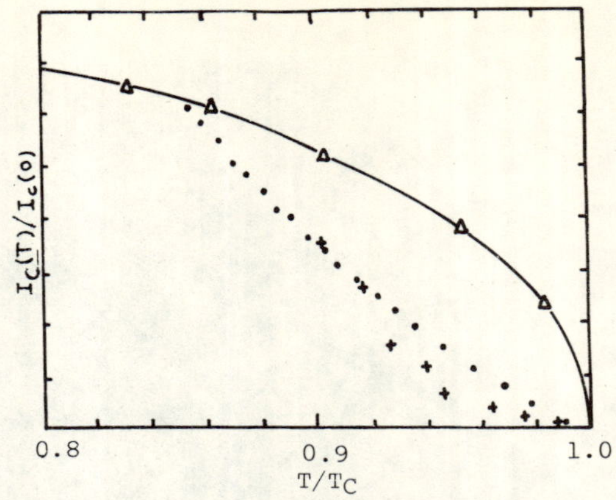

Fig.4 Ic(T) curve of $GdBa_2Cu_3O_{7-y}$ sample in zero magnetic field. Solid circle represents the experimental result of ours; cross represents the result of Chen et al.[5]; delta represents the result of Lu etal.[4]; solid line is the theoretical curve of Ambegaokar-Baratoff theory[2].

$$Ic(T)=(\pi/2)(\Delta(T)/R_N)\tanh(\Delta(T)/2k_BT) \quad (3)$$

where R_N is the junction resistance and $\Delta(T)$ is the energy gap. When T approaches T_c, $I_c(T)$ is proportional to (T_c-T) i.e. it has linear behavior. If the insulative layer (I) vanishes, it degenerates to S-N-S junction, whose critical current Ic(T) is described by the following equation:

$$I_c(T)=Io(1-(T/T_c))^2\exp(-2d_N/\xi_N) \quad (T\to Tc) \quad (4)$$

where ξ_N is the coherent length of the electron in normal state, d_N is the thickness of the normal metal layer. If $d_N>l$, its Ic(T) characteristic must be carefully described by McMillan theory [3] with notable feature as:

$$d^2I_c(T)/dT^2>0 \qquad (T\to T_c) \qquad (5)$$

Our experimental results rightly follow the behavior of S-N-I-N-S junction with $d_N>l$, which may reflect that the thicker grain boundaries play an important role in restricting the critical current density. However, due to different preparational conditions, the thickness and properties of the N layers or the I layers may vary greatly which leads to the change of the behavior of Ic(T) among those described by eq.(3)--(5). The results of Lu et al.[4] follow eq.(3)(A-B theory) very well. This indicates that the N layers in their sample are relatively thinner, and the weak-link junctions are of the feature of S-I-S junctions. The results of Chen et al.[5]

are similar to ours, i.e. showing the feature described by eq.(5). We believe that it is due to the thicker grain boundaries and that the main contribution to the Ic(T) of the bulk sample is from the S-N-I-N-S junctions.

b. Single crystalline materials
Since it is one of the most efficient methods to determine the intrinsic critical current of superconducting materials by measuring that on the single crystal, many researches have been carried out in this field in the world.

Fig.5 shows the I-V curve we measured on single crystalline sample $YbBa_2Cu_3O_{7-y}$ We adopted the criterion of Ic as the value of the current at the appearance of voltage on the sample. About the result shown, special attention should be paid to the fact that the crystal has two critical currents, Ic_1 and Ic_2 which shows the nonideality of the sample.When $I<Ic_1$ (=52mA), no voltage is measured on the sample. At $I>Ic_1$, voltage appears on the sample and increases non-linearly with the current I until $I \approx I'$ (=92mA), but its slope much smaller than the resistance of the sample at normal state. When $I=Ic_2$, the voltage increases drastically and the sample enters the normal state. This phenomenon is similar to that observed on bulk polycrystalline sample which is related to the SGS and we attribute the two critical currents to the weak links and crystal respectively. The weak links in single crystalline material may originate from two sources as follow: first, imperfection of structure which mainly concluds the twin boundary between crystals, structural deviation and mechanical defects; second, the inhomogeneity of composition which concludes the varied stoichiometry(mainly for oxygen) and impurities. Although it is almost impossible to decide the Jc of the singlecrystalline material by above dc transporting measurement for not knowing the effective section, we think the Jc cannot reach a high value since so many weak links exist in the non-ideal samples. In the other single crystals we grow with only one Ic, its critical current seems somewhat larger. This may indicate that the weak links are also the major hindrance to the raising of Ic in the single crystals.

Many laboratories have measured the magnetization Jc in single crystal. By measuring the magnetization curve (hysteresis) and from the critical model of Bean, they estimate the Jc of single crystal. The typical values of Jc are as follow[6,7]: at T=4.2K, $Jc^{\perp} \approx 2x10^7 A/cm^2$

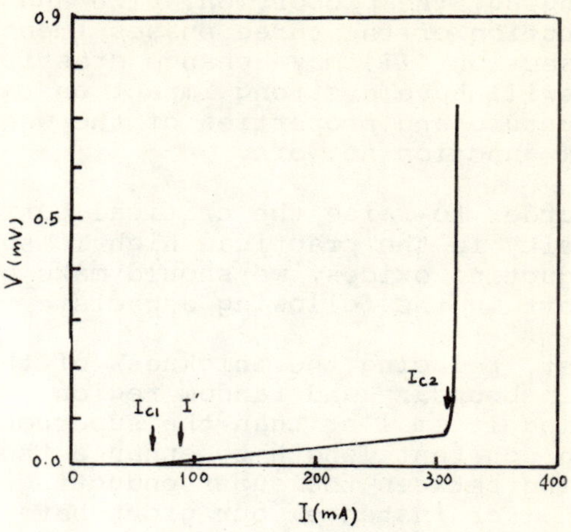

Fig. 5 the I-V curve of $YbBa_2Cu_3O_{7-y}$

$Jc^{\parallel} \approx 2x10^6 A/cm^2$ and they degrades by one order applying magnetic field of 4 Tesla.

c. Thin film materials
It is commonly believed that the high quality thin films are vital for fundamental physical studies as well as for device application; so many groups in the world have devoted themselves to the improvement of techniques to prepare (new high Tc) superconducting oxides in the form the of thin film.

In order to get high quality thin film, the selection of substrate and the temperature plays the key role. It has been widely-accepted that the single crystalline $SrTiO_3(001)$ and MgO(001) are the best choice and 600-700°C is the most suitable temperature of the substrates. As for the methods to grow thin film, it seems almost all methods, such as epitaxy, dc or rf magnetron-sputtering, plasmon-sputtering, laser-melted evaporation, etc., can get good film under suitable preparational conditions. The best thin film updating is prepared by Itozaki et al.(Sumitomo Electric) in epitaxial singlecrystalline Ho-Ba-Cu-O films grown on MgO(001) substrates using magnetrosputtering. Its $Jc=3.54x10^6 A/cm^2$ in zero field and $Jc=1.5x10^6 A/cm^2$ in 1.0 Tesla at 77.3K. This result is consistent with the intrinsic critical current density measured in the single crystal mentioned above.

SOME PROPOSALS ON RAISING Jc

According to all the experimental results and theoretical analysis mentioned above the key to the raise of Jc of polycrystalline $LnBa_2Cu_3O_{7-y}$ materials lies in the adoption of the most reasonable factors of original stoichiometry and preparational condition. With these factors changing, the proportion, size and distribution of the three phases (mentioned in section III) may change drastically. Thiswill have a strong impact on the structure and properties of the weak-link junction network.

In order to raise the critical current density in the practical high Tc superconducting oxides, we should make great effort in the following aspects.

First, reducing the thickness of the grain boundary and random region and making it smaller than the superconducting coherent length to enhance the coupling between the superconducting grains. For instance, our group have studied the effect of adding Ag and Pb in the $YBa_2Cu_3O_{7-y}$ sample and found that the Jc(77K) of $YBa_2Cu_3Pb_xO_{7-y}$ sample is three times as much as that of $YBa_2Cu_3O_{7-y}$ sample.

Second, making the grains orient in a certain direction, because of the strong anisotropy in the superconductivity of $LnBa_2Cu_3O_{7-y}$. If the grains are connected in a-b plane, then d(thickness of the grain boundary)<30-70Å is requested; while connected along c-axis or randomly oriented, d<7-10Å is requested. In addition, the random region and grain boundary in the whole wire (or ribbon and thin film) need to be less than the (superconducting) coherent length everywhere. Under these restrictions, the random regions and grain boundary may no longer be limit factor on Jc, but serve as the pinning centers to help increse Jc. The AT&T Bell Laboratory have made the $YBa_2Cu_3O_{7-y}$ samples, by the method of melted texture growth, which Jc (77K)=7400A/cm^2, and still Jc$\sim 10^3$A/cm^2 under 1 Tesla.

Third, making the superconductivity the same everywhere in the grains. In order to ensure the homogeneity of the number and ordering degree of the oxygen vancancies, we'd better turn the materials into thin wire or ribbon.

Fourth, if the thickniss of the random region or grainboundary in the polycrystalline materials is impossible to be reduced to the order of coherent length, it would better to grow single crystalline fibre or thin film, and in the mean time introduce the pinning centers with the size of coherent length and separated by the penetrating length.

From the above discussion, we can see the difficulties in raising the critical current density of the high Tc superconducting oxides. As the practical superconducting materials, it is also necessary to adopt the method of stablization like that used in conventional materials.

PROSPECT OF APPLCATIONS

There is no doubt about the great potential of the new high Tc superconducting materials in applications. It may bring out revolutionary changes in the fields of electrical industry, space science, electronics, computer and transducer science, traffic, medical science and high-energy physics, etc. According to the current level and tendency of development of the materials, it is estimated that its application in superconducting electronics can be realized quite soon, but that in electrical power still need 10-15 years. In the near future, some applications like SQUID apparatus, infrared transducer and superconducting magnetic screen may come into use. While the large scale applications such as the superconducting magnet, superconducting computer and transmission of electricity are the goal for a long time.

As mentioned above, how to raise Jc of the high Tc superconducting bulk material is still a key problem in the development of electrical power applications. But there seems no reason to feel pessimism about this. Hingoroni, from the Electric Research Laboratory of America once estimated the intrinsic Jc as high as 10^7 A/cm^2 in $LnBa_2Cu_3O_{7-y}$ supercondutors. So it is quite possible to prepare practical wire or ribbon materials by using novel techniques. This is the major target of research in a certain period from now on. In the meanwhile, researches on the intensity, flexibility , stability and reliability of working at high temperature should carried out.

REFERENCES

[1]K.A. Müller, M.Takashige, J.G.Bednorz, Phys.Rev.Lett.,58(1987) 1143
[2]V. Ambegaokar, A. Baratoff, Phys. Rev. Lett.,10(1963) 486; 11(1963) 104
[3]W.L.McMillan, Phys. Rev. 175(1968)537
[4]L. Lu, H.M. Dun, D.L. Zhang, preprint
[5]K.Y. Chen, B.C. Mao, Y.M. Cai, J.Luo, B. Zhou, Y.J. Qian, preprint
[6]P.R. Dinger, T.K. Worshington, W.J. Gallagher and R.L. Sandstrom, Phys. . Rev.Lett., 58(1987) 2687
[7]D.E. Farrel, B.S. Chandrasekhar, M.R. DeGuire, M.M. Fang, V.G. Kogan, J.R. Clem and D.K. Finnemore, Phys. Rev. B36(1987) 4025

CONSIDERATIONS ON AND EXPERIENCE WITH HELIUM FORCED-FLOW-COOLING TECHNIQUES FOR SUPERCONDUCTING FUSION MAGNETS

W. Lehmann
Kernforschungszentrum Karlsruhe (KfK), Institut für Technische Physik
Postfach 3640, D 7500 Karlsruhe, F.R. Germany

ABSTRACT

Helium forced-flow-cooling (ffc) through internally cooled superconductor cables has proven true as a concept superior to helium bath cooling in powerful and large-volume fusion magnets. Manifold experience with He I ffc under single phase "supercritical" conditions has been acquired with the Euratom LCT Coil at KfK and with three LCT Coils altogether within the LCT Test Facility at ORNL in USA. Complementary alternative ffc modes have been investigated at KfK for other actual applications: two-phase He I ffc for pulsed TOKAMAK Poloidal Field Coils and He II ffc for TOKAMAK Coils in general to enhance the field of NbTi-Coils up to 11 T.
This report covers the main aspects on He ffc technology, summarizes the results of the development work conducted at KfK and outlines the forseen further investigations in this field and at this place.

1. REASONS FOR FORCED-FLOW-COOLING

Large and powerful superconducting magnets are best cooled by forced-flow-cooling (Table 1). The advantages become particularly attractive in large superconducting magnets for fusion applications, for instance, for the toroidal and poloidal coils of planned next step TOKAMAK reactors (NET, ITER) with approximate dimensions of 30 m circumference for the D-shaped toroidal field coils and up to 70 m circumference for the circular poloidal field coils.

Table 1: Advantages and disadvantages of He forced-flow-cooling compared to He bath cooling

Advantages

- powerful cooling also for long channels
- possibility to adapt cooling by routing He flow
- effective cooling irrespective of the physical location (1 phase cooling)
- specific und uniform cool down, warm up
- low He-volume within the magnets
- better mechanical stability of the winding
- better electrical insulation of the winding

Disadvantages

- higher He mass flow ($\dot{m}$) required
- energy requirement for He circulation or heat load at low temperature, respectively
- instabilities (also in this case) possible if there are big changes in density

2. SINGLE PHASE OR TWO PHASE FFC ?

Figure 1 shows the useful enthalpy differences of helium available under single-phase and two-phase operating conditions. In case of single-phase cooling at so-called supercritical cooling conditions ($p > p_c$), cooling capacities equal to those of evaporation cooling are available only if a relatively high temperature increase between the inlet and the outlet of the cooling channel and also a low pressure level can be permitted. On the other hand, two-phase cooling, offering much less room for temperature and pressure variations, allows very attractive cooling capacities to be achieved at relatively low LHe I temperatures and at high permitted evaporation fraction. However, depending on the structure/separation of the two phases, high vapor fractions may result in highly increased pressure losses, flow and heat transfer instabilities, which may be difficult to calculate. Especially in complex superconducting magnets with long, wound, parallel cooling channels only a very small part of the heat of evaporation can therefore be utilized without any problems (e.g., $x_{outlet} \leq$ 10-20 %). The enthalpy differences, which can be used for magnet cooling, in that case are comparable to those associated with single-phase cooling, which can be calculated accurately. Moreover, single-phase ffc allows operation at higher He temperatures, which may be permitted in some superconducting magnets, for instance, with Nb_3Sn conductors.

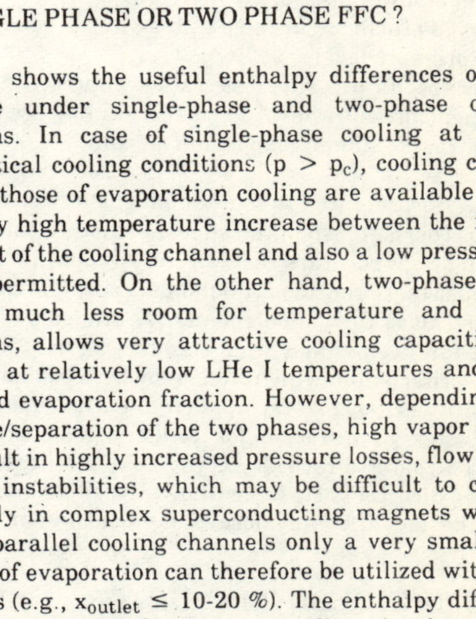

Fig. 1: Helium pressure-enthalpy diagram. Field of condition.

3. OPERATING DATA FOR SINGLE PHASE FFC (HE I). LCT EXPERIENCE

Operating data

For cooling of superconducting magnets in TOKAMAK fusion reactors, He I cooling temperatures of 4.2-4.5 K are suitable for reasons of economy and reliability in the cooling supply. This also applies because the Nb_3Sn- or other A15-conductors to be used there will have higher transition temperatures than NbTi. However, in magnet test facilities, it is reasonable to set magnet operating temperatures also below 4.2 K so that the superconductor and the coil elements can be operated at the extremes of their capability, i.e., even with low safety margins or at the highest possible fields. The pressure level within the magnet must be optimized on the basis of the following aspects:

- High useful enthalpy differences of He for cooling the magnet at lower pressure levels (see Fig. 1).
- Improved efficiency in the JT stage of the refrigerator at lower pressure levels.
- Higher specific pressure losses at lower pressure levels.
- Sufficient pressure level for direct cooling systems with downstream "far end" expansion machine .

The mass flow required to cool the magnets results from the continous thermal load of the superconducting cable and the permissible temperature increase of the helium within the cable under these continous and under transient disturbance conditions.Mass flow supplements for transient heat loads have so far been taken into account by use of so-called stability criteria following criteria for bath cooled coils. But the stability criteria so far applied to bath cooled magnets, such as Stekly's criterion for full cryogenic stability, or Maddock's cold-end-recovery cryostability criterion, cannot be applied directly to He forced-flow- cooled magnets. This is due to the fact that any local heat load in an internally forced-cooled superconducting cable will affect the whole downstream region as well.

For this reason, all forced-flow-cooled LCT coils were dimensioned on the basis of so-called critical-energy stability considerations [1]. A problem associated with this approach was the correct assumption of possible disturbances and of the intensity and propagation of the heat released in that case. In the conductor concepts of the three forced-cooled LCT coils shown in Fig. 2, e.g., different disturbance probabilities caused by conductor movements have been anticipated because of the different cable designs. On the other hand, also the cooling reserves (enthalpy change potentials) in the conductor material and in the coolant differ. as do the transient heat transfer conditions because of the different perimeters and surfaces of the conductor wetted by the coolant. The extra mass flows calculated for safe removal of the probable critical energy therefore differed, also on account of the pressure drop and, last, but not least, because of different extra safety margins (Fig. 2).

LCT experience

In the late seventies KfK has performed initial experiments with single phase He I ffc in long channel geometries with relevance to fusion magnets within the HELITEX Test Facility [2]. On that basis, KfK decided in favor of this cooling mode for the European toroidal field coil of the IEA Large Coil Task (LCT) project.

Prior to the delivery of the European LCT (LCT-EU) coil to USA, it was tested within the KfK TOSKA Test Facility in spring of 1984 [3]. This test made it possible to get first operating experience with supercritical He I ffc within the 28 parallel conductor cooling channels under field conditions up to 5.6 T and within a closed Secondary Cooling Circuit (SCC, chapter 6) with a cold piston pump as He circulator.

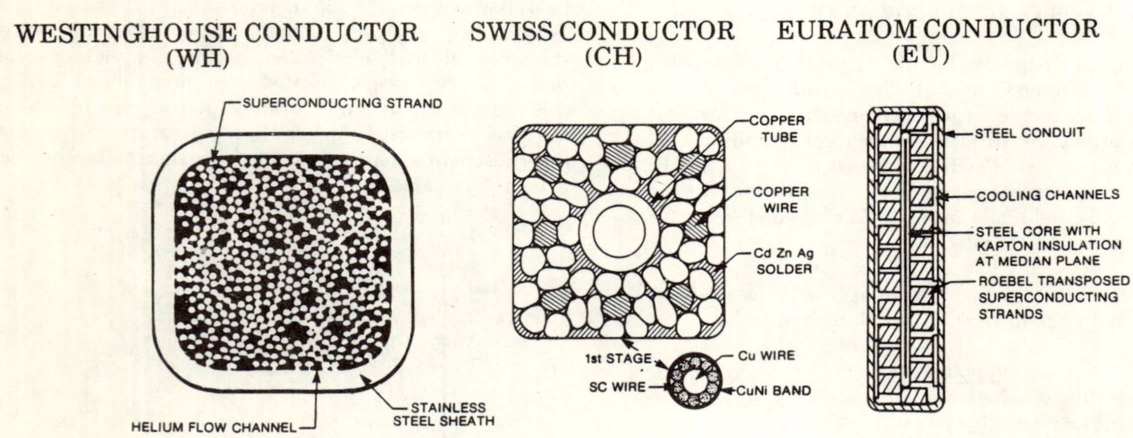

	WH	CH	EU
Areas:	Nb_3Sn : 1,87 cm^2	NbTi : 0,43 cm^2	NbTi : 0,29 cm^2
	Cu : 1,38 cm^2	Cu : 1,79 cm^2	Cu : 1,38 cm^2
	He : 0,91 cm^2	He : 0,159 cm^2	He : 0,95 cm^2
overall dimension	: 20,8 x 20,8 mm	18,5 x 18,5 mm	40 x 10 mm
hydraulic dia d_h	= 0,032 cm	0,45 cm	0,154 cm
cooled perimeter P	= 112,9 cm	1,41 cm	16,5 cm (24,6 wetted)
cooled length L	= 120 m x 46 75 x 2	250 x 22	250 x 28
Nominal current I	= 17760 A	13000 A	11400 A
$\dot{m}_{total}$, design [9]	= 109 g/s	121 g/s	≦ 280 g/s
$\dot{m}_{total}$,Standard I Test	= 120 g/s	70→33 g/s	150→33 g/s
($\dot{m}_{case}$,Standard I Test	= --	140→120 g/s	50→33 g/s)

Fig. 2: Design and cooling parameters of the LCT Forced-Flow-Conductors [6]

The feasibility and safe operation of s.c. toroidal field coils in a torus structure was demonstrated in 1986/87[4]. This was done within the International Fusion Superconducting Magnet Test Facility IFSMTF [5] at Oak Ridge/USA. In this exclusive magnet technology experiment, TOKAMAK reactor conditions for the coils were simulated under steady and non-steady state as well as symmetrical and asymmetrical mechanical, electrical and thermal loads within the six-coil torus. Three of the six coils (D-shaped with an inner bore of 3,5 x 2,5 m) were forced-flow-cooled by supercritical He I. The coil conductor configurations can be seen from Fig. 2. Ffc was done in a PCC mode. (chapter 6). The maximum mass flow of the screw-type compressors of the refrigerator was 300 g/s. Subcooling and recooling of the supercritical helium was achieved by a 3,5 K/0,5 bar heat exchanger bath. The maximum pressure difference available at the magnet was 5 bar, as the far end expansion turbine required a minimum inlet pressure of 10 bar.

The "thermohydraulic" behavior" of the forced-flow-cooled coils under steady state nominal conditions is apparent from Fig. 3. The helium mass flow $\dot{m}$ required for coil cooling and stabilization was considerably lower in all coils than had originally been assumed. This drastically reduced the power required for He circulation through the magnets and the heat load generated at LHe temperature level ($\sim \dot{m}^3 \cdot C$).

Stability tests were carried out with all three coils to verify the amount of deviation from the original nominal conditions (Fig. 2) which is permissible in the thermal load (additional loads) and in the mass flow (part load), respectively:

- "Current sharing threshold tests" were performed with the EU and the WH coils in which the helium was heated electrically before entering specific superconducting channels and the point at which a voltage generated could be observed (approaching the operating limits) [7, 8].
- "Nuclear heating tests" were performed with the CH coil: The heat load was generated by means of 11 full-turn constant ohmic heaters [9].
- Tests at mass flow rates reduced from the nominal data (Fig. 2) as originally defined were run with all three forced-flow-cooled coils under Standard I Conditions (test coil under 100 %, neighbouring coils at 80 % nominal current).

The most important features experienced with supercritical He I ffc of the LCT-coils may be summarized as following.

Flow Characteristics

- Pressure losses for He operating conditions can be calculated with great accuracy on the basis of model experiments at ambient temperature ($\Delta p(Re)$).
- Because of its special winding technique ("two-in-hand, He inlet in the innermost layer"), the LCT EU coil has a particularly low specific pressure loss (cf. Fig. 3).
- The necessary coolant mass flows in all three forced-flow-cooled coils were (far) below assumptions, corresponding to the critical energy stability considerations. Consequently, the pressure loss in the winding of the LCT EU coil dropped from an expected level of 1 bar (at an expected 150 g/s) to $\leq$ 0.1 bar at $\leq$ 33 g/s.
- The distribution of the coolant flow over many parallel channels of equal lengths (CH and EU coils) was uniform and easy to manage without the assistance of control valves under all operating conditions. In the EU coil, this was true even at the extremly low mass flow of 8 g/s through the seven double pancakes ($\rightarrow$ flow velocity in the SC cable $\approx$ 0.02 m/s).

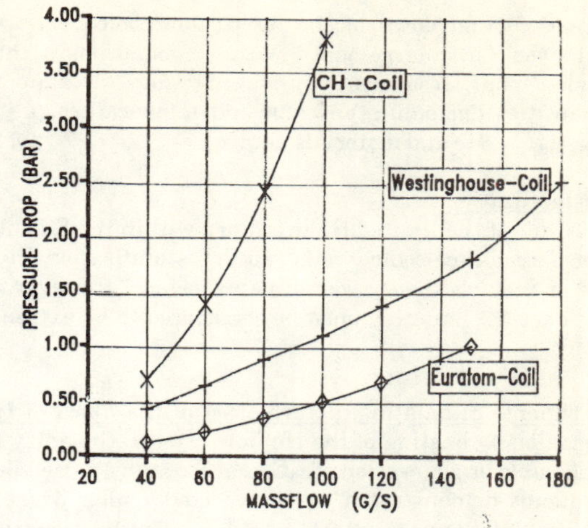

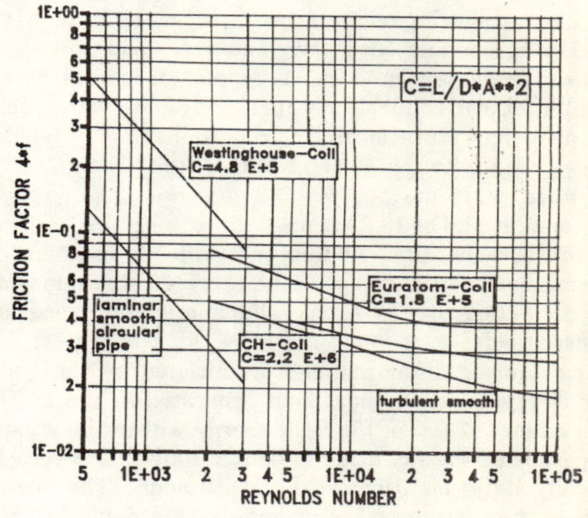

Fig. 3: Pressure loss characteristics and friction factors $\lambda = 4 \cdot f$ of the LCT-Conductors

Heat Transfer

- The heat transfer to He meets expectations in accordance with the Dittus-Boelter correlation. This means, e.g. a heat transfer coefficient α = 17 mW/cm^{2}K for a mass flow of 33 g/s through the LCT-EU-winding. The α-values required in the NbTi coils for steady state cooling of the thermal loads (static heat loads on the winding and nuclear heat in accordance with fusion reactor conditions with $\leq$ 150 µW/cm^2) are far below this level.
- Related to the total volume of the CH-coil the simulated nuclear heat load was 0,12 mW/cm^3, which corresponds to twice the level of nuclear heat load expected in NET. This heat load could be easily carried away during charging of the coil to 13 kA/6 T and operating at that level with mass flow rates of 70 g/s through the winding and 140 g/s through the casing at 12 bar inlet pressure and 3.8 K and 4.1 K inlet temperature, respectively [9].
- Under transient conditions (experiments with pulsed fields causing AC heat loads in the conductor), the maximum thermal loads estimated by calculation at the LCT EU conductor were 0.3 mW/cm^2. This corresponds to a heat flux approx. 100 times lower than what can be removed by a He mass flow of only 1 g/s per cooling channel in accordance with the Dittus-Boelter relation. Consequently, no stability problems whatsoever arose when the He-temperature was raised to 0.3 K below the current sharing temperature at nominal current.

Dosed cooling down of the forced-flow-cooled coils was achieved in agreement with calculations (three-dimensional treatment by taking into account the quantities depending on time and temperature, i.e. $\dot{m}$, $T_{He\text{-}Supply}$, He and materials data, α).

Cryostabilization
As a result of the limited He inventory within the SC cables of the forced-flow-cooled coils, no cryostabilization in the sense of Stekly's theory was to be expected. However, the experience accumulated must be considered to be extremely positive.

Experiments Simulating disturbances in the Form of Heat Generation by heating of the He flow prior to the entry into the SC cable or generation of AC heat losses by pulsed fields and dumps demonstrated an unexpectedly high degree of stability sufficient for all TOKAMAK opeating conditions. The "current sharing temperatures" measured at the EU coil were in agreement with the conductor load line as determined in single-strand (short-sample) measurements [7].
The LCT-EU coil could be operated safely close to the current sharing temperature locally at conditions "far outside the cryogenic stability region according to Stekly" ($\alpha_{St} > 1$; e.g. $\alpha_{St} = 18$ with $\dot{m}_{Winding} = 30$ g/s, $\alpha_{Dittus\ Bolter} = 16$ mW/cm^{2}K at 11.4 or 15.2 kAmps).
Due to the monolithic winding and principle used for force transmission the EU-coil could be operated stable for hours with currents above the rated value and including exposure to pulsed fields.
During dumps, either planned or initiated by the quench detection systems, the heat load generated in the LCT-EU coil was low (< 1 % of the field energy within the winding) and permitted the s.c. mode to be maintained and recooling to steady state conditions within 1 - 2 hours. The pressure increase within the coil was approx. 4 - 6 bar within 6 seconds. The amount of He blown down at 15 bar was below 15 % (equal approx. 90 Nm^3) of the total He volume of this coil under IFSMTF conditions.

Experiments Simulating Impaired Coolant Supply also demonstrated an unexpectedly high degree of stability of the forced-flow-cooled NbTi coils:
The He mass flow through the CH winding could easily be reduced from approx. 70 to 33 g/s and, through the EU coil, from 150 to 33 g/s (corresponding to $c \approx 0.1$ m/s and $Re \approx 4{\cdot}10^3$ within the EU coil conductors). The coils were operated safely under these cooling conditions.
Closing an inner cooling channel of the LCT-EU coil winding produced neither an electric resistance nor heat in the coil operated at nominal current.
An experiment simulating total failure of the cryosupply to the LCT-EU coil (winding and casing) for more than 1 h did not produce a quench, although the coil was discharged from the nominal current of 11.4 kA at 5 A/s only after 10 minutes. This demonstrated that a well designed forced-flow-cooled coil, even with additional losses during fast ramping (here 7 W) can be brought safely from rated current to zero current over a period of at least 1 h without ffc. (Even after this period, the temperature of the He outlet was ≤ 4.2 K).

Power required for He circulation
The power input required to circulate the helium through the forced-flow-cooled NbTi coils was reduced drastically, compared to expectations, because of the low level of disturbance energies.
In the winding of the LCT-EU coil, this allowed the He mass flow, $\dot{m}$, to be reduced to 1/5, the pressure drop, Δp, to approx. 1/20 and, consequently, the ideal circulation power, P, of the quasi-incompressible supercritical helium, $P \sim \dot{m}{\cdot}(\Delta p/\bar{\rho})$, to 1/100.
The ideal total circulation power for the mass flows through the winding and the casing of the LCT-EU coil thus amounted to less than half the total static thermal load of the coil within the IFSMTF.
Probably, a tailormade complete design of the magnet system and the refrigerator system, or a SCC magnet system (chapter 6), would allow $\dot{m}$ and thus P to be reduced even further.
A separate cooling channel system for the coil casing has been found to be advantageous. It reduces the thermal load on the winding, the cool down time, and the recovery time after a dump, provided that there is sufficiently good thermal conduction between the casing and the winding. Moreover, a solid connection between the casing and the winding can be ensured at any time in case the casing/winding are cooled down/warmed up more quickly. The pressure loss, and thus the recirculation power in and for cooling the casing, can be kept low because of the large possible cooling channel cross sections.

4. FFC WITH TWO-PHASE HE I. KFK EXPERIENCE

The advantages of two-phase ffc (positive JT coefficient, high latent cooling reserve, and high cooling potential relative to the mass flow, respectively) seem to be attractive especially in a poloidal NbTi coil operated in the pulsed mode.
The pressure drop of the two-phase flow must be minimized,
- on the one hand, in order to be able to keep the LHe bath at a slight overpressure relative to the atmosphere, if possible, and
- on the other hand, to ensure a sufficient temperature reserve for the superconductor.

This applies both to the PCC and the open and closed SCC modes. In the "S.C. Poloidal Field Coils for NET" development project carried out at KfK a s.c. model coil is under construction, which will be tested within the KfK TOSKA Test Facility. For this model coil, which also acts as a model for the first s.c. Poloidal Field Coil of TORE SUPRA, a decision was taken in favor of a superconductor with dual cooling principle [10]. In that version, the ring shaped arrangement of the superconducting subcables is directly surrounded by stagnant He I which, in turn, is in thermal contact with the He I flowing through the inner tube of the superconducting cable in a two-phase mode.

This two-phase forced flow has to remove the continous but pulsed heat caused by AC losses due to the plasma control in TORE SUPRA. Beside this it has to take over heat resulting from brief local heat pulses or in extreme cases from a plasma disruption by further use of reserve cooling capacity in form of latent heat.
In order to ensure stable and effective quasi-isothermal cooling of this superconducting cable, cryogenic investigations with two-phase He I were carried out at two stages in the KfK TOSKA Test-Facility [11,12] and have been prepared, respectively.
(A) Experiments with a quasi-horizontal single test channel (copper pipe of 9.9 mm dia., 200 m length, 3 m winding dia.) to investigate the flow behavior and heat transfer characteristics.

(B) Thermohydraulic investigations of four parallel quasi-horizontal test channels (each 50 m long, otherwise of the same geometry). The four pipe channels are equipped with heaters and mass flow control circuits to simulate the four pancakes of the poloidal model coil under fabrication [10].

The experiments conducted so far (A) have produced the following positive experience:

- Optical observation of the flow in the single channel within two viewing sections made of Pyrex glass 5 cm long, indicated as a typical flow pattern "wavy flow" and "stratified flow", respectively (operating conditions: $\dot{m}$ = 1 - 8 g/s, G = 13 - 108 kg/m^{2}s, x = 0.05 - 0.8; p/p_{crit} = 0.38 - 0.63).
- Despite separately flowing phases ("each phase continuous within itself"), the homogeneous computation model turned out to be sufficiently accurate for estimates of the pressure drop. The agreement between the experiment and calculation, under the assumption of the pipe friction coefficient of a smooth pipe, was within a range of ± 15 % [12].
- As a consequence of the favorable flow characteristics of both phases, a large amount of the heat of evaporation can be used and cooling is possible with small mass flow at a low pressure drop without stability problems. The investigations permitted stable operating conditions to be achieved under thermal loads of 0 - 0.4 W/m and 0 - 1.3 mW/cm^2, respectively, with vapor contents at the outlet of the heated pipe of x = 0 - 1 (x_{inlet} = 0, p/p_{crit} = 0.63). The mass flow density was G ≈ 33 - 108 kg/m^{2}s. This result is valid for steady state heat load conditions!
- The heat transfer coefficients measured under steady-state operating conditions and at Re numbers of approx. 10^4 - 10^5 and vapor contents of approx. 4 - 57 % are around 0.03 - 0.05 W/cm^{2}K in agreement with data measured by other authors, e.g., the data determined at NBS in a short vertical test section [13].
- For the simulation of the AC losses due to plasma control in TORE SUPRA, experiments were run with pulsed thermal loads (especially 0.1 and 0.2 W/m, respectively, ≈ 0.32 and 0.64 mW/cm^2, respectively) with pulse periods of 30 s interrupted by breaks of 120 s. In a flow generated by a "natural" pressure gradient, these pulsed loads resulted in dangerous operating conditions with complete stop of the flow and flow reversal, respectively. The use of a piston pump to drive the flow permitted a sufficiently stable forward flow of the two-phase helium. Consequently, such a pump will be used to cool the poloidal field coil. However, even with a pump, the continuous pulsed mode of operation requires a higher coolant flow rate than steady-state load operation does, if sufficient forward flow is to be achieved at all times. Consequently, it allows only a more modest use to be made of latent heat (cf. Fig. 4).

In summary, it can be said from the experimental results that the worst case pulsed thermal heat load to be expected in the Poloidal Field Coil for TORE SUPRA (0.1 W/m over 30 sec with 240 sec pause) can be cooled safely by use of a LHe pump with a mass flow of 3 g/s per superconductor cooling channel. This is valid for the assumptions of 400 m long SC-channels, x_{inlet} = 0, G = 32 kg/m^{2}s and the vapor content x_{outlet} to be expected is ≤ 0.15. The pressure drop will be less than 60 mbar.

For simulation of the behavior of several poloidal field coil pancakes with different thermal loads individual mass flow control loops with venturis, controllers and controlling valves have been installed within the test arrangement for further investigations (B).

Finally, it must be emphasized that this experience cannot be extrapolated to nonhorizontal cooling channels. In three-dimensional cooling channel arrangements with risers and downcomers, especially low He mass flows coupled with high vapor contents can create dispersed flow conditions and, in the worst case, slug and plug flow conditions, with the consequence of flow instabilities and greatly increased pressure losses and heat transfer degradation compared to homogenous flow conditions.

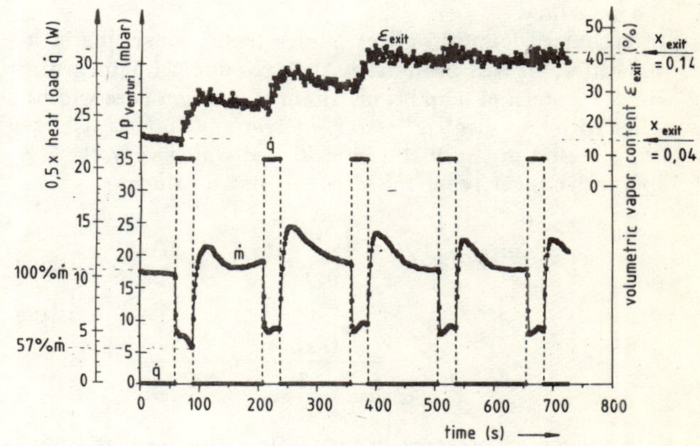

Fig. 4: 2phase massflow ($\dot{m}$) and quality (ε,x) response on pulsed heat load ($\dot{q}$)

Cooling channel with 9,9 mm inner dia and 100 m heated length. Heat pulses of 30 s length with 120 s pause. He circulation by 3 cylinder piston pump. $\dot{m}_{100\%}$ = 2,8 g/s, T_{in} = 4,55 K, p ≈ 1,4 bar.

5. ATTRACTIVITY OF HE II FFC . KFK EXPERIENCE

At an operating temperature of 1.8 K, also NbTi superconductors with internal ffc are able to produce the peak field strength of 11 T required for TOKAMAK toroidal fields coils (e.g., NET). Consequently, this technology is an alternative to the Nb_3Sn coils, which are more difficult to design and manfacture.

Within the framework of the European Fusion Technology Program, extensive experimental investigations of He II ffc have been conducted at KfK [14 - 17]. On this basis, studies have been performed concerning He II cooling technology for NET Toroidal Field Coils [18, 19].

The investigations carried out so far have been run in a test system with a vertical test section (Cu pipe) using fountain effect pumps to drive the subcooled He II flow. They were concentrated especially on the He II flow characteristics, heat transfer to He II, and the characteristics and efficiency of fountain effect pumps. The most important results will be summarized below as follows:

- The pressure drop of the He II flow within the circular copper pipes follows the well known relation for turbulent flows of Newtonian fluids in smooth pipes, which is in agreement with the experience accumulated by other authors [20, 21]. In general, it can be said that the pipe system of the He II cooling circuit can be designed as for a normal fluid medium, because the flow velocities are above the "critical velocity" (transition from a laminar no-loss superflow to a turbulent dissipative superflow).

- The heat transfer coefficient, α, of heated copper pipes to the subcooled He II flow is independent of the pressure and the flow velocity, but dependent on the heat flux, $\dot{q}$, and the fluid temperature, T_{fl}. Increasing $\dot{q}$ and T_{fl} will increase α. When the λ-temperature is exceeded, there is a major drop in the heat transfer coefficient from 100 to some 30 %.
- Independent of the channel cross section, the critical heat flow through stagnant He II in the cooling channels is confirmed to be

$$\dot{q}_{crit} = F_{\lambda}\,(p,T) \cdot L^{-1/3}$$

for dissipative counter flow of He II with F_{λ} (p,T) for the "degree of subcooling" and L for the length of the cooling channel [22]. In the case of (macroscopic) He II forced flow in long channels this heat flux can be exceeded by adapting the mass flow.
- In optimized fountain effect pumps (FEP) consisting of a superfilter (at KfK preferably Al_2O_3 powder of 1.5 μm grain size, or a stack of membrane filters of 0.01 μm pore width) followed by an electric heater, deliveries (Δp (P_{heater}), $\dot{m}$ (P_{heater}) and $\dot{m}$ (Δp)), can be achieved which very closely approach or even reach the ideal calculated values:

$$\Delta p_{FEP\,ideal} = \bar{p} \int_{T_{in}}^{T_{out}} s(\bar{p},T) \cdot dT;$$

$$\dot{m}_{FEP\,ideal} = \frac{P_{heater}}{s_{out} \cdot T_{out}} = \frac{P_{heater}}{\Delta h_{heater}}$$

Both quantities, pressure head and flow rate are highest at low system pressure. But there is only little degradation when such loops are operated at pressures of up to about 7 bar.
- Different types of heat exchangers prove to be necessary to achieve optimum performance of such pumps. At the cold end where all heat is rejected to the saturated bath (He II_s) a fin type device with large surface areas is needed. For many applications it proves to be advantegous to use the "free" waste heat of the loop (e.g. the heat picked-up in a large superconducting coil) to activate the pump instead of doing that by electrical heating. In this case of self-sustained operational mode a shell and tube type heat exchanger is needed at the warm end of the pump [17, 19].

The studies on He II ffc of NET NbTi Toroidal Field Coils, in which use of self-sustained FEP (SSFEP) was supposed, indicated that both types of s.c. cables developed by KfK together with Vacuumschmelze GmbH for LCT-EU and the Poloidal model coil can be used for this purpose. At cooling channel lengths of 500 m, the thermal loads due to nuclear heat and AC losses (a total of 500 W in 22 pancakes) can be safely carried away. Because of the advantageous hydraulic data of these types of conductors (d_h, A), the pressure head attainable in single-stage SSFEP is sufficient [19].
The other activities prepared at KfK for He II ffc of fusion magnets will be carried out in the following two stages:

(1) Comissioning of a SSFEP cooling circuit designed for a He II mass flow of 6 g/s within an internally cooled NbTi conductor (L = 30 m, I = 3000 A) placed and to be tested in the 10 T background field of the KfK High Field Test Facility HOMER.
(2) 1.8 K test of the LCT-EU coil using several parallel SSFEP for He II ffc of the 7 NbTi double pancakes with 28 cooling channels 250 m long. This is to produce a maximum field at the conductor of 11 T from a power supply of 16 kA. This background field can be used to test subsequent NET pancakes in the TOSKA-UP Test Facility.

6. COOLING CIRCUIT CONFIGURATIONS

Figure 5 shows the basic principles of primary circuit cooling (PCC) and secondary circuit cooling (SCC) in superconducting magnets. In the case of PCC, the helium compressor of the refrigeration plant has the dual functions of He compression and He circulation through the magnet followed by liquefaction. In the SCC mode of operation, the forced flow through the magnet is achieved by means of a cold circulator ("pump") in a separate, self-contained circuit, which is thermally coupled by means of heat exchangers with the LHe bath of the refrigerator primary circuit.
Naturally, there are a number of possible variations in the two techniques. The PCC design shown in Fig. 5a) has the special characteristic of the whole JT mass flow being passed through the magnet and then expanded to LHe bath conditions. The SCC variant sketched in Fig. 5b) has the special feature of the circulation unit operating at LHe operating temperature. The advantages listed in Table 2 show the reasons for favoring the SCC technique (as shown in Fig. 5 b)) over a PCC approach for a large superconducting magnet. These advantages count especially when existing He standard refrigerators with limited JT mass flows are used and in research and prototype systems with a variety of operating conditions, some of which are not even known in advance.

As NbTi magnets normally allow an enthalpy difference, Δh_M, to be used which is small relative to the heat of evaporation, the SCC approach needs a much lower JT refrigerator mass flow, as long as heat generation in the circulation pump, $\dot{Q}_{cir}$, is not much higher than the heat load on the magnet, $\dot{Q}_M$. Consequently, the SCC process sketched in Fig. 5b), is particularly advantageous in the generation of high He mass flow through the magnet when using a standard refrigerator with limited JT mass flow. When designing or purchasing a new system complete with the magnets and refrigerator, alternative refrigerator variants will be selected in the case of the PCC principle. The reason for that is that the intermediate cooling required for better utilization of the cooling capacity of a standard refrigerator after partial sections of the superconducting magnets needs sophisticated measures to be taken and, consequently, is expensive and risky. The alternative refrigerator variants are characterized by the fact that a large part, or all, of the mass flow delivered by the He compressor is being passed through the magnet. But downstream of the magnet, only a part is expanded into the LHe heat exchanger bath, while the bulk at supercritical pressure condition is passed back to a medium-pressure stage of the He compressor. Fig. 5c) shows such a PCC configuration for a medium pressure level in the magnet [23].

For the reasons mentioned above, KfK favors the SCC method. This applies not only to single-phase supercritical He I ffc, but also to subcooled He II ffc.
In the case of two-phase He I ffc of the poloidal field coil an "open" SCC method is used. This means that in differences to the Fig. 5b) the cold recirculation pump will work as an LHe immersion pump. It extracts liquid helium straight from the LHe heat exchanger bath and pumps it to the coil. After partial evaporation in the coil, the liquid-vapor mixture is recirculated into the LHe heat exchanger bath (LHe-HEX).

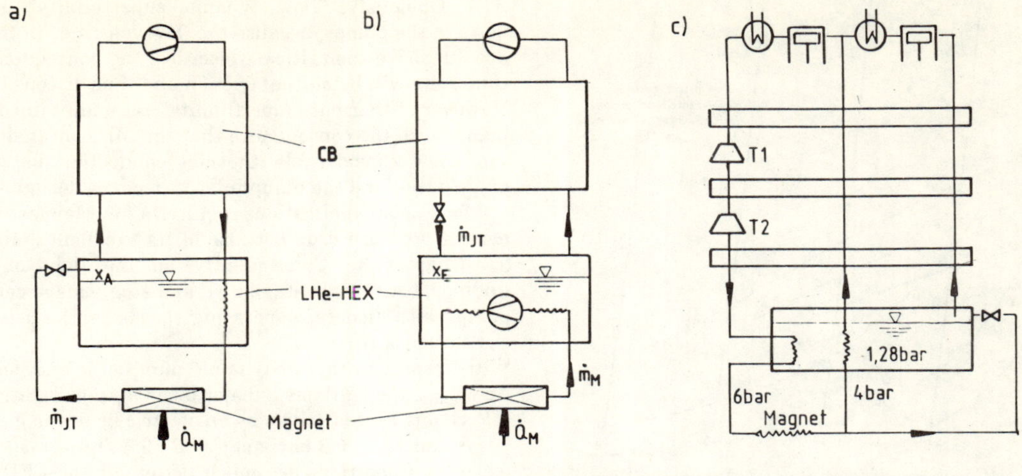

Fig, 5: Fundamental cooling principles/circuits. a) PCC Primary Cooling Circuit. b) SCC secondary Cooling Circuit. c) Alternative PCC from [23]. CB = Coldbox of 4.5 K standard refrigerator. LHE-HEX = LHe-heat exchanger bath. $\dot{m}_{JT}$ = JT mass flow. $\dot{Q}_M$ = Magnet heat load.

Table 2: Characteristics of the Secondary (SCC) and the Primary Cooling Circuit (PCC) principles

SCC-Advantages

Flexibility of Operation

- pressure p_M and mass flow $\dot{m}_M$ within the magnet can be set to optimum heat transfer, stability and cooling conditions. $\dot{m}_M > \dot{m}_{Refrigerator}$ possible. Reserve cooling capacity in form of LHe storage.

Decoupling of Refrigerator and Magnet

- separate circuits for refrigeration and magnet cooling with no $\dot{m}$- and p-interference
- $\Delta h_{Refrigerator\ low\ temp.}$ independent on p_M-level → better $\eta_{Refrigerator}$ and $\eta_{Cooling\ System}$ possible

SCC-Disadvantage

- additional cold machine with risk of failure and additional heat generation $\dot{Q}_{Circ.} \sim \dot{Q}_M \cdot \Delta p_M / \rho$.

PCC-Constraints

Coupling of Refrigerator and Magnet Circuits

- Δh_M and Δh_{Refr} limited by technical principle. Δh_{Refr} decreasing with increasing p_M-level. High $\dot{m}_{JT}$ for 1stage magnet cooling → oversized refrigerator (Variant 5a)

Circulators for supercritical or subcooled He

The need to develop cryogenic He circulators resulted from the KfK decision in favor of ffc of the LCT EU coil (required mass flow up to 150 g/s according to the 1979 design) and the wish for an in-house test at KfK prior to delivery of the coil to the USA. The underlying reasons for this development were these:

(1) The maximum available He mass flow (= JT flow) of the existing KfK refrigerators amounted to 30 g/s.
(2) High-capacity cryogenic He circulators were neither available commercially nor under development anywhere in the world at that time.

Together with two industrial companies, therefore, two helium piston pumps, one three-cylinder pump, and one single-cylinder pump, were developed for a peak performance of 150 g/s and Δp = 3 bar [24, 25]. This was done also in view of the advantages of the SCC method with cryogenic circulators listed in Table 2.

These pumps have been operated during the domestic test of the LCT-EU coil within the TOSKA-Testfacility.

Because of the much more uniform delivery of the three-cylinder pump in comparison to the one-cylinder pump, the three-cylinder pump was used almost exclusively. Under a variety of operating conditions, it worked most satisfactorily in this circuit without any damper volumes. As the mass flows to be pumped were far lower than originally thought necessary, the pump in this application was run mainly in the part load regime ($\dot{m} \leq$ 25 - 40 g/s; $\Delta p <$ 0.2 bar) at a low efficiency, but also with less absolute heat generation. When first installed in the TOSKA facility, the one-cylinder pump could not be fitted out with the required damper elements. For this reason, it could be operated only at speeds ≤ 140 rpm, which corresponds to ≤ 50 g/s mass flow rate. This confirmed that the one-cylinder piston pump cannot be operated free from risk at high speeds in a closed He cooling circuit at cryogenic temperatures (4 - 5 K) in the absence of effective damping systems. At insufficiently damped pressure amplitudes there may be excitation of the transfer lines if the pressure wave frequencies are close to, or equal to, the eigenfrequencies of the piping. The advantage mentioned above of the flexibility of the SCC mode with the refrigerator and the magnet cooling circuits decoupled from each other, were confirmed in very easy and quick matching to different operating conditions. Within the 1000 h of test operation of the LCT-EU coil in the TOSKA facility, single-phase He I ffc and the SCC cooling principle have proved to be satisfactory and have demonstrated the expected advantages.

In the light of the high levels of stability attainable in NbTi coils as demonstrated in the LCT project, the specific coolant flows in general can be kept at a lower level in future systems and, consequently, the required pumping heads can be minimized considerably. Besides the use of single-stage piston pumps, this would also allow single-stage centrifugal pumps to be used.

The advantages of rotating machines are: only one moving part (rotor), uniform delivery, less wear, and expected longer service life. One drawback is the delivery characteristics with the strong dependence of $\dot{m}$ on Δp (cf. Fig. 6).

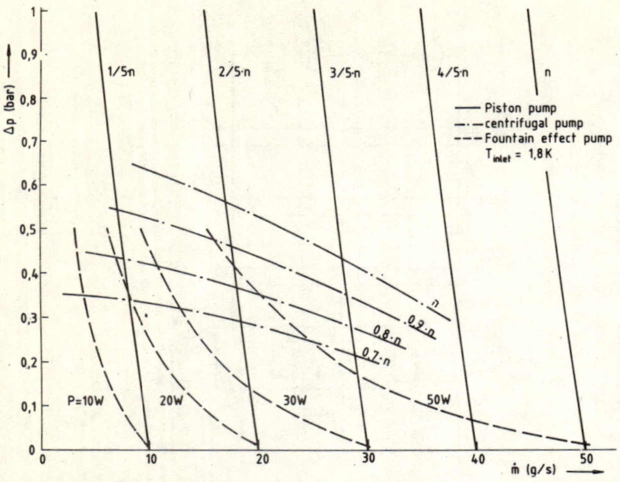

Fig. 6: Typical delivery characteristics of 1stage helium pumps.

The latter point was one of the reasons why a three cylinder piston pump was planned again for LHe pumping through the poloidal model coil needing two-phase helium flow.

The forced flow of subcooled He II, like the forced flow of supercritical He I can be achieved both in the direct PCC mode or in the indirect closed SCC mode. KfK favoring this SCC mode, works on two He II circulation methods at the same time:

circulation of the He II by fountain effect pumps (FEP)
- circulation of the He II by mechanical pumps (MP).

The table below is a summary of specific problems and characteristics of the two types of pumps.

Principal questions regarding the functioning of both pump systems have meanwhile been answered positively. Also the mechanical pump is able to pump superfluid "no-viscous" He II without any degradation. It behaves in the normal fluid mode with respect to its delivery characteristics and efficiency ($\eta_{th} = \Delta p \cdot \dot{V}/P_{pump}$). It can be designed with the methods employed for media of normal fluidity. This has been demonstrated experimentally both with rotating MP, e.g., at NBS/Boulder and UW. and with a piston pump at KfK. Obviously, "flow dynamic superfluidity" is broken down in the pumps, because the flow velocities in the pumps are far above the critical velocities, the consequence being that there will be mutual friction and thus viscous behavior. However, "thermal superfluidity" is maintained in the pump, with the consequence that the MP operated on He II can show considerable thermal conduction between the pump outlet and the pump inlet. A modern design of a NbTi coil for Fusion applications requires a low He mass flow at a modest pressure drop because of its excellent stabilization (cf. Chapter 3). Consequently, in the light of current findings, both single-stage FEP and single-stage centrifugal pumps with moderate operating speeds can be used for the circulation of He II.

With respect to thermodynamic pump efficiency, $\eta_{th} = \dot{V} \cdot \Delta p/P_{pump}$, the FEP has a major disadvantage relative to the MP as long as it is driven by an electric heater (e.g., for $\dot{m}$ = 10 g/s and $\Delta p \approx 0.3$ bar: $\eta_{FEP} \approx 5$ - 6 %, $\eta_{MP} \approx 50$ - 60 % is feasible). Conditions are much better for the FEP when it can be integrated in a self-sustained system (SSFEP). This is possible in He II ffc of Fusion Magnets, in which the heat load of the magnet, transferred to the recirculated He, can be used to drive the FEP.

In view of the major advantages existing for both types of pumps (FEP and MP), and also of the open questions associated with large-scale application and long-therm reliability under all operating conditions in fusion magnet facilities, KfK is preparing the He II test of the LCT EU coil in such a way that both types of pumps will be operated side by side for comparison.

7. SUMMARY OF EXPERIENCES AND CONCLUSIONS FOR FUTURE APPLICATIONS

7.1 He I ffc at supercritical conditions. (LCT-Project)

The advantages of coils with ffc due to fewer movements, gaps, stresses and acoustic noise resulted in a high mechanical stability of these coils.

The high quality electric insulation within the coils was confirmed by finding no short circuits in any of three coils with ffc under steady state and non-steady state operating conditions.

Table 3: Comparison of Fountain Effect Pumps (FEP) and Mechanical Pumps (MP) for He II ffc.

	Self-sustained FEP (SSFEP)	MP
Principle	Simple, no moving parts	Rotating and/or oscillating parts
Functioning + operating behavior	Operating behavior? Matching of driving heat in SSFEP?	Pumping of superfluid helium?
Limits of operation	$\Delta p_{max\ 1stage} \approx 0.45$ bar (1.8 - 2.15 K) $(\dot{m}/A_{SF})_{max} \approx 0.5 - 2$ g/s · cm² 1) 1.8 - 2.15 K	"None", as a range of MP types are avalaible
Structural volume (cold)	Rel. large with heat exchangers required at SF-inlet and outlet	Compact version possible (high speed CP)
Capital cost	Low	Higher
Efficiency, ηs	Low, e.g., 5-8 % (FEP) 2)	Approx. 50-80 % is possible.
Operating cost	High	Rel. low
Risks/drawbacks Reliability	Long-term stability of SF? No problems over 700 h within KfK cryogenic test section. In principle no limitation in life.	Wear parts limited in life. Irregular pumping in PP.
Experience	KfK, on a laboratory scale	E.g., NBS/Boulder and UW (CP), KfK (PP).

CP = centrifugal pump PP = piston pump SF = superfilter FEP = fountain effect pump SSFEP = self-sustained FEP

1) Theoretically no limition ($\dot{m} \sim A_{SF}$); pratically limited because of SF contrations → leak flow problems possible. $A_{max} \approx 10$ cm² (KfK recommendation). 2) Better in case of SSFEP, if "waste heat" is used.

The positive expectations with regard to a good superconductor stabilization were exceeded. An enormous reduction in cooling mass flow and subsequent in power for helium circulation was possible. The LCT-EU type coil does not need any mass flow for stabilization (stagnant He is sufficient). A small amount of mass flow is necessary for cooling the heat load on the coil from outside.

An excellent and calculable cool down behavior and stable low temperature cooling conditions in complex cooling channel systems with 22/28/48 parallel channels without flow control in all three coils were demonstrated.

Mechanical disturbances in the rigid coils were so small that operation outside the "cryogenic stability region" (Stekly) was possible.

Because of the low He inventory of coils with ffc and due to the low heat generation and He density changes during dumps and quenching, no He-losses occur when compact cold buffer vessels will be used.

The good calculability and scaling characteristics with respect to mechanical, electric and flow dynamics behavior was confirmed.

7.2 He I Two-phase ffc.

In a (quasi)-horizontal cooling channel, such as an internally cooled superconducting cable of a poloidal field coil, the advantages of this cooling technology (positive JT effect, utilization of heat of evaporation) can be exploited up to high vapor contents without any flow instabilities as long as the heat load is constant. The flow pattern is separated, but ordered, thus causing slight pressure losses only which can be estimated from the homogeneous model.

In a pulsed mode operation of the type occurring in TORE SUPRA, which is amplified in comparison to future TOKAMAK Reactors, the mass flow can be kept relatively small only if a piston pump enforces the "directed" circulation of the helium.

7.3 He II ffc under subcooled conditions.

He II ffc produces not only the advantages listed under 7.1 above, but also offers a raise of the maximum fields in coils from approx. 8 T (4 K) to approx. 11 T (1.8 K). In parallel, He II ffc in long cooling channels allows much more heat to be removed than by a He I ffc or by stagnant (subcooled) He II. The design of the He II cooling circuit piping and mechanical pump can follow the design employed for a medium of normal fluids. Studies on the base of experimental investigations confirm that He II cooling of NbTi Toroidal Field Coils for NET, with 500 m long SC-cables similar to the LCT-EU cable, e.g., can be made by use of self sustained fountain effect pumps.

7.4 Cooling Circuit Configurations.

Both basic cooling principles PCC and SCC have proved to be satisfactory and have their specific pros and cons. We feel that the SCC mode is particularly suitable for tests with research and prototype facilities undergoing manifold, sometimes unknown, operating conditions and having low circulation capacities. The very positive LCT-experience with respect to the stability in cooling (with low mass flow and low heat loads of He circulation) and with respect to mechanical and electrical behavior of ffc magnets confirmed that the SCC mode would be a proper concept for future reactor magnets 2,5 times the LCT size and their superconductor concepts, e.g. those under development for NET.

Overall can be summarized that the He ffc for large-volume and powerful fusion magnets with internally cooled s.c. cables has demonstrated its attractive features by studies, cryogenic tests and extensive operating phases of the three He I ffc LCT coils in the six-coil torus system at ORNL/USA. These experience are a sound basis for future projects like NET or ITER either with He I-or He II ffc techniques.

REFERENCES

1 P. Komarek, Stability problems in Supercondutors, Proc. of Conf. on Cryog.Fundamentals, Warschau (Polen)1983.

2 G. Krafft et al., IIR Saclay (1981).

3 W. Herz et al., Proc. ICEC 10, Helsiniki (1984), 83-86, 382-385.

4 Fusion Engineering and Design, Vol.7 (1988),Sept. 1988.

5 T.L. Ryan et al., Proc. 8th Symp. on Eng. Probl. of Fusion Res., IEEE N.Y. (1979), p. 1187.

6 J.L. Young et al., The forced flow cooled coils for the IEA LCT, Adv. Cryog. Eng. Vol. 27, 11-19.

7 W. Herz et al., Testing of the EURATOM LCT Coil, a forced-flow-cooled NbTi Coil in the IFSMTF, IEEE Trans. on Magn. Vol. MAG-23 No. 2 (1987) 1693-1696.

8 Dresner et al., First results of the Westinghouse coil in the IFSMTF, same IEEE Vol., 1701-1705.

9 G. Vecsey et al., Preliminary test results of the Swiss LCT-Coil, same IEEE Vol., 1697-1700.

10 S. Förster et al., A 2MJ 150 T/s pulsed ring coil. Status of the design and the test arrangements, Proc. MT 9, Zürich, 1985, 32-35.

11 H. Katheder, M. Süßer, Pressure drop of adiabatic and nonadiabatic horizontal two phase helium flow, Proc. ICEC 11, Berlin 1986.

12 H. Katheder, M. Süßer, Results of flow experiments with two phase helium for cooling of superconductors, Proc. ICEC 12, Southampton 1988.

13 P.J. Giarratano et al., Forced convection heat transfer so subcritical Helium I, Adv. Cryog. Eng. Vol. 19 (1974).

14 A. Hofmann, Thermomechanically driven He II flow, an option for fusion magnets with internally cooled conductors, Proc. ICEC 11, Berlin 1986, 306-311.

15. A. Hofmann et al., Operational characteristics of loops with He II flow driven by fountain effect pumps, Adv. Cryog. Eng. Vol. 23, 1988, 471-478.

16 R. Srinivasan and A. Hofmann, Investigations on cooling with forced flow of He II. Part 1, Cryogenics 25 (1985), 641-649 and 650-657.

17 A. Hofmann et al., Heat exchangers for He II cooling loops with self sustained fountain effect pumps. Cryogenics 27 (1987), 682-688.

18 A. Hofmann and K. Khalil, Considerations on magnet design based on forced flow of He II in internally cooled cables, Proc. SOFT 14, 1986.

19 A. Hofmann, A study on nuclear heat load tolerable for NET/TF Coils cooled by internal flow of He II, KfK Report No. 4365, Febr. 1988.

20 W. Johnson and Mc Jones, Measurement of axial heat transport in He II with forced convection, Adv. Cryog. Eng. Vol. 23 (1978), 363-370.

21 P.L. Walstrom et al., Turbulent flow pressure drop in various He II transfer system components, Cryogenics 1988, Vol. 28, Febr. 101-109.

22 G. Bon Mardion et al., Helium II in low-temperature and superconductive magnet engineering, Adv. Cryog. Eng. Vol. 23, 1978, 358-362.

23 H. Quack, 7th Symp. Eng. Probl. Fus. Res., Knoxville (1977).

24 W. Lehmann, J. Minges, Adv. Cryog. Eng. Vol. 29 (1984), 813-820.

25 J. Hübener et al., Proc. ICEC 10 Helsinki (1984), 574-577.

SUPERCONDUCTING THIN FILM BY EVAPORATION

Jacob Azoulay
Center for Technological Education Holon,
P.O.B. 305, Holon 58102, Israel.

Dan Goldschmidt
School of Physics and Astronomy
Raymond and Beverly Sackler Faculty of Exact Sciences
Tel Aviv University, Tel Aviv 69978, Israel.

ABSTRACT

A new evaporation method of high-temperature superconducting films, the resistive evaporation from a single source, is reported here for the first time. The source material, that was inserted into a tungsten boat in a conventional vacuum system, consisted of a pulverized mixture of Cu, YF_3 and BaF_2. The handling of source material required only grinding and mixing of the raw materials. Its deposition onto $SrTiO_3$ substrates yielded superconducting properties very similar to those obtained in a layer-by-layer resistive evaporation of these materials. In particular, resistive transition onset at 75K and zero-resistance at ~40K, with critical currents of $2000A/cm^2$ at ≈ 10K have been measured. The broad transition may be attributed to a copper concentration gradient, as measured by Auger depth profiling, or to a residual flourine-rich phase.

INTRODUCTION

Superconducting oxide thin-films have a high potential for applications in fields such as microelectronics and detectors, and have attracted much attention in the last two years. Applications in the microelectronics field require probably high-quality films, that have high critical temperatures (T_c) and large critical currents (I_c). In other fields such as light detectors, less perfect films can be utilized and may even be of advantage. For instance, use of thin-film weak-links for direct detection of light[1] (i.e. by directly altering I_c) may benefit from low critical currents improving the signal to noise ratio. In this application a relatively low T_c may not be a big disadvantage either, since in any case, light sensors operate at low temperatures. Thus, thin film preparation methods that are simple and amenable for mass production, and that yield moderate quality superconducting properties, as compared to epitaxial films prepared by elaborate methods, are interesting in their own right.

In a previous publication,[2] we have described an evaporation procedure that used solely resistive evaporation from tungsten boats. It consisted of three boats containing separate powders of YF_3, Cu and BaF_2 that were evaporated layer-by-layer in this order onto a heated $SrTiO_3$ substrate. The resulting films had the onset of the resistive transition at 75K with zero-resistance at ~50K and critical current densities of $3000A/cm^2$ at ≈ 10K. Here we report on a new procedure that utilizes the same components now mixed together and evaporated resistively from a single boat. As far as we know,[3] this is the first reported success of resistive evaporation of a high T_c superconducting film from a single source and, as will be described below, is probably the most simple vacuum deposition method requiring almost no previous handling of the source material. The crucial point is the choice of starting materials; those chosen have relatively close vapor pressure curves. Cu on the one hand, and Y_2O_3, BaO and other oxides on the other, have widely different melting points and vapor pressure, and cannot be utilized in this approach.

EXPERIMENTAL

Many experimental details are similar to those employed in the layer-by-layer evaporation.[2] Prior to evaporation we have flushed the diffusion-pumped vacuum chamber with O_2. No evaporation control other than evaporation to completion was attempted. The handling of the source material included only the mixture and grinding of the starting materials weighted in appropriate amounts to yield the 1:2:3 cation composition. We have used pulverized Cu and aggregated BaF_2 and YF_3 for this purpose. The powder mixture was inserted into a regular tungsten boat and evaporated at ~1100°C with an average rate of ~7A/sec. During evaporation, the $SrTiO_3$ substrate was kept at 500°C. However, in one evaporation we left the substrate at room temperature; no substantial change in film properties was observed.

The post evaporation treatment included an anneal at 880°C in flowing O_2 for 10 minutes and then quenching by removing the sample from the quartz-tube furnace. Usually the oxygen was bubbled through a water beaker at the outlet of the quartz tube. Since we suspected existence of Flourine-rich phases in our films, we have also tried annealing in water vapor by flowing the O_2 through a heated water beaker at the tube's inlet.[4] All our attempts at various water vapor pressures and concentrations acted adversely on film properties.

Electrical contacts were employed by pressing In. Resistance vs. temperature [R(T)] and current-voltage (I-V) measurement were carried out employing conventional 4-pole dc techniques.[5]

RESULTS AND DISCUSSION

We have examined about a dozen different films. The results showed some scatter: about half of the films had a resistive transition quite similar to that reported earlier on the layer-by-layer films. The remaining films had inferior properties, particularly a lower zero-resistance temperature, and a less metallic behaviour in the normal state.

In one case we tried Yttrium stabilized ZrO_2 as a substrate. That film had a very low zero-resistance temperature as well, the same onset temperature (~75K) and a sharp semiconducting behaviour in the normal state. Another film, evaporated onto $SrTiO_3$ held at room-temperature showed results similar to those shown below. The results reported here were all obtained on one film. Thus even though the resistive transition is typical of our better films, the microstructural characterization has been carried out on only one sample. This film's thickness, measured by an interference microscrope, was 0.45 μm. The resistivity curve is shown in Fig.1. The

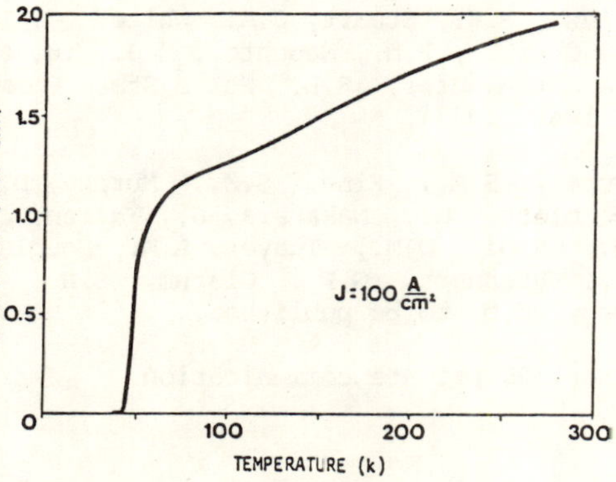

Fig.1 Resistive transition of a thin-film evaporated resistively from a single source.

normal resistance shows a metallic behaviour with resistivity values similar to those observed in bulk ceramic $Y Ba_2 Cu_3 O_{7-\delta}$. The resistance above the onset (20Ω) is about $\frac{1}{4}$ of the resistance in the layer-by-layer films,[2] reflecting a resistivity that is twice smaller. The transition onset at 75K and zero-resistance at ~40K are very similar to those reported earlier on the layer-by-layer films[2] (here, as well, there is a systematic error in the reading of low temperatures). In Fig.2 we show a typical I-V curve taken at ≃ 10K and corresponding to a critical current density of ~2000A/cm^2, again very similar to the layer-by-layer films.

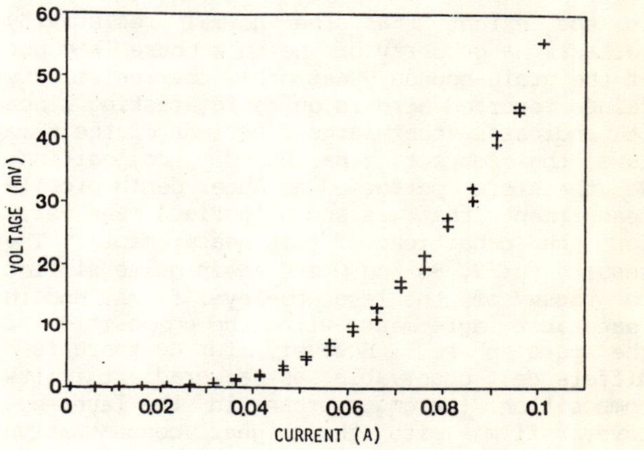

Fig.2 Current-voltage characteristic at ≃ 10K corresponding to a critical current of 20mA.

The striking similarity between both kinds of films suggests similar microstructure in both. In Fig.3 we show a scanning electron microscope picture of the film. Most of the grains are elongated and connected in a head-to-tail morphology inclined one with respect to the other. This, indeed, was the case also in the layer-by-layer films. As explained previously,[2] this may indicate conduction along a preferred (a,b) crystallographic plane aligned parallel to the long grain axis.

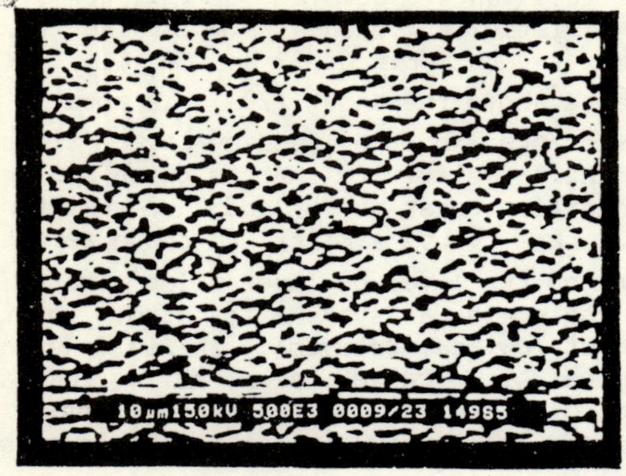

Fig.3 Scanning electron microscope picture of the same film showing elongated superconducting grains.

To the extent that the normal resistivity reflects a property of the bulk phase (and not of the grain-boundary network), the resistivity value reported here is quite interesting since it indicates that large fractions of the film have the correct $Y Ba_2 Cu_3 O_{7-\delta}$ composition. We therefore performed an Auger depth profile measurement that is shown in Fig.4 (see Ref.2 for the conditions of this measurement). The results for Y, Ba and O are again quite similar to those of the layer-by-layer films, and in reasonable agreement with the composition of the pure phase. However, for Cu there is a difference, observable as a gradient in its composition (stronger than in the layer-by-layer film) with the higher concentration values facing the front surface. Note that the average Cu concentration has the approriate value as required by the 1:2:3 composition. The Cu rich surface in this film may explain the lower resistivity in this film.

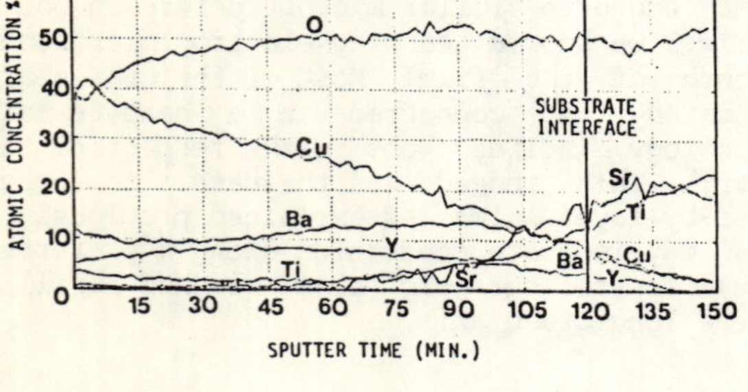

Fig.4 Auger depth profile.
Note the strong gradient in Cu concentration.

The gradient in Cu concentration may be one reason for the broad transition. Fluorine may be another reason for the transition broadening.[6] It has also been shown that the fluorine is not incorporated into the Y-Ba-Cu-O phase, but rather is present in a second amorphous phase.[7] Note, that no trace of fluorine has been detected in the Auger spectra. This in itself does not definitely mean that there is no fluorine in the film, since in many cases fluorine is unstable and decomposes under the electron beam.[8] However, it possibly means that the fluorine concentration is not very large since in superconducting films prepared by other methods and measured under the same conditions, fluorine traces could be observed.[8] A secondary fluorine phase may have a strong effect on the transition width particularly if it is present as a thin amorphous coating on the superconducting grains.[7]

ACKNOWLEDGEMENT

WE would like to thank R. Brener for the Auger measurement and J. Larea for the SEM measurement. This work was partially supported by the U.S.-Israel Binational Science Foundation.

REFERENCES

1 Nishino, T. et al., Jap. J. Appl. Phys. 26, L1320 (1987).

2 Azoulay, J. and Goldschmidt, D. submitted to Appl. Phys. Lett.

3 Laibowitz, R.B. private communications.

4 Gupta, A., Jagannathan, R., Cooper, E.I., Giess, E.A., Landman, J.I., and Hussey, B.W. Appl. Phys. Lett. 52, 2077 (1988).

5 Goldschmidt, D. Phys. Rev. B. in press.

6 Davies, P.K., Stuart, J.A., White, D., Lee, C., Chaikin, P.M., Naughton, M.J., Yu, R.C. and Ehrenkaufer, R.L. Solid State Commun. 64, 1441 (1987).

7 Zahurak, S.M., Fine, S.M., Murphy, D.W., Greenblatt, M., Nakahara, S., Warren, Jr., W.W., Krol, D.M., Thayer, A.M., Douglass, D.C., Brennert, G.F., Glarum, S.H., and Giorgy, E.M. to be published.

8 Brener, R. private communication.

9. EDUCATION AND MANAGEMENT ON CRYOGENICS

TECHNICAL EDUCATION AND SCIENTIFIC RESEARCHES ON CRYOGENICS IN UNIVERSITIES IN CHINA

Zhang Zhiyou
Power-Machinery Eng. Dept. of Xi'an Jiaotong University
Xi'an, Shaanxi Province, PRC

ABSTRACT

In the paper, the development of the full-time universities in China was described simply, and it is pointed out that the government of the People's Republic of China pays attention consistently to the technical education for industries.
In China, the technical education on the speciality "refrigeration equipments and cryogenic technique" has a history of more than 30 years. In the paper, teaching plan of the speciality and outlines of the teaching programs and textbooks of its curricula were introduced briefly and analytically. And also, brief introduction of the cultivation of postgraduates and of scientific research work was given. With these introductions, some necessary description and brief comments were presented. At last, several private views of the author were centributed.

INTRODUCTION

Since the founding of the People's Republic of China, the government pays attention consistently to the technical education for industries and the cultivation of engineering technicians. Because the socialist economic construction requires a large professional contingent in technical fields.
Up to 1949, there were 207 institutions* of high education in China, of which only 18 colleges belong in engineering course. In the same year, the total number of students in China was about 150000, and 17.8% of which (27500 students) was studing in engineering colleges. From 1952 to 1953, a nationwide regulation for institutions and departments had been carried out. The policy of the regulation was "to develop specialized colleges and to reorganize and emphasize comprehensive universities with a focal point to cultivate technicians for industrial construction and teachers".[1] After the regulation, the total number of universities (colleges) was reduced to 184 but that for engineering course was increased to 38 with a proportion of 20.6%. In 1953, there were 212000 students, in which 80000 students were in the area of engineering course taking a proportion of 37.7%. In the past period of more than 30 years, the higher educational undertaking developed more rapidly. For example, in 1980, there existed 675 universities /colleges and an enrollment of 1.14 millions students, in which 203 universities /colleges and 0.38 millions students belong in engineering course taking proportions of 30.1% and 33.5% respectively; and in 1986, the corresponding sums are 1054 and 271 universities /colleges, 1.88 and 0.65 millions students with proportions of 25.7% and 34.3% respectively. In recent future, the proportions for engineering course may be decreased slightly.
Since 1952 the system to cultivate students in light of speciality was emploied. Such a system was erected according to the experience of USSR. In 1953, 215 kinds of specialities were installed in all the universities and colleges in which 107 kinds (taking 49.8%) are refered to engineering course. Thereafter, the kinds of specialities expanded continuously to 323 in 1957, to 627 in 1962 and to 1039 in 1980, and about one half of them are refered to engineering course.
After the beginning of the eighth decade, the education system began to be reformed along with the reformation of social economy. The practices of the education reformation are: to reduce the kinds of specialities, to reduce the proportion of engineering specialities, to reform and merge the original specialities and to install new specialities of rising courses. The purpose of the reformation is to strengthen the ability of the higher education to adapt the changing situation of the society and economy. Up to now, the kinds of engineering specialities had been reduced from 537 to 204[2] and there still exist a decreasing tendency.
In China, the speciality concerning cryogenics is installed together with refrigeration and named "refrigeration equipments and cryogenic technique". It is firstly seted up in 1957 in Xi'an Jiaotong University and the first class was graduated in 1961. Now there are five uni-

*Refers only to full-time universities and colleges in the whole paper.

versities installed with this speciality, the general situation of them is tabulated below.

Name of universities	Founding year of the specility	Amount of graduated students	Amount of students in study
Xi'an Jiaotong University	1956	1380	260
Shanghai Jiaotong University	1958	450	250
Huazhong University of Science and Technology	1961	600	270
Shanghai Institute of Mechanical Eng.	1970	790	230
Zhejiang University	1958	290	170

FUNDAMENTAL TRAINING TO STUDENTS

In the institutions of higher education the fundamental training to students is carried out according to the teaching plan of specilities.
Through the past decades, though about the charater and content of the teaching plan there were many changes, such as from five-year system to four-year-system, from unified plan to whole China to guiding plan and then to reference plan, the general goal of cultivation did not changed essentially. The general goal of cultivation for engineering specialities is to enable the students to develop morally, intellectually and physically, and to help the students to undergo fundamental training for engineers in order that they may get good basis for becoming technical specialists.
Now in the five universities previously mentioned, four-year system teaching plan is adopted for the speciality "refrigeration equipments and cryogenic technique". The teaching plan stipulates 36 to 40 courses with 2500 to 2600 lecture hours in four years. The courses may be divided into five categories:
1) Common courses: Include theoretical courses for politics, physical education, foreign languages, business management, basis of laws and military theory. There are generally 3 courses for politic theory, the aim of which is to carry on the education in Marxism-Leninism and the policy of the communist Party of China. The common courses possess about 760 lecture hours and take a proportion of about 29% in the total hours of the teaching plan.
2) Basic courses: Include higher mathematics, chemistry, physics and experiment of physics. The aim of basic courses is to enable students to obtain knowledges about the fundamentals of nature sciences and to undergo preliminary training on technical ability about computation and experiment. The basic courses possess about 420 lecture hours and take a proportion of about 16%.
3) Technical basic courses: By virtue of the technical basic courses, the fundamental theories of applied science and technical knowledges of industrial production are instructed to students with the aim to found a strong base for them to study specilized courses. This category plays most important role in cultivating students and has comparatively more courses. For this category of courses there may be some differences between different universities, the typical ones for the speciality "refrigeration equipments and cryogenic technique" are: programming language, linear algebra, mathematic-physical equation, engineering drawing, theoretical mechanics, mechanics of materials, mechanism, machine member design, metallography and heat-treatment, engineering thermodynamics, heat transfer, fluid mechanics, electrical engineering and electronics, metalworking, tolerance and technical measurement and interface technique for microcomputer. This category of courses has a total sum of lecture hours more than 1000 with a proportion about 40%.
4) Specialized courses: Although the specialized courses reflect the cultivating goal for the speciality, yet they possess lesser courses and lecture hours (about 10%). The typical specialized courses on cryogenic technique are principle of cryogenic technique,cryogenic installations, turboexpander, reciprocating expansion machine (or combined as cryogenic machines), measuring technique for cryogenics.
5) Elective courses: The function of this category is to widen the domain of knowleges or to enhance the theoretical or specialized knowledges for students. Each student selects only a few courses with lecture hours about 150, but the category possesses a big sum of subjects differ from university to university and are related to the scientific researches of the teachers. As follows are the examples of lective courses taken by the five universities in the field of cryogenic technique: technology of manufacture, separation of gases,cryogenic refrigerator, heat exchanger for cryogenic technique, cryogenic vacuum technique, three dimensional and two-phase flow in turboexpander, cryogenic heat transfer, cryogenic thermodynamics, cryogenic insulation, introduction to low temperature physics, magnetic refrigerator, materials for cryogenic engineering, superconductivity and its application, cryobiological engineering.
In addition to the five categories of courses, some measures of practicality are stipulated in the teaching plan. They are as follows.
1. Courses for exercise and seminar: They are used as auxiliary measures for lectures to enable the integration of theory with practice.
2. Experiments: Experiments are stipulated for minority of the courses with the aim to train the students with experimental methods and measuring techniques.

3. Teaching and productive practices: Two or three times of teaching and productive practices are stipulated with a total time of 10-15 weeks. Teaching practice is carring out in the practising factories of the universities and it is claimed that students must operate machines as workers. Productive practice is carring out in specialized factories with the aim to enable students to gain some practical productive knowledge and skill.
4. Course designs and graduation design: Every student in his academic life must complete two or three course designs, one for the course machine member design and the others for specialized courses. And before graduation, the graduation design should be fulfilled in the peroid of 8-14 weeks. By means of the course and graduation designs it is capable to train the students for the ability to analyse and solve practical problems with the theoretical knowledges learnt from the courses. During the procession of graduation design (or graduation thesis), the student will conclude and operate synthetically the knowledges learnt and will study independenty, therefore it is a more effective measure to cultivate the students for capability.

In the universities in China, examination for courses and oral examination of graduation design are to be taken seriously. In the teaching plan, the total times for examination is stipulated not less than 12 weeks in four years.

The formulation of teaching plan is one of the most important affairs. The guiding or reference plan is usually formulated by the teach-guiding committee of the spciality which is an interinstitutional academic committee composed of experienced professors. According the guiding or reference teaching plan the executive plan for each university is formulated. On formulation of teaching plan five important relations must be deal with excellently, which are relation between moral, physical and intellectual educations, relation between the knowledges of natural, technical sciences and humanities, relation between theory and practice and between theoretical lecturing and practising measures, relation between basic courses and specialized courses and relation between the demand of the plan and enduring ability of the students.

Nww in China, reformations are going on. In order to meet the requirements for the development of commodity economy, the reformation of the education system and teaching substance is necessary. The tendentious views on the reformation of higher education are to widen the field of speciality, to increase the numbers of elective courses and to pay great attention to the cultivation of capability for students.

PROGRAM AND TEXTBOOK FOR COURSES

Each course has its own teaching program. It stipulates the scope and profundity for the course and provides chief basis to the teacher for teaching. The teaching programs of courses are formulated in accordance with the teaching plan and are emploied to ensure the implementation of the plan.

The teaching programs of all courses stipulated in the teaching plan are drafted by experienced teachers and are examined and approved by the teach-guiding committee of the speciality with the exception of that for elective courses.

To the formulation of the teaching program for each course, it is claimed to carry through the following principles:
1. To conform the demand of the teaching plan and coincide with the lecture hours stipulated;
2. To apply theory to reality and avoid from theory to theory;
3. To guarantee the scientific and systematic nature of the course and reserve as far as possible its completeness under the principle "fewer but better";
4. To avoid disjointment and unnecessary repition with other courses;
5. To mirror the advances in its field.

The teaching program of course is also the basis to select or to author textbooks. In China, for majority of the courses (except elective courses), national textbooks are adopted. National textbooks are authored in a planned way and under the sponsorship of teach-guiding committee of the speciality. The national textbooks must fulfill the following requirements:
1. Consistent with the teaching program of the course;
2. Handling materials with correct academic viewpoint and logical method;
3. Reflecting the principle that theory combines with practice;
4. No any principled error;
5. Possessing higher scientific level and mirroring new achievements in the field;
6. Having a suitable volume to mirror the principle "fewer but better";
7. Writing in a popular and fluent style and easy to study.

In recent year, more than ten national textbooks were authored for the specialized courses of the speciality "refrigeration equipments and cryogenic technique" and were published by the Press of Machine-Building Industry. The national textbooks (and their chief editors) are: Refrigeration Principles and Equipments (by Zhang Zhiyou), Reciprocating Compressors for Refrigeration (Miao Daoping), Centrifugal Compressors for Refrigeration (Chang Hongshu), Rotary Compressors for Refrigeration (Li Wenlin), Refrigerating Installations and its Automation (Zhou Gennan), Principles and Installations of Cryogenic Technique (2 volumes, Zhang Zhiyou and Shi Bingsan), Turboexpanders (Ji Guanghua), Reciprocating expansion machines (Chou Yiming), Cryogenic Refrigerators (Bian Shaoxong), Cryogenic Heat Transfer (Guo Fangzhong) and Cryogenic Heat Exchangers (Chen Changqing).

In addition, there were many kinds of textbooks

for elective courses which are authored and published separately within the universities. These textbooks possess they own style differing from foreign textbooks including that of Soviet Union which we have utilized in the early years. They also came up to advanced standard. But it seems that there are some common tendencies with them: description is somewhat detailed and more data for design are introduced. Along with the reformation of education system, textbooks with new style will be authored and published in the recent futhre.

CULTIVATION OF POST GRADUATES

The cultivation of postgraduates in the field of refrigeration and cryogenic technique started 30 years ago. In the period 1957-1962 the Xi'an Jiaotong University enrolled 7 postgraduates four of them were cultivated collaborately with a Soviet specialist engaged by the University. Mass-cultivation of postgraduates in the field of cryogenics was carried on from 1978, and progressed quickly after the promulgation and execution of <<Ordinances of the People's Republic of China for Academic Degrees>> in 1980. The table shows the general situation of the five universities for cultivation of postgraduates in recent ten years.

Name of Universities	Postgraduates for master degree		Postgraduates for doctor degree	
	Graduated	In study	Graduated	In study
Xi'an Jiaotong University	23	24		2
Shanghai Jiaotong University		22		2
Huazhong University of Science & Technology	27	13		
Shanghai Institute of Mechanical Eng.	19	15		4
Zhejiang University	6	10		3

In China, the cultivating standard for postgraduates is to master substantial basic theory and systematical knowledge in the field of the speciality and to possess great ability for researching and teaching work. It is certain that the standard for postgraduates of doctor degree must be higher than that of master degree. In past, majority of the postgraduates for master degree was enrolled from the students graduated in that very year. In recent years, the proportion of postgraduates enrolled from working personel increased year by year. The principles for enrollment are measuring in an all-round way, adimitting by selecting the better ones, and putting quality before quantity.

Every postgraduate has his/her own cullivating plan formulated by his/her teacher and approved by the head of department. The postgraduate of master degree, as an example, in the period of 2 years and 8 months, should study about 10 courses and complete graduation thesis with a load of about 2500 working hours. The graduation thesis is consistent with the studing orientation (realm). In the five universities mentioned above, there are a lot of studing orientations in the field of cryogenic technique, such as: liquefaction and separation of gases, heat transfer and heat-exchangers at cryogenic temperature, adsorption at lower temperature, cryogenic machines, three dimensional and two-phase flow in turboexpander, gas-lubricating technique for cryogenic machine, cryogenic insulation technique, cryogenic refrigerator, cryogenic crushing, cryobiological engineering, etc.

In the universities in China, the courses for postgraduates are offered independently and not interrelated with the courses for students. In the past 10 years, a series of courses for the postgraduates in the field of cryogenic technique is offered in the five universities, the principal ones of which are: thermodynamics for cryogenic technique, cryogenic heat transfer and heat-exchanger, mass transfer at lower temperature, gas-liquid two-phase flow of cryogens, heat transfer with phase transition for cryogens, dynamic theory of adsorption, gas separation, cryogenic insulation technique, positive displacement expansion machine, three dimensional and two-phase flow in turboexpander, gas-lubricating technique for cryogenic machines, theory of cryogenic machines, cryobiological engineering, theory of solidifing transition, cryogenic vacuum technique, cryogenic refrigerator. And in the same time several reference books for teaching have been authored. Therefore, in universities, cultivation of postgraduates impels the teachers to raise their own academic level and promotes the development of scientific researches.

The universities often cultivate postgraduates in cooperation with research institutes. This is a more effective way for the postgraduate cultivation. In this way, the postgraduates study courses in university and does research work for his/her graduation thesis in institute, so strong points of both are sufficiently utilized. In addition, the universities also cultivate postgraduates in cooperation with universities or research institutes of other countries, or foreign advisory professors are invited for postgraduate cultivation. This is also useful for cultural and technical interchanges between two universities and between two countries.

SCIENTIFIC RESEARCHES

Carring on scientific researches is one of the important tasks for institutions of higher education. In China, as demanded by the government, every key university should be established as dual centers: teaching and researching. In fact, in the past forty years, a lot of research works on basic theory, application and development have been done and gratifying results have been achieved.

Scientific researches on cryogenic technique began at 1958 firstly in Xi'an Jiaotong university, and the first subject to research was the development of single-stage stirling-cycle refrigerator. Massive research work spreaded in the late seventies. According to incomplete statistics more than 80 research subjects on cryogenic technique were accomplished within the five universities mentioned above of which the following subjects are awarded ones or the ones of great significance: adsorptive purification of liquid oxygen by self-circulation in air separating plant, recovery of argon from low pressure air separating plant, medium pressure turboexpander with gas-lubricating bearings, miniature two-stage regenerative refrigerator, magnetic free refrigerator, closed-cycle cryogenic installation for superconductive magnetic separation, productive installation for superfluid helium of atmospheric pressure, cryogenic installation in the vicinity of 1 K by He^4 depressing, effective compact cryogenic heat exchanger of new type, large microscope system for cryobiology, cryogenic preservation of medical biomaterials, cryogenic crushing of waste rubber and plastic articles, cryogenic operator for ophthalmology using liquid nitrogen, magnetic free Dewar reservoir for liquid helium, large Dewar reservoir for liquid hydrogen, 100 liter liquid oxygen reservoir with new type insulation, heat-separator and its application in recovery plants of light hydrocarbons, etc.

At present, more and more research subjects on cryogenics are being carried on in the universities. They are proceeding to new areas of cryogenics and will meet the expectation that new higher aim shall be achieved in future years.

In general, the research subjects carried on in universities are stemed from three ways. Some subjects are supported and donated by the National Foundation of Natural Sciences, some are assigned by the National Committee of Education, and some are contracted from factories or institutes as crosswise cooperation. In recent years, crosswise cooperation develops more quickly.

Carring on scientific researches in universities is of deep significance. Firstly, its results can contribute directly to the economic construction and social development. Also, through the research work the academic level and teaching quality may be elevated, and batches of postgraduates and young researchers may be growing well. We will go this way continuously.

SEVERAL PRIVATE VIEWS

With relation to the above description and statement several private views are provided as concluding remarks of the paper.

1. In the first half of this century the full-time institutions of higher education in China imitated chiefly west countries, from system to curricula and even using their textbooks without translation. In fifties, to learn the experience of Soviet Union was underlined, system, curricula, teaching plans and teaching programs of Soviet universities were copied indiscriminately and immense Soviet textbooks were being translated. Starting from late fifties, we tried to reform higher education and tried to find out our own way. Despite the serious destruction in the Great Cultural Revolution, the higher educational undertaking restored quickly and made great progress in the past ten years. Now, comprehensive reformation, inclusive of the reformation of higher education, is carring on. It is believed that, through the reformation and development in the future ten years, at the end of this century, a socialist higher educational system with Chinese distinguishing feature shall be established.

2. To cultivate students in light of speciality is a practically effective measure in universities. General speaking, this will remain unchanged in the future reformation. Therefore the speciality "refrigeration equipments and cryogenic technique" will exist continuously, but it will possess a widened field and incremental content.

3. Even though a lot of teaching and research work on cryogenic technique have been done in the five universities they are still in the primary developing stage. The socialist construction in China demands a large quantity of specialists to cultivate and a great number of technical problems to solve. In the last decade of this century mighty advances concerning cryogenic technique will be inevitable. For this, international cooperation shall be necessary.

ACKNOWLEDGMENT

The author would like to spread heartfelt thanks to Profs. Chen Changqing, Chen Guobang, Chen Zijiu, Ji Guanghua, Jiang Nengzhao and Zhang Bingyuan for their kindness to provide materials for the paper.

REFERENCES

[1] <<Yearbook of China's Education>> 1949-1981 (in Chinese), 1984

[2] <<Yearbook of Statistics 1987>> (in Chinese) 1988.

10. LATE ARRIVED PAPERS

ANALYSIS OF TEMPERATURE PROFILE IN LOW TEMPERATURE HEAT EXCHANGER

Guangsan Sun
Shanghai Jiao Tong University, P.R.China

ABSTRACT

Heat transfer at low temperature, both heat capacity and heat conductivity are strong temperature dependent. Moreover, substantial temperature different are established between hot and cold ends. Sometimes, very small temperature different is required at hot end. Then temperature profile in low temperature heat exchanger can be overlaped. The overlaped temperature profile will violate the minimum energy comsuming in order to reach the target temperature or violate the energy recovery at low temperature. This paper presents the appearance of overlaped temperature profile. How can it be result from and what it'll result in? The judgement formula is derived.

INTRODUCTION

Design of heat exchanger involves lots of sophisticated schedules. Many designers assume they have performed their function adequately if the design meets their requirments specified. However, many instances lead to the deviation of actual heat exchanger parameters apart from the designed values. This paper only aimed to avoid the temperature profile of heat exchanger overlaped at its early design stage.
For design heat exchanger to meet the special requirments, such as heat duty, flow rate, exit temperature, maximum pressure drops and velocity are specified. Sometimes due to the necessity of energy recovering, the temperature difference at hot end has to be very small, say, less than 0.5° C.
The design conditions can be expressed as follows:

$$Q_{act} \geqq Q_{min}$$
$$(\Delta p)_{act} \leqq (\Delta p)_{max}$$
$$(\Delta T)_{hot} \leqq (\Delta T)_{max}$$

Under these cases, it's necessary to make a Q-T sketch to calculate the average temperature difference and point out the possible trouble.

Q-T DIAGRAM

If neglect the cold leakage and pressure drop, the foundamental equations of heat exchanger are as follows:

$$Q = m'c'_p(T'_2-T'_1) = m''C''_p(T''_2-T''_1)$$

or $$q = Q/m' = I'_2-I'_1 = m(I''_2-I''_1)$$

and $$Q = KF\Delta T_{cp}$$

or $$q = Q/m' = K(F/m')\Delta T_{cp}$$

Here $$m = m''/m'$$

Q-T diagram is the curve of heat transfered versus temperature, as shown in Fig.1 Obviously, the difference between two curves represents the temperature difference in each section. The profile of two curves depends on the properties of fluids and parameters of p'_1, p''_2, T'_1, T''_2, F and m.

TWO CHANNELS COUNTERFLOW HEAT EXCHANGER

We can draw temperature profile B when m, p'_1, T''_1 are specified. Curve A can be moved up and down along the linear T''_2 = const. In another words, we can change F/m to control two exit temperarure T''_1 and T'_2, as shown in Fig.2.

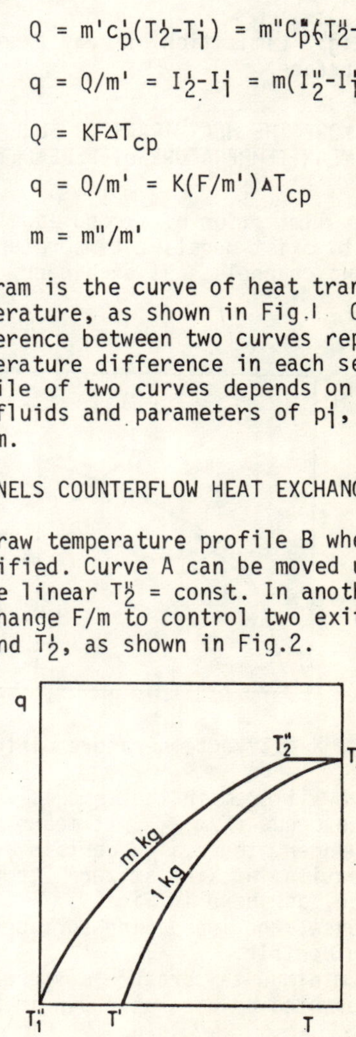

Fig.1 Q-T Diagram

If we increase F/m, then ΔT_{cp} decreased. For a certain surface, we get a correspondent ΔT_{cp}. Fig.2 shows the limitation of heat transfer surface. Two curves intersected at point C, i.e., at section C, drawing force equals zero. If F/m continuously increased, $F/m = (F/m)_{lim}$ $\Delta(F/m)$, what will be happened?

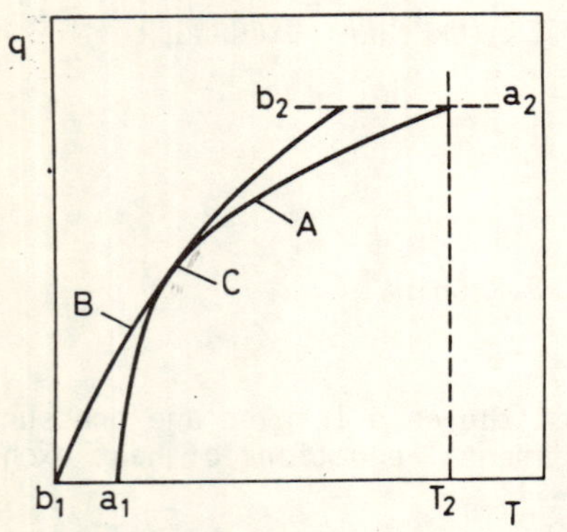

Fig.2 Limitation of heat exchanger surface

DOES THE MORE THE HEAT TRANSFER SURFACE, THE LESS THE MEAN TEMPERATURE DIFFERENCE?

Due to the limitation of second law. It's impossible to exist negative temperature difference in two channels heat exchanger, as shown in Fig.3.

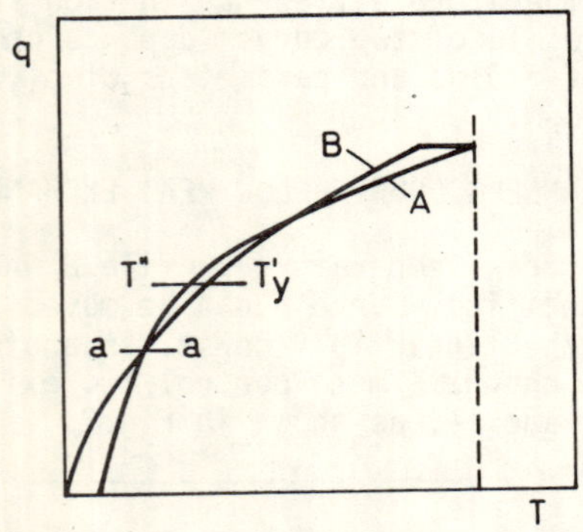

Fig.3. Negative temperature difference

So curve A will remain the same as in Fig.2 when F/m exceeds $(F/m)_{lim}$. It means in between heat exchanger, there is a certain area, not only a section, working at zero temperature difference, as shown in Fig.4.
In this case, the increasing surface has no operating benifit.
Also the minimun temperature difference at hot end represented by a_2b_2 as shown in Fig.2.

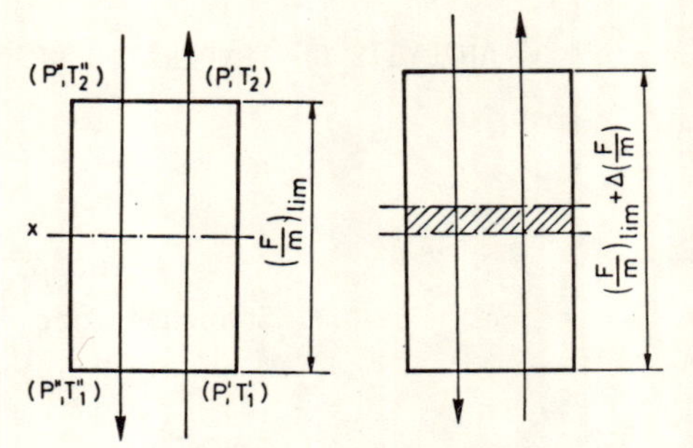

Fig. 4 Zero temperature difference

JUDGEMENT FORMULA

The necessary condition for producing "negative temperature difference" in two channels heat exchanger is:
The slope of curve A

$$(\partial Q/\partial T)_P = C_p'$$

or

$$(\partial i'/\partial T)_{P'} = C_p'$$

The slope of curve B

$$(\partial mi''/\partial T)_{P''} = mC_p''$$

As shown in Fig.3, at first, curve A is on the right side of curve B, ΔT = positive. As the section moves to the hot end, ΔT decreased and finally becomes negative, i.e. goes to the left side. Because the temperature difference at hot end is supposed to be positive, so curve A goes back to the right side again.
It means, C_p' first greater than mC_p'' and then smaller than mC_p''.
Thisis the necessary condition to produce "negative temperature difference" in two channels counterflow heat exchanger, but not sufficient condition. In addition, the temperature difference at hot end is small enough, then it can be happened at the early design stage.

FORU CHANNELS HEAT EXCHANGER

Fig.5 is the four channels parallel flow plate heat exchanger. Temperature profile is in dimensioness temperature versus dimensioness length.

Here $\theta = (T-T_1')/(T_2''-T_1')$

The temperature curves of the heated channel number 3 and the cold channel number 2 intersect one another at a certain position y_{int}.

This point depends on the parameter C. C is the ratio of $(NTU)_c/(NTU)_h$. After y_{int}, the direction of heat flow will be reversed.

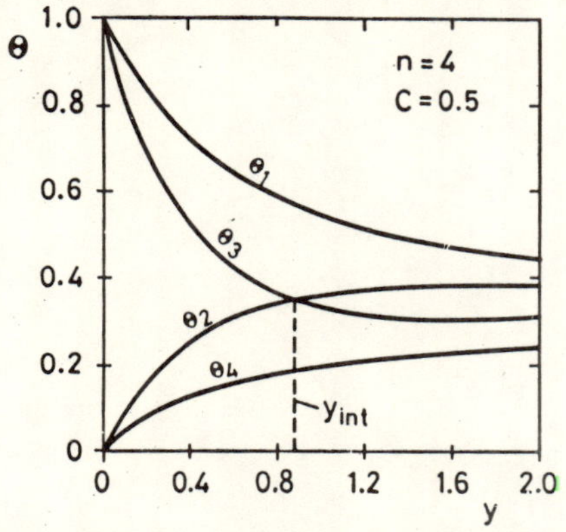

Fig.5 Four parallel plate heat exchanger

Fig.6 shows the variation of the intersection point with parameter C. In all position above the curve, a smaller amount of heat will be withdraw from the second to the third stream. C equals $m''C_p''/m'C_p'$ when K and F equal corstant.

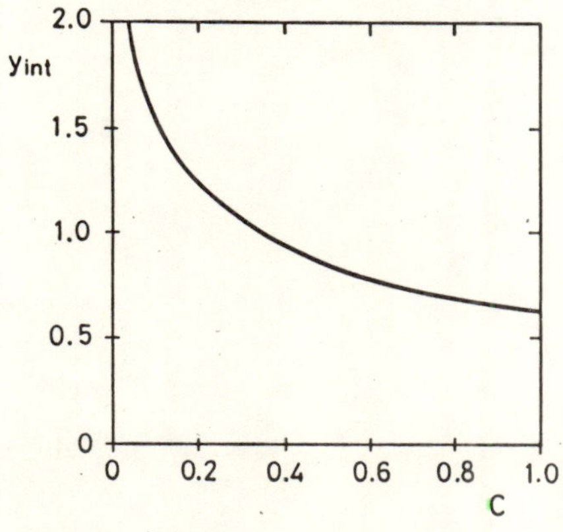

Fig.6 Variation of the intersection poirt with the heat capacity ratio

The more the intersection of temperature profile, the less the thermodynamics efficiency of heat exchanger. With the same principle mentioned above, we can change F/m to avoid the severe temperature profile overlaped. Fig.7 is the four channels counterflow plant heat exchanger. A large amount of area is working at zero temperature difference when C is rather high.

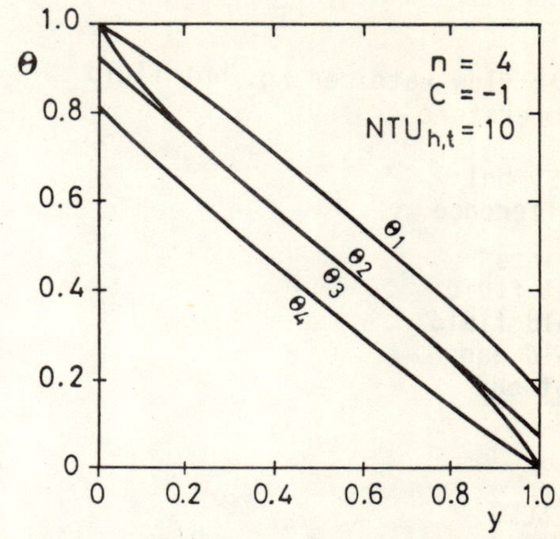

Fig.7 Four counterflow plate heat exchanger

Here

$NTU_{h,t} = 3/2\ NTU_h$

CONCLUSION

A. In some special cases, the increasing of heat exchanger surface does not correspondent to the decreasing of mean temperature difference.
B. Overlaped temperature profile should be always avoided both in two channels or more than two channels heat exchanger.
C. For two channels heat exchanger, the direction of heat transfer can not be changed. So overlaped temperature profile can only be exist in early design stage. If so, part of the surface will be working under zero temperature difference.
D. For more than two channels heat exchanger, the temperature profile may intersect one another. After this point, the direction of heat flow will be reversed. The more the intersection, the less the thermodynamic efficiency.
E. Counterflow arrangment should always be considered, also for the reason of avoiding severe temperature profile overlaped.

ACKNOWLEDGEMENT

One of the authors Guangsan Sun would like to acknowledge the financial support from the "ICTP Programme for Training and Research in Italian Laboratories, Trieste, Italy". Also for the assistance of CNR, ITEF Padova, Italy and Mr. B.Toniolo for his careful drawing.

NOMENCLATURE

C_p - heat capacity at constant pressure
F - heat transfer surface
K - heat transfer coefficient
m',m" - flow rate of hot and cold stream
Q - heat flow rate

q - heat flow rate per Kg. hot fluid
T - temperature

Greak Symbol
Δ - difference

Subscripts
' - hot fluid
" - cold fluid
1 - cold end
2 - hot end

REFERENCES

1. Walter Frost, Heat Transfer at Low Temperature, Plenum Press New York and London
2. M.Khalil Bassiouny, H.Martin, "Temperature Distribution in a Four Channels Plate Heat Exchanger" Heat Transfer Eng. Vol.6 NO.2, 1985
3. E.S.R.Gopal, Specific Heat at Low Temperature, Heywood Books, London
4. Jerry Taborek, Evolution of Heat Exchanger Design Techniques Heat Transfer Eng. Vol.1, NO.1, July - Sep. 1979

HEAT TRANSFER COEFFICIENT IN HELIUM VAPOUR FILM AROUND A HEATED WIRE

F. JEBALI, G. DEFRESNE, M.C. DULUC, M. FOUAIDY, J. WANG, M.X. FRANCOIS
Laboratoire de Thermodynamique des Fluides - PARIS VI
Campus Universitaire, Bat. 502TER, 91405 ORSAY Cedex, France

ABSTRACT

Heat transfer from a solid to liquid helium in presence of vapor film is studied by using a thin horizontal metallic wire electrically heated. In this paper we present some analysis of heat transfer coefficient measurements for different film boiling regimes, for which, in a previous work (4,5,6) we have shown that three typical modes of the liquid-vapor interface are to be observed.
The influence of the internal convection of the helium II bath seems to be important for the onset of the natural convection in the vapor film. We note that, for all these regimes, the film boiling heat transfer coefficient is high (1500 $W/m^2.k$) and varies with surface wire temperature.

INTRODUCTION

Although many experimental measurements (1,2,3) of the film boiling heat transfer have already been made in saturated helium II, it appears that the heat transport mechanisms in the vapor and in the vapor-liquid helium II interface are not yet truly understood even for stationnary conditions. This question is of practical interest and more of fundamental one as for example the correlation between interface stability and heat flux density and the heat transfer coefficient.
In a previous work (4,5,6) we performed a very sensitive optical system which allows to get photographs or high speed film (2000 Fr/s) of the different film boiling regimes. It was confirmed that the results depend on several parameters as well as the heat transport properties of the superfluid bath and the vapor layer, the wire diameter and the overhead hydrostatic pressure which governs, by a non linear equation, both the helium II thermal conductance and the vapor layer thickness. Thus the experiments must be driven very carefully and the optical measurements become necessary for a better understanding of the phenomenous

In this paper, we start from the precise description of the heat transfer regimes given in other publications (4,5,6,7) and focus our attention on the heat transfer coefficient measurement and its interpretation.
It seems actually well admitted that for stationnary heat flux applied to a thin wire being immersed in a saturated helium II bath, and bigger than the so-called peak heat flux, (q_p) the nucleated vapor layer can exist in roughly three different configurations.
The first one (fig. 1.a) with no apparent interfacial mass flux ; the second one (fig. 1.b), with stationnary or slowly variable deformations leads to a quasi classical film boiling. As said above, the observed configuration for a given heat flux Q depends on the bath temperature T_o, the immersion depth H of the wire under the free surface leading to a heat transfer coefficient given by :

$$h = h\ (Q,\ H,\ T_o)$$

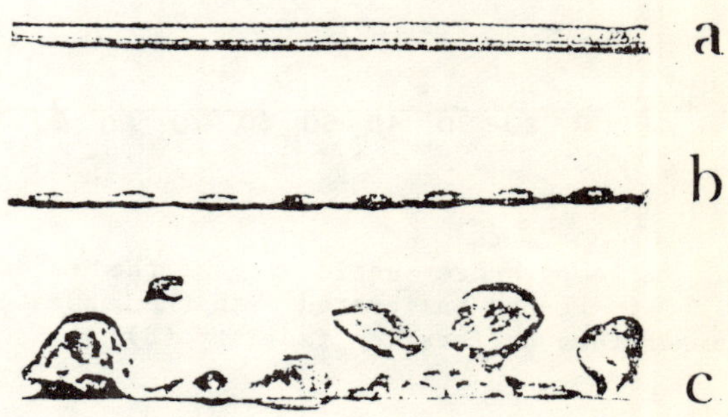

Fig. 1 Three typical modes of the vapor liquid He II interface submitted to a heat flux and a small pressure over saturation condition. For explanation see text.

(a) Experimental system

The experimental apparatus and sample are described in (5) ; we used a 32 μm diameter, 3 cm length thermoresistive alloy which has been carefully calibrated by the laboratory calibration facility (8). The temperature sensitivity $1/R(dR/dT)$ (fig. 2), R is the electrical wire resistance, is rather high at low temperature ($4.10^{-1}K^{-1}$ at T=1.9 K) due to the presence of lead in the alloy, but falls drastically in the range 5-15 K to be nearly constant until 300 K at about few 10^{-3} K^{-1}. Such a temperature behaviour is convenient for Kapitza resistance measurement but impedes a very precise determination of the first wire temperature step associated with the vaporization. The horizontal wire, tightened between two vertical pure copper posts, soldered to the two voltage V leads, is connected to a current supply by two other leads. The controlled parameters are the heat power $Q_W = RI^2$ injected in the wire, or the wire heat flux density $q_W = Q_W$: dL, the bath temperature T_o and the immersed depth H. The measured values are the thickness of the pseudo-cylindrical vapor layer, the mean temperature <T> of wire and H. All measurements are made in quasi-stationnary conditions of the controlled parameters.

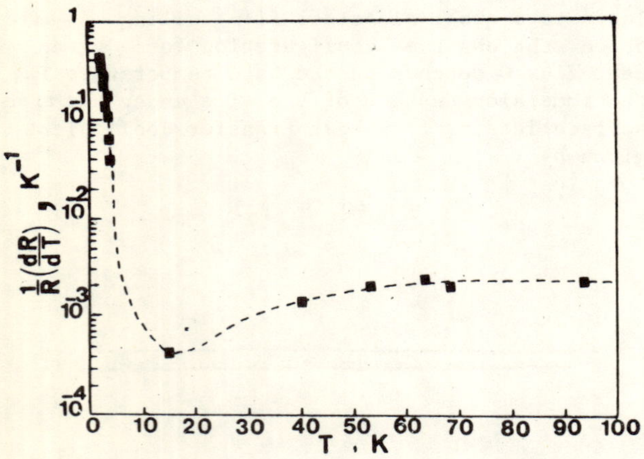

Fig. 2 Temperature sensitivity of the heated wire (Cu-Pb-Sn) calibrated with the Laboratory temperature calibration facility (8).

(b) Results

The results given in this paper have been got from about 50 runs made in a temperature range close to the point where the heat transport properties of helium II are strongly dependent on temperature. They can be summarized as follows :

1. At fixed T_o, several values of H, we measure the heat transfer characteristic of the wire. A typical example of the results obtained for two different depths H is reported (fig. 3.a.). The regions marked K, a, b, c, refer respectively to the Kapitza, first, second and third film boiling regimes described above and are determined by optical observation. The interfacial heat flux density q_i, which has been verified in good agreement with Ametistov theory, (1, 2) can be written as

$$q_i = A(T_o, d, H).H.\sqrt{T_o}$$

where A is a rather complicated fonction of the parameters in brackets. Thus, at H and T_o fixed, q_i has got an unique value and any increase of q_W will increase in size the vapor layer thickness. We have successfully checked the simple relation :

$$q_i = q_W \cdot d/e$$

for the a and b regions where there is no interfacial mass transfer.

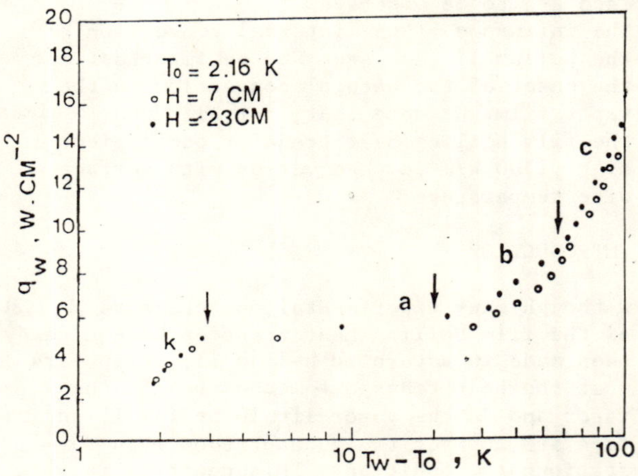

Fig. 3a Heat flux density as a function of the temperature difference between the wire (T_W) and the bath temperature T_o.

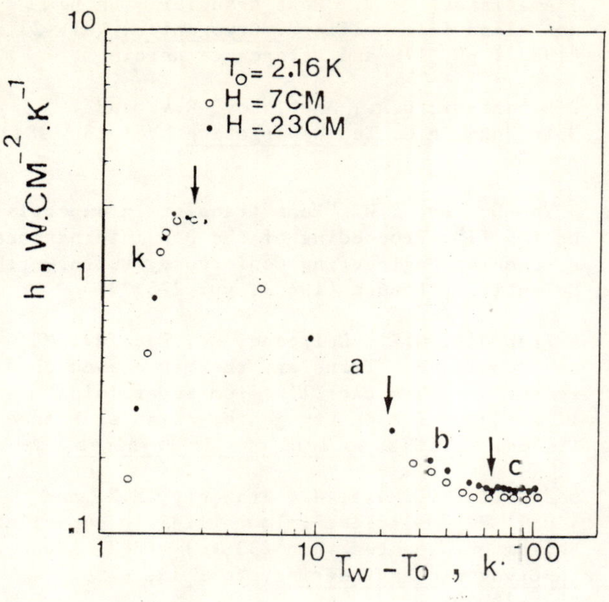

Fig. 3b Heat transfer coefficient deduced from fig. 3a.

2. The heat transfer coefficient can now be taken out from the results of fig. 3.a and is presented fig. 3.b as $h=h(T_W-T_o)$. We choose To which is close to the interfacial temperature on the liquid side as the reference temperature to calculate the overall heat transfer efficiency. A useful comparison can be made between the h values in the different regions K, a, b, c.

3. Such experiment is done for ten values of H and the corresponding h are plotted fig. 4. This heat transfer coefficient corresponds to the regime b with $t_W \sim 40$ K. Notice that the slope $(\partial h/\partial H)q_W$, obtained here, is in good agreement with the results of Bett and Leonard (9) for a flat surface as we got

$$(\partial h/\partial H)q_W = 1200 \text{ W.m}^{-3}\text{K}^{-1}$$

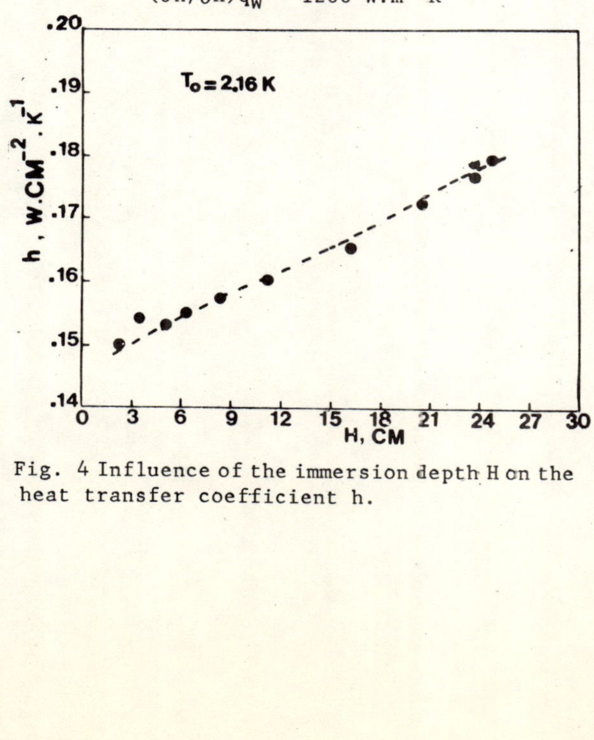

Fig. 4 Influence of the immersion depth H on the heat transfer coefficient h.

4. As the helium II heat transport properties are well known to depend on the heat flux densities in a rather simple way, the questionned part of the process are the heat transfer mechanisms in the vapor layer and at the vapor-liquid interface which includes its vapor and liquid boundary layers. But for T_o and H fixed, one can expect that a qualitative behaviour of h(q) or h(ΔT) allows to choose between a diffusive or a convective heat transfer in the vapor. We showed in a previous paper (7) that for small and constant values of e, let say 10^{-4} m, the wire temperature T_W varies versus the heat flux density q_W in agreement with the diffusive process.

In order to confirm this conclusion for different values of e, we approximate the vapor layer to a cylindrical cover of the wire and define again an equivalent thermal resistance h_v of the layer which actually cannot be distinguished of h in that measurement.

$$h_v = \frac{q_W \Pi L d}{T_W - T_o} = \frac{2 \Pi L \bar{k}}{\text{Log } \frac{e}{d}}$$

where $\bar{k}$ is the heat conductivity of the gas at its mean temperature, with :

$$\bar{k} = \frac{5}{2} C_v \eta \ , \quad \eta = 5.02\ T^{0.647} . 10^{-6} \text{ Cgs}$$

$$C_v = 3.117 \text{ J.g}^{-1}.\text{K}^{-1}$$

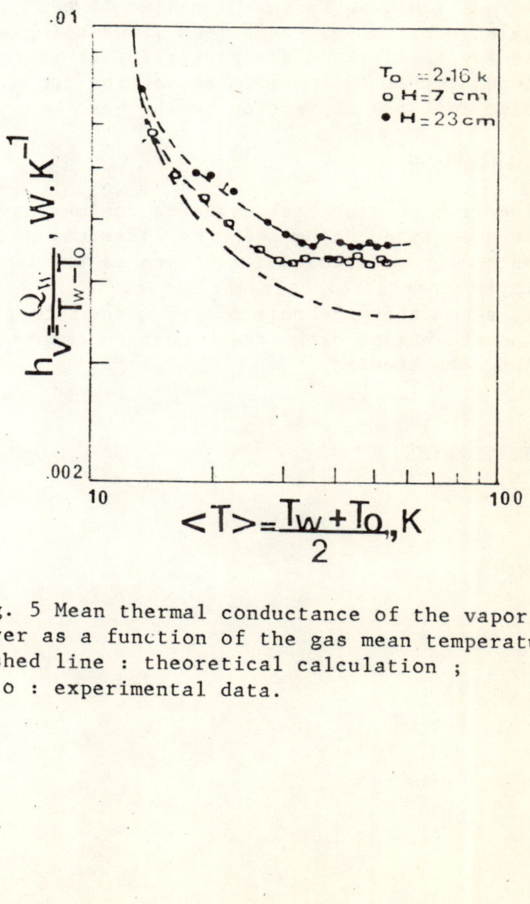

Fig. 5 Mean thermal conductance of the vapor layer as a function of the gas mean temperature dashed line : theoretical calculation ;
•, o : experimental data.

Plotted in (Fig. 5) are the experimental results of Q_W/ T versus <T> for H=7 cm and H=23 cm, and the theoretical calculations after (1). It can be noticed that a good agreement between both is only obtained for small values of T_W. Moreover every transition from a regime to another induces a more pronounced deviation for the theoretical diffuse suggested process.

5. An other way to test the onset of a convection mechanism in the heat transfer process, is to plot the Nusselt number versus the Grashoff or Rayleigh number defined as

$$Nu = \frac{he}{\bar{k}}, \quad Gr = \frac{g\beta(T_W - T_o)e^3\rho^2}{\eta^2}$$

$$Pr = \frac{\eta C_p}{\bar{k}}, \quad Ra = Pr.Gr$$

where the characteristic length could be the thickness e. For low values of $T_W - T_o$, Nu is nearly independent of Rayleigh as expected. But the change in slope associated with a new behaviour given by

$$Nu \propto Ra^{0,5}$$

occurs for critical Rayleigh number value of about 10^{-3}. This very small magnitude is not in agreement with a classical natural convection, unless it is due to the presence of the large horizontal temperature gradient in the gas. Moreover large convection streams of normal fluid $\rho_N.V_N$ in the helium II bath were able to be taken, the stream lines being marked by small solid air particles. It is thus thought that this helium convection could also initiate convection in the gas.

CONCLUSION

The film boiling heat transfer regimes have been analyzed using both the classical thermal method and spatial observations of the vapor layer geometrical characteristics.
It seems that the role played by the interface and the helium bath itself must be emphasized and more studied.

REFERENCES

1 Ametistov, Ye.V. "Heat transfer with He II noiseless film boiling". Cryogenics, Vol. 23 (1983), pp. 179 and references herein.

2 Labuntzov, D.A., Ametistov, E.V. and Spiridona, A.G. Teploenergetica, Vol 28 (1981), pp 18.

3 Van Sciver, S.W. "Heat transfer in superfluid helium II". Proceeding of the 8th International Cryogenics Engineering Conference, Butterworth Scientific, London (1980), pp. 228-237.

4 François, M.X., Defresne, G., Fouaidy, M. and Jebali, F. "Transfert thermique avec changement de phase dans l'helium superfluide" IIèmes journées de Cryogénie, Aussois, France (1986), pp 119. Les Editions de Physique.

5 Jebali, F., Masa, J., François, M.X. and Vidal, F. "Noiseless film boiling in superfluid helium on a heated thin (32 μm) wire". Advances in Cryogenic Engineering, Vol. 33, (1988), pp. 425-430.

6 Jebali, F., Fouaidy, M. and François, M.X. "Comportement de l'interface vapeur-helium superfluide crée autour d'un fil chauffé" Annales de Physique, Colloque n° 2, supplément au n° 3, Vol.13 (Juin 1988), pp 89-90.

7 Jebali, F., Masa, J., François, M.X. and Vidal, F. "Temperature and heat flux measurements in noiseless film boiling with cylindrical symmetry in superfluid helium" J. Phys. D : Appl. Phys., Vol. 21 (1988), pp 1508-1511.

8 Defresne, G., Fouaidy, M. and François, M.X. "Thermométrie Cryogénique (1.3 K - 40 K) : étalonnage et loi de comportement" IIèmes journées de cryogénie, Aussois, France (1986), pp 151.

9 Betts, K.R., Leonard, A.C. "Free convection film boiling from a flat horizontal surface in saturated helium II" Adv. Cry. Eng. vol. 21 (1975), pp 282.

HEAT TRANSFER STUDY CF SHUTTER CONDENSER OF REFRIGERATOR

Liu Shuren and Wu Yezheng
Division of Refrigeration Department of Power Machinery
Engineering Xi'an Jiaotong University, Xi'an, China

ABSTRACT

This paper makes an intensive study of heat transfer problem on the shutter condenser of refrigerator. The influence of miniblade number on heat transfer process is analysed. During the study, other factors are considered such as heat transfer on the plate, thermal radiation, heat conduction, and heat transfer with phase change. In the end, overall heat transfer coefficient of shutter condenser is presented. After comparing theoretical solution with experimental result, satisfactory agreement is found between them.

INTRODUCTION

Shutter condenser is widely applied in refrigerator. Since people use refrigerator more and more in the present, it is necessary to make the intensive study of shutter condenser. Overall thermal resistance Ro of shutter condenser is consisted of three parts, the outside resistance Rn, the conduction resistance Rc, and the inside resistance Rp.

$$Ro = Rn + Rc + Rp \qquad (1)$$

Resistance Rn can also be devided into three independent parts, the free convection resistance of inclining miniblade Ri, the free convection resistance of vertical plate Rv, and the equivalent radiation resistance Rr.

ANALYSIS

A. Ri

It is hypothesized that the miniblade is an isothermal plate. The mathematical model of the finite length two dimensional isothermal plate inclined at an angle ϕ to the direction of the gravitational acceleration can be set up. In the present analysis the inclined angle ϕ has a positive sign for the region "above" the plate, and negative sign for the region "below" the plate. The coordinate system, velocity direction and orientation of plate are shown in Fig. 1.

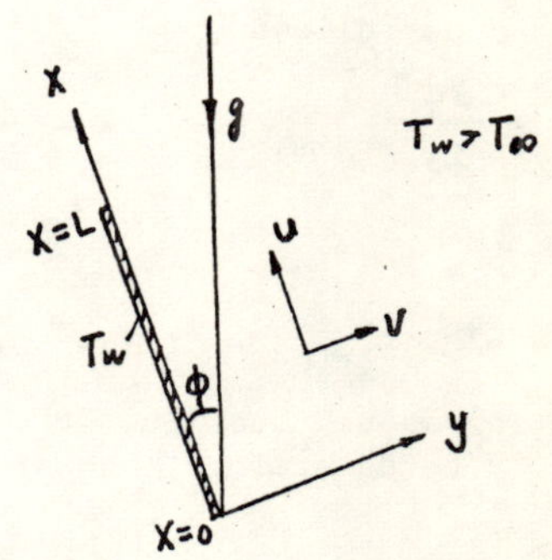

Fig. 1

With the assumptions that all physical properties of the fluid, except density, are constants. The governing Navier-Stokes, continuity and energy equations are as follows

$$u\frac{\partial u}{\partial x} + v\frac{\partial u}{\partial y} = -\frac{1}{\rho}\frac{\partial}{\partial x}(P - P_\infty) + \gamma\left[\left(\frac{\partial^2 u}{\partial x^2}\right) + \left(\frac{\partial^2 u}{\partial y^2}\right)\right] + g\beta(T - T_\infty)\cos\phi \qquad (2)$$

$$u\frac{\partial v}{\partial x} + v\frac{\partial v}{\partial y} = -\frac{1}{\rho}\frac{\partial}{\partial y}(P - P_\infty) + \gamma\left(\frac{\partial^2 v}{\partial x^2} + \frac{\partial^2 v}{\partial y^2}\right) + g\beta(T - T)\sin\phi \qquad (3)$$

$$\frac{\partial u}{\partial x} + \frac{\partial v}{\partial y} = 0 \qquad (4)$$

$$u\frac{\partial}{\partial x}(T - T_\infty) + v\frac{\partial}{\partial y}(T - T_\infty) = a\left[\frac{\partial^2}{\partial x^2}(T - T_\infty) + \frac{\partial^2}{\partial y^2}(T - T_\infty)\right] \qquad (5)$$

The boundary conditions associated with above equations are

$$y = 0, \quad u = 0, \quad v = 0, \quad T = Tw$$
$$y \to \infty \quad u \to 0, \quad v \to 0, \quad P \to P \quad T \to T \qquad (6)$$

After introducing new dimensionless variables in the form, we have

$$x=x^+L, \quad u=u^+\frac{\gamma}{L}Gr, \quad y=y^+L,$$

$$v=v^+\frac{\gamma}{L}\,Gr$$

$$T - T_\infty = T^+(Tw - T_\infty), \; P - P_\infty = P^+\frac{\rho\gamma^2}{L^2}\,Gr$$

Where Grashof number Gr is recognized as

$$Gr = \frac{g\beta(T_W - T_\infty)\,L^3\cos\phi}{\gamma^2}$$

Because the dimensionless steam function satisfies the continuity equation, U^+ and V^+ can be expressed as follows

$$u^+ = \frac{\partial\psi^+}{\partial y^+}, \qquad v^+ = -\frac{\partial\psi^+}{\partial x^+}$$

Let

$$X = x^+ \qquad Y = y^+\,Gr^{-\frac{1}{4}}$$

After solving above equations, the temperature profiles may be determined, and are given by

$$\frac{T - T_\infty}{T_W - T_\infty} = H(\eta) + Gr^{-\frac{1}{4}} \times \frac{3F(\infty)}{\pi(4x)^{\frac{1}{4}}} \times -\left[\sum_{m=0}^{\infty}\frac{X^{\frac{1}{4}}}{m+\frac{1}{4}}\,h_m(\eta)\right] \tag{7}$$

The local heat transfer coefficient will be defined as

$$\alpha(Tw - T_\infty) = -\lambda_w\left(\frac{\partial T}{\partial y}\right)_w$$

It may be expressed in terms of a Nusselt number Nu

$$Nu = [-H'(o)]\left(\frac{Gr}{4}\right)^{\frac{1}{4}}(\cos\phi)^{\frac{1}{4}} + \frac{3F(\infty)}{4\pi}\left[\sum_{m=0}^{\infty}\frac{X^{(m+\frac{1}{4})}}{m+\frac{1}{4}}\,h'_m(o)\right] \tag{8}$$

The mean Nusselt number is

$$Nu = \frac{4}{3}[-H'(o)]\left(\frac{Gr}{4}\right)^{\frac{1}{4}}(\cos\phi)^{\frac{1}{4}} + \frac{3}{4}\,\frac{F(\infty)}{\pi}\left[\sum_{m=0}^{\infty}\frac{X^{(m+\frac{1}{2})}}{(m+\frac{1}{4})(m+\frac{1}{2})\,L^{m+\frac{1}{2}}}\,h'_m(o)\right] \tag{9}$$

After simplizing equation (9), it may be written

$$Nu = 0.668\left(\frac{Gr\cos\phi}{4}\right)^{\frac{1}{4}} - 0.16 \tag{10}$$

With equation (10), the value of Ri can be obtained.

B. Rv

The area of plate between miniblade can be regarded as vertical isothermal plate. The formulation of vertical isothermal plate may be recommended as [1]

$$Nu = 0.59\,(Gr \cdot Pr)^{\frac{1}{4}} \tag{11}$$

The value of the Rv can be yielded from equation (11), The heat flows Q1, Q2 and Q12 are the result of natural convection, and can be given by

$$Q1 = F1\,(Tw1 - T_\infty)/R_i \tag{12}$$

$$Q2 = F2\,(Tw2 - T_\infty)/Rv \tag{13}$$

$$Q12 = Q1 + Q2 \tag{14}$$

Where F1 —— area of miniblate
F2 —— area of vertical plate
Tw1 —— temperature of miniblate
Tw2 —— temperature of miniblate

C. Rr

Assuming the condenser as a grey body, the heat flow of thermal radiation between condenser and ambient can be shown by

$$Q3 = (F1 + F2)\varepsilon\sigma(Tw^4 - T_\infty^4) \tag{15}$$

Hence we have equivalent radiation resistance Rr

$$Rr = \frac{1}{\varepsilon\sigma(Tw+T_\infty)(Tw^2+T_\infty^2)} \tag{16}$$

Then the natural convection resistance Rn can be given by means of follow method

$$Q = Q1 + Q2 + Q3$$

$$Q = F\alpha\Delta T$$

$$Rn = \frac{F\Delta T}{Q} \tag{17}$$

Where F = F1 + F2

Q —— total heat transfer coefficient

D. Rp

When refrigerant is condensed in the condenser, heat transfer coefficient can be shown as [2]

$$\alpha = 6\left(\frac{\beta}{q\,d_i}\right)^{\frac{1}{3}} \tag{18}$$

Where q —— specific rate of heat flow
d_i —— internal diameter of tube
c,β —— constants

Rp can be obtained form equation (18)

E. Rc

Formulation of conduction resistance is

$$Rc = \frac{F}{2\pi l\lambda / \ln(d_2/d_1)} \quad (19)$$

Since Ro=Rn+Rp+Rc, the overall heat transfer coefficient is

$$K = \frac{1}{Ro} \quad (20)$$

CONCLUSION

A study on the heat transfer of shutter consenser has been carried out. The mean Nusselt number of shatter condenser of refrigerator can be obtain

$$Nu = 0.9\left(\frac{Gr\cos\phi}{4}\right)^{\frac{1}{4}} - 0.22 \quad (21)$$

The formulation can be applied in project calculation when the inclined angle is smaller then 30 °C, the rate of length to width is not more than 8, the area of miniblade is approximation equal to the area of part of vertical plate. The results obtained from both the calculation and experiment data are listed in Table 1.

Table 1 Comparison of overall coefficient of heat transfer

Condition	Temperature (°C)	Refrigerant temperature (°C)	Overall heat transfer coefficient		deviation (%)
			Theoretical ($w/m^2.k$)	Experiment ($w/m^2.k$)	
1	29.96	41.33	9.547	9.222	4.5
2	34.27	56.75	10.68	9.51	10.9
3	30.31	51.36	10.48	9.77	6.7
4	34.59	63.36	11.13	10.43	6.3
5	30.22	58.72	10.63	10.05	5.4
				Mean deriation	6.76%

According to Table 1., the mean deviaticn is only 6.8%, it means the good agreement between the theoretical solution and experimental result.

NOMENCLATURE

u	Velccity component in x direction (m/s)
v	velocity component in y direction (m/s)
p	static pressure (N/m^2)
T	temperature (K)
g	acceleration due to gravity (m/s^2)
L	length of the miniblate (m)
m	dimensionless exponent introduced
F,H,h_m	dimensionless functions introudced
Pr	Prandtl number
ρ	density of fluid (kg/m^3)
β	coefficient of thermal expansior of fluid (1/K)
α	local heat transfer coefficient ($w/m^3.k$)
γ	kinematic viscosity of fluid (m^2/s)
λ	thermal conductiving of fluid (w/m.k)
φ	plate surface inclination angle
a	thermal diffusivity of fluid (m^2/s)
x	coordinate along the plate surface (m)
y	coordinate normal to the plate surface (m)
ψ	dimensionless steam function
∞	ambient condition
w	plate condition

REFERENCES

[1] Jame Sucec Heat Transfer 1975.

[2] Wu Ye-zheng and Han Bao-qi "Principle of Refrigeration and Equipments" Xi'an Jiaotong University Press, 1987.

NEW RESULTS ON THE MUTUAL FRICTION COEFFICIENT NEAR THE CRITICAL TEMPERATURE T_λ IN SUPERFLUID HELIUM

M. FOUAIDY, G. DEFRESNE, M.C. DULUC, F. JEBALI and M.X. FRANCOIS
Laboratoire de Thermodynamique des Fluides, PARIS VI
Campus Universitaire, Bat. 502TER, 91405 ORSAY Cedex, FRANCE

ABSTRACT

To study the phase transition of a pressurized (e.g $P > P_\lambda$) He II layer submitted to a heat flux, it is necessary to know with a large precision the experimental value of the Gorter-Mellink $A_{GM}(T)$ coefficient and the m exponent of Gorter-Mellink law, in the vicinity and as close as possible of the λ-temperature. We have perform an experiment to get such results in the temperature range between 2K and $T_\lambda - T$ = one millikelvin. The test cell was an He II filled vertical channel where we can move and localize the He II - He I interface by varying the heat flux density, and perform a quasi-local measurement of $A_{GM}(T)$.
Our data are in well agreement with the previous work of Vinen (1957), Brewer and Edwards (1961), Dimotakis and Broadwell (1973), Childers and Tough (1976) for T < 2,155 K, but not above this temperature especially with Dimotakis'results.

INTRODUCTION

The growing interest in superfluid helium (He II) as cooling medium for superconducting devices (1) in different applications as energy storage, fusion reactors which require a large temperature uniformity and stability against uncontrolled perturbations demands extension of the understanding of its transport characteristics.
In these applications, He II channel is used to connect the heat source to the refrigerator and heat load can be large enough to induce a local or broad phase transition into helium I or helium vapor at the heated surface. To study such phase transition of He II layer at pressure around P_λ (P_λ=5038 $N.m^{-2}$) and submitted to stationnary heat flux, it is necessary to know with large precision the experimental value of the Gorter-Mellink $A_{GM}(T)$ coefficient the heat conductivity function f(T) in turbulent conterflowing He II and the m exponent of Gorter-Mellink law in the vicinity and as close as possible of the λ-temperature (T_λ=2,172 K)

THEORY

Heat flow regimes of conterflowing He II have been extensively investigated in narrow ($D < 10^{-3}$cm) and wide ($D \geqslant 10^{-1}$cm) flow channels (2-4). An excellent review of this subject have been given recently by Tough (5) and it is established that He II channels exhibit mainly two distincts flow regimes :
- the well know Landau regime observed at sufficiently low relative velocities (e.g heat flux q) where no turbulence is present and the only dissipation is provided by the normal fluid viscosity η_n ; in this linear regime, the temperature gradient ∇T is related to the heat flux by the equation :

$$\vec{q} = - \frac{(\rho.S)^2.T.D^2\vec{\nabla T}}{32\eta_n}$$

Where ρ is the fluid density, S the entropy per unit mass and D is channel flow diameter.
- At a certain critical heat flux q_c depending on temperature and channel diameter ($q_c \sim 1/D$), one observes the transition to a non-linear supercritical regime (e.g turbulent regime $q > q_c$).
In the most applications (D > 1 mm), the supercritical or Gorter-Mellink turbulent regime is of concern. To account for the non linear dependance of the temperature gradient on the heat flux, in 1949 Gorter and Mellink (2) postulated an internal frictionnal force F_{sn} between the normal fluid and a tangled mass of vortex lines in the superfluid.
This mutual friction resulting of normal-fluid excitations scattering from the vortex lines is related to the relative velocity $w = |\vec{v}_n - \vec{v}_s|$ where $\vec{v}_n$ and $\vec{v}_s$ are respectively velocities fields of normal and superfluid component (3) :

$$F_{sn} = A_{GM}(T)\rho_s\rho_n (w-w_o)^{m-1}.w$$

for $w > w_c$

$$F_{sn} = 0 \quad \text{for} \quad w < w_c$$

In this expression $A_{GM}(T)$ and m are two independant experimentally determined parameters. Note that in the original theory (5), the turbulent state is assumed homogeneous and isotropic leading to m = 3.
However some experiments shows that the turbulence can be non isotropic (6) and experimental evidence (7) shows that the Gorter-Mellink m exponent is higher than 3 especially in the vicinity of the λ-temperature. In order to explain the increase in the power law of the Gorter-Mellink relation-ship between temperature gradient and heat flux near the λ-point, Swanson and Donnelly (8) have developped recently a theory deduced from dynamical rules.
In the fully developped turbulent state, the mutual friction is the dominant dissipative mechanism as compared with the viscous dissipation in the normal fluid hence the heat flow is governed by the Gorter-Mellink law :

$$q^{m-1} \cdot \vec{q} = - \frac{\vec{\nabla T}}{f(T)}$$

Where f(T) is the heat conductivity function in turbulent He II given by :

$$f(T) = \frac{A_{GM}(T) \cdot \rho_n}{S(\rho_s S \cdot T)^m}$$

In our knowledge few experimental results on f(T) have been reported close to T_λ. So it is interesting to measure f(T) and $A_{GM}(T)$ in the neighbourhood of the λ-point where these parameters diverge in such strong way that they cannot be seriously predicted by extrapoling earlier results.
We have perform an experiment to get such results in the temperature range between T = 1,8 K and T_λ - T = 1 mk for heat flux ranging from 10^3 to 10^4 w/m^2.

EXPERIMENTAL METHOD AND APPARATUS

A. APPARATUS

The experimental cell has been described elswere (9) : the flow tube is a stainless steel cylindrical channel with an inside diameter of 10 mm and a lenght of 150 mm. This vertically oriented channel is closed at its bottom by a copper block and connected to the helium II regulated bath at is upper end. Heat is supplied to the copper block by two heaters, the constantan wire (H_1) wrapped around and the constantan strip (H_2) stuck to its bottom. These two differents locations of the heater have been used in order to detect the influence of heat flux uniformity in the copper block on the heat flow in the He II channel. The power supply to the heater can be driven with various positive or negative slope rate ranging from 10μW/s to 0.1W/s. A stand-by position is available for each value of the heat flux. Five calibrated carbon thermometers (Allen-Bradley) are located in the channel at 5, 25, 50, 100, 135 mm respectively from the copper surface whose temperature is measured by two Allen-Bradlen carbon resistors.
The bath temperature which is monitored to within $\pm 50 \mu K$ is measured by a calibrated Germanium thermometer n°19937 of the lake shore Cryotronics Inc. and by a precise measurement of the vapor pressure (Texas Instrument Fused-quartz-gauge, model 145-01).
The test channel and the copper block are placed into a vacuum insulated can, and thus all the heat supplied to the copper is transfered to the He II channel ; the minimum heat leak (< 1%) by conduction between the copper and the stainless steel tube can be detected by another thermometer attached to the tube.

B. PRINCIPE OF THE METHOD

In one dimensional conterflow channel with zero mass flux, the temperature field is related to the local heat flux by the Gorter-Mellink law :

$$q^m = - \frac{1}{f(T)} \cdot \left(\frac{dT}{dx} \right)$$

Taking into account the heat flux variation at thermometers locations (constriction effect), this equation can be integrated from the location of the jth thermometer (xj) to the channel exit and hence (10) :

$$q^m (L - x_j + X(j)m) = \int_{T_o}^{T_j} \frac{dT}{f(T)}$$

where :

xj : location of the jth thermometer in the channel (1 < j < 5)
L : channel lenght
T_j : local temperature at xj
T_o : bath temperature
X(j) : is a geometric factor depending on the thermometer location in the channel (constriction effect)

A small variation of few percent of the heat flux induce at xj a local temperature variation δTj, so at first order (in δTj) we can write :

$$\left((q+\delta q)^m - q^m\right)(L - x_j + mX(j)) \simeq \frac{\delta T_j}{f(T_j)}$$

Finally one obtains the system of equations :

$$\left((q+\delta q)^m - q^m\right)\left(L - x_j + mX(j)\right) \simeq \frac{\delta T_j}{f(T_j)}$$

$$q^m\left(x_j - x_k + m\left(X(k) - X(j)\right)\right) \simeq \frac{T_k - T_j}{2}\left(\frac{1}{f(T_j)} + \frac{1}{f(T_k)}\right)$$

The left hand side of the last equation can be linearized with relative error less than 1 %.

The final system is resolved by an iterative method (10).

C. EXPERIMENTAL PROCEDURE

The detailled description of instrumentation is given elsewhere (10).
The experiment has been carried out for different bath temperature 1,8 K < T_o < 2,171 K at near saturated vapour pressure. Once the bath temperature is set to the desired value and regulated by means of AC-Bridge regulator; the temperatures in the channel are recorded as function of heat flux in the following way:

(1) a supercritical heat power Q_j is supplied at the bottom of the channel (and the temperature field Tj(Qj) in He II is measured once the stationnary state is reached.

(2) a small heat pulse δQj($\sim$ 5 %) is added to the actual value and the new values Tj + δTj are recorded.

(3) these value are stored in the disk for numerical treatment at the end of the experiment and so on.

IV - RESULTS AND DISCUSSION

The mean measured value ($\sim$ 200 data points) of the m exponent in the temperature range between 1.8 k and 2.1 k is found to be 3.05 with an R.M.S. of 8 %. We can conclude that in this temperature range and for heat flux ranging from 3 10^3 w/m^2 to 10^4 w/m^2, the turbulent state in He II is essentially homogeneous (e.g m=3). Using this value in our measurements we can compare them to earlier results (Fig. 1) : the data are in an excellent agreement with measurement done by quite different methods (e.g transverse second sound attenuation in rectangular schaped channel (VINEN), local temperature measurement with a single small thermometer on a traversing assembly in cylindrical duct (Dimotakis (11)).

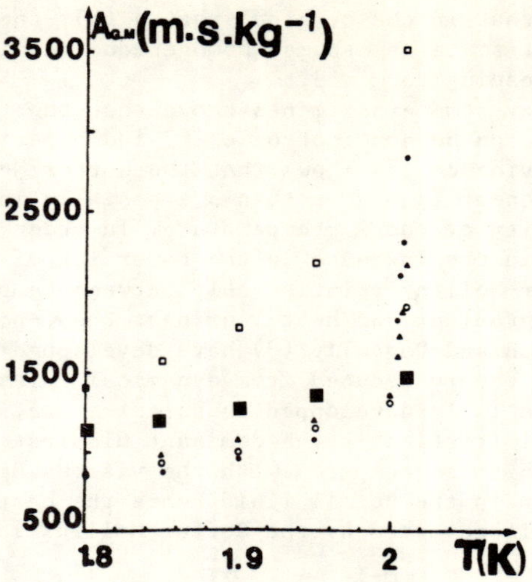

Fig. 1 : Gorter-Mellink coefficient A_{GM}(T) calculated with m=3
● : our results (D=10mm)
■ : Childers and Tough (1976) (12)
▲ : Dimotakis and Broadwell (1973) (11)
□ : Brewer and Edwards (1962) (4)
○ : Vinen (1957)

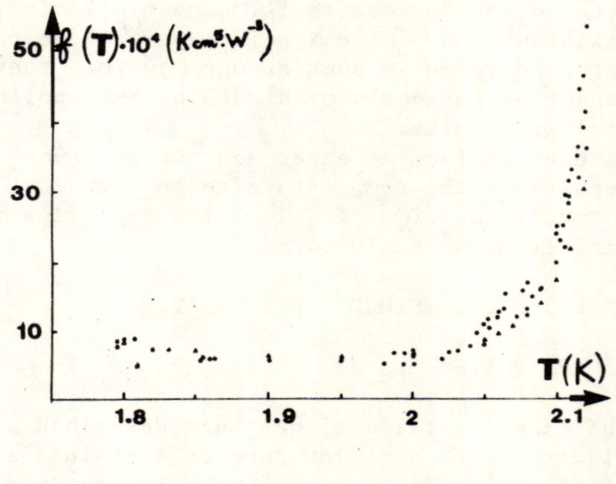

Fig. 2 : heat conductivity function f(T)
● : thermometer $T_{h1}(\delta T_1)$ } our results
* : thermometer $T_{h2}(\delta T_2)$ }
▲ : results of Dimotakis et al.

We note also a strong divergence of A_{GM}(T) as λ-point is approached. Also are presented (Fig. 2-4) the results on the heat conductivity function f(T) using m=3 in calculation. As compared with Dimotakis results, the agreement is very good up to T=2,155 k, but our results shows a less strongler divergence of f(T) in the neighbourhood of T_λ . Note that only few results have been reported by the author in this temperature range (e.g T_λ - T < 12 mK).

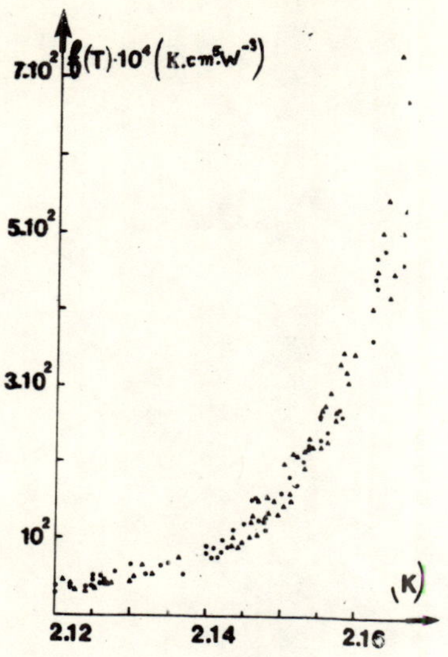

Fig. 3 Heat conductivity function f(T)
▲ : thermometer $Th_1(\delta T_1)$
* : thermometer $Th_2(\delta T_2)$

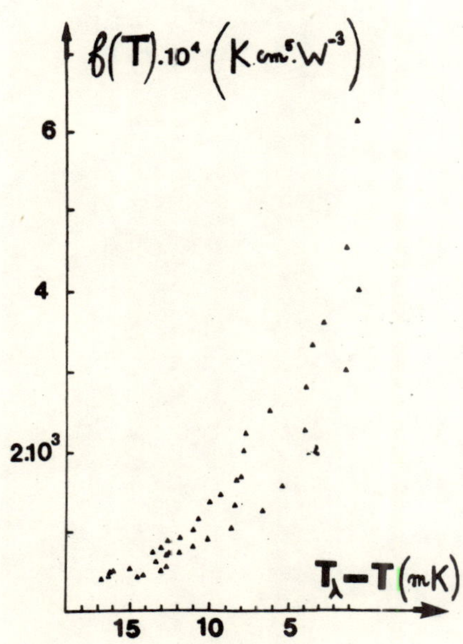

Fig. 4 Heat conductivity function in the vicinity of T_λ

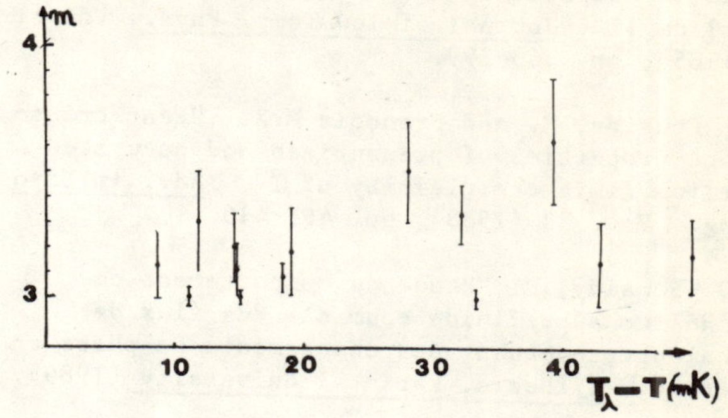

Fig. 5 experimental values of the Gorter-Mellink m exponent

Some works on superfluid turbulence (7) shows an increase of the power low for large heat flux (> 10 q_c) or in the vicinity of the λ-point.

The experimental results are shown in Fig. 5 : the m exponent measured (2mk < T_λ - T < 50 mk ; $10^3 < q < 10^4$ w/m^2) range from 3 to 3,62 giving a mean value in this temperature range of 3,18 with an R.M.S. of 6 %.

REFERENCES

1 Aymar, R. et al IEEE Transactions on Magnetics, Vol MAG 17, N° 1, pp 38.

2 Gorter, CJ. and Mellink J.H.. "on the irreversible processes in liquid Helium II". Physica, Vol XV (1949), pp. 285-304.

3 Vinen, W.F. "Theory of the mutual friction" Proc. of the Royal Society, A 240 (1957), pp 493-515.

4 Brewer, D.F. and Edwards "Heat conduction by liquid Helium II in capillary tubes, III. Mutual friction". The Philosophical Magazine, Vol. 7 (1962), pp. 721-735.

5 Tough, J.T. Prog. Low Temp. Phys., Vol. VIII (1982), p. 1330.

6 Ashton, R.A. and Northby J.A. "Ion currents in turbulent conterflow of helium II". Physical Review Letters, Vol. 30 (1973) pp. 1119-1122.

7 Ahlers, G. "Mutual friction in He II near the superfluid transition". Physical Rev. Let. Vol. 22 (1969), pp. 54-56.

8 Swanson, C.E. and Donnelly, R.J. "Vortex dynamics and scaling in turbulent conterflowing Helium II". Journal of low Temp. Phys., Vol.61 (1985), pp. 363-399.

9 Fouaidy, M. and François M.X., "Heat transport properties of pressurized and saturated Helium II in the vicinity of T . Adv. in Cryo Eng., Vol. 33 (1988), pp. 407-416.

10 Fouaidy, M. "Etude du comportement de l'Hélium superfluide soumis à des flux de chaleur provoquant des changements de phase en paroi", PhD Thesis. Paris VI University (1989).

11 Dimotakis, P.E. and Broadwell J.E. "Local temperature measurements in supercritical conterflow in liquid helium II". The Physics of Fluids, Vol.16 (1973), pp 1787-1795.

12 Childers, R.K and Tough J.T. "Helium II thermal counterflow : temperature and pressure difference an analysis in term of the Vinen theory". Physical Review B, Vol.13 (1976), pp 1040-1055.

EXERGETIC ANALYSIS FOR NITROGEN LIQUEFIER DESIGN

R.G. Gates and M.J. Andrecovich
The BOC Group Technical Center
100 Mountain Ave.
Murray Hill, N.J. 07974

J. Marshall
BOC Limited
The Priestley Center
Guildford, Surrey GU2 5XY

Z.Q. Sun
BOC Technologies, Ltd.
100 Yan An Road (E), Shanghai
The People's Republic of China

ABSTRACT

This paper presents an overview of some of the methods through which The BOC Group, p.l.c., has designed large, high-performance, nitrogen liquefiers, and illustrates these methods relative to parallel-expander plants. Exergetic analysis was used throughout, providing a means of evaluating either a whole liquefier or a working cycle sub-system relative to the thermodynamic ideal.

Computer simulations of component cycles were conducted to provide efficiency maps, and the interaction of these cycles was investigated both graphically, using heat integration concepts, and more exactly with whole-plant simulations.

Parallel-expander liquefiers, where each turbine exhausts at a common pressure and temperature, were explored and found to offer both efficient operation and, compared to the more common series topology, lower heat exchanger duty.

INTRODUCTION

Nitrogen liquefiers associated with large air separation plants must perform their function with maximal efficiency to be economic. Continuing improvement in compressors and expansion turbines has been of benefit to these plants, but concomitant upgrading of process design, to take full advantage of these improvements, is also required. This article describes the use of exergetic analysis by The BOC Group, p.l.c., and outlines, by way of example, the development of new designs for large liquefiers with superior characteristics.

Nitrogen liquefiers usually comprise several interacting refrigeration cycles. Conventional design techniques allow the assessment of the value of process modifications to the cycle as a whole, but offer little information on how such changes influence the individual refrigeration cycles or their interaction. Exergetic analysis of each of the cycles in a liquefier shows what trade-offs have been made in a process modification and may suggest further improvements.

The approach of this process study was; first, to computer-simulate individual refrigeration circuits to identify parameter adjustments leading to the highest exergetic efficiency; second, to assemble these improved circuits into a whole using heat integration techniques; and finally, to simulate the entire liquefier to allow performance assessment and parameter tuning.

EXERGETIC EFFICIENCY

Exergy, like entropy, is a defined thermodynamic quantity, but it has the dimensions of energy. Exergy is a state function, and for nitrogen refrigeration circuits where there are no composition changes it may be defined as;

$$X \equiv H-H_o-T_o(S-S_o); \qquad (1)$$

where the subscript o refers to the reference state.

Once the reference state is defined, the value of exergy at any other state point is unique. If the exergies of two states, 1 and 2, are subtracted to calculate the difference obtained in moving from 1 to 2, the enthalpies and entropies of the reference state cancel.

$$X_2-X_1 = X_{1\to 2} = H_2-H_1-T_o(S_2-S_1) \qquad (2)$$

Here the ideal work in moving from state 1 to state 2 is calculated. If, for example, state 2 involves an isothermal compression from state 1, the exergy difference is the least work which would be required to produce this compression. Similarly, if cooling takes place at constant pressure from state 1 to state 2, equation (2) gives the least possible refrigeration work needed for this cooling.

The time rate of change of exergy may also be defined for a material stream flowing in some process where its thermodynamic state is altered.

$$d(X_{1\to 2})/dt = \dot{M}(x_{1\to 2}) = \dot{M}[h_2-h_1-T_o(S_2-S_1)]; \quad (3)$$

where $\dot{M}$ is the flow rate, mass/time, and x, h, and s are specific values pertaining to a unit mass of the working fluid.

The values for h and s are, of course, taken directly from thermodynamic data tables. T_o is commonly set at the ambient air or water temperature. In this paper a value of 300K was used.

Exergetic efficiency is defined as:

$$\eta \equiv \text{desired output/necessary input} \quad (4)$$

In this nitrogen liquefier analysis "necessary input" is the shaft power required to maintain steady operation, and "desired output" is the time rate of change in useful exergy imparted to a fluid stream through cooling. Exergetic efficiency, then, compares a real liquefier cycle to a thermodynamically ideal one and is thus a universal measure of its performance.

TOPOLOGIES CONSIDERED

Two liquefier topologies were considered in this work, denoted as "series" and "parallel". A typical configuration of a large series liquefier is shown in Fig. 1.

Power input is through a multi-stage compressor which provides nitrogen under pressure to the product stream and the working cycles (defined as those which provide the refrigeration). Four heat exchangers are depicted, HE-1 through HE-4. A Freon refrigeration system removes heat from the highest-temperature exchanger, HE-1. Two expanders are used in this example, a warm and a cold. These are connected in series; i.e., the exhaust temperature of the warm turbine is approximately equal to the inlet temperature of the cold turbine. A sequence of J-T expansions and separators completes the liquefaction of the product stream.

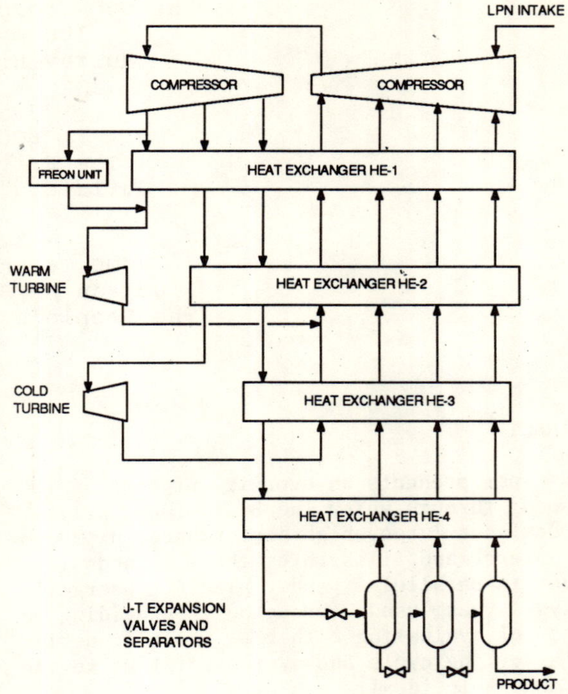

Fig. 1 Series Liquefier Schematic

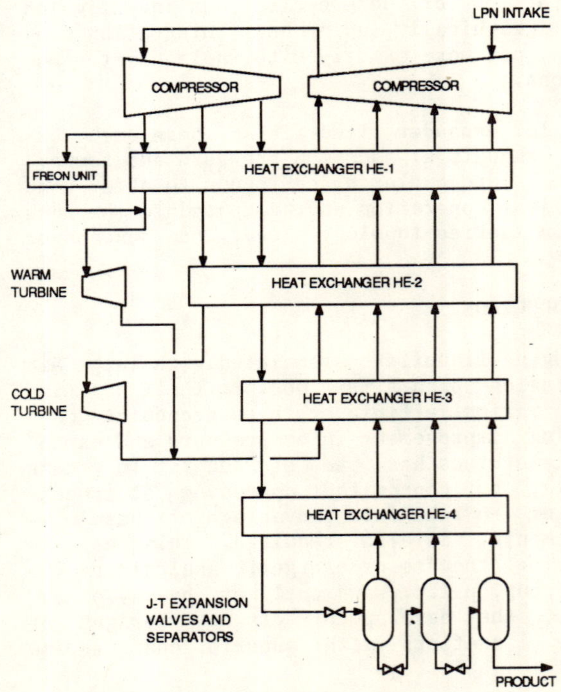

Fig. 2 Parallel Liquefier Schematic

The parallel liquefier, shown in Fig. 2, is similar to the series liquefier except for an alternative arrangement of the expanders. Both turbines exhaust at the same temperature and pressure into a common line, but each has a different inlet temperature and pressure from the other. This novel topology has certain advantages, and these will be explored in this article.

ANALYTICAL TOOLS

(1) Computer simulations

Computer simulations were written to provide performance data for both individual working cycles of a liquefier and for liquefiers as a whole. Both commercially available and inhouse process flowsheeting programs were used for this work.

The working cycle simulations were relatively straightforward due to the simplified process models upon which they were based. The warm turbine circuit, for example, involved only a compressor, a Freon system, an expander and a heat exchanger. Where whole-plant simulations were performed, the process model had to consist of a number of interacting elements; therefore the simulation had to be appropriately sophisticated.

The whole-liquefier simulations were topology-specific; one was formulated for each of the arrangements which was to be treated. From the given input data, these programs calculated the flow rates, temperatures, pressures and enthalpies of all streams in the flowsheet. Heat duties, EMTD and UA values for each heat exchanger were also compiled, as were the power produced by each expander, the power required by the Freon unit (if present), and the powers for the booster and recycle compressors. An additional feature was the calculation of exergetic efficiency for each of the working fluid cycles. The programs also provided an overall efficiency figure, based on the net work required by the plant, and determined the exergy increase experienced by the product stream.

The specifications for each simulation included:

a) For each heat exchanger; minimum allowable internal pinch temperature, warm or cold end temperature differences, and pressures and temperatures for certain streams where a priori choices of operating conditions had to be made.
b) For each expander; isentropic efficiency, generator efficiency, the isentropic efficiency of the associated booster compressor, some subset of inlet and outlet temperatures and pressures, and the vapor fraction or superheat of the outlet stream of the coldest expander.
c) For the recycle compressors; stage efficiency in two groups, outlet and suction pressures, seal losses, motor efficiency and pressure drops in associated piping.
d) For the liquefier as a whole; temperature of the Freon system outlet stream, thermodynamic state of the entering stream and product and the flow rate of the product.

The program included a feature which iterated certain stream variables until each of the exchangers satisfied both the conditions for pinch and for preselected temperature differences between warming and cooling streams. Again, the optimization was specific to the topology selected. Once the operability conditions for the heat exchangers were met within a user-defined precision, the program terminated and printed the calculated outputs.

Frequently, optimization was desired over a wider range of variables than the program could handle. In such cases, the program was run repetitively, with specifications altered between runs. Thus a series of optimized solutions was generated, and the best of the group was then selected.

(2) The Heat Availability Diagram

The Heat Availability Diagram (acronym "HAD") provides the conceptual basis upon which individual circuit parameters are adjusted to allow the various working cycles to be assembled into a whole refrigeration plant. A sample HAD is shown in Fig. 3 for a liquefier operating at 50 atm head pressure. The ordinate is linear in a transformed temperature, $(T_o/T)-1$, but scaled in terms of T, and the abscissa shows the total enthalpy difference. Notice that the ordinate scale is the inverse of the Carnot Coefficient of Performance at the temperature in question and therefore represents the ideal work to pump a unit quantity of heat from a temperature T to the reference level. It is a property of the diagram, then, that the area included between any segment of a constant pressure process curve and T_o is equal to the the exergy difference between the segment end points.

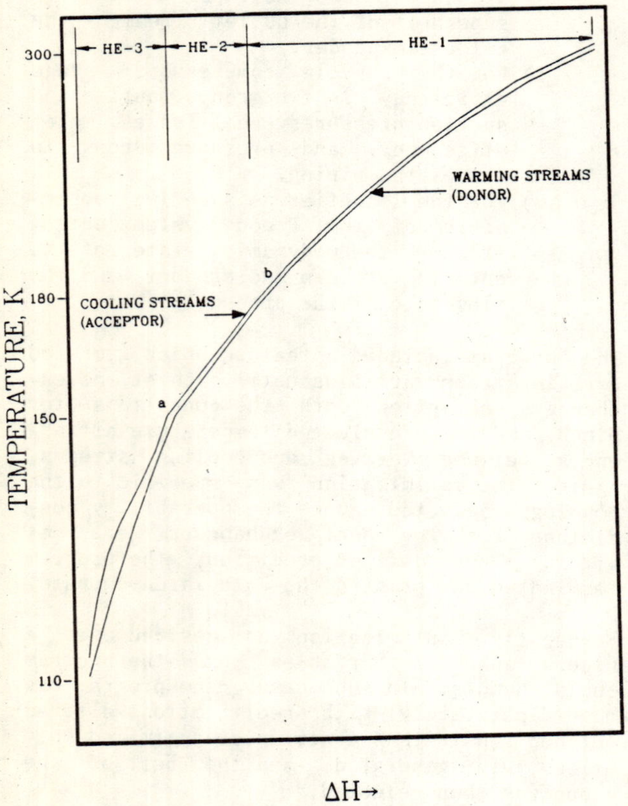

Fig. 3 Typical HAD, Series Liquefier

All constant (or nearly constant) pressure processes in the liquefier can be depicted on the HAD by the two curves shown in the Figure. The upper curve is a composite cooling curve. All the cooling flows in the main heat exchangers are represented here. This "acceptor" curve represents streams accepting refrigeration or exergy. The lower curve, the "donor", represents the total enthalpy change of all of the return streams including those from the separators. These return streams, of course, provide refrigeration to the main heat exchangers.

The boundaries of the warmer heat exchangers, HE-1 through HE-3, are shown as vertical lines in Fig. 3. First law balance requires that the donor and acceptor streams in each exchanger experience the same enthalpy change between inlet and outlet. The curves representing the streams are so positioned that their beginning and end points fall directly above one another, as shown. The area above the donor curve in an exchanger is the exergy supplied; the area above the acceptor is the exergy absorbed. The difference between the two areas is the lost work in the exchanger engendered by acceptor-to-donor heat flow through a finite temperature difference.

Note that the acceptor must always be above the donor to provide the temperature differential needed for heat flow. The minimum temperature difference shown in each exchanger is the pinch point.

Since thermal lost work in an exchanger is represented by the area between its donor and acceptor curves, these curves should be brought as close together as possible while respecting the minimum pinch conditions. This is done, as a liquefier is designed, by adjusting expander flow rates and the temperature at which these working cycle flows enter and leave the main exchangers.

The slope of a donor or acceptor curve is a function of the flow that it involves. If flow is increased, maintaining the same average pressure, the slope diminishes; i.e., more heat is absorbed or rejected between any two temperatures. Thus, if a donor curve rises so steeply as to cross an acceptor, the donor flow must be increased to reduce its slope. If the donor curve falls away from the acceptor, the return flows can be reduced. These considerations form the basis of a step-by-step procedure through which the HAD may be used to heat-integrate a liquefier, and were used in the conceptual and penultimate design phases of the parallel turbine plants.

ANALYSIS OF INDIVIDUAL CIRCUITS

As a preliminary to whole-liquefier design, individual working-fluid cycles were simulated and their exergetic efficiencies mapped as key parameters were altered. This effort provided canons for cycle optimization which were used to set initial operating conditions before the final adjustments were made in the whole-plant computer simulations.

(1) The Freon Circuit

Freon-circuit performance was derived in a separate study based on both manufacturer's data and theoretical considerations. It was taken into account in the whole-plant simulations, but only approximated in the working-cycle data to follow.

In general, Freon system exergetic efficiency increased as the outlet temperature was lowered, provided, of course, that the system complexity (the number of stages) was increased as well to reflect the widening of the temperature span.

(2) The Warm Turbine Circuit

The studies of warm turbine cycle performance used the process model shown in Fig. 4.

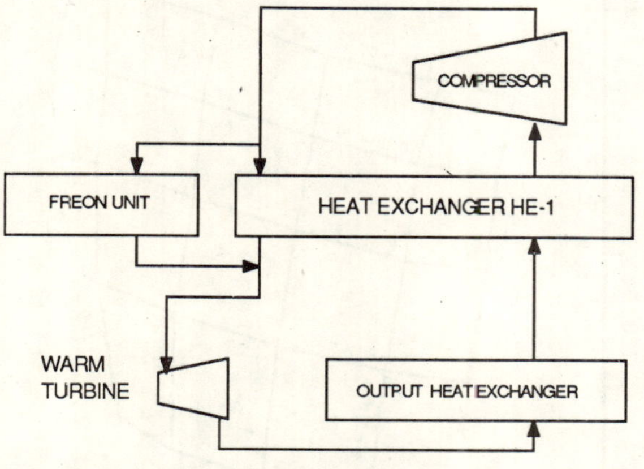

Fig. 4 Process Model, Warm Turbine Circuit

A Freon system with a 29% exergetic efficiency provided balancing refrigeration down to the inlet temperature of the expander. The return stream exchanged with the supply stream in HE-1 in which warm- and cold-end approaches of 5K and 3K respectively were assumed. Pressure drops of 0.2 atm were imposed in supply and exhaust lines. An isothermal efficiency of 72.5% was assumed for the compressor and an adiabatic efficiency of 85% was applied to the expander. Refrigeration was supplied by the circuit through an output heat exchanger. Two performance charts are shown in Figs. 5 and 6. Each Figure shows the loop performance at a specific turbine inlet temperature. Exergetic efficiencies can be read along constant inlet or constant outlet pressure lines as a function of outlet temperature, and a line is included showing efficiency when the expander is run with a saturated outlet.

A comparison of the two Figures shows the pronounced effect of inlet temperature on efficiency, which increased as this temperature was lowered. At an 80 atm inlet pressure and 33.6 atm back pressure, for example, the displayed efficiency at a 210K inlet was 37.9%, while at 180K it was 44.0%.

For a given inlet temperature, inlet and outlet pressures significantly influenced efficiency. If inlet pressure was held fixed, efficiency rose as outlet pressure was lowered; an effect especially pronounced at the lower inlet temperatures. If outlet pressure was constant, efficiency rose as inlet pressure was increased. The best cycles were the ones with low inlet temperature, high inlet pressure and low outlet pressure. For all combinations, (except for those with supercritical exhaust) the best efficiency was found when the exhaust was at the dew point.

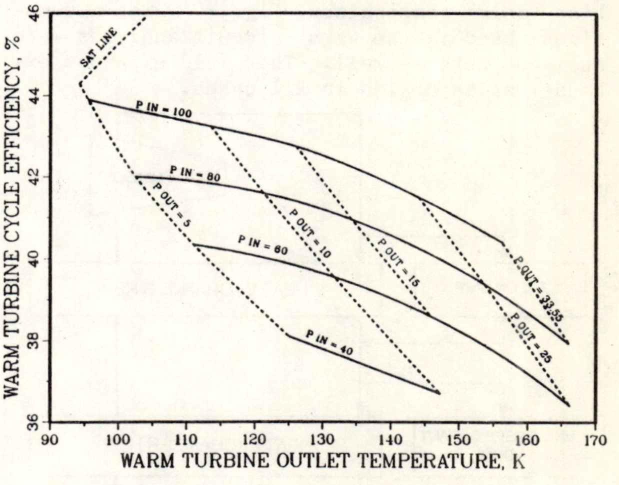

Fig. 5 Performance Curves for Warm Turbine Circuit, 210K Inlet

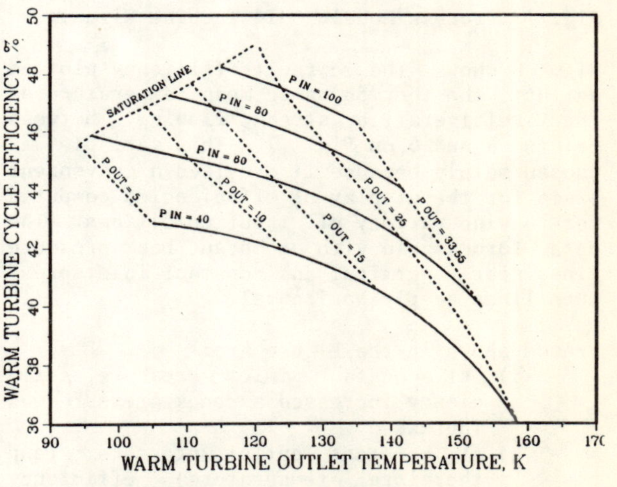

Fig. 6 Performance Curves for Warm Turbine Circuit, 180K Inlet

To summarize: the main effects seen were that warm turbine circuit efficiency was improved by raising inlet pressure, lowering inlet temperature, and lowering outlet temperature.

(3) The Cold Turbine Circuit

The process model analyzed to describe cold turbine performance was similar to that of the warm turbine, but contained an additional element, a 44% exergetic efficiency refrigeration system operating as a constant-efficiency warm turbine circuit between 210K and the cold turbine inlet temperature, Fig. 7. The assumptions used in the warm circuit analyses were applied here as well. The cold expander exhaust was saturated in all cases.

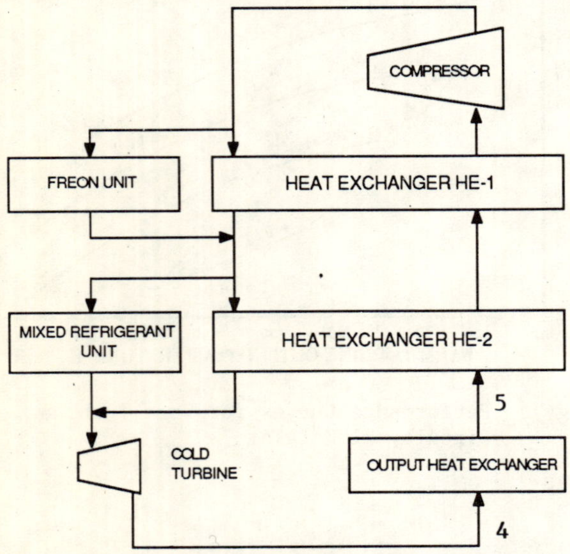

Fig. 7 Process Model, Cold Turbine Circuit

Fig. 8 shows the circuit efficiency plotted against the thermodynamic mean temperature of the refrigeration stream flowing between points 4 and 5 on Fig. 7. This variable was chosen mainly because it provided a convenient basis for the display of efficiencies computed for a wide variety of input conditions. The data form a grid with constant back pressure lines nearly vertical and constant inlet pressure lines nearly horizontal.

Trends shown in the Figure are:

a) At constant inlet pressure, efficiency increased as back pressure was increased.
b) At constant outlet pressure (and therefore temperature) efficiency (usually) decreased slowly as inlet pressure or temperature was increased. Exceptions are seen along the 20 atma back pressure line at intermediate inlet pressures and on the 15 atma line at low inlet pressures.
c) Thermodynamic mean temperature rose as inlet pressure or temperature was increased.

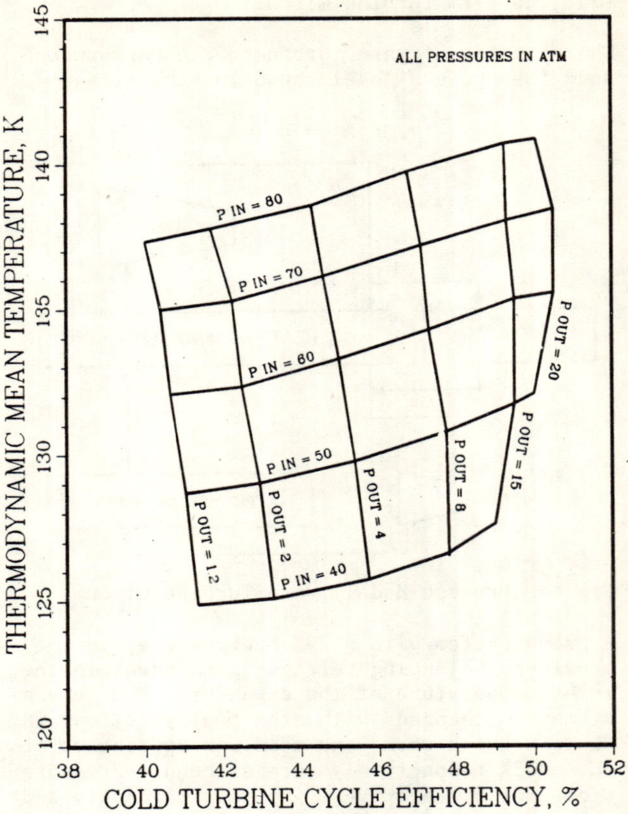

Fig. 8 Performance Curves for Cold Turbine Circuit

Summary of Turbine Circuit Efficiency Calculations:

a) Cold turbine circuits with saturated outlets improved in efficiency as back pressure was raised. Other factors had lesser effects.
b) Warm turbine circuit efficiency was improved by raising inlet pressure, lowering inlet temperature and lowering outlet temperature.

Therefore, decreasing the warm turbine outlet temperature to that of the cold expander, as is done in the parallel case, should not impair the efficiency of the warm turbine circuit. On this basis, then, parallel operation of the expansion turbines seemed viable, especially, as will be shown, at the higher product stream pressures.

(4) The J-T Valve and Separator Group

Relative to this last stage of the liquefaction process, where the product stream is expanded through J-T valves and flash separators in tandem, a few generalizations about exergetic efficiency can be made. The overall efficiency of the expansion sequence was improved both by increasing the number of expansions, (up to a point) and increasing the temperature of the final product.

The efficiency of the first expansion step was lower than that in the subsequent steps, but it was optimal if the temperature drop across the first J-T valve was about 30K. In the following expansion stages, efficiency (usually) decreased as the pressure differential was increased.

The influence of the product stream pressure and temperature entering the first J-T valve on efficiency is complex; it was, however, taken into account in the whole-plant computer simulations.

TOPOLOGIES COMPARED USING THE HAD

As indicated, the HAD can be used to establish expander circuit flows in a liquefier. It also displays heat exchanger duty and the lost work due to heat flow across a finite temperature difference. These quantities are influenced by liquefier topology, and the nature of this influence will be explored for two arrangements, series and parallel.

For the sake of simplicity, the illustrations of the method will involve only two expanders. The method is, of course, applicable to liquefiers with any number of working fluid circuits.

(1) The Series Arrangement

The first step in establishing a HAD for any topology is the plotting of the cooling curve for the stream(s) supplying the product and separator return flows. Generally, the product flow is set at unity; thus the curve to be drawn has a total flow of about 1.5 units. Separator inlet pressure and temperature must be defined, as well as the topology of the separators and the associated heat exchangers. After this is done, the return flows may be calculated from heat and mass flow balances for these units.

Once the cooling curve is drawn on the HAD as shown in Fig. 9, the inlet temperature of the cold turbine and the outlet temperature of the warm turbine are selected. In the case at hand they are arbitrarily set at 152K and 150K respectively. The donor curve is then plotted, beginning at the cold end of the exchanger associated with the cold turbine, (HE-3) and ending at 150K. The donor stream in this temperature range involves the cold turbine return flow, at a rate to be determined, and the flows from all separators at rates already calculated. This segment of the donor curve is positioned by changing the turbine return flow rate so that the curve is as close to the acceptor curve as possible, but not so close that the temperature difference at any point on the curve is less than the minimum pinch allowed. Placing this curve establishes the cold turbine flow.

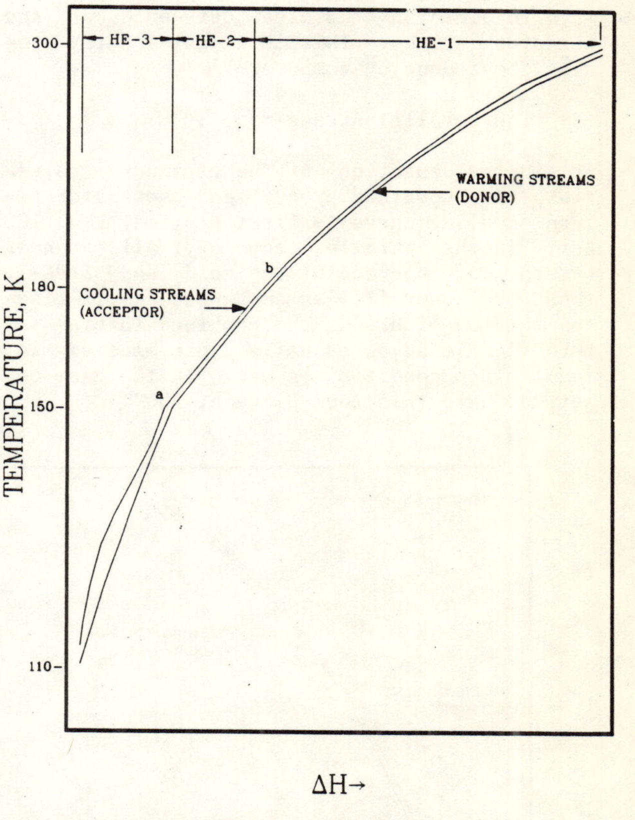

Fig. 9 Typical HAD, Series Liquefier (repeat of Fig. 3)

The acceptor curve must include the supply flow for the cold turbine. This supply is, after all, a stream cooled between ambient and (in this case) 152K, thus beginning at 152K the slope of this curve must be decreased to correspond to the added heat duty that this supply flow imposes. This is shown as the segment between points a and b on Fig. 9.

The donor stream above 150K carries the warm turbine returns as well as the returns from the cold turbine and the separators. Its slope is adjusted by changing the warm turbine loop flow so that it approaches the acceptor curve without coming too close.

Above 180K the acceptor flow is diminished by the flow that is diverted into the Freon evaporators. The greater this subtraction, the steeper the acceptor curve becomes. This is the third and final flow to be established by HAD concepts.

Fine adjustment of heat integrations can be performed by changing heat exchanger approaches (and therefore loop flow rates) to try to diminish the area between donor and acceptor curves. This is a cut-and-try process, best done by computer.

(2) The Parallel Arrangement

As in the depiction of the expanders in series, the combined product and separator return cooling curve is first plotted, Fig. 10. But, in the parallel topology, all expander returns are blended at the cold end of HE-3, thus the donor flow is unchanged as it rises to the top of HE-1. The combined turbine return flow is adjusted until a preselected minimum pinch condition is met. In the case described here this took place at 150K.

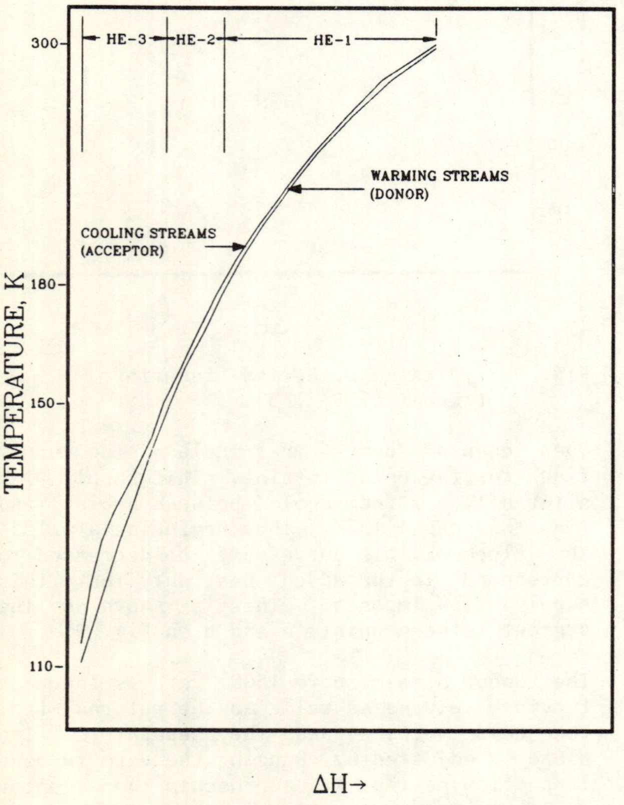

Fig. 10 Typical HAD, Parallel Liquefier

Above 150K, acceptor and donor curves diverge. Since there is no alteration in flow in the donor stream, its representation in the HAD is fixed at all points. It is the acceptor curve, then, which must be bent downward towards the donor by adding the supplies for the turbines to the flow of the combined product cooling curve at appropriate points, Fig. 10. This added flow cannot, of course, exceed the summed turbine returns. The cold turbine receives part of the total flow at 150K; the balance is supplied to the warm expander at 180K. Obviously, these temperatures and partial flows can be adjusted to minimize heat exchanger lost work. The Freon duty is derived here as it was in the series example.

The essential difference between the series and parallel topologies in now clear. In the series arrangement, the acceptor stream carried the entire cold turbine flow down to 152K. With the turbines in parallel, some of the supply flow leaves the acceptor at 180K, the rest at 150K. In the series case, therefore, downward inflection of the acceptor curve was relatively severe, and began at a relatively low temperature in contrast to the parallel case. The consequences of this to heat duty in the two warmest exchangers can be seen when Figs. 9 and 10 are compared. The parallel case generally involves less total expander flow and less duty in the warmest heat exchangers than does the series.

PARALLEL TURBINE TOPOLOGY, PARAMETRIC STUDY

Because of the advantages of the parallel turbine topology for nitrogen liquefiers described in the prior section, a simulation study was made of its performance. A comparable series liquefier was simulated as well, so that the efficiency of those parallel liquefiers which were treated could be related to the performance of a well-designed series unit. This latter plant, operating at a supply pressure of 50 atma, had been optimized earlier through a succession of computer studies.

Parallel-topology plants were investigated with supply pressures of 40, 50 and 60 atma. Efficiency was calculated in up to six cases at each pressure as a function of the product stream temperature entering the low-temperature exchanger associated with the J-T valve and separator group (HE-4).

(1) Results and Discussion

The results of the calculations are plotted in Fig. 11 and summarized in Table 1. A judgment was made, based first on the HAD and then upon calculations, that a third expander was necessary in the 40 and 50 atma cases to obtain appropriate low-temperature heat integration. This was included in the results.

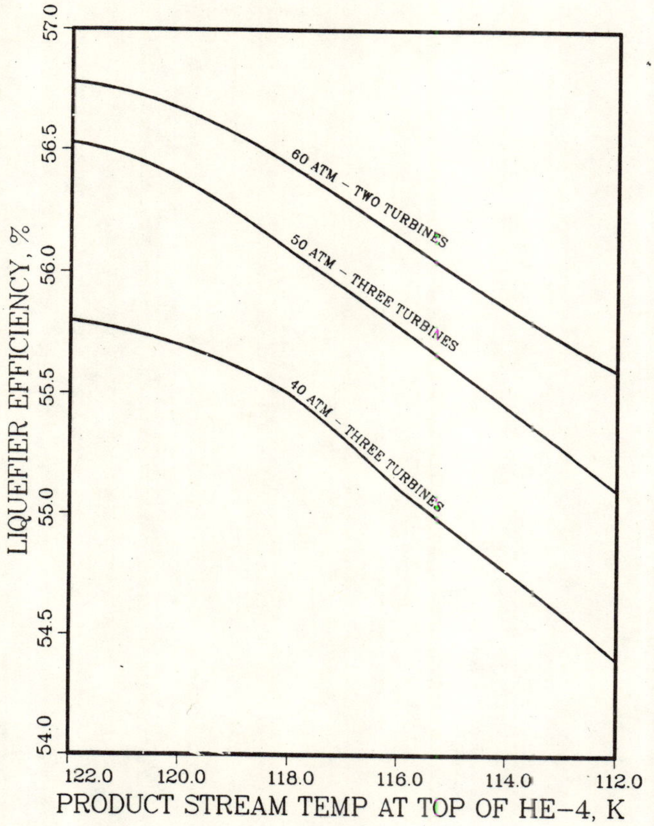

Fig. 11 Performance Curves for Parallel Liquefiers

Fig. 11 is a performance plot for the liquefiers investigated, showing liquefier exergetic efficiency as a function of an optimizing parameter, the temperature of the product stream entering the coldest exchanger (HE-4).

Liquefier efficiency was significantly improved by raising the product stream pressure and increasing the product stream temperature where it enters HE-4. However, referring to Table 1, where the product stream temperature was high, the inlet pressure to the warmest turbine was excessive. Thus, the more practical cases are those where this pressure was relatively low. Picking the lowest of each group, the liquefier efficiencies calculated were 55.6%, 55.1% and 54.4% for the 60, 50 and 40 atma product pressures respectively. These numbers compare favorably to the 53.7% efficiency calculated for a series version with two turbines operating with a 50 atma supply pressure, and with a warm turbine inlet temperature (180K) identical to the parallel cases.

Table 1. Data for Parallel Liquefiers (Compared to Series)

Run #	Prod. Pres. (atma)	No. of Expdrs.	Temp. of Product Strm.(K)	WT Inlet Press. (atma)	Rel. HE-1 Duty	Eff. (%)
1	60	2	122	122	0.53	56.8
6	60	2	112	86	0.53	55.6
8	50	3	122	122	0.62	56.5
13	50	3	112	86	0.62	55.1
15	40	3	122	122	0.85	55.8
20	40	3	112	86	0.76	54.4
28	52	2	–	52	0.72	53.6
Sers.	50	2	–	50	1.00	53.7

Case #28 is of special interest; this is a two-turbine arrangement with certain adjustments made to the topology to allow operation at a maximal pressure of 52 atma. Here the calculated efficiency was nearly identical to the series case, 53.6%.

The relative heat exchanger duty is of significance. As indicated in Table 1, and as predicted by HAD analysis, the overall exchanger duty is lower in the parallel cases than in the series. For example, when case #13 is compared to the 50 atma series plant; the duty of the warmest exchanger, (HE-1) in the parallel liquefier was found to be 62% of that in the series case, while the other exchangers were more nearly equivalent. In case #28, the HE-1 duty was 72% of that in the series liquefier.

(2) A Special Case at 80 atma

Two trends seen in the previous section, the increase in efficiency with increase in supply pressure, and the reduction in the number of expanders needed as this pressure was increased, lead to an exploration of still higher pressures. Calculations were made for plants operating at 80 atma, and it was found that one turbine could be eliminated without impairing heat integration. This single-turbine liquefier was explored and compared to a two-turbine series unit also operating at 80 atma.

As in the lower-pressure cases, the parallel (now single-expander) liquefier showed an efficiency nearly identical to the two-expander series case, but the warmest heat exchanger duty was cut almost in half. Reductions in duty were seen for the colder exchangers as well; overall, the single-turbine plant required only 56% of the exchanger area needed in the series.

SUMMARY

This article has attempted at least to outline the analytical and computational methods used by the BOC Group in designing large nitrogen liquefiers, and to show how these methods lead, in two specific cases, to the identification of new and interesting topologies. U.S. Patents have been awarded to the Group in relation to these topologies.

REFERENCES

1 Gates, R., "Gas Liquefaction Method and Apparatus", U.S. Patent No. 4,740,223 of 04/26/88 (to BOC Group, Inc.).

2 Gates, R. and Marshall, J., "Gas Liquefaction Method and Apparatus", U.S. Patent No. 4,758,257 of 07/19/88 (to BOC Group, Inc.).

SUPERCONDUCTIVITY RESEARCH AT THE NAVAL RESEARCH LABORATORY

R.J. Soulen, Jr., M.S. Osofsky, L.E. Toth, L.A. Allen, W.W. Fuller, W.T. Elam
J.H. Claassen, V. LeTourneau, P.R. Broussard, , S.B. Qadri, E.F. Skelton,
W.L. Lechter, R. Rayne, R. Gossett, F.J. Rachford, H.A. Hoff, B. Das,
M.Z. Harford, P.Lubitz, S.A. Wolf

Naval Research Laboratory
Washington, DC 20375-5000

ABSTRACT

We are engaged in a program to develop and study the properties of bulk and thin film samples of high transition temperature superconductors. We report on the progress made in these areas over the past several months. In the case of thin films, we have deposited films of YBaCuO on a variety of substrates (MgO, $SrTiO_3$) by coevaporation and have found the critical current density for the best films (deposited on $SrTiO_3$) to be $1x10^6 A/cm^2$ at 4.2K, and $4x10^3$ A/cm^2 at 77K. These films were superconducting below 88K. We also discuss a new technique for rapid preparation of thin films which has been successfully been used to prepare several high T_c materials. A new technique for measuring the critical current density in thin films has also been developed which does not require the films to be patterned. This measurement technique also is a factor of 100 times more sensitive than conventional methods. In the area of bulk superconducting materials, we have discovered a new material, TlSrCaCuO, which is made under conditions of high temperature and pressure. This material enters the superconducting state below 109 K, and completes the transition below 104 K. Finally we have developed techniques to cast molten BiSrCaCuO into a wide variety of geometrical shapes which can be made superconducting below 80 K with the proper anneal.

INTRODUCTION

Many scientists at the Naval Research Laboratory are engaged in a broadbased program to develop and study the properties of thin film and bulk samples of high-transition temperature (high-T_c) superconductors. This article will treat some of these activities, concentrating on those with which the authors have had direct experience. The topics reported in this article fall into two categories: Thin Films and Bulk Samples.

THIN FILMS

A variety of methods have been used to prepare thin films of all of the high T_c materials. Using a single target, thin films have been evaporated by dc and rf sputtering, by laser ablation, and by "flash evaporation". Thin films have also been prepared by coevaporation from multiple sources. We will report here on the last two methods since they have been the ones most extensively developed to date at NRL.

Since coevaporation requires a source for each component, it is best suited for preparation of thin films comprised of only a few elements. We have therefore used this technique only for the YBaCuO superconducting system (1) Films of YBaCuO were prepared by coevaporation of Y, Cu and BaF_2 onto substrates of $SrTiO_3$, MgO, and Yttria stabilized Zirconia (YSZ) . The substrates were cleaned before evaporation and were nominally at room temperature during deposition. Y and Cu were evaporated from electron beam sources with feedback rate control, while BaF_2 was evaporated from a temperature-controlled Knudsen cell without feedback rate control. Oxygen was introduced during the evaporation at a pressure of 10^{-6} Torr. Deposition rates were typically 1-10Å/sec and film thicknesses are between 5000 and 10,000 Å after annealing. Samples were annealed after deposition by removing them from the evaporation chamber and placing them in an oven at 850 °C in flowing moist O_2. The chemical composition of the deposited films was determined by elastic backscattering of alpha particles (RBS), while the lattice spacing and crystallinity were characterized by x-ray diffraction and scanning electron microscopy.

We have found that the characteristics of coevaporated films depend significantly on the substrate, and that the best results are achieved with $SrTiO_3$. For this substrate material, the superconductive transitions for stoichiometric YBaCuO films begin at a temperature of 91 K and are complete below 88 K. The transition is not markedly degraded by off-stoichiometric concentrations and is not strongly dependent on annealing procedures. The films are highly ordered and polycrystalline; the majority of the crystals have their c axis perpendicular to the plane of the substrate , and there is long range order in the plane of the film. For thin films grown on MgO, the onset of T_c is depressed by 6 K, the transitions are wider and are much more sensitive to annealing and composition, and there is very little inplane order. For films on YSZ, the transitions are even broader, with the onset at 90 K. This study of the interaction of superconductor and substrate is not complete and will continue so that improvements may be made.

The critical current density, J_c, was measured for a stoichiometric and a nonstoichiometric film of YBaCuO deposited on $SrTiO_3$ (2) . The films were patterned into a four-probe geometry and J_c was measured by a conventional transport method. The results as a function of temperature are shown in Fig. 1. The J_c for the stoichiometric film was rather high, even at liquid nitrogen temperatures, whereas the T_c and J_c values for the off-stoichiometric film were depressed. The dependence of J_c on an external, transverse magnetic field was also measured for the stoichiometric film at 77 and 4.2 K. At the former temperature, J_c was depressed very rapidly with the application of the magnetic field (approximately tenfold per 3 kG) , whereas at 4.2 K, J_c was depressed only by a factor of 10 at 100 kG. These results suggest that pinning forces are not large in the vicinity of the boiling point of liquid nitrogen, but that they are at lower temperatures.

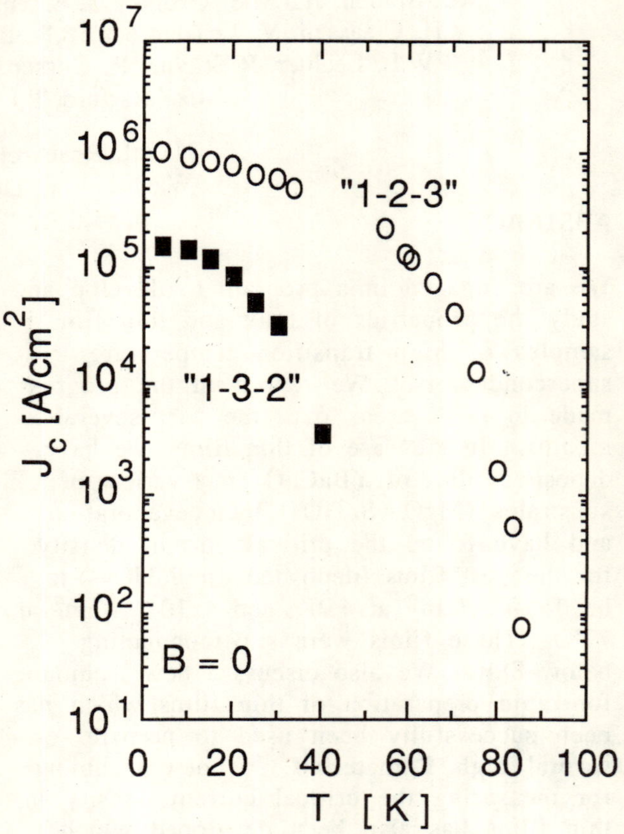

Fig. 1: J_c vs. T for two YBaCuO films. The "1-2-3" films has the stoichiometric composition, and the "1-3-2" film is barium-rich and copper-poor.

The second method for preparation of thin films used at NRL is that of "flash evaporation", (3) in which a small pellet of material of a specific composition is completely evaporated onto a substrate by an electron beam, thereby preserving the stoichiometry of the target. Since a single target is used, and since the composition is preserved (verified by EBS), this method is better suited than code-position to making superconducting compounds with more components, such as the BiSrCaCuO and TlBaCaCuO systems. Thin films may also be prepared very rapidly by this technique making it especially well suited for surveying new materials or for varying composition. To demonstrate the utility of this method, we made several samples of Bi- and Tl -based superconducting compounds. Quite satisfactory superconducting materials can be made by this method.

The critical current density is one of the most important physical parameters used to characterize superconducting materials. Unfortunately it is very inconvenient to measure J_c by conventional methods since the film must be patterned and leads must be attached to it. Aside from these inconveniences, there is the additional concern that the processes used for patterning and lead attachment might influence the superconducting properties. Accordingly, J. Claassen (4) has developed an inductive method for determining J_c for thin films in which the losses in a pancake shaped coil placed in proximity to an unpatterned thin film are detected and related to J_c. It is not necessary to pattern the film nor to attach contacts to it in order to employ this contactless method. It offers the further advantage of providing approximately 100 times greater sensitivity than the traditional method. This inductive method has been used to measure the J_c of samples of Nb and YBaCuO and the results compare well with values obtained by the conventional method. An unexpected discovery revealed by the greater sensitivity of the inductive method is that losses in YBaCuO were found to persist to much lower currents than had been previously expected. Indeed, the absence of a dissipationless current flow down to the very small currents studied suggests that J_c is not a well defined parameter in this material. The results are interpreted as providing evidence for flux creep mechanisms at work in this material. We intend to extend this technique by studying several more superconducting films and comparing the results to the conventional method.

BULK SAMPLES

We have prepared bulk superconducting samples for study by the following methods: ceramic powder processing, melting (5) ,hot isostatic pressing (6), and casting from a melt (7). The preponderance of our bulk samples (including targets for thin film preparation) were prepared by the first method but, since the technique has been described elsewhere and is in widespread use throughout the world, we will not treat it here.

We have used the second technique sparingly to produce a few, small single crystals of $ErBa_2Cu_3O_7$ (5) . The largest of these crystals, grown at the surface of molten material, was moderately large (0.75mm x 4mm x 4mm) and was the best sample we have produced. We submitted this crystal to many measurements. The resistive transition showed T_c to be centered at 92.7K and and to be less than 1 K wide. Study of the dc Meissner effect with a SQUID magnetometer confirmed the results of the resistance measurements. Magnetization curves taken with the same apparatus were reversible above 77K, indicating that flux pinning was small. Since many high frequency electronic applications envisioned for high T_c materials depend on the quality of the surface impedance,R_s, we measured it for this crystal since the other measurements showed that it was of exceptional quality. We found that R_s did indeed decrease rapidly below T_c, but the temperature dependence did not follow that expected from the BCS prediction. We did show, however, that R_s at a frequency of 9.2GHz was less than that of copper at temperatures below 50 K.

We have used hot isostatic pressing (HIP) to produce a new composition of a Tl-based superconductor (6) . We calcined $SrCO_3$ and CuO powders, then added Tl_2O_3, pressed the powders into a rod which was wrapped in a Pt foil and then placed in a sealed and evacuated stainless steel envelope. The sample was reacted at 850°C for 1 hour at a pressure of 30,000 psi (the HIP process). The resulting material was analyzed and found to have the composition $Tl_{1.7}Sr_3Ca_{0.7}Cu_{4.3}O_y$, indicating an accidental contamination of Ca. Nevertheless, we found that the sample exhibited zero electrical resistance below 105 K. The susceptibility indicated the presence of a second phase, since it went through two distinct transitions at temperatures of 100 K and 70 K. By using this method of sample preparation we were able to make a Ba-free Tl-based superconductor which exhibited superconducting behavior above 100 K.

Turning to the final topic, preparation of materials by casting, we have found that the material $Bi_4Sr_3Ca_3Cu_6O$ may be readily cast and machined into a number of useful shapes. For instance, we have cast a cylinder 8 cm long and 1.9 cm in diameter, bored out a 1 cm hole, annealed it and have used it as a superconducting magnetic shield. We have poured a sputtering target disk 7.6 cm in diameter and 1 cm thick. Finally, we have machined some thin wafers (1cm x 1 mm x 0.1 mm) for J_c measurements. The as-cast material is oxygen deficient and not superconducting, but we have shown that a 30 hour anneal in air at 800 ^oC produces good superconducting material. Thus it is easy to cast the material, machine it to the approximate shape, anneal it, and give it a finish machining treatment (since some distortion occurs during anneal). The samples are superconducting below ~80 K, and studies of their surface impedance and Jc are underway.

During the course of studying the optical properties of several bulk, high-T_c materials, H. Hoff noticed a correlation between T_c and the color of the material in reflected polarized light at room temperature. This correlation has been extended to a preliminary study and report (8). We have studied samples of the following materials (T_c given in parentheses): $YBa_2Cu_3O_7$(90K), $ErBa_2Cu_3O_7$(90K), $NdBa_2Cu_3O_7$(90K), $La_{1.9}Sr_{0.15}CuO_4$(38K), $Bi_2Sr_2CaCu_2O_8$(85K), $Tl_{1.7}Sr_3Ca_{0.7}Cu_{4.3}O_y$(109K), $Bi_{0.7}Pb_{0.3}SrCaCu_{1.8}O_y$(114K), and $Tl_2Ba_2Ca_2Cu_3O_{10}$ (120K). We have found that all the cuprate materials which have superconducting phases have the same characteristic golden brown color,while non-superconducting phases have a grayish brown color. We also performed the following experiment: we took a sample of $ErBa_2Cu_3O_{6.5}$ and, by successively removing oxygen,varied T_c and registered the color. The sample started with a T_c of 60K and a golden color. As the T_c was reduced in four steps to a temperature below 4K, the color of the sample gradually changed from golden brown to grayish brown. We speculate that the color may be due to the presence of free carriers in the material, but the explanation is no quantitative. Nevertheless, this method appears to provide a very simple test for the presence of superconductivity in cuprate superconductors without requiring the samples to be cooled.

REFERENCES

1. P. R. Broussard, J. H. Claassen, C. R. Gossett, M. S. Osofsky, and W.T. Elam, J. Superconductivity**1** (1988).

2. L. H. Allen, P. R. Broussard, J. H. Claassen, and S. A. Wolf, Appl. Phys. Lett.,**53**,1338 (1988).

3. M. S. Osofsky, P. Lubitz, M. Z. Harford, A. K. Singh, S. B. Qadri, E. F. Skelton, W. T. Elam, R. J. Soulen, Jr., W. L. Lechter, and S. A. Wolf, "Superconductivity in Thin Films of Bi-Sr-Ca-Cu-O Prepared by a Simple Flash Evaporation Technique", Appl. Phys. Lett. **53**,1663 (1988)

4. J. H. Claassen, "Inductive Measurements of Critical Current Density in Superconducting Thin Films". IEEE Trans. on Mag., to be published.

5. M.S. Osofsky, D.U. Gubser, B. Das, L.E. Toth, V. LeTourneau, E.F. Skelton, S.B. Qadri, S.A. Wolf, A.K. Singh, H.A. Hoff, C.S. Pande, R.J. Soulen Jr., J.E. Schooley, "Properties of Single Crystal and Nearly Bulk Density High Temperature Superconductors", Proc. of Conf. on High-Tc Superconductors, Materials, and Mechanisms, Feb.29-March 4, 1988, Zurich, Switzerland

6. W.L. Lechter, M.S. Osofsky, R.J. Soulen, Jr., V.M. LeTourneau, E.F. Skelton, S.B., Qadri, W.T. Elam, H.A. Hoff, R.A. Hein, "Bulk Superconductivity Above 100K in the Ti-Sr-Ca-Cu System",Solid State Communications **68**, 519-521(1988).

7. L. E. Toth, B. N. Das, R. J. Rayne, B. A. Bender, W. L. Lechter, M. S. Osofsky, R. J. Soulen, Jr., "Chemical Casting of High Tc Superconducting BiSCCO", to be published,

8. H. A. Hoff, M. Rubinstein, M. S. Osofsky, A. K. Singh, L. E. Richards, W. L. Lechter, L. E. Toth, B. N. Das, and C. S. Pande, " Color Indicator of Cuprate Superconductivity by Polarized Light Microscopy", to be published.

HEAT TRANSFER COEFFICIENT MEASUREMENTS IN FORCED FLOW HE II

J. G. Weisend II and S. W. Van Sciver
Applied Superconductivity Center
University of Wisconsin, 1500 Johnson Drive
Madison, WI 53706, USA

ABSTRACT

Recently there has been increased interest in using forced flow He II to cool superconducting magnets. However, in order to produce reliable designs a thorough knowledge of the heat transfer coefficient, h, in forced flow He II is necessary. Early work suggested the heat transfer process would be Kapitza conductance limited and as a result h would be independent of flow rate. In the only significant measurements to date, Kramer (1) found h to be a function of the fluid temperature and applied heat flux but not flow rate. The purpose of the work described here is to verify and expand upond Kramer's data providing a better understanding of the dependence of h on various parameters in forced flow He II.

Two different heat transfer test sections are used in this experiment. The sections are copper tubes with and I.D. of 1.09 cm and two different lengths, 2.54 cm and 0.3175 cm. He II flows through the tubes at rates up to 40 gm/s. Heat is applied via manganin wire heaters wrapped on the outside of the test sections. Ge resistance thermometers are imbedded within the test sections to allow determination of the wall temperature and additional thermometers are immersed in the helium to provide a bulk fluid temperature. The maximum heat flux tested in these experiments is 3.15 W/cm2 and data is taken at Reynolds numbers as high as $3x10^6$. The helium enters the test sections at 1.9K and the helium pressure is kept at $3x10^5$ Pa. The design of these test sections allows measurements of the local heat transfer coefficient using the short test section and the average heat transfer coefficient using the long test section. The test sections are included as part of a 5 meter horizontal test loop operated at the UW Liquid Helium Flow Facility.

REFERENCE

1. Kramer, H. P., "Heat Transfer to He II Flow," Presented at the 12th International Cryogenic Engineering Conference, Southampton, UK, July, 1988.

COMPUTER AIDED DESIGN OF MULTISTREAM COILED-TUBE HEAT EXCHANGERS FOR AIR SEPARATION SYSTEMS

Sunil Sarangi, P.S.R. Aithal, G. Venkatranam and K.V. Ravi Kumar
Process Equipment and Design Laboratory, Cryogenic Engineering Centre
India Institute of Technology, Kharagpur 721 302, India

ABSTRACT

Coiled tube (Hampson type) heat exchangers are extensively used in cryogenic process plants, particularly in medium and high pressure systems. While design methods and heat transfer correlations can be found in literature, no standardised design procedure is available.

COILHX5 is a user-friendly interactive program to design multistream (up to 5 streams) coiled tube heat exchangers written for personal computers. The program has 3 sections:

1) Input of customer specifications,
2) Calculation of exchanger geometry,
3) Output of results.

The input and output segments are written in (complied) Turbo-BASIC for better screen handling and graphic output, whereas the main body of the program is written in MS-FORTRAN for higher computational speed and access to different physical property programs. The three segments are called sequentially from a batch file.

The program determines the full geometry of the exchanger using variable specific heats of fluids or fluid mixtures, geometrical and configurational constraints and any other information provided by the designer. The output provides:

(a) geometrical specifications by the designer.
(b) relevant intermediate results,
(c) temperapure profiles.
(d) a graphic representation of the exchanger.

A modified version of this program has been used in a local industry with significant saving in design time. The benefit of better accuracy and higher flexibility is even greater.

THE VOID FRACTION MEASUREMENT IN THE TWO PHASE FLOW OF HELIUM (4.2k)

J. Wang*, G. Defresne, M.X. Francois, B. Allegri
Laboratoire de Thermodynamique des Fluides, Campus Universitaire, Bat 502 Ter
91405 ORSAY cedex, France

* Visiting Scholar, Department of Mechanical Engineering
Wuhan Institute of Textile Technology
Wuhan, China

ABSTRACT

The present work concerns the measurement of the void fraction in two-phase flow of cryogenic fluids (Helium and Nitrogen) by means of a capacitance sensor.

In the first section of this paper, the concept of the measurement is described. It is based on the variation of the capacitance of a cylindrical condenser in which the two-phase system is flowing.

The performances of the electronics (capacitance bridge with synchrone detection) are described. The sensitivity is about 1V/pF for nominal capacitance of 1 to 100 pF. This allows to detect a void fraction variation of new percent in a Helium flow.

The sensor(length=15cm;diameter=28/30mm) has been calibrated in vertical and horizontal orientations, where it was used as level detector hin Helium bath, in order to study the influence of flow patterns on void fraction measurements. The experimental results are in good agreement with theoretical predictions.

The effect of geometry on the flow pattern and velocity field of the liquid and gas phases has been studied experimentally using an optical system; the velocity of small bubble of gas gragged by the liquid has been measured by a high speed camera (2000 fr/s). The results are presented in this paper.

PREDICTION OF TENSILE STRENGTH IN FIBER REINFORCED COMPOSITES USING MONTE CARLO METHOD

Hisham H. Abdel Mohsen
Applied Superconductivity Center
University of Wisconsin-Madison
Madison, WI 53706, USA

ABSTRACT

The standard shear lag model is modiffied to account for the stresses developed between the fiber and the matrix during cooldown in unidirectional fiber composites. Utilizing this micromechanics model with the conjuction with the chain of bundles probability model and Monte Carlo simulation method, unidirectional composites with different ratios of the constituents and different specimen sizes are analyzed. The effect of cryogenic enviroment on the tensile strength of unidirectional fiber composites is also discussed.

CRYOGENIC SYSTEM OF NEW FUSION REACTOR

Junya Yamamoto
Plasma Physics Laboratory
Kyoto University
Gohasho, Uji, Kyoto 611, Japan

ABSTRACT

Ministry of Education, Science, and Culture is planning to construct large helical type fusion research system. This machine is made from a superconducting coil, which energy is 2 GJ. Large cryogenic system including 3000 L/h to 5000 L/h helium liquefier will be prepared for the purpose. Design scheme of the system is presented at the conference.

RESEARCH AND DEVELOPMENT IN PROGRESS TO OVERCOME FIELD EMISSION IN SUPERCONDUCTING ACCELERATOR

Q.S. Shu, K. Gendreau, W. Hartung, J. Kirchgessner, D. Moffat
R. Noer, H. Padamsee, D. Rubin, J. Sears
115 Newman Lab, Cornell University
Ithaca, N.Y. 14853, USA

ABSTRACT

Systematic studies are in progress for understanding, controlling and eliminating field emission (FE) in superconducting cavities so that higher gradients can be achieved for application of superconducting cavities to linear colliders. A key tool used to improve our understanding is a high speed temperature mapping diagnostic systxich can locate and characterize individual emitters as well as other sources of RF losses. The influence of various treatments on the properties of over 100 emitters analysed will be presented. We find that effectiveness of HE processing continues as higher RF power levels are used. Systematic exposure studies show that the predominant source of emitters is not the dust-free air (class 100) used for drying cavities, nor the pure methanol used for final rinsing. On the other hand, we have shown that gases condensed on the cavity cold wall can drastically increase emission at pre-existing sites. In exploration of potential cures, we have applied UHV heating above 1100 C as the final surface treatment on 1-cell, 1.5 GHz cavities. The average accelerating fields increased by 75 present in comparison with similar chemically prepared cavities. The best cavity reached surface electric (magnetic) field of mv/m (1260 OE).

THE ROLE OF ENGINEERS IN ENHANCING COMPETITIVENESS THROUGH TECHNOLOGY

Nirmal Chatterjee
Air products & Chemicals, Inc.
Allentown, PA 18195, USA

ABSTRACT

In virtually all product lines in the chemical industry, technology has always been the key to profitability and success. This is even more true today with intense world-wide competition. It is very clear that technology is changing rapidly and we must be able to incorporate these changes in new and existing plants in minimum time and at minimum cost. This requires engineers to be at the leading edge of technology; to be knowledgeable about the fundamentals of engineering; to have an appreciation of costs; and to know how cost-benefit analyses are carried out.

In the segment of the cryogenic process industry that yields liquid oxygen and nitrogen products, improvements on the order of 40% have taken place over the last ten years. This is particularly impressive because this technology has been considered "mature". Many of these improvements have also been retrofitted to existing plants to reduce their costs. The discussion will include: specific technology improvements, their timely implementation, the thermodynamic analyses that were crucial in identifying opportunities, and the concept of cost-benefit analysis. The discussion will clearly show how future engineers will have to be even more flexible, be prepared to take calculated risks, and have an excellent grasp of the fundamentals.

Development of a Compact Dilution Refrigerator for Zero Gravity Operation

PAT R. ROACH

NASA Ames Research Center
Moffett Field, California 94035

BEN HELVENSTEIJN

Sterling Federal Systems, Inc.
Palo Alto, California

Infrared and X-ray detectors for deep-space astronomy can benefit greatly from having detectors cooled to 0.1 K or below. The most-used technique for reaching these temperatures in the laboratory is the He-3-He-4 dilution refrigerator. Its potential for continuous operation and the fact that it doesn't need a magnetic field with a heavy magnet and power supply are advantages that must be balanced against the need to properly control the liquids of the dilution refrigerator in a zero-gravity environment.

A compact dilution refrigerator of unique design using internal charcoal adsorption pumps has been modified and is being tested for operation in zero gravity. This commercially-available refrigerator is self-contained with no external pumps or gas handling system and provides very reliable operation since it has no moving parts. All operations are performed with heaters and are completely computer controlled. The refrigerator provides many hours of operation at very low temperature before the charcoal pumps must be recycled.

This single-cycle system has had porous sintered metal added to its liquid chambers in order to confine and control the liquid phases at low temperatures. A detailed analysis has shown that operation down to 0.05 K should be possible. Results of the operation of the refrigerator with this sinter are presented. Successful operation with the sinter in a normal configuration will be followed by testing with the liquid chambers in an inverted geometry to simulate −1 G acceleration. Success of these tests will demonstrate the gravity-independence of the design and will guarantee the ability to operate in zero G. Future modifications to achieve continuous operation will then be implemented.

Category: 2

Pat R. Roach
MS 244-10
NASA-Ames Research Center
Moffett Field, CA 94035
(415) 694-3191

A RANGE OF AUTOMATIC STANDARD HELIUM LIQUEFIERS AND REFRIGERATORS

Guy GISTAU
Refrigeration group, **L'AIR LIQUIDE,**
38360 SASSENAGE FRANCE

1 THE NEEDS

1.1 RECAPITULATION OF THE ROLE OF A CENTRALISED LIQUEFACTION

A centralised liquefaction facility must feed one or several laboratories (its customers) :

- as and when demanded,
- at minimum cost.

A flow diagram of a typical liquefaction facility is shown in fig. 1.

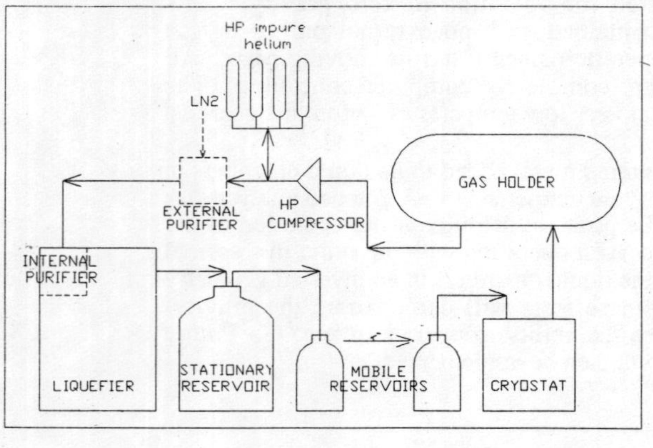

FIGURE 1
A CENTRALIZED LIQUEFACTION FACILITY

1.2 HOW BEST TO FILL THIS ROLE ?

To supply the customers as as when they demand implies :

- sufficient liquefaction capacity,
- possibility to increase output to meet occasional, consumption peaks,
- a high reliability liquefier,
- a storage tank of sufficient capacity,
- good organisation.

To supply customers at minimum cost implies :

- a thermodynamically efficient liquefier,
- a reduced labour time,
- machines with minimum maintenance requirement,
- minimal occupation of floor space.

2 THE IMPORTANT FACTORS

2.1 PRODUCTION CAPACITY

The number of litres of liquid helium produced per hour is a convenient way of expressing the production capacity of a liquefier. The numerical value of the production capacity of a liquefier feeding a storage tank can be influenced by the means of measurement employed.

The level of the liquid in the storage tank can be :

- rising,
- constant.

When the level of liquid rises in the tank it displaces cold gas which passes through the heat exchangers thus contributing to cooling of incoming gas to be liquefied. The production capacity determined by measuring the rate of rise of the liquid level is approximately 15 % greater than if the measureament is made by maintaining the level constant i.e. withdrawing liquid as it is produced.

If, now, liquid is withdrawn faster than the rate at which it is produced, the liquefier must replace the volume of liquid withdrawn by an equal volume of cold fluid. At a withdrawal rate normally employed, there is no production of liquid. In fact, for practical purposes the production capacity of a liquefier connected to a storage dewar from which regular liquid withdrawals are made, is close to that measured at **constant level**.

It is therefore the liquefaction rate at constant level which should be taken into account when estimating the required operating time.

The performance is affected by the purity of the helium to be liquefied since the adsorption or the freeze-out of impurities imposes an additional load except when the purification is carried out by an independant external purifier, generally liquid-nitrogen cooled.

The performance of a liquefier can be boosted either by pre-cooling with liquid nitrogen or by increasing the available process gas flow by using additional compressors.

Finally, consideration may be given to the periods during which the liquefier may function :

- Is it possible to operate through the night or at weekends ?
- Can it operate unattended ?
- Is it capable of restarting after an interuption of utilities ?

The autonomy represented by the useful volume of the storage reservoir must be such that it permits plant shutdown for prescribed preventive maintenance.

Do not forget to plan for a normal growth of demand for liquid helium.

2.2 PEAK DEMANDS FOR LIQUID OUTPUT

Liquid nitrogen pre-cooling which backs the warmer expansion machine, permits the liquefaction rate to be practically doubled.

With this in mind, the machine's constructor builds a liquid nitrogen boiler and the associated additional head exchange surface into the heat exchanger in order to warm the vaporised liquid to ambient temperature.

To take advantage to this additional pre-cooling, one simply opens the liquid nitrogen inlet valve. This boosted mode of operation is established very rapidly, and can be used as and when required, according to the relative costs of liquid nitrogen or electrical power.

2.3 THE RELIABILITY OF THE LIQUEFIER

The reliability of the machine depends on the reliability of each component. Indeed a liquefier may be prevented from operating by a faulty check valve or by a compressor.

The constructor must seek a compromise between quality and price of all components.

Independently of components, the operating procedures can influence reliability ; mistakes can cause unscheduled stops which reduce output.

Finally, preventive maintenance limits downtime due to breakdown.

2.4 THE EFFICIENCY OF THE LIQUEFIER

The specific energy consumption (number of kWh of electricity to liquefy 1 litre of helium) is a practical way of expressing the efficiency of a liquefier.

The efficiency of system components may degrade with operating time, depending on their basic technology (for example, reciprocating compressor mass throughput may diminish as wear takes place).

If, as it is the case in many countries, the price of electrical energy is lower at certain times, it is an advantage to be able to shutdown the liquefier in peak periods.

2.5 LABOUR COSTS

The largest part of labour costs is incurred by operation, and supervision of the liquefier when in operation. Following this, are the regular liquid helium withdrawals.

Finally, there are maintenance labour costs.

2.6 MACHINE MAINTENANCE

It is desirable to select components which require a minimum of maintenance. Machines such as oil lubricated screw compressors and gas lubricated expansion turbines are the most obvious examples.

When more-frequent maintenance work is required it should be simple to perform and, if possible, by personnel who are not specially-trained.

3 L'AIR LIQUIDE PROPOSES A SOLUTION : THE HELIAL FAMILY

3.1 THE HELIAL LIQUEFIERS

The HELIAL family of liquefiers has been created to meet the foregoing needs.

Let us take a closer look at where HELIAL stands in respect of each of these parameters.

The liquefaction rates, level rising, of HELIAL when fed with pure helium are indicated in table 1.

	LIQUEFACTION RATE l/h	
	without LN2 precooling	with LN2
HELIAL 7	7	15
HELIAL 15	15	30
HELIAL 20	20	40
HELIAL 50	50	100

TABLE 1
HELIAL'S LIQUEFACTION RATES

HELIAL 7, 15 and 20 are based on identical cryogenic components (the "coldbox").

Rated capacity is obtained by connecting 1, 2 or 3 identical helium compressors which correspond to 7, 15 or 20 l/h. The range of liquefaction rates shown in table 1 meets the requirements of most laboratories.

HELIAL 7 : COLD BOX MODULE

HELIAL is normally fitted with an internal "freeze-out" purifier, cooled by the liquefier cycle gas. Liquid nitrogen is not necessary for its operation.

The reliability of the HELIAL family is derived from :

- the oil-lubricated screw compressor with its oil removal system (to date, screw compressor installations supplied by L'AIR LIQUIDE have accrued a total of more than 160 000 running hours.
- cryogenic expansion turbines with aerostatic bearings with which HELIAL 50 is equipped (as well as 60 others liquefiers or refrigerators throughout the world)
- the choice of minor components, proven in service in many liquefiers
- the totally-automatic control system which avoids all accidental operating errors.

The expansion engine fitted to the HELIAL 7, 15, 20 models has been designed for maximum operating time between maintenance periods. We must await operational experience to see if our specification is met.

The efficiency of the HELIAL machines compares favourably with other available liquefiers. See table 2.

The mass throughput of the HELIAL screw compressors does not degrade in time since there is no sliding contact between moving parts producing wear. The overall efficiency of the liquefier is unaffected.

HELIAL 50 : COLD BOX MODULE

The automatic control system permits the operators to pre-programm shutdown and re-start in order to avoid operation in peak periods of high-priced electricity. It is one way of reducing operating costs.

	SPECIFIC ENERGY CONSUMPTION	
	without LN2	with LN2 pre-cooling
HELIAL 7	4	1,86
HELIAL 15	3,73	1,86
HELIAL 20	4,20	2,10
HELIAL 50	**2,93**	**1,54**
X1	7,50	2,64
X2 + 1C	4,50	2,57
X2 + 2C	7,20	2,40
X3	4,50	2,25

TABLE 2
SPECIFIC ENERGY CONSUMPTION COMPARISONS

It is clear that the automated operation reduces operator labour cost. The principle of automatic operating should be emphasised ; on HELIAL machines this is total. Not only does this encompass everyday operations such as startup, cooldown, liquefaction, re-start afterutility failure, but also the less-frequent procedures such as purging compressor or coldbox, or regenerations.

The VDU terminal and printer are not necessarily placed near the liquefier but can be located at the most convenient place, eg. in the office of the person in charge of the installation.

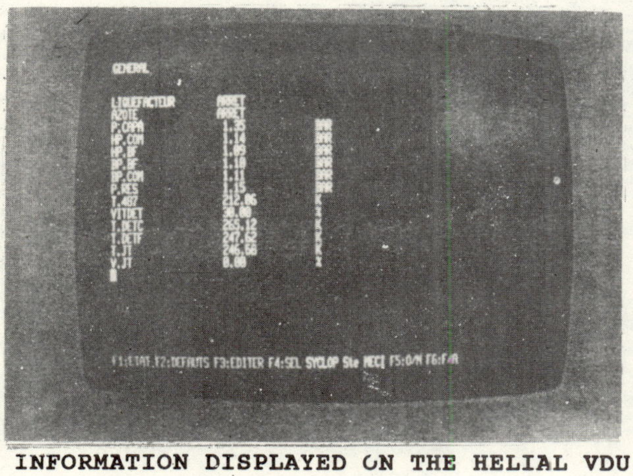

INFORMATION DISPLAYED ON THE HELIAL VDU

Remote supervision and alarm can be achieved if a conventional telephone line is available. This facility is particularly attractive at night or weekends. In France, the widespred and free MINITEL system can be used in such a way.

Maintenance of HELIAL is minimal and easy.

Minimal in the case of components such as screw compressors or gas-lubricated turbines, and easy in the case of the reciprocating expansion engine (the replacement of all wearing components is done in a few hours and no special tools are required). If the user has neither the time nor the competence for maintenance work, he can consider a preventive maintenance contract offered by L'AIR LIQUIDE. In this case programmed maintenance is undertaken by L'AIR LIQUIDE's representatives.

The HELIAL family meets the expectations of liquefaction centre operators. Comparing the needs of a centralized liquefaction facility with the characteristics of the HELIAL family shows that these liquefiers meet present needs satisfactorily.

3.2 THE HELIAL REFRIGERATORS

With minor changes (cryogenic connections, cool-down by-passes) the HELIAL liquefiers can be as well operated as 4.5 K or 20 K refrigerators.

THE HELIAL 7 COMPRESSOR MODULE

4 FUTURE

Liquid helium production is currently achieved automatically and in a satisfactory manner. t can now be established that labour costs are inevitably being focussed on task which remain manual :

- liquid helium withdrawal
- helium stock control

It is in these areas where development effort should be concentrated.